Argument-Driven Inquiry
in
Fifth-Grade Science
Three-Dimensional Investigations

Argument-Driven Inquiry
in
Fifth-Grade Science

Three-Dimensional Investigations

Victor Sampson, Todd L. Hutner,
Jonathon Grooms, Jennifer Kaszuba,
and Carrie Burt

nsta Press
National Science Teaching Association
Arlington, Virginia

nsta Press®
National Science Teaching Association

Claire Reinburg, Director
Rachel Ledbetter, Managing Editor
Andrea Silen, Associate Editor

ART AND DESIGN
Will Thomas Jr., Director

PRINTING AND PRODUCTION
Catherine Lorrain, Director

NATIONAL SCIENCE TEACHING ASSOCIATION
1840 Wilson Blvd., Arlington, VA 22201
www.nsta.org/bookstore
For customer service inquiries, please call 800-277-5300.

Copyright © 2021 by Argument-Driven Inquiry, LLC.
All rights reserved. Printed in the United States of America.
24 23 22 21   4 3 2 1

NSTA is committed to publishing material that promotes the best in inquiry-based science education. However, conditions of actual use may vary, and the safety procedures and practices described in this book are intended to serve only as a guide. Additional precautionary measures may be required. NSTA and the authors do not warrant or represent that the procedures and practices in this book meet any safety code or standard of federal, state, or local regulations. NSTA and the authors disclaim any liability for personal injury or damage to property arising out of or relating to the use of this book, including any of the recommendations, instructions, or materials contained therein.

PERMISSIONS
Book purchasers may photocopy, print, or e-mail up to five copies of an NSTA book chapter for personal use only; this does not include display or promotional use. Elementary, middle, and high school teachers may reproduce forms, sample documents, and single NSTA book chapters needed for classroom use only. E-book buyers may download files to multiple personal devices but are prohibited from posting the files to third-party servers or websites, or from passing files to non-buyers. For additional permission to photocopy or use material electronically from this NSTA Press book, please contact the Copyright Clearance Center (CCC) (*www.copyright.com*; 978-750-8400). Please access *www.nsta.org/permissions* for further information about NSTA's rights and permissions policies.

Library of Congress Cataloging-in-Publication Data
Names: Sampson, Victor, 1974- author. | Hutner, Todd, 1981- author. | Grooms, Jonathon, 1981- author. | Jordan-Kaszuba, Jennifer, author. | Burt, Carrie, author.
Title: Argument-driven inquiry in fifth-grade science : three-dimensional investigations / by Victor Sampson, Todd L. Hutner, Jonathon Grooms, Jennifer Kaszuba, and Carrie Burt.
Description: Arlington, VA : National Science Teaching Association, [2020] | Includes bibliographical references and index.
Identifiers: LCCN 2020026348 (print) | LCCN 2020026349 (ebook) | ISBN 9781681405230 (paperback) | ISBN 9781681405247 (adobe pdf)
Subjects: LCSH: Science--Study and teaching (Elementary)--Activity programs. | Science--Experiments. | Inquiry-based learning. | Fifth grade (Education)
Classification: LCC Q164 .S2544 2020 (print) | LCC Q164 (ebook) | DDC 372.35--dc23
LC record available at *https://lccn.loc.gov/2020026348*
LC ebook record available at *https://lccn.loc.gov/2020026349*

Contents

Preface ix
Acknowledgments xiii
About the Authors xv
Introduction xvii

SECTION 1 - The Instructional Model: Argument-Driven Inquiry

Chapter 1. An Overview of Argument-Driven Inquiry 3

Chapter 2. The Investigations 33

SECTION 2 - Matter and Its Interactions

Investigation 1. Movement of Matter Into and Out of a System: What Causes the Water to Come Out of the Tube in the Balloon Water Dispenser?
- Teacher Notes 42
- Investigation Handout 71
- Checkout Questions 80

Investigation 2. Movement of Particles in a Liquid: Why Do People Use Hot Water Instead of Cold Water When They Make Tea?
- Teacher Notes 82
- Investigation Handout 111
- Checkout Questions 120

Investigation 3. States of Matter and Weight: What Happens to the Weight of a Substance When It Changes From a Solid to a Liquid or a Liquid to a Solid?
- Teacher Notes 123
- Investigation Handout 150
- Checkout Questions 159

Investigation 4. Chemical Reactions: Which Pairs of the Available Liquids Produce a New Substance When They Are Mixed Together?
- Teacher Notes 161
- Investigation Handout 189
- Checkout Questions 198

Contents

Investigation 5. Reactions and Weight: What Happens to the Total Weight of a Closed System When a Chemical Reaction Takes Place Within That System?
    Teacher Notes 200
    Investigation Handout 229
    Checkout Questions 238

Investigation 6. Physical and Chemical Properties: What Are the Identities of the Unknown Powders?
    Teacher Notes 240
    Investigation Handout 266
    Checkout Questions 275

SECTION 3 - Motion and Stability

Investigation 7. Gravity: What Direction Is the Gravitational Force Exerted by Earth on Objects?
    Teacher Notes 278
    Investigation Handout 304
    Checkout Questions 313

SECTION 4 - Ecosystems: Interactions, Energy, and Dynamics

Investigation 8. Plant Growth: Where Do the Materials That Plants Need for Growth Come From in the Environment?
    Teacher Notes 318
    Investigation Handout 346
    Checkout Questions 355

Investigation 9. Energy in Ecosystems: How Do We Best Model the Transfer of Energy Into and Within the Living Things That Are Found in the Arctic Ocean?
    Teacher Notes 356
    Investigation Handout 382
    Checkout Questions 392

Contents

Investigation 10. Movement of Carbon in Ecosystems: How Does the Amount of Dissolved Carbon Dioxide Gas Found in Water Change Over Time When Aquatic Plants and Animals Are Present?

    Teacher Notes 394

    Investigation Handout 424

    Checkout Questions 434

SECTION 5 - Earth's Place in the Universe

Investigation 11. Patterns in Shadows: How Does the Location of the Sun in the Sky Affect the Direction and Length of Shadows?

    Teacher Notes 438

    Investigation Handout 463

    Checkout Questions 472

Investigation 12. Daylight and Location: Why Does the Duration of Daylight Change in Different Locations Throughout the Year, and Why Isn't the Pattern the Same in All Locations on Earth?

    Teacher Notes 474

    Investigation Handout 505

    Checkout Questions 515

Investigation 13. Stars in the Night Sky: How Do the Number and Location of the Constellations That We Can See in the Night Sky Change Based on the Time of Year?

    Teacher Notes 519

    Investigation Handout 549

    Checkout Questions 559

Investigation 14. Star Brightness: How Does Distance Affect the Apparent Brightness of a Star?

    Teacher Notes 561

    Investigation Handout 587

    Checkout Questions 596

Contents

SECTION 6 - Earth's Systems

Investigation 15. Geographic Position and Climate: Why Is the Climate in Western and Eastern Washington State So Different?

Teacher Notes 600
Investigation Handout 627
Checkout Questions 636

Investigation 16. Water Reservoirs: How Are Farmers Able to Grow Crops That Require a Lot of Water When They Live in a State That Does Not Get Much Rain?

Teacher Notes 638
Investigation Handout 665
Checkout Questions 674

SECTION 7 - Appendixes

Appendix 1. Standards Alignment Matrixes 679

Appendix 2. Overview of *NGSS* Crosscutting Concepts and Nature of Scientific Knowledge and Scientific Inquiry Concepts 691

Appendix 3. Some Frequently Asked Questions About Argument-Driven Inquiry 695

Appendix 4. Peer-Review Guide and Teacher Scoring Rubric 699

Appendix 5. Safety Acknowledgment Form 701

Appendix 6. Checkout Questions Answer Guide 703

Image Credits 709
Index 711

Preface

There are a number of good reasons for teaching children about science in elementary school: to give students a strong foundation in the basics to prepare them for what they will be expected to know or do in middle and high school; to get students interested in science early so that more people will choose to go into a science or science-related career; and to take advantage of children's natural curiosity about how the world works, using the information included as part of the science curriculum to answer many of their questions. We, however, think that one of the most important reasons to teach children about science in elementary school is that it will be useful for them in everyday life.

Science is useful because it, along with engineering, mathematics, and the technologies that are made possible by these three fields, affects almost every aspect of modern life in one way or another. For example, people need to understand science to be able to think meaningfully about policy issues that affect their communities or to make informed decisions about what food to eat, what medicine to take, or what products to use. People can use their understanding of science to help evaluate the acceptability of different ideas or to convince others about the best course of action to take when faced with a wide range of options. In addition, understanding how science works and all the new scientific findings that are reported each year in the media can be interesting, relevant, and meaningful on a personal level and can open doors to exciting new professional opportunities. The more a person understands science, which includes the theories, models, and laws that scientists have developed over time to explain how and why things happen and how these ideas are developed and refined based on evidence, the easier it is for that person to have a productive and fulfilling life in our technology-based and information-rich society. Science is therefore useful to everyone, not just future scientists.

A Framework for K–12 Science Education (NRC 2012; henceforth referred to as the *Framework*) is based on the idea that all citizens should be able to use scientific ideas to inform both individual choices and collective choices as members of a modern democratic society. It also acknowledges the fact that professional growth and economic opportunity are increasingly tied to the ability to use scientific ideas, processes, and ways of thinking. From the perspective of the *Framework*, it is important for children to learn science because it can help them figure things out or solve problems. It is not enough to remember some facts and terms; people need to be able to use what they have learned while in school. We think this goal for science education is not only important but represents a major shift in what should be valued inside the classroom.

The *Framework* asks all of us, as teachers, to reconsider what we teach in grades K–5 and how we teach it, given this goal for science education. It calls for all students, over multiple years of school, to learn how to use disciplinary core ideas (DCIs),

Preface

crosscutting concepts (CCs), and scientific and engineering practices (SEPs) to figure things out or solve problems. The DCIs are key organizing principles that have broad explanatory power within a discipline. Scientists use these ideas to explain the natural world. The CCs are ideas that are used across disciplines. These concepts provide a framework or a lens that people can use to explore natural phenomena; thus, these concepts often influence what people focus on or pay attention to when they attempt to understand how something works or why something happens. The SEPs are the different activities that scientists engage in as they attempt to generate new concepts, models, theories, or laws that are both valid and reliable. All three of these dimensions of science are important. Students not only need to know about the DCIs, CCs, and SEPs but also must be able to use all three dimensions at the same time to figure things out or to solve problems. These important DCIs, CCs, and SEPs are summarized in Table P-1.

When we give students an opportunity to learn how to use DCIs, CCs, and SEPs to make sense of the world around them, we also provide an authentic context for students to develop fundamental literacy and mathematic skills. Students are able to develop literacy and mathematics skills in this type of context because doing science requires people to obtain, evaluate, and communicate information. Students, for example, must read and talk to others to learn what others have done and what they are thinking. Students must write and speak to share their ideas about what they have learned or what they still need to learn. Students can use mathematics to make measurements and to discover trends, patterns, or relationships in their observations. They can also use mathematics to make predictions about what will happen in the future. When we give students opportunities to do science, we give students a reason to read, write, speak, and listen. We also create a need for them to use mathematics.

To help students learn how to use DCIs, CCs, and SEPs to figure things out or solve problems while providing a context for them to develop fundamental literacy and mathematical skills, elementary teachers will need to use new instructional approaches. These instructional approaches must give students an opportunity to actually do science. To help teachers who teach elementary school make this instructional shift, we have developed a tool called argument-driven inquiry (ADI). ADI is an innovative approach to instruction that gives students an opportunity to use DCIs, CCs, and SEPs to construct and critique claims about how things work or why things happen. As part of this process, students must talk, listen, read, and write in order to obtain, evaluate, and communicate information. ADI, as a result, creates a rich learning environment for children that enables them to learn science, language, and mathematics at the same time.

Preface

TABLE P-1
The three dimensions of *A Framework for K–12 Science Education*

Scientific and engineering practices (SEPs)	Crosscutting concepts (CCs)
• SEP 1: Asking Questions and Defining Problems • SEP 2: Developing and Using Models • SEP 3: Planning and Carrying Out Investigations • SEP 4: Analyzing and Interpreting Data • SEP 5: Using Mathematics and Computational Thinking • SEP 6: Constructing Explanations and Designing Solutions • SEP 7: Engaging in Argument From Evidence • SEP 8: Obtaining, Evaluating, and Communicating Information	• CC 1: Patterns • CC 2: Cause and Effect: Mechanism and Explanation • CC 3: Scale, Proportion, and Quantity • CC 4: Systems and System Models • CC 5: Energy and Matter: Flows, Cycles, and Conservation • CC 6: Structure and Function • CC 7: Stability and Change

Disciplinary core ideas (DCIs)		
Earth and Space Sciences (ESS)	**Life Sciences (LS)**	**Physical Sciences (PS)**
• ESS1: Earth's Place in the Universe • ESS2: Earth's Systems • ESS3: Earth and Human Activity	• LS1: From Molecules to Organisms: Structures and Processes • LS2: Ecosystems: Interactions, Energy, and Dynamics • LS3: Heredity: Inheritance and Variation of Traits • LS4: Biological Evolution: Unity and Diversity	• PS1: Matter and Its Interactions • PS2: Motion and Stability: Forces and Interactions • PS3: Energy • PS4: Waves and Their Applications in Technologies for Information Transfer

Source: Adapted from NRC 2012.

This book not only describes how ADI works and why it is important but also provides 16 investigations that can be used in the classroom to help students reach the performance expectations found in the *Next Generation Science Standards* (NGSS Lead States 2013) for fifth grade.[1] The 16 investigations described in this book will also enable students to develop the disciplinary-based literacy skills outlined in the *Common Core State Standards for English Language Arts* (NGAC and CCSSO 2010)

---

[1] See *Argument-Driven Inquiry in Third-Grade Science* (Sampson and Murphy 2019b) and *Argument-Driven Inquiry in Fourth-Grade Science* (Sampson and Murphy 2019a) for additional investigations for students in elementary school.

Preface

because ADI gives students an opportunity to make presentations to their peers; respond to audience questions and critiques; and then write, evaluate, and revise reports as part of each investigation. In addition, these investigations will help students learn many of the mathematical ideas and practices outlined in the *Common Core State Standards for Mathematics* (NGAC and CCSSO 2010) because ADI gives students an opportunity to use mathematics to collect, analyze, and interpret data. Finally, and perhaps most important, ADI can help emerging bilingual students meet the *English Language Proficiency (ELP) Standards* (CCSSO 2014) because it provides a language-rich context where children can use receptive and productive language to communicate and to negotiate meaning with others. Teachers can therefore use these investigations to align how and what they teach with current recommendations for improving science education.

References

Council of Chief State School Officers (CCSSO). 2014. *English language proficiency (ELP) standards*. Washington, DC: NGAC and CCSSO. *https://elpa21.org/wp-content/uploads/2019/03/Final-4_30-ELPA21-Standards_1.pdf*.

National Governors Association Center for Best Practices and Council of Chief State School Officers (NGAC and CCSSO). 2010. *Common core state standards*. Washington, DC: NGAC and CCSSO.

National Research Council (NRC). 2012. *A framework for K–12 science education: Practices, crosscutting concepts, and core ideas*. Washington, DC: National Academies Press.

NGSS Lead States. 2013. *Next Generation Science Standards: For states, by states*. Washington, DC: National Academies Press. *www.nextgenscience.org/next-generation-science-standards*.

Sampson, V., and A. Murphy. 2019a. *Argument-driven inquiry in fourth-grade science*. Arlington, VA: NSTA Press.

Sampson, V., and A. Murphy. 2019b. *Argument-driven inquiry in third-grade science*. Arlington, VA: NSTA Press.

Acknowledgments

We would like to thank the following individuals for piloting the investigations and giving us feedback about ways to make them better.

Coppell Independent School District (ISD), Coppell, Texas

Shari Hamam
Teacher
Richard J. Lee Elementary

Shelby Holbrook
Teacher
Richard J. Lee Elementary

Laura Jennings
Teacher
Richard J. Lee Elementary

Samira Khan
Teacher
Richard J. Lee Elementary

Pricilla Shaner
STEM Specials Designer
Coppell ISD

May Voltz
Teacher
Richard J. Lee Elementary

Frisco ISD, Frisco, Texas

Lauren Tilley Clark
Teacher
Scott E. Johnson Elementary

Tricia Elrod
Teacher
McSpedden Elementary

Alicia Jahnke
Teacher
Sem Elementary

Laura Lee McLeod
Coordinator of Elementary Science
Frisco ISD

Alyson Prior
Teacher
Sonntag Elementary

Hoover City Schools, Birmingham, Alabama

Lincoln Clark
Teacher
Hoover City Schools

Julie Erwin
Teacher
Bluff Park Elementary

Polly Ohlson
Teacher
Deer Valley Elementary

Stacey Rush
Teacher
Greystone Elementary

Patrick Woody
Teacher
Gwin Elementary

Peoria Unified School District, Glendale, Arizona

Justin Henry
Teacher
Heritage Elementary STEAM School

Jan Ogino
Teacher
Heritage Elementary STEAM School

Curtis Smith
Principal
Heritage Elementary STEAM School

Acknowledgments

Round Rock ISD, Round Rock, Texas

Vanessa DeBecze
Teacher
Joe Lee Johnson Elementary School

Ashley Hentges
Teacher
Joe Lee Johnson Elementary School

Lauren Lightfoot
Teacher
Joe Lee Johnson Elementary School

Jamie Nair
STEAM Coordinator
Joe Lee Johnson Elementary School

Gina Picha
Instructional Coach
Joe Lee Johnson Elementary School

Kirsten Prud'Homme
Project Lead the Way Launch Coordinator
Joe Lee Johnson Elementary School

Vanessa Stenulson
Librarian
Joe Lee Johnson Elementary School

Brian Wolfe
Instructional Technology Specialist
Joe Lee Johnson Elementary School

Kate Wood
Teacher
Joe Lee Johnson Elementary School

About the Authors

Victor Sampson is an associate professor of STEM (science, technology, engineering, and mathematics) education at The University of Texas at Austin (UT-Austin). He received a BA in zoology from the University of Washington, an MIT from Seattle University, and a PhD in curriculum and instruction with a specialization in science education from Arizona State University. Victor also taught high school biology and chemistry for nine years. He is an expert in argumentation and three-dimensional instruction in science education, teacher learning, and assessment. Victor is also an NSTA (National Science Teaching Association) Fellow.

Todd L. Hutner is an assistant professor of science education at the University of Alabama in Tuscaloosa. He received a BS and an MS in science education from Florida State University (FSU) and a PhD in curriculum and instruction from UT-Austin. Todd's classroom teaching experience includes teaching chemistry and physics in Texas and Earth science and astronomy in Florida. His current research focuses on the impact of both teacher education and education policy on the teaching practice of secondary science teachers.

Jonathon Grooms is an assistant professor of curriculum and pedagogy in the Graduate School of Education and Human Development at The George Washington University. He received a BS in secondary science and mathematics teaching with a focus in chemistry and physics from FSU. Upon graduation, Jonathon joined FSU's Office of Science Teaching, where he directed the physical science outreach program Science on the Move. He also earned a PhD in science education from FSU. His research interests include student engagement in scientific argumentation and students' application of argumentation strategies in socioscientific contexts.

Jennifer Kaszuba is a doctoral student in the STEM education program at UT-Austin. She received an MS in curriculum and instruction from Our Lady of the Lake University and a BA in biology from Texas A&M University. Jennifer taught high school chemistry, physical science, and biology in Texas for 10 years before serving as a science specialist assisting districts in central Texas with curriculum and instruction.

Carrie Burt is a former elementary and middle school teacher, curriculum writer, and instructional coach. She taught third grade for three years and sixth through eighth grades for five years in Frisco, Texas. Carrie received a BA in psychology and an MAT from Austin College.

Introduction

A Vision for Science Education in Elementary School

The current aim of science education in the United States is for *all* students to become proficient in science by the time they finish high school. *Science proficiency,* as defined by Duschl, Schweingruber, and Shouse (2007), consists of four interrelated aspects. First, it requires an individual to know important scientific explanations about the natural world, to be able to use these explanations to solve problems, and to be able to understand new explanations when they are introduced to the individual. Second, it requires an individual to be able to generate and evaluate scientific explanations and scientific arguments. Third, it requires an individual to understand the nature of scientific knowledge and how scientific knowledge develops over time. Finally, and perhaps most important, an individual who is proficient in science should be able to participate in scientific practices (such as planning and carrying out investigations, analyzing and interpreting data, and arguing from evidence) and communicate in a manner that is consistent with the norms of the scientific community. These four aspects of science proficiency include the knowledge and skills that all people need to have to be able to purse a degree in science, be prepared for a science-related career, and participate in a democracy as an informed citizen.

This view of science proficiency serves as the foundation for *A Framework for K–12 Science Education* (NRC 2012). The *Framework* calls for all students to learn how to use disciplinary core ideas (DCIs), crosscutting concepts (CCs), and scientific and engineering practices (SEPs) to figure things out or solve problems as a way to help them develop the four aspects of science proficiency. The *Framework* was used to guide the development of the *Next Generation Science Standards* (*NGSS;* NGSS Lead States 2013). The goal of the *NGSS,* and other sets of academic standards that are based on the *Framework,* is to describe what *all* students should be able to do at each grade level or at the end of each course as they progress toward the ultimate goal of science proficiency.

The DCIs found in the *Framework* and the *NGSS* are scientific theories, laws, or principles that are central to understanding a variety of natural phenomena. An example of a DCI in Earth and space sciences is that the solar system consists of the Sun and a collection of objects that are held in orbit around the Sun by its gravitational pull on them. This DCI not only can be used to help explain the motion of planets around the Sun but also can be used to explain why we have tides on Earth, why the appearance of the Moon changes over time in a predictable pattern, and why we see eclipses of the Sun and the Moon.

The CCs are ideas that are important across the disciplines of science. The CCs help people think about what to focus on or pay attention to during an investigation. For example, one of the CCs from the *Framework* is Energy and Matter: Flows,

Introduction

Cycles, and Conservation. This CC is important in many different fields of study, including astronomy, geology, and meteorology. This CC is equally important in physics and biology. Physicists use this CC when they study how things move, why things change temperature, and the behavior of circuits or magnets. Biologists use this CC when they study how cells work, the growth and development of plants or animals, and the nature of ecosystems. It is important to help students see the value of the CCs as a tool for developing an understanding of how or why things happen.

The SEPs describe what scientists do as they attempt to make sense of the natural world. Some of the SEPs include familiar aspects of what we typically associate with "doing" science, such as Asking Questions and Defining Problems, Planning and Carrying Out Investigations, and Analyzing and Interpreting Data. More important, however, some of the SEPs focus on activities that are related to developing and sharing new ideas, solutions to problems, or answers to questions. These SEPs include Developing and Using Models, Constructing Explanations and Designing Solutions, Engaging in Argument From Evidence, and Obtaining, Evaluating, and Communicating Information. All these SEPs are important to learn because scientists engage in different practices, at different times, and in different orders depending on what they are studying and what they are trying to accomplish at that point in time.

Few students in fifth grade have an opportunity to learn how to use DCIs, CCs, and SEPs to figure things out or to solve problems. Instead, most students are introduced to facts, concepts, and vocabulary without a real reason to know or use them. This type of focus in fifth grade does little to promote and support the development of science proficiency because it emphasizes "learning about" science rather than learning how to use science to "figure things out." This type of focus also reflects a view of teaching that defines *rigor* as covering more topics and *learning* as the simple acquisition of more information.

We must think about rigor in different ways before we can start teaching science in ways described by the *Framework*. Instead of using the number of different topics covered in a particular grade level as a way to measure rigor in our schools (e.g., "we made fifth grade more challenging by adding more topics for students to learn about"), we must start to measure rigor in terms of the number of opportunities students have to use DCIs, CCs, and SEPs to make sense of different phenomena (e.g., "we made fifth grade more challenging because students have to figure out where plants get the materials they need to grow"). A rigorous class, in other words, should be viewed as one where students are expected to do science, not just learn about science. From this perspective, our goal as teachers should be to help our students learn how to use DCIs and CCs as tools to plan and carry out investigations, construct and evaluate explanations, and question how we know what we know instead of just ensuring that we "cover" all the different DCIs and CCs that are included in the standards by the end of the school year.

To better promote and support the development of science proficiency, we must also rethink what learning is and how it happens. Rather than viewing learning as an individual process where children accumulate more and more information over time, we need to view learning as both a social and an individual process that involves being exposed to new ideas and ways of doing things, trying out these new ideas and practices under the guidance of more experienced people, and then adopting the ideas and practices that are found to be useful for making sense of the world (NRC 1999, 2008, 2012). Learning, from this perspective, requires children to "do science" while in school not because it is fun or interesting (which is true for many) but because doing science gives children a reason to use the ideas and practices of science. When children are given repeated opportunities to use DCIs, CCs, and SEPs as a way to make sense of the world, they will begin to see why these ideas and practices are valuable. Over time, children will then adopt these ideas and practices and start using them on their own. We therefore must give our students an opportunity to experience how scientists figure things out and share ideas so they can become "socialized to a greater or lesser extent into the practices of the scientific community with its particular purposes, ways of seeing, and ways of supporting its knowledge claims" (Driver et al. 1994, p. 8).

It is important to keep in mind that helping children learn how to use the ideas and practices of science to figure things out by giving them an opportunity to do science is not a "hands-off" approach to teaching. The process of learning to use the ideas and practices of science to figure things out requires constant input and guidance about "what counts" from teachers who are familiar with the goals of science, the norms of science, and the ways things are done in science. Thus, learning how to use DCIs, CCs, and SEPs to figure things out or to solve problems is dependent on supportive and informative interactions with teachers. This is important because students must have a supportive and educative learning environment to try out new ideas and practices, make mistakes, and refine what they know and how they do things before they are able to adopt the ideas and practices of science as their own.

The Need for New Ways of Teaching Science in Elementary School

Science in fifth grade has historically been taught through a combination of direct instruction and hands-on activities. A typical lesson often begins with the teacher introducing students to a new concept and related terms through direct instruction. Next, the teacher will often illustrate the concept by giving a demonstration or asking students to complete a hands-on activity. The purpose of including a demonstration

Introduction

or a hands-on activity in the lesson is to provide the students with a memorable experience with the concept. If the memorable experience is a hands-on activity, the teacher will often provide his or her students with a step-by-step procedure to follow to help ensure that no one in the class "gets lost" or "does the wrong thing" and everyone "gets the right results." The teacher will usually assign a set of questions for the students to answer on their own or in groups after the demonstration or the hands-on activity to make sure that everyone in the class "reaches the right conclusion." The lesson usually ends with the teacher reviewing the concept and all related terms to make sure that everyone in the class learned what they were "supposed to have learned." The teacher often accomplishes this last step of the lesson by leading a whole-class discussion, by assigning a worksheet to complete, or by having the students play an educational game.

Classroom-based research, however, suggests that this type of lesson does little to help students learn key concepts (Duschl, Schweingruber, and Shouse 2007; NRC 2008, 2012). This finding is troubling because, as noted earlier, one of the main goals of this type of lesson is to help students understand an important concept by giving them a memorable experience with it. In addition, this type of lesson does little to help students learn how to plan and carry out investigations or analyze and interpret data because students have no voice or choice during the activity. Students are expected to simply follow a set of directions rather than having to think about what data they will collect, how they will collect it, and what they will need to do to analyze it once they have it. These types of activities can also lead to misunderstanding about the nature of scientific knowledge and how this knowledge is developed over time due to the emphasis on following procedure and getting the "right" results. These hand-on activities, as a result, do not reflect how science is done at all.

Many fifth-grade teachers started using more inquiry-based lessons in the late 1990s and early 2000s to help address shortcomings of these typical science lessons. Inquiry-based lessons that are consistent with the definition of *inquiry* found in *Inquiry and the National Science Education Standards* (NRC 2000) share five key features:

1. Students need to answer a scientifically oriented question.
2. Students must collect data or use data collected by someone else.
3. Students formulate an answer to the question based on their analysis of the data.
4. Students connect their answer to some theory, model, or law.
5. Students communicate their answer to the question to someone else.

Many fifth-grade teachers also changed the traditional sequence of science instruction when they started using more inquiry-based lessons. Rather than introducing students to an important concept or principle through direct instruction and then having students do a hands-on activity to demonstrate or confirm it, they used an inquiry-based lesson as a way to give students a firsthand experience with a concept before introducing terms and vocabulary (NRC 2012). Inquiry-based lessons, as a result, are often described as an "activity before content" approach to teaching science (Cavanagh 2007). The focus of these "activity before content" lessons, as the name implies, is to help students understand the core ideas of science. Inquiry-based lessons also give students more opportunities to learn how to plan and carry out investigations, analyze and interpret data, and develop explanations. These lessons also give students more voice and choice so they are more consistent with how science is done.

Although classroom-based research indicates that inquiry-based lessons are effective at helping students understand core ideas and give students more voice and choice than typical science lessons, they do not do as much as they could do to help students develop all four aspects of science proficiency (Duschl, Schweingruber, and Shouse 2007; NRC 2008, 2012). For example, inquiry-based lessons are usually not designed in a way that encourages students to learn how to use DCIs, CCs, and SEPs because they are often used to help students "learn about" important concepts or principles (NRC 2012). These lessons also do not give students an opportunity to participate in the full range of SEPs because these lessons tend to be designed in a way that gives students many opportunities to learn how to ask questions, plan and carry out investigations, and analyze and interpret data but few opportunities to learn how to participate in the practices that focus on how new ideas are developed, shared, refined, and eventually validated within the scientific community. These important sense-making practices include developing and using models; constructing explanations; arguing from evidence; and obtaining, evaluating, and communicating information (NRC 2012). Inquiry-based lessons that do not focus on sense-making also do not provide a context that creates a need for students to read, write, and speak, because these lessons tend to focus on introducing students to new ideas and how to design and carry out investigations instead of how to develop, share, critique, and revise ideas. These types of inquiry-based lessons, as a result, are often not used as a way to help students develop fundamental literacy skills. To help address this problem, teachers will need to start using instructional approaches that give students more opportunities to figure things out.

This emphasis on "figuring things out" instead of "learning about things" represents a big change in the way we have been teaching science in fifth grade. To figure out how things work or why things happen in a way that is consistent with how science is actually done, students must have opportunities to use DCIs, CCs, and SEPs at the same time to make sense of the world around them (NRC 2012). This

Introduction

focus on students using DCIs, CCs, and SEPs at the same time during a lesson is called *three-dimensional instruction* because students have an opportunity to use all three dimensions of the *Framework* to understand how something works, to explain why something happens, or to develop a novel solution to a problem. When teachers use three-dimensional instruction inside their classrooms, they encourage students to develop or use conceptual models, develop explanations, share and critique ideas, and argue from evidence, all of which allow students to develop the knowledge and skills they need to be proficient in science (NRC 2012). A large body of research suggests that all students benefit from three-dimensional instruction because it gives all students more voice and choice during a lesson and it makes the learning process inside the classroom more active and inclusive (NRC 2012).

We think investigations that focus on making sense of how the world works are the perfect way to integrate three-dimensional science instruction into elementary classrooms. Well-designed investigations can provide opportunities for students to not only use one or more DCIs to understand how something works, to explain why something happens, or to develop a novel solution to a problem but also use several different CCs and SEPs during the same lesson. A teacher, for example, can give his or her students an opportunity to figure out if the mass of a substance changes when it melts. The teacher can then encourage them to use what they know about Matter and Its Interactions (a DCI) and their understanding of Scale, Proportion, and Quantity (a CC) to plan and carry out an investigation to figure out how to measure the total weight of a piece of wax, a chunk of ice, and a few milliliters of vegetable shortening before and after they are heated to the point that they melt. In addition to planning and carrying out an investigation, they must also ask questions; analyze and interpret data; use mathematics; construct an explanation; argue from evidence; and obtain, evaluate, and communicate information (seven different SEPs).

Using DCIs, CCs, and SEPs at the same time is important because it creates a classroom experience that parallels how science is done. This, in turn, gives all students who participate in the investigation an opportunity to deepen their understanding of what it means to do science and to develop science-related identities (Carlone, Scott, and Lowder 2014; Tan and Barton 2008, 2010). In the following section, we will describe how to promote and support the development of science proficiency through three-dimensional instruction by using an innovative instructional model called argument-driven inquiry (ADI).

Argument-Driven Inquiry as a Way to Promote Three-Dimensional Instruction While Focusing on Literacy and Mathematics

The ADI instructional model (Sampson and Gleim 2009; Sampson, Grooms, and Walker 2009, 2011) was developed as a way to change how science is taught in our schools. Rather than simply encouraging students to learn about the facts, concepts, and terms of science, ADI gives students an opportunity to use DCIs, CCs, and SEPs to figure out how things work or why things happen. ADI also encourages children to think about "how we know" in addition to "what we have figured out." The ADI instructional model includes eight stages of classroom activity. These eight stages give children an opportunity to *investigate* a phenomenon; *make sense* of that phenomenon; and *evaluate and refine* ideas, explanations, or arguments. These three aspects of doing science help students learn how to figure something out and make it possible for them to develop and refine their understanding of DCIs, CCs, and SEPs over time.

Students will use different SEPs depending on what they are trying to accomplish during an investigation, which changes as they move through the eight stages of ADI. For example, students must learn how to ask questions to design and carry out an investigation in order to investigate a phenomenon, which is the overall goal of the first two stages of ADI. Then during the next stage of ADI, the students must learn how to analyze and interpret data, use mathematics, develop models, and construct explanations to accomplish the goal of making sense of the phenomenon they are studying. Students then need to evaluate and refine ideas, explanations, or arguments during the last five stages of this instructional model. Students therefore must learn how to ask questions; obtain, evaluate, and communicate information; and argue from evidence. These three goals provide coherence to a three-dimensional lesson, create a need for students to learn how to use each of the SEPs (along with one or more DCIs and one or more CCs), and will keep the focus of the lesson on figuring things out. We will provide a more detailed discussion of what students do during each stage of ADI in Chapter 1.

ADI also provides an authentic context for students to develop fundamental literacy and mathematics skills. Students are able to develop these skills during an ADI investigation because the use of DCIs, CCs, and SEPs requires them to gather, analyze, interpret, and communicate information. Students, for example, must read and talk to others to learn to gather information and to find out how others are thinking. Students must also talk and write to share their ideas about what they are doing and what they have found out and to revise an explanation or model. Students, as a result, "are able to fine-tune their literacy skills when they engage in science investigations because so many of the sense-making tools of science are consistent with, if not identical to, those of literacy, thus allowing a setting for additional practice and refinement that can enhance future reading and writing efforts" (Pearson, Moje, and Greenleaf 2010, p. 460). Students must also use mathematics during an ADI investigation to measure what they are studying and to find patterns in their observations, uncover differences between groups, identify a trend over time, or

Introduction

confirm a relationship between two variables. They also use mathematics to make predictions. Teachers can therefore use ADI to help students develop important literacy and mathematics skills as they teach science. We will discuss how to promote and support the development of literacy and mathematics skills during the various stages of ADI in greater detail in Chapter 1. We will also describe ways to promote and support productive talk, reading, and writing during ADI in the Teacher Notes for each investigation.

ADI investigations also provide a rich language-learning environment for emerging bilingual students who are learning how to communicate in English. A rich language-learning environment is important because emerging bilingual students must (1) interact with people who know English well enough to provide both access to this language and help in learning it and (2) be in a social setting that will bring them in contact with these individuals so they have an opportunity to learn (Lee, Quinn, and Valdés 2013). Once these two conditions are met, people are able to learn a new language through meaningful use and interaction (Brown 2007; García 2005; García and Hamayan 2006; Kramsch 1998). ADI, and its focus on giving students opportunities to use DCIs, CCs, and SEPs to figure things out, also provides emerging bilingual students with opportunities to interact with English speakers and opportunities to do things with language inside the classroom (Lee, Quinn, and Valdés 2013). Emerging bilingual students therefore have an opportunity to use receptive and productive language to communicate and to negotiate meaning with others inside the science classroom. Teachers can promote and support the acquisition of a new language by using ADI to give emerging bilingual students an opportunity to *investigate* a phenomenon; *make sense* of that phenomenon; and *evaluate and refine* ideas, explanations, or arguments with others. Teachers can then provide support and guidance as students learn how to communicate in a new language. We will provide a more detailed discussion of how teachers can use ADI to promote language development in Chapter 1. We will also provide advice and recommendations for supporting emerging bilingual students as they learn science and how to communicate in a new language at the same time in Chapter 1 and Appendix 3.

Organization of This Book

This book is divided into seven sections. Section 1 includes two chapters. The first chapter describes the ADI instructional model. The second chapter provides an overview of the information that is associated with each investigation. Sections 2–6 contain the 16 investigations. Each investigation includes three components:

- Teacher Notes, which provides information about the purpose of the investigation and what teachers need to do to guide students through it.
- Investigation Handout, which can be photocopied and given to students at the beginning of the lesson. The handout provides the students with a phenomenon to investigate, an overview of the DCI(s) and CC(s) that students can use during the investigation, and a guiding question to answer.
- Checkout Questions, which can be photocopied and given to students at the conclusion of investigation. The Checkout Questions consist of items that target students' understanding of the DCI(s) and the CC(s) addressed during the investigation.

Section 7 consists of six appendixes:

- Appendix 1 contains several standards alignment matrices that can be used to assist with curriculum or lesson planning.
- Appendix 2 provides an overview of the CCs and nature of scientific knowledge (NOSK) and nature of scientific inquiry (NOSI) concepts that are a focus of the different investigations.
- Appendix 3 lists some frequently asked questions about ADI.
- Appendix 4 provides a peer-review guide and teacher scoring rubric, which can be photocopied and given to students.
- Appendix 5 provides a safety acknowledgment form, which can also be photocopied and given to students.
- Appendix 6 provides an answer guide for the Checkout Questions.

References

Brown, D. H. 2007. *Principles of language learning and teaching.* 5th ed. White Plains, NY: Longman.

Carlone, H., C. Scott, and C. Lowder. 2014. Becoming (less) scientific: A longitudinal study of students' identity work from elementary to middle school science. *Journal of Research in Science Teaching* 51: 836–869.

Cavanagh, S. 2007. Science labs: Beyond isolationism. *Education Week* 26 (18): 24–26.

Driver, R., Asoko, H., Leach, J., Mortimer, E., and Scott, P. 1994. Constructing scientific knowledge in the classroom. *Educational Researcher* 23: 5–12.

Duschl, R. A., H. A. Schweingruber, and A. W. Shouse, eds. 2007. *Taking science to school: Learning and teaching science in grades K–8.* Washington, DC: National Academies Press.

García, E. E. 2005. *Teaching and learning in two languages: Bilingualism and schooling in the United States.* New York: Teachers College Press.

Introduction

García, E. E., and E. Hamayan. 2006. What is the role of culture in language learning? In *English language learners at school: A guide for administrators*, eds. E. Hamayan and R. Freeman, 61–64. Philadelphia, PA: Caslon Publishing.

Lee, O., H. Quinn, and G. Valdés. 2013. Science and language for English language learners in relation to *Next Generation Science Standards* and with implications for *Common Core State Standards* for English language arts and mathematics. *Educational Researcher* 42 (4): 223–233. Available online at *http://journals.sagepub.com/doi/abs/10.3102/0013189X13480524*.

Kramsch, C. 1998. *Language and culture*. Oxford, UK: Oxford University Press.

National Research Council (NRC). 1999. *How people learn: Brain, mind, experience, and school*. Washington, DC: National Academies Press.

National Research Council (NRC). 2000. *Inquiry and the National Science Education Standards*. Washington, DC: National Academies Press.

National Research Council (NRC). 2008. *Ready, set, science: Putting research to work in K–8 science classrooms*. Washington, DC: National Academies Press.

National Research Council (NRC). 2012. *A framework for K–12 science education: Practices, crosscutting concepts, and core ideas*. Washington, DC: National Academies Press.

NGSS Lead States. 2013. *Next Generation Science Standards: For states, by states*. Washington, DC: National Academies Press. *www.nextgenscience.org/next-generation-science-standards*.

Pearson, P. D., E. B. Moje, and C. Greenleaf. 2010. Literacy and science: Each in the service of the other. *Science 328* (5977): 459–463.

Sampson, V., and L. Gleim. 2009. Argument-driven inquiry to promote the understanding of important concepts and practices in biology. *American Biology Teacher* 71 (8): 471–477.

Sampson, V., J. Grooms, and J. Walker. 2009. Argument-driven inquiry: A way to promote learning during laboratory activities. *The Science Teacher* 76 (7): 42–47.

Sampson, V., J. Grooms, and J. Walker. 2011. Argument-driven inquiry as a way to help students learn how to participate in scientific argumentation and craft written arguments: An exploratory study. *Science Education* 95 (2): 217–257.

Tan, E., and A. Barton. 2008. Unpacking science for all through the lens of identities-in-practice: The stories of Amelia and Ginny. *Cultural Studies of Science Education* 3 (1): 43–71.

Tan, E., and A. Barton. 2010. Transforming science learning and student participation in sixth grade science: A case study of a low-income, urban, racial minority classroom. *Equity and Excellence in Education* 43 (1): 38–55.

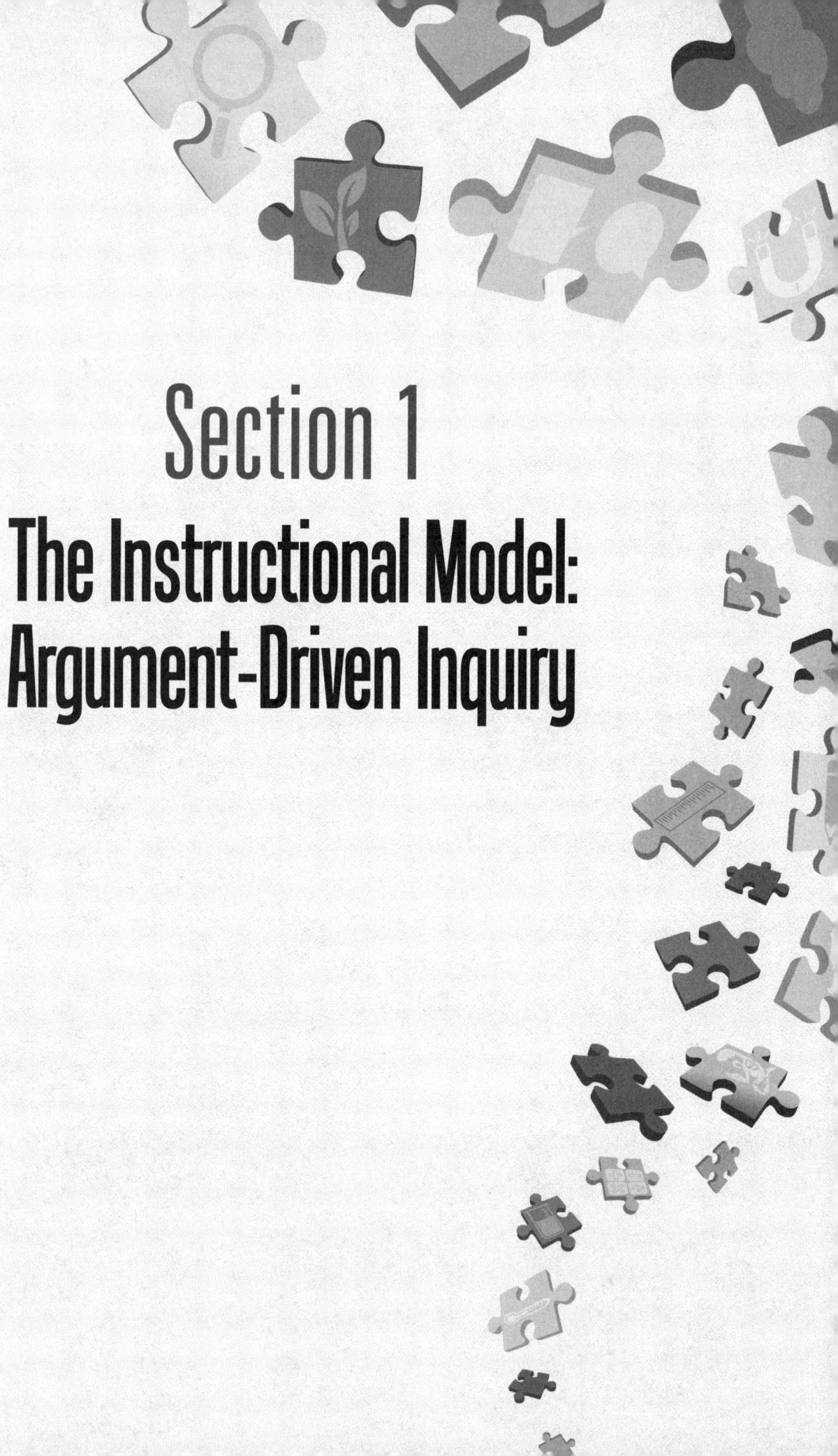

Section 1
The Instructional Model: Argument-Driven Inquiry

Chapter 1
An Overview of Argument-Driven Inquiry

The Argument-Driven Inquiry Instructional Model

The argument-driven inquiry (ADI) instructional model (Sampson and Gleim 2009; Sampson, Grooms, and Walker 2009, 2011) was created to help change the way science is taught in our schools. This instructional model includes eight stages of classroom activity. As children participate in each stage of ADI, they have an opportunity to *investigate* a phenomenon; *make sense* of that phenomenon; and *evaluate and refine* ideas, explanations, or arguments. These eight stages of the instructional model provide a structure that supports children in fifth grade as they learn to plan and carry out an investigation in order to figure something out and enables them to develop and refine their understanding of the disciplinary core ideas (DCIs), crosscutting concepts (CCs), and scientific and engineering practices (SEPs) over time (NGSS Lead States 2013; NRC 2012). ADI also provides an authentic context for students to develop fundamental literacy skills and to learn or apply mathematical concepts and practices, and it provides a social setting that enables emerging bilingual students acquire a new language as they learn science.

Rigor and equity are two hallmarks of ADI. A rigorous learning experience gives students an opportunity to figure out how or why something happens or to develop a solution to a meaningful problem by engaging them in the intellectual activities and practices of science. For us, rigor does not mean covering more content or decreasing the amount of time that is spent on a topic. Rigor is about learning from mistakes, thinking in new ways, and encouraging the use of ideas and practices from science to make sense of the world around us. Rigor, in other words, presses learners to go beyond what they currently know and can do. Equity means providing all students, rather than a select few, with a fair opportunity to learn the ideas and practices of science. It also means doing everything possible to ensure that every student inside a classroom feels like their ideas and participation are valued because they have unique life experiences and ways of talking or thinking that are useful for figuring out how or why something happens or how to develop a solution to a problem. Equitable instruction teaches all students to see themselves, other students, and the teacher as knowers and doers of science, mathematics, or engineering and as an important component of a supportive learning community.

In this chapter, we will explain what happens during each of the eight stages of ADI. These eight stages are the same for every ADI investigation. Students, as a result, quickly learn what is expected of them during each stage and can focus their attention on learning how to use DCIs, CCs, and SEPs to figure out how something works or why something happens. Figure 1 (p. 4) provides an overview of the eight stages of the ADI instructional model.

Chapter 1

FIGURE 1

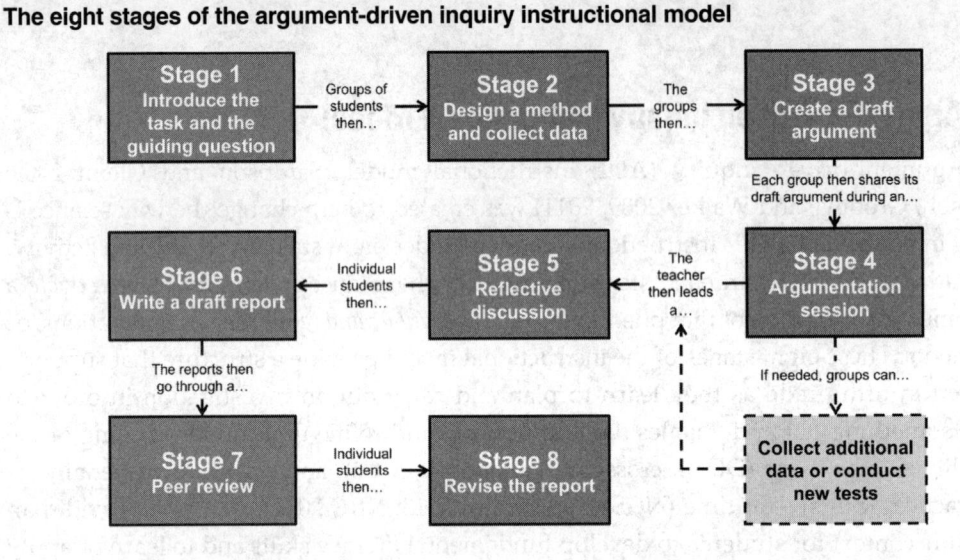

The eight stages of the argument-driven inquiry instructional model

```
Stage 1                  Stage 2                  Stage 3
Introduce the  Groups of  Design a method   The    Create a draft
task and the   students → and collect data  groups → argument
guiding question  then...                   then...
                                                    │
                                          Each group then shares its
                                          draft argument during an...
                                                    ↓
Stage 6        Individual  Stage 5           The    Stage 4
Write a draft  students ←  Reflective      teacher ← Argumentation
report         then...     discussion      then leads  session
                                           a...        │
   │                                         ↑        │
The reports then                             │   If needed, groups can...
go through a...                              │        ↓
   ↓                                         │   ┌─────────────┐
Stage 7        Individual  Stage 8           └───│Collect additional│
Peer review    students →  Revise the report     │data or conduct │
               then...                           │new tests       │
                                                 └─────────────┘
```

We will also provide hints (in boxes) for implementing each stage as part of our explanation of the ADI instructional model. To supplement our explanation and hints found in this chapter, Appendix 3 provides answers to frequently asked questions about ADI. These answers will help you encourage productive talk among students, support emerging bilingual students, improve students' reading comprehension, and improve the quality of feedback during the peer-review process. The answers in Appendix 3 also include techniques for helping students when they get stuck as they are developing their draft argument or writing their report.

Stage 1: Introduce the Task and the Guiding Question

An ADI activity begins with the teacher identifying a phenomenon to investigate and offering a guiding question for the students to answer. The goal of the teacher during this stage of the model is to create a need for students to use DCIs, CCs, and SEPs to figure something out. First, you will provide each student with a copy of an Investigation Handout (or if using the student workbook, ask them to turn to the Investigation Log). Read the first paragraph of the "Introduction" in the Investigation Handout or the Investigation Log aloud and then direct the students to explore a phenomenon for a few minutes. This exploration can be a firsthand or secondhand experience with a phenomenon. Examples of firsthand experiences include watching a tea bag being added to a cup of hot water, adding a few drops of vinegar to baking soda and watching what happens over time, or tracking how the direction and length of a shadow changes over the course of an afternoon. A secondhand experience, in contrast, typically involves watching a video of something that happens. An example of a video that provides a good secondhand experience with a phenomenon

An Overview of Argument-Driven Inquiry

might be a time-lapse video of the night sky or a plant growing from a seed.

This brief exploration of a phenomenon is designed to encourage students to ask questions and create a need for them to figure something out. Students should be encouraged to record what they observe and any questions they might have during this brief exploration (see Figure 2). Emerging bilingual students should be allowed to use English, their native language, or some combination of the two (see Figure 2). You should then give the students an opportunity to share their observations and questions with the rest of the class. At this point, students are interested and want to know more about the phenomenon.

FIGURE 2

A student recording observations of a phenomenon in the Investigation Handout

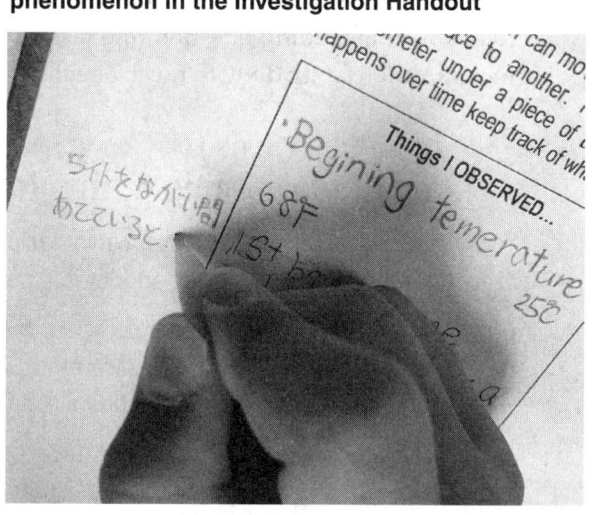

You can then have the students read the rest of the "Introduction" *or* ask them to follow along as you read it aloud. (In some investigations, this reading is divided into two parts—for example, reading the next two paragraphs before you do a demonstration, and then reading the rest of the "Introduction.") The remainder of the text included in the "Introduction" provides a brief overview of one or more DCIs and CCs or some other important ideas that the students can use to make sense of what they are seeing and doing during the rest of the investigation. The ideas found in the "Introduction" are *not* what the students need to figure out during the investigation; instead, these ideas are designed to function as tools for students to use during an investigation to help them make sense of a phenomenon. Your goal at this stage is to *put these ideas on the table* so students can work with them and learn more about them as they design and carry out an investigation (SEP 3), analyze and interpret the data they collect (SEP 4), construct an explanation (SEP 6), and argue from evidence (SEP 7). The goal here is *not* to front-load vocabulary or to tell students what they "should learn" by the end of the investigation. Be sure to give students the opportunity to discuss any ideas that they think are important or useful with their classmates after you read the "Introduction" aloud or have the students read it on their own.

It is also important for the teacher to hold a "tool talk" (Blanchard and Sampson 2018) during this stage, taking a few minutes to explain how to use the available materials and equipment. The tool talk is important because fifth-grade students are often unfamiliar with these materials and equipment. Even if the students are familiar with them, they may use them incorrectly or in an unsafe manner unless they are reminded about how they work and the proper way to use them. You should therefore review specific safety protocols and precautions as part of the tool talk. Including a tool talk during this stage is useful because students often find it difficult to design a method to collect the data needed to answer the guiding question (the task of stage 2) when they do not understand what they can and cannot do with the available materials and equipment.

> **Keep the following points in mind during stage 1 of ADI:**
> - The initial activity is important. It is designed to provide a phenomenon to explain, trigger students' curiosity, and "create a need to read." Do not skip it.
> - Students will likely use some or most of the information that they include in the "Things we KNOW from what we read…" box to help justify their evidence in their arguments during stage 3 of the investigation.
> - Don't worry if students "don't get it" yet or struggle to comprehend what they are reading at this point; they will revisit the text later in the lesson.
> - There are many supports for helping students comprehend what they read (i.e., activating prior knowledge, providing a shared experience, making connections, synthesizing, and talking with peers) already embedded into this stage. You might not need to provide much extra support.

Once all the students understand the goal of the investigation and how to use the available materials, you should divide the students into small groups (we recommend three or four students per group) and move on to the second stage of the instructional model.

Stage 2: Design a Method and Collect Data

Small groups of students develop a method to gather the data they need to answer the guiding question and then carry out that method during this stage of ADI. The overall intent of this stage is to provide students with an opportunity to use one or more DCIs and one or more CCs to plan and carry out an investigation (SEP 3). It also gives students an opportunity to learn how to use appropriate data collection techniques, which include but are not limited to controlling variables, making multiple observations, and quantifying observations, and how to use different types of data collection tools, such as rulers to measure length, scales to measure mass, and graduated cylinders to measure volume. This stage of ADI also gives students a chance to see why some approaches to data collection and tools work better than others and how the method used during a scientific investigation is based on the nature of the question and the phenomenon under investigation. Students even begin to learn how to deal with the uncertainties that are associated with all empirical work as they discover the importance of attending to precision when they take measurements and attempt to eliminate factors that may change the results of their tests.

This stage begins with students discussing two questions that are designed to encourage the students to use one or two CCs as a lens to determine what data they need to collect and how they should collect it (see the "Plan Your Investigation" section of the Investigation Handout or the Investigation Log). For example, students might be asked to discuss a question such as:

- What types of *patterns* might we look for to help answer the guiding question?
- What information do we need to find a relationship between a *cause* and an *effect*?
- What measurement *scale* and *units* might we use as we collect data?
- How might *energy transfer into, within, or out of the system?*
- What are the *components of the system* we are studying and how do they *interact*?
- How might the *structure* of what you are studying relate to its *function*?

Once students have discussed these questions in their small groups and shared their ideas with the rest of the class, each small group can begin to plan their investigation. To facilitate this process, you should direct the students to fill out the graphic organizer in the "Plan Your Investigation." The graphic organizer helps guide students through the process of planning an investigation by encouraging them to think about what type of data they will need to collect and how to collect it. Figure 3 shows a group of students working together to plan an investigation using the graphic organizer.

FIGURE 3

A group of students working together to plan an investigation

If the students get stuck as they are planning their investigation, you should not tell them what to do. Instead, bring over some of the available materials and ask them probing questions, such as "I have all these materials, what data do I need to collect?" or "Now that we know what data we need, what should we do first?" Once the students finish filling out the investigation proposal, look it over and either approve it or offer suggestions about how to improve their plan for collecting data during the investigation. If you identify a flaw in their plan, ask probing questions to help the students identify and correct the flaw rather than just telling them what to fix. Examples of probing questions to help students identify a flaw in the investigation proposal include "I'm not sure what you mean here, could you explain that another way?" or "Do you think you have all the information you need to answer the guiding question?"

> **Keep the following points in mind during stage 2 of ADI:**
> - Not all investigations in science are experiments. Scientists use different methods to answer different types of questions.
> - The graphic organizer found in the "Plan Your Investigation" section of the Investigation Handout or the Investigation Log is designed to help students think about how to answer the guiding question of the investigation, what data they need to collect, and how they will need to analyze it; it is not a scientific method.
> - The graphic organizer makes student thinking about investigation design visible so teachers can use it as an embedded formative assessment.
> - Students can use some or most of the information that they include in the graphic organizer to help write their investigation report in stage 6.

FIGURE 4

A group of students working together to collect data

It is important to remember that all the student-designed investigations do not need to be the same. The students will learn more about how to do science during the later stages of the instructional model when the groups use different methods to collect and analyze data. The groups should then carry out their plan and collect the data they need to answer the guiding question once you approve their proposal (see Figure 4).

Stage 3: Create a Draft Argument

The next stage of the instructional model calls for each group to create a draft argument. To accomplish this task, the students must first analyze the measurements (e.g., temperature and mass) and/or observations (e.g., appearance and location) they collected during stage 2 of ADI. Once the groups have analyzed and interpreted the results of their analysis (SEP 4), they will need to make sense of the phenomenon based on what they found out. They can then develop an answer to the guiding question based on what they figured out. Often, but not always, developing an adequate answer to the guiding question requires the students to construct an explanation (SEP 6). The students can then create a draft argument to share what they have learned with the other students in the class. The intent of having students craft arguments is to encourage them to focus not only on "what they figured out" but also on "how they know what they know." The investigations that give students opportunities to analyze and interpret measurements that they collected during stage 2 also enable students to learn concepts and practices outlined in the *Common Core State Standards for Mathematics* (NGAC and CCSSO 2010).

The argument that the students create consists of a claim, the evidence they are using to support their claim, and a justification of their evidence. The *claim* is their answer to the guiding question and thus is often an explanation for how or why something happens. The *evidence* consists of an analysis of the data they collected and an interpretation of the analysis. The evidence often includes a graph that shows a difference between groups, a trend over time, or a relationship between variables; it also includes a statement or two that explains what the analysis means (but not why it matters). The *justification of the evidence* is a statement that explains why the evidence matters. The justification of the evidence, in other words, is used to defend the choice of evidence by making the DCIs, CCs, and/or assumptions underlying the collection of the data, the analysis of the data, and interpretation of the analysis explicit so other people can understand why the evidence is important and relevant. The components of a scientific argument are illustrated in Figure 5 (p. 10). Crafting an argument that consists of these three components helps students learn how to argue from evidence (SEP 7).

It is not enough for students to be able to include all the components of argument when they have an opportunity to argue from evidence. It is also important for students to understand that, in science, some arguments are better than others. Therefore, an important aspect of arguing from evidence in science involves the evaluation of the various components of the arguments put forward by others. The framework provided in Figure 5 highlights two types of criteria that students can and should be encouraged to use to evaluate an argument in science: empirical criteria and theoretical criteria. *Empirical criteria* include

- how well the claim fits with all available evidence,
- the sufficiency of the evidence,

- the relevance of the evidence,
- the appropriateness and rigor of the method used to collect the data, and
- the appropriateness and soundness of the method used to analyze the data.

Theoretical criteria refer to standards that allow us to judge how well the various components are aligned with the DCIs and CCs of science; examples of these criteria are

- the sufficiency of the claim (i.e., Does it include everything needed?);
- the usefulness of the claim (i.e., Does it help us understand the phenomenon?);
- how consistent the claim is with accepted theories, laws, or models (e.g., Does it fit with our current understanding of motion and forces?); and
- how consistent the interpretation of the results of the analysis is with accepted theories, laws, or models (e.g., Is the interpretation based on what we know about the relationship between structure and function?).

FIGURE 5

The components of a scientific argument and some criteria for evaluating an argument

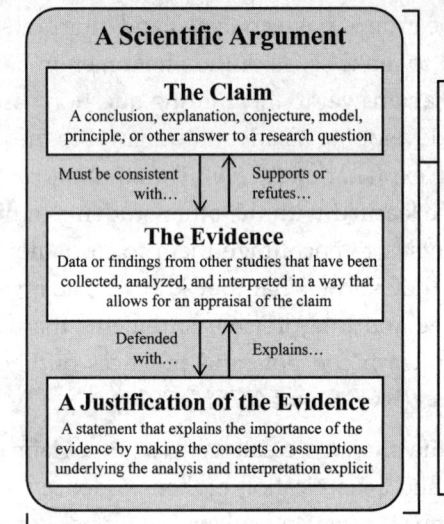

A Scientific Argument

The Claim
A conclusion, explanation, conjecture, model, principle, or other answer to a research question

Must be consistent with… ↕ Supports or refutes…

The Evidence
Data or findings from other studies that have been collected, analyzed, and interpreted in a way that allows for an appraisal of the claim

Defended with… ↕ Explains…

A Justification of the Evidence
A statement that explains the importance of the evidence by making the concepts or assumptions underlying the analysis and interpretation explicit

The quality of an argument is evaluated using…

Empirical Criteria
- The claim is consistent with the evidence.
- The amount of evidence is sufficient.
- The evidence is relevant.
- The method used to collect the data was appropriate and rigorous.
- The method used to analyze the data was appropriate and sound.

Theoretical Criteria
- The claim is sufficient (includes everything needed).
- The claim is useful (helps us understand the phenomenon we are studying).
- The claim is consistent with current theories, laws, or models.
- The interpretation of the data analysis is consistent with current theories, laws, or models.

The generation and evaluation of an argument are shaped by…

Discipline-Specific Norms and Expectations
- The theories and laws used by scientists within a discipline
- The methods of inquiry that are accepted by scientists within a discipline
- Standards of evidence shared by scientists within a discipline
- How scientists communicate with each other within a discipline

What counts as quality in terms of these different criteria varies from discipline to discipline (e.g., biology, geology, physics, chemistry) and within the specific fields of each discipline (e.g., the fields of ecology, botany, and zoology within the discipline of biology). This variation in what counts as quality is due to differences in the types of phenomena that the scientists within these disciplines or fields investigate (e.g., changes in populations over time, how a trait is inherited in a plant, an adaptation of an animal), the types of methods they use (e.g., descriptive studies, experimentation, computer modeling), and the different DCIs that they use to figure things out. It is therefore important to keep in mind that what counts as a quality argument in science is discipline and field dependent.

An Overview of Argument-Driven Inquiry

Each group of students should create their draft argument in a medium that can easily be viewed by the other groups. This is important because each group will share their draft argument with the other students in the class during the next stage of ADI (stage 4). We recommend that students use dry-erase markers to create their draft argument on a 2' × 3' whiteboard during this stage (see Figure 6). Students can also create their draft arguments using presentation software such as Microsoft's PowerPoint, Apple's Keynote, or Google Slides and devote one slide to each component of an argument. They can then share their arguments with others using a tablet or a laptop (see Figures 11 and 12 later in this chapter). The choice of medium is not important as long as students are able to easily modify the content of their argument as they work and other students can easily see each component of their argument. Students should include the guiding question of the investigation and the three main components of an argument on the whiteboard or other medium. Figure 7 shows the general layout for a presentation of an argument.

Figures 8 and 9 (p. 12) provide examples of an argument crafted by students. Notice that the argument in Figure 8 is written in English and the argument in Figure 9 is written in Spanish. To help support emerging bilingual students in learning science and English at the same time, it is important to allow students to communicate their ideas in their first language, English, or a combination of the two. This is important because (a) it

FIGURE 6

A group of students creating a draft argument on a whiteboard

FIGURE 7

The components of an argument that should be included on a whiteboard (outline)

The Guiding Question:	
Our Claim:	
Our Evidence:	Our Justification of the Evidence:

Note: This outline is referred to as the "Argument Presentation on a Whiteboard" image in stage 3 of each investigation.

FIGURE 8
An example of a student-generated argument on a whiteboard

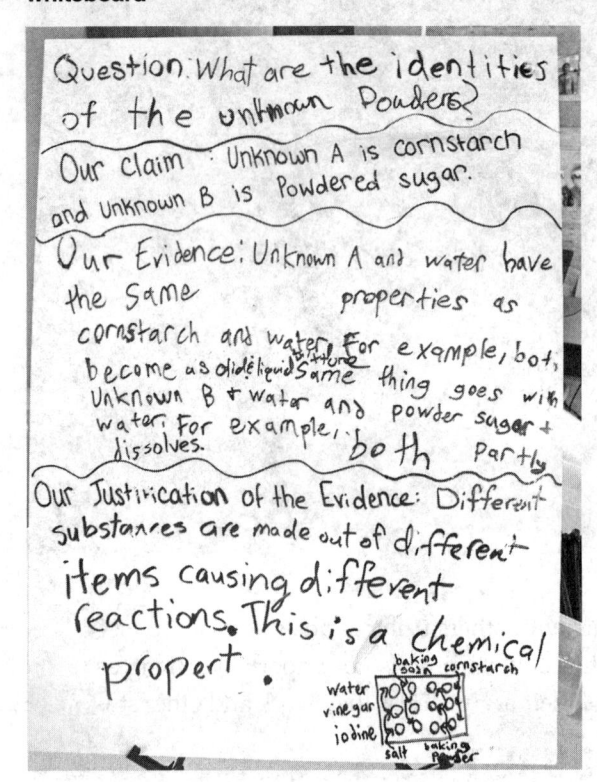

FIGURE 9
An example of a student-generated argument written in Spanish

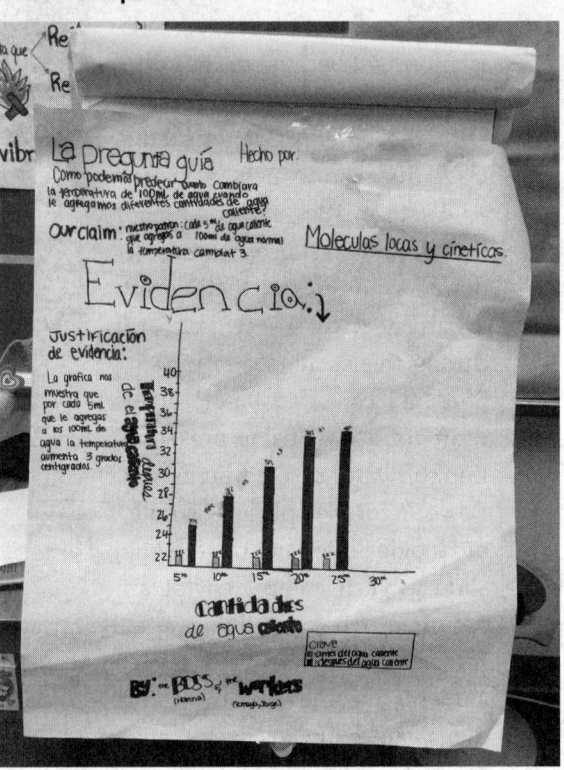

allows students to use their home language and culture as a resource for making sense of the phenomenon (Escamilla and Hopewell 2010; Goldenberg and Coleman 2010; González, Moll, and Amanti 2005), (b) students learn new languages through meaningful use and interaction (Brown 2007; García 2005; García and Hamayan 2006), and (c) students' academic language development in their native language facilitates their academic language development in English (Escamilla and Hopewell 2010; García and Kleifgen 2010; Gottlieb, Katz, and Ernst-Slavit 2009).

This stage of the model can be challenging for students in fifth grade because they are rarely asked to make sense of a phenomenon based on raw data, so it is important for teachers to actively work to support their sense-making. In this stage, you should circulate from group to group to act as a resource person for the students, asking questions that prompt them to think about what they are doing and why. To help students remember the goal of the activity, you can ask questions such as "What are you trying to figure out?" You can also ask them questions such as "Why is that information important?" or "Why is

An Overview of Argument-Driven Inquiry

that analysis useful?" to encourage them to think about whether or not the data they are analyzing are relevant or the analysis is informative. To help them remember to use rigorous criteria to determine if a claim is acceptable or not, you can ask, "Does that fit with all the data?" or (for an investigation concerning matter and interactions) "Is that consistent with what we know about the properties of matter?"

> **Keep the following points in mind during stage 3 of ADI:**
> - If you uncover a flaw or if something important is missing in an argument as you move from group to group, ask them probing questions such as "I see you put a table here, but you did not really explain what we should pay attention to. Is there a way to help your classmates understand what is really important in this table?" or "I'm not sure what you mean here, could you explain that another way?"
> - All the arguments should not be the same (or perfect) at this point in the lesson. The argumentation session during stage 4 will be more interesting (and students will learn more about how to do science) if each group has a different claim, evidence, and justification of the evidence so there is something to discuss.
> - The students will have an opportunity to revise their draft argument at the end of the argumentation session (stage 4 of the lesson).
> - One of the best ways to ensure that important ideas spread through the class is to help one or two groups develop a very strong component of their argument. Other groups will see these examples and add a version of them to their own arguments during the argumentation session. For example, help one group make a perfect graph, another group write out a perfect interpretation of their analysis, and a third group include a core idea that the other groups are missing in their justification.

It is important to remember that at the beginning of the school year, students will struggle to develop arguments and will often rely on inappropriate criteria such as plausibility (e.g., "That sounds good to me") or fit with personal experience (e.g., "But that is what I saw on TV once") as they attempt to make sense of the phenomenon, construct explanations, and support their ideas with sufficient evidence and an adequate justification of the evidence. However, as students learn why it is useful to use evidence in an argument, what makes evidence valid or acceptable from a scientific perspective, and the importance of providing a justification for their evidence through repeated practice, *students will improve their ability to argue from evidence* (Grooms, Enderle, and Sampson 2015; Strimaitis et al. 2017). This is an important principle underlying the ADI instructional model.

Stage 4: Argumentation Session

The fourth stage of ADI is the argumentation session. In this stage, each group is given an opportunity to share, evaluate, and revise their draft arguments. The process of sharing their arguments with their classmates requires students to communicate their explanations for the phenomenon under investigation (SEP 8) and support their ideas with evidence (SEP 7). Other students in the class are expected to listen to presentations of these arguments, ask questions as needed (SEP 1), evaluate the arguments based on empirical and theoretical criteria (SEP 4 and SEP 7), and then offer critiques (SEP 8) along with suggestions for improvement. At the end of this stage the students are expected to revise their draft arguments based on what they learned from interacting with other students and seeing examples of other arguments.

This stage is included in the ADI instructional model because research indicates that students develop a better understanding of DCIs and CCs, learn how to argue from evidence, and acquire better critical-thinking skills when they are exposed to alternative ideas, respond to the questions and challenges of other students, and are encouraged to evaluate the merits of competing ideas (Duschl, Schweingruber, and Shouse 2007; NRC 2012). Research also suggests that students learn how to distinguish between alternative ideas using rigorous scientific criteria and are able to develop scientific habits of mind (such as treating ideas with initial skepticism, insisting that the reasoning and assumptions be made explicit, and insisting that claims be supported by valid evidence) when they have an opportunity to participate in these argumentation sessions (Sampson, Grooms, and Walker 2011). This stage provides the students with an opportunity to learn from and about the practice of arguing from evidence.

FIGURE 10

A group of students presenting their arguments to the other groups in the class during a whole-class presentation argumentation session

There are three types of formats that teachers can use during the argumentation session. One format is the *whole-class presentation format*, in which each group gives a presentation to the whole class (see Figure 10). This format is often useful when students are first learning how to propose, support, evaluate, challenge, and refine

An Overview of Argument-Driven Inquiry

ideas in the context of science, because the teacher can provide more support as students interact with each other. This format begins with the teacher asking a group to present their argument to the class. The teacher can then encourage students in the class to ask questions, offer critiques, and give the presenters suggestions about ways to improve. The teacher can also ask questions as needed. We also recommend that the presenters keep a record of the critiques made by their classmates and any suggestions for improvement. The students who listen to the presentation should also be encouraged to keep a record of good ideas or potential ways to improve their own arguments by recording them in the "Argumentation Session" section of their Investigation Handout or Investigation Log.

Another format is called the *gallery walk*. In this format, each group sets up their argument so others can see it at their workstation. The entire group then moves to a different workstation. While at a different workstation, a group is expected to read the argument at that workstation (which was created by a different group) and offer critiques and suggestions for improvement (see Figures 11 and 12). Notice in Figures 11 and 12 that the students in this classroom created their draft argument on tablets and then used the tablets to share their argument with their classmates. The students should move to a different workstation after a few minutes so they can read and critique an argument written by a different group. Students should also be encouraged to keep a record of good ideas or potential ways to improve their own arguments as they move from workstation to workstation to offer critiques and feedback.

After repeating this process in the gallery walk three or four times, every group will have had an opportunity to read and critique three or four different arguments and receive feedback from three or four different groups. Students often use sticky notes to provide feedback to each other during a gallery walk argumentation session (see Figures 11 and 12). This type of format is useful because everyone in the classroom will be actively engaged during the

FIGURE 11
Students critiquing arguments and providing feedback during a gallery walk argumentation session. Students in this class created draft arguments using a tablet.

FIGURE 12
Some of the feedback students gave to each other during a gallery walk argumentation session using sticky notes

Argument-Driven Inquiry in **Fifth-Grade Science**: Three-Dimensional Investigations

argumentation session and will have a chance to see different arguments. This format, however, does not give students a chance to support, critique, and challenge the ideas, explanations, and arguments of other groups through talk, because two different groups are never at the same workstation at the same time.

A third format is called the *modified gallery walk*. We recommend that teachers use this format rather than a whole-class presentation format or gallery walk format whenever possible because it provides more opportunities for student-to-student talk and ensures that all ideas are heard and that all students are actively involved in the process. This is especially important for helping students develop speaking and listening skills. It also provides a context for emerging bilingual students to use both productive (speaking and writing) and receptive (listening and reading) language and to learn science through meaningful interactions (Brown 2007; García and Hamayan 2006). In the modified gallery walk format (see Figure 13), one or two members of the group stay at their workstation to share their group's ideas (we call these students *presenters*) while the other group members go to different groups one at a time to listen to and critique the arguments developed by their classmates (we call the students who go to different groups *travelers*). We recommend that travelers visit at least three different workstations during the argumentation session. We also recommend that the presenters keep a record of the critiques made by their classmates and any suggestions for improvement (see Figure 14). The travelers should also be encouraged to keep a record of good ideas or potential ways to improve their own arguments as they travel from group to group. The presenters and the travelers can record this information in the "Argumentation Session" section of the Investigation Handout or the Investigation Log.

It is important to note, however, that supporting and promoting productive interactions between students inside the classroom can be difficult because the practices of arguing from evidence (SEP 7) and obtaining, evaluating, and communicating information in science (SEP 8) are foreign

FIGURE 13

An example of a modified gallery walk argumentation session. Notice that in this format there are multiple discussions going on at the same time.

FIGURE 14

An example of a modified gallery walk argumentation session. Notice that the presenter and a traveler are recording good ideas that they can use to improve their argument.

to most students when they first begin participating in ADI. In a non-ADI classroom, students are not typically expected to think about or engage with the ideas, explanations, or arguments of their classmates, so students are often reluctant to ask questions (SEP 1) in this type of context. To encourage more productive interactions between students, materials, and ideas, the ADI instructional model requires students to generate their arguments in a medium that can be seen by others. By looking at the ideas that are presented on whiteboards, tablets, or paper, students tend to focus their attention on evaluating the claim, evidence, and justification and how well the various components of the argument align with the DCI(s) and CC(s) that students are using to figure things out rather than attacking the source of the ideas. This strategy often makes an argumentation session more productive and makes it easier for students to identify and weed out faulty ideas. It is also important for the students to view the argumentation session as an opportunity to learn. The teacher, therefore, should describe the argumentation session as an opportunity for students to collaborate with their peers and as a chance to give each other feedback so the quality of all the arguments in the classroom can be improved, rather than as an opportunity to determine who is right or wrong. This is why we ask students to keep track of what they see and hear so they can use the ideas of other to make their own argument better.

Keep the following points in mind during stage 4 of ADI:

- Make sure to stress that the goal of the argumentation session is to share ideas and to help each other out. The goal of the argumentation session is not to see who has the best argument. Students need to see value in the ideas of others and view their classmates as a resource. This does not happen when students are trying to show others that they have the best argument.

- Be sure to remind students before the argumentation session starts that they will have an opportunity to revise their arguments based on what they learn from their peers at the end of the argumentation session. This is important because students need to have a reason to engage with the ideas of others.

- Encourage students to talk to each other during this stage. It is really important to give students voice and choice during the argumentation session.

- Use a note card to keep track of interesting ideas you hear or see, good examples, or important contributions that were made by different students when the students are sharing and critiquing the arguments. You can bring these ideas, examples, or contributions up at the end of this stage (before students begin revising their arguments) or during stage 5 (the reflective discussion).

Chapter 1

Just as is the case in earlier stages of ADI, it is important for the classroom teacher to be involved in (without leading) the discussions during the argumentation session. Once again, you should move from group to group or from student to student to keep everyone involved and model what it means to argue from evidence. You can ask the presenters questions such as "Why did you decide to analyze the available data like that?" or "Were there any data that did not fit with your claim?" to encourage students to use empirical criteria to evaluate the quality of the arguments. You can also ask the presenters to explain how the claim they are presenting fits with a DCI or a CC or to explain why the evidence they used is important based on a DCI or a CC. In addition, you can ask the students who are listening to a presentation questions such as "Do you think their analysis is accurate?" or "Do you think their interpretation is sound?" or even (in the case of an investigation related to matter and its interactions) "Do you think their claim fits with what we know about the properties of matter?" These questions can serve to remind students to use empirical and theoretical criteria to evaluate an argument during the discussions. Overall, it is the goal of the teacher at this stage of the lesson to encourage students to think about how they know what they know and why some claims are more valid or acceptable in science. This stage of the model, however, is not the time to tell the students that they are right or wrong.

FIGURE 15

Students discussing what they learned from other groups as they revise their draft arguments at the end of the argumentation session

At the end of the argumentation session, it is important to give students time to meet with their original group so they can discuss what they learned by interacting with individuals from the other groups and they can revise their initial arguments (see Figure 15). This process can begin with the presenters sharing the critiques and the suggestions for improvement that they heard during the argumentation session. The students who visited the other groups during the argumentation can then share their ideas for making the arguments better based on what they observed and discussed at other workstations. Students often realize that the way they collected or analyzed data was flawed in some way at this point in the process. You should therefore encourage students to collect new data or reanalyze the data they collected as needed, and you can also give students time to conduct additional tests of ideas or claims. At the end of this stage, each group should have a final argument that is much better than their draft argument.

Stage 5: Reflective Discussion

You should lead a whole-class reflective discussion during stage 5 of ADI. The goals of this discussion are to give students an opportunity to think about and share what they know and how they know it and to ensure that all students understand the DCI(s) and the CC(s) they used during the investigation to make sense of the phenomenon. This discussion also encourages students to think about ways to improve their participation in scientific practices such as planning and carrying out investigations, analyzing and interpreting data, and arguing from evidence. At this point in the instructional sequence, you should also encourage students to think about a nature of scientific knowledge (NOSK) or nature of scientific inquiry (NOSI) concept (see Appendix 2).

It is important to emphasize that the reflective discussion is not a lecture; it is an opportunity for students to think about important ideas and practices and to share what they know or do not understand. The more students talk during this stage, the more meaningful the experience will be for them and the more you can learn about their thinking. Table 1 (p. 20) lists some important talk goals for students (e.g., encourage students to support an idea, encourage students to think with others) and some talk moves (e.g., press for support, invite contributions) that you can use to make progress toward these goals during the reflective discussion. Table 1 also includes examples of questions that you can ask students as you facilitate the discussion. These questions are organized by talk goal and move (Michaels and O'Connor 2012; Schwarz, Passmore, and Reiser 2017). For example, if you have a student talk goal of "encourage students to share, expand on, or clarify an idea" at some point during the reflective discussion, you can use the "share an idea" talk move by asking the class a question such as "How might we explain that?" or the "expand on an idea" talk move by asking the class a question such as "Can you tell me more about that?" These talk goals and moves are only a suggestion; however, we find them useful when we facilitate a whole-class discussion.

Begin the discussion by asking students to share what they know about the DCI(s) and the CC(s) they used to figure things out during the investigation. You can show several images as prompts and then ask questions to encourage students to share how they are thinking about the DCI(s) and the CC(s) they used earlier to explain the phenomenon under investigation and to provide a justification of the evidence in their arguments. You should *not* tell the students what results they should have obtained or what information should be included in each argument. Instead, you should focus on the students' thoughts about the DCI(s) and the CC(s) by providing a context for students to share their views and explain their thinking. Remember, this stage of ADI is a *discussion*, not a presentation about what the students "should have seen" or "should have learned." We provide recommendations about what teachers can do and the types of questions that teachers can ask to facilitate a productive discussion about the DCI(s) and the CC(s) during this stage as part of the Teacher Notes for each investigation.

TABLE 1

Student talk goals and teacher talk moves

Talk goal	Talk move	Examples
Encourage students to put forward an idea	Share an idea	• What do you see going on here? • What do you think is happening? • How might we explain that? • What do we need to do to solve this problem? • What are you thinking about?
	Expand on an idea	• Can you tell me more about that? • What do you mean by that? • Can you give me an example of that? • Is that all you wanted to say?
	Clarify an idea	• Can you say that in another way? • I think you are saying [rephrase]. Is that what you mean? • What do you mean by that?
Encourage students to think about the acceptability of an idea	Press for support	• Why do you think that? • What is your evidence for that? • How did you arrive at that conclusion? • What makes you say that? • How do you know?
	Critique an idea	• Do you agree or disagree with that idea and why? • What do we think about that idea? • Does that fit with what we have figured out so far? • Does it always work that way? • Is there another way to do it?
Encourage students to think with others	Invite contributions	• Would anyone like to add to that idea? • What should we keep or change about that idea? • Can anyone else give me some examples of that? • Can anyone take that idea and push it a little further? • Does anyone want to respond to that idea?
	Re-voice an idea	• Who can explain that to me in another way? • Can anyone share what [name] said again for those who might have missed it?
Encourage students to think about what they know and what they don't know yet	Assess progress	• I'm confused now. What have we figured out so far? • So, what do we all agree on? • What are we sure and not so sure about? • What do you think are the most important ideas that we have been talking about so far?
	Move forward	• What do we need to talk about next? • What are we still trying to figure out? • What questions do we have now?

Source: Adapted from Michaels and O'Connor (2012) and Schwarz, Passmore, and Reiser (2017).

An Overview of Argument-Driven Inquiry

Next, you should switch the topic of discussion to a NOSK or NOSI concept, using what the students did during the investigation to help illustrate one of these important concepts (NGSS Lead States 2013). This stage provides a golden opportunity for explicit instruction about NOSK and how this knowledge develops over time in a context that is meaningful to the students. For example, you can use the investigation as a way to illustrate a NOSK concept such as the differences between observations and inferences, data and evidence, or theories and laws. You can also use the investigation as a way to highlight a concept related to NOSI. For example, you might discuss the following concepts:

- How scientists investigate questions about the natural world
- The wide range of methods that scientists can use to collect data
- How science is a way of knowing
- The assumptions that scientists make about order and consistency in nature

Research in science education suggests that students only develop an appropriate understanding of NOSK and NOSI concepts when teachers *explicitly* discuss them as part of a lesson (Abd-El-Khalick and Lederman 2000; Lederman and Lederman 2004; Schwartz, Lederman, and Crawford 2004). In addition, by embedding a discussion of a NOSK concept or a NOSI concept into each investigation, teachers can highlight these important concepts over and over again throughout the school year rather than just focusing on them during a single unit. This type of approach makes it easier for students to learn these abstract and sometimes counterintuitive concepts. As part of the Teacher Notes for each investigation, we provide recommendations about which concepts to focus on and examples of questions that teachers can ask to facilitate a productive discussion about these concepts during this stage of the instructional sequence.

Keep the following points in mind during stage 5 of ADI:
- Keep this stage short—about 15 minutes.
- Make sure that students are doing most of the talking during this stage.
- Your goal during the reflective discussion is to figure out what students are thinking and build on their ideas as needed.
- Make sure that every student has an opportunity to contribute to the discussion and that they feel like their ideas and contributions are valued.

You should end this stage by encouraging the students to think about what they learned about the practices of science and what they should do when they engage in these practices in the future. This is important because students are expected to design their own investigations, decide how to analyze and interpret data, and support their claims with evidence in every ADI investigation. These practices are complex, and students cannot be expected to master them without being given opportunities to try, fail, and then learn from their mistakes. To encourage students to learn from their mistakes during an investigation, students must have an opportunity to reflect on what went well and what went wrong. You should therefore encourage the students to think about what they did during their investigation, how they chose to analyze and interpret data, how they decided to argue from evidence, and what they could have done better. You can then use the students' ideas to highlight what does and does not count as quality or rigor in science and to offer advice about ways to improve in the future. Over time, students will gradually improve their abilities to participate in the practices of science as they learn what works and what does not. To help facilitate this process, we provide questions that teachers can ask students to help elicit their ideas about the practices of science and set goals for future investigations in the Teacher Notes for each investigation.

Stage 6: Write a Draft Report

Stage 6 is included in the ADI model because writing is an important part of doing science. Scientists must be able to read and understand the writing of others as well as evaluate its worth. They also must be able to share the results of their own research through writing. In addition, writing helps students learn how to articulate their thinking in a clear and concise manner, encourages metacognition, and improves student understanding of DCIs and CCs (Wallace, Hand, and Prain 2004). Finally, and perhaps most important, writing makes each student's thinking visible to the teacher (which facilitates assessment) and enables the teacher to provide students with the educative feedback they need to improve.

In stage 6, each student is required to write an investigation report using his or her Investigation Handout or Investigation Log and his or her group's argument as a starting point. The report should address three fundamental questions:

1. What were you trying to figure out and why?
2. What did you do to answer your question and why?
3. What is your argument?

The format of the report is designed to emphasize the persuasive nature of science writing and to help students learn how to communicate in multiple modes (words, figures, tables, and equations). The three-question format is well aligned with the components of research articles (i.e., introduction, method, results, and discussion) but allows students to see the important role argument plays in science. We have included sentence starters in the "Draft Report" section of the Investigation Handout and the Investigation Log to help facilitate the writing process (see Figure 16). The sentence starters are intended to act a guide for students as they learn to write in the context of science and should make the assignment less intimidating for students. We recommend that you use these sentence

An Overview of Argument-Driven Inquiry

starters to encourage students to write in a clear and concise manner. Over time, you may decide that you no longer need to use them.

Stage 6 of ADI is important because it allows students to learn how to construct an explanation (SEP 6), argue from evidence (SEP 7), and communicate information (SEP 8). It also enables students to master the disciplinary-based writing skills outlined in the *Common Core State Standards for English Language Arts* (*CCSS ELA;* NGAC and CCSSO 2010). As discussed in "Stage 3: Create a Draft Argument," allowing emerging bilingual students to write and communicate what they have learned in their first language, English, or a combination of the two will help support these students in learning science and English at the same time. It is important to scaffold this process for the emerging bilingual students based on their current level of English language proficiency because (a) students' development of academic language and academic content knowledge are interrelated processes (Echevarría, Short, and Powers 2006; Gottlieb, Katz, and Ernst-Slavit 2009) and (b) students' access to instructional tasks that require complex thinking is enhanced when linguistic complexity and instructional support match their levels of language proficiency (Gottlieb, Katz, and Ernst-Slavit 2009).

FIGURE 16
Students writing a draft report to share what they figured out during the investigation

Keep the following points in mind during stage 6 of ADI:

- There are many writing supports (i.e., sentence starters, pre-writing, graphic organizers) already embedded into this stage. You might not need to provide much extra support for your students.

- Don't worry if students do not produce a perfect report at this point. All the reports will be reviewed and revised before they are turned in to you. The reports should be viewed as a starting point. Encourage students to "just get something down on paper" so "we can work on it together."

- Writing is an important component of this model. Do not skip this stage. When a student writes the report on his or her own, it not only helps each student *learn to write* in the context of science (an important literacy skill) but also gives each student an opportunity to *write to learn* (develop a better understanding of the core ideas of science by writing about them).

- Do not overscaffold the writing progress. Mistakes are opportunities to learn. Do not take these opportunities away from your students.

Stage 7: Peer Review

The students have an opportunity to review the reports in pairs using the peer-review guide and teacher scoring rubric (PRG/TSR; see Appendix 4) during the seventh stage of ADI. The PRG/TSR contains specific criteria that are to be used by a pair of students as they evaluate the quality of each section of the investigation report as well as the quality of the writing. There is also space for the reviewers to provide the author with feedback about how to improve the report. Once a pair of students finishes reviewing a report as a team, they are given another report to review. When students are grouped together in pairs, they only need to review two different reports. Be sure to give students only 15 minutes to review each report (we recommend setting a timer to help manage time). When students are grouped into pairs and given 15 minutes to complete each review, the entire peer-review process can be completed in 30 minutes (2 different reports × 15 minutes = 30 minutes).

> **Keep the following points in mind during stage 7 of ADI:**
> - The goal of the peer-review process is not to assign grades. Rather, it gives students an opportunity to give and receive feedback about their writing.
> - The peer-review process is one of the best ways for students to learn how to write. When students read and review two different reports, they have an opportunity to (1) see examples of texts written by others for the same purpose, (2) discuss what counts as quality and why some reports are stronger that others, (3) discuss ways to strengthen a report, and (4) pick up things that they can do to improve their own reports.
> - You can review the writing of the emerging bilingual students in your class depending on their English language proficiency, but make sure that your emerging bilingual students participate in the peer-review process.
> - Don't skip this step—you will see tremendous growth in your students' writing skills the more they participate in the peer-review process.

Reviewing each report as a pair using the PRG/TSR is an important component of the peer-review process because it provides students with a forum to discuss what counts as high quality or acceptable and, in so doing, forces them to reach a consensus during the review process. This method also helps prevent students from checking off "yes" for each criterion on the PRG/TSR without thorough consideration of the merits of the report. It is also important for students to provide constructive and specific feedback to the author when areas of the report are found to not meet the standards established by the PRG/TSR. The peer-review process provides students with an opportunity to read good and bad examples of the reports. This helps the students learn new ways to organize and present information, which in turn will help them write better on subsequent reports. It also provides an opportunity and a mechanism for all students in the classroom, including emerging bilingual students, to develop shared norms for what counts as high-quality writing in the context of science.

This stage of the model is intended to give students opportunities to ask questions (SEP 1) and obtain and evaluate information (SEP 8). This stage also gives students an opportunity to develop the reading skills that they need to be successful in science. Students must be able to determine the central ideas or conclusions of a text and determine the meaning of symbols, key terms, and other domain-specific words. In addition, students must be able to assess the reasoning and evidence that an author includes in a text to support his or her claim and compare or contrast findings presented in a text with those from other sources when they read a scientific text. Students can develop all these skills, as well as the other discipline-based reading skills found in the *CCSS ELA*, when they are required to read and critically review reports written by their classmates. This stage is also beneficial for emerging bilingual students because students learn language through meaningful use, such as reading and reviewing a report, and interaction with others (Brown 2007; García 2005; García and Hamayan 2006).

Stage 8: Revise the Report

The final stage in the ADI instructional model is to revise the report. Each student is required to rewrite his or her report using the reviewers' comments and suggestions as a guideline. The author is also required to explain what he or she did to improve each section of the report in response to the reviewers' suggestions in the author response section of the PRG/TSR.

Once the report is revised, it is turned in to the teacher for evaluation. If the students are using the Investigation Handout, they write the final report on a separate piece of paper and then turn that in together with the Investigation Handout, which contains the original draft report, and the PRG/TSR. If the students are using the workbook, they write both the draft report and the final report in the Investigation Log, which also contains the PRG/TSR, so they turn in the entire workbook. The teacher can then provide a score on the PRG/TSR in the column labeled "Teacher Score" and use these ratings to assign an overall grade for the report. This approach provides all students with a chance to improve their writing mechanics and develop their reasoning and understanding of the content. This process also offers students the added benefit of reducing academic pressure by providing support in obtaining the highest possible grade for their final product.

> **Keep the following points in mind during stage 8 of ADI:**
> - Students don't always use the feedback they receive from others to improve the quality of their report. Make sure you move from student to student and encourage them to make changes based on the feedback.
> - If students get stuck as they are writing their final report, model how to use feedback. For example, you might say, "Okay, based on the peer-review guide, I need to do more to describe what I did to collect data. I bet if I add information about _____, readers will have a better idea of what I did."

The PRG/TSR is designed to be used with any ADI investigation, thus allowing you to use the same scoring rubric throughout the entire year. This is beneficial for several reasons. First, the criteria for what counts as a high-quality report do not change from investigation to investigation. Students therefore quickly learn what is expected from them when they write a report, and you do not have to spend valuable class time explaining the various components of the PRG/TSR each time you assign a report. Second, the PRG/TSR makes it clear which components of a report need to be improved next time, because the grade is not based on a holistic evaluation of the report. Students, as a result, can see which aspects of their writing are strong and which aspects need improvement. Finally, and perhaps most important, the PRG/TSR provides you with a standardized measure of student performance that can be compared over multiple reports across semesters, thus allowing you to track improvement over time.

The Role of the Teacher During Argument-Driven Inquiry

If the ADI instructional model is to be successful and student learning is to be optimized, the role of the teacher during an investigation that was designed using this model must be different from the teacher's role during a more traditional science lesson. The teacher *must* act as a resource for the students, rather than as a director, as students work through each stage of the activity; the teacher must encourage students to think about *what they are doing* and *why they made that decision* throughout the process. This encouragement should take the form of probing questions that teachers ask as they walk around the classroom, such as "Why do you want to set up your equipment that way?" or "What type of data will you need to collect to be able to answer that question?" Teachers must also restrain themselves from telling or showing students how to "properly" conduct the investigation. However, teachers must emphasize the need to maintain high standards for a scientific investigation by requiring students to use rigorous standards for what counts as a good method or a strong argument in the context of science.

An Overview of Argument-Driven Inquiry

Finally, and perhaps most important for the success of these investigations, teachers must be willing to let students try and fail, and then help them learn from their mistakes. Teachers should not try to make the investigations included in this book "student-proof" by providing additional directions to ensure that students do everything right the first time. We have found that students often learn more from an ADI investigation when they design a flawed method to collect data during stage 2 or analyze their results in an inappropriate manner during stage 3, because their classmates quickly point out these mistakes during the argumentation session (stage 4) and it leads to more teachable moments.

Because the teacher's role during an ADI investigation is different from what typically happens in a classroom, we have provided a chart describing teacher behaviors that are consistent and inconsistent with each stage of the instructional model (see Table 2). This table is organized by stage because what the students and the teacher need to accomplish during each stage is different. It might be helpful to keep this table handy as a guide when you are first attempting to implement the investigations found in the book.

TABLE 2

Teacher actions during the stages of the ADI instructional model

Stage	What the teacher does that is...	
	Consistent with ADI model	**Inconsistent with ADI model**
1: Introduce the task and the guiding question	• "Creates a need" for students to design and carry out an investigation by introducing a phenomenon • Reads the introduction aloud to the students • Supplies students with the materials and equipment they will need • Holds a "tool talk" to show students how to use the materials and equipment • Reviews relevant safety precautions and protocols • Allows students to tinker with the equipment they will be using later	• Provides a list of vocabulary terms • Tells students what they will figure out • Tells students that there is one correct answer • Makes students do everything on their own • Skips going over the safety precautions • Skips introducing the phenomenon to save time

Continued

Table 2 *(continued)*

Stage	What the teacher does that is...	
	Consistent with ADI model	**Inconsistent with ADI model**
2: Design a method and collect data	• Encourages students to ask questions as they design their investigations • Encourages students to use the crosscutting concept(s) (CC[s]) to decide what data are important to collect • Asks groups probing questions about their investigation plan (e.g., "Why did you do it this way?") and the type of data they expect from that design • Checks over the investigation proposal and offers feedback as needed • Assists students as they get stuck • Ensures that all students are safe	• Gives students a procedure to follow • Does not question students about the method they design or the type of data they expect to collect • Approves vague or incomplete investigation proposals • Makes everyone in the class collect data the same way
3: Create a draft argument	• Reminds students of the guiding question and what counts as appropriate evidence in science • Requires students to generate an argument that provides and supports a claim with genuine evidence • Encourages students to use the disciplinary core idea(s) (DCI[s]) and the CC(s) in the justification of the evidence • Encourages all student in the group to make equal contributions • Provides just-in-time instruction as students get stuck	• Requires only one student to be prepared to discuss the argument • Moves to groups to check on progress without asking students questions about why they are doing what they are doing • Tells students the right answer • Has the class create a single argument together • Allows one student to do all the work for a group
4: Argumentation session	• Establishes and maintains classroom norms for discussions • Encourages student-to-student talk • Keeps the discussion focused on the elements of the argument • Encourages students to use appropriate criteria for determining what does and does not count	• Allows students to negatively respond to others • Asks questions about students' claims before other students can ask • Allows students to discuss ideas that are not supported by evidence • Allows students to use inappropriate criteria for determining what does and does not count
5: Reflective discussion	• Encourages students to discuss what they learned about the DCI(s) and the CC(s) • Encourages students to discuss what they learned about a nature of scientific knowledge (NOSK) or nature of scientific inquiry (NOSI) concept • Encourages students to think of ways to design better investigations in future • Asks students what they think	• Provides a lecture on the content • Skips over the discussion about a NOSK or NOSI concept to save time • Tells students what they "should have learned" or what they "should have figured out"

Continued

An Overview of Argument-Driven Inquiry

Table 2 (*continued*)

Stage	What the teacher does that is... Consistent with ADI model	What the teacher does that is... Inconsistent with ADI model
6: Write a draft report	• Reminds students about the audience, topic, and purpose of the report • Provides an example of a good report and an example of a bad report • "Chunks" the writing process into manageable pieces • Provides just-in-time instruction as students get stuck	• Has students write only a portion of the report • Moves on to the next activity/topic without providing support • Expects students to complete the entire report with little or no assistance
7: Peer review	• Establishes and maintains classroom norms for the review process • Encourages students to remember that while grammar and punctuation are important, the main goal is an acceptable scientific claim with supporting evidence and justification • Reminds students of what counts as specific and useful feedback • Holds the reviewers accountable	• Allows students to make critical comments about the author (e.g., "This person is stupid") rather than their work (e.g., "This claim needs to be supported by evidence") • Allows students to just check off "Yes" on each item • Allows students to skip giving feedback during the peer-review process • Has students review reports on their own
8: Revise the report	• Requires students to edit their reports based on the reviewers' comments • Requires students to respond to the reviewers' feedback	• Allows students to turn in a report without a completed peer-review guide • Allows students to turn in a report without revising it first

How to Keep Students Safe During ADI Investigations

It is important for all of us to do what we can to make school science investigations safe for everyone in the classroom. We recommend four important guidelines to follow. First, we need to have proper safety equipment such as, but not limited to, fire extinguishers and an eye wash in our classrooms. Second, we need to ensure that students use appropriate personal protective equipment (PPE; e.g., sanitized indirectly vented chemical-splash goggles and nonlatex gloves) during all parts of the investigations (i.e., setup, the hands-on investigation, and cleanup) when students are using potentially harmful supplies, equipment, or chemicals. At a minimum, the PPE we provide for students to use must meet the ANSI/ISEA Z87.1D3 standard. Third, we must review and comply with all safety policies and procedures, including but not limited to appropriate chemical management, that have been established by our place of employment. Finally, and perhaps most important, we all need to adopt legal safety standards and better professional safety practices and enforce them inside the classroom.

We provide safety precautions for each investigation and recommend that all teachers follow these safety precautions to provide a safer learning experience inside the classroom. The safety precautions associated with each investigation are based, in part, on the use of the recommended materials and instructions, legal safety standards, and better professional safety practices. Selection of alternative materials or procedures for these investigations may jeopardize the level of safety and therefore is at the user's own risk. Remember that an investigation includes three parts: (1) setup, which is what the teacher and/or the students do to prepare the materials for the investigation; (2) the actual investigation, which involves students using the materials and equipment; and (3) the cleanup, which includes cleaning the materials and putting them away for later use. The safety procedures and PPE we recommend for each investigation apply to all three parts.

We also recommend that you go over the 11 safety rules that are included as part of the safety acknowledgment form with your students before beginning the first investigation. Once you have finished going over these rules with your students, have them sign the safety acknowledgment form. You should also send the form home for a parent or guardian to read and sign to acknowledge that they understand the safety procedures that must be followed by their child. The safety acknowledgment form can be found in Appendix 5. Another elementary science safety acknowledgement form can be found on the National Science Teaching Association Safety Portal at *http://static.nsta.org/pdfs/SafetyAcknowledgmentForm-ElementarySchool.pdf*.

References

Abd-El-Khalick, F., and N. G. Lederman. 2000. Improving science teachers' conceptions of nature of science: A critical review of the literature. *International Journal of Science Education* 22: 665–701.

Blanchard, M. and V. Sampson. 2018. Fostering impactful research experiences for teachers (RETs). *Eurasia Journal of Mathematics, Science and Technology Education* 14 (1): 447–465.

Brown, D. H. 2007. *Principles of language learning and teaching.* 5th ed. White Plains, NY: Pearson.

Duschl, R. A., H. A. Schweingruber, and A. W. Shouse, eds. 2007. *Taking science to school: Learning and teaching science in grades K–8.* Washington, DC: National Academies Press.

Echevarría, J., D. Short, and K. Powers. 2006. School reform and standards-based education: A model for English-language learners. *Journal of Educational Research* 99 (4): 195–210.

Escamilla, K., and S. Hopewell. 2010. Transitions to biliteracy: Creating positive academic trajectories for emerging bilinguals in the United States. In *International perspectives on bilingual education: Policy, practice, controversy,* ed. J. E. Petrovic, 69–94. Charlotte, NC: Information Age Publishing.

García, E. E. 2005. *Teaching and learning in two languages: Bilingualism and schooling in the United States.* New York: Teachers College Press.

García, E. E., and E. Hamayan. 2006. What is the role of culture in language learning? In *English language learners at school: A guide for administrators,* eds. E. Hamayan and R. Freeman, 61–64. Philadelphia, PA: Caslon Publishing.

García, O., and J. Kleifgen. 2010. *Educating emergent bilinguals: Policies, programs, and practices for English language learners.* New York: Teachers College Press

Goldenberg, C. and R. Coleman. 2010. *Promoting academic achievement among English learners: A guide to the research.* Thousand Oaks, CA: Corwin Press.

González, N., L. Moll, and C. Amanti. 2005. *Funds of knowledge: Theorizing practices in households, communities and classrooms.* Mahwah, NJ: Erlbaum.

Gottlieb, M., A. Katz, and G. Ernst-Slavit. 2009. *Paper to practice: Using the English language proficiency standards in PreK–12 classrooms.* Alexandria, VA: Teachers of English to Speakers of Other Languages.

Grooms, J., P. Enderle, and V. Sampson. 2015. Coordinating scientific argumentation and the *Next Generation Science Standards* through argument driven inquiry. *Science Educator* 24 (1): 45–50.

Lederman, N. G., and J. S. Lederman. 2004. Revising instruction to teach the nature of science. *The Science Teacher* 71 (9): 36–39.

Michaels, S., and C. O'Connor. 2012. Talk science primer. The Inquiry Project, TERC. *http://inquiryproject.terc.edu/shared/pd/TalkScience_Primer.pdf.*

National Governors Association Center for Best Practices and Council of Chief State School Officers (NGAC and CCSSO). 2010. *Common core state standards.* Washington, DC: NGAC and CCSSO.

National Research Council (NRC). 2012. *A framework for K–12 science education: Practices, crosscutting concepts, and core ideas.* Washington, DC: National Academies Press.

NGSS Lead States. 2013. *Next Generation Science Standards: For states, by states.* Washington, DC: National Academies Press. *www.nextgenscience.org/next-generation-science-standards.*

Sampson, V., and L. Gleim. 2009. Argument-driven inquiry to promote the understanding of important concepts and practices in biology. *American Biology Teacher* 71 (8): 471–477.

Sampson, V., J. Grooms, and J. Walker. 2009. Argument-driven inquiry: A way to promote learning during laboratory activities. *The Science Teacher* 76 (7): 42–47.

Sampson, V., J. Grooms, and J. Walker. 2011. Argument-driven inquiry as a way to help students learn how to participate in scientific argumentation and craft written arguments: An exploratory study. *Science Education* 95 (2): 217–257.

Schwartz, R. S., N. Lederman, and B. Crawford. 2004. Developing views of nature of science in an authentic context: An explicit approach to bridging the gap between nature of science and scientific inquiry. *Science Education* 88: 610–645.

Schwarz, C. V., C. Passmore, and B. J. Reiser. 2017. *Helping students make sense of the world using next generation science and engineering practices.* Arlington VA: NSTA Press.

Strimaitis, A., S. Southerland, V. Sampson, P. Enderle, and J. Grooms. 2017. Promoting equitable biology lab instruction by engaging all students in a broad range of science practices: An exploratory study. *School Science and Mathematics* 117 (3–4): 92–103.

Wallace, C., B. Hand, and V. Prain, eds. 2004. *Writing and learning in the science classroom.* Boston: Kluwer Academic Publishers.

Chapter 2
The Investigations

How to Use the Investigations

This book includes 16 argument-driven inquiry (ADI) investigations. These investigations are not designed to replace an existing science curriculum, but rather are designed to function as a tool that teachers can use to integrate more three-dimensional instruction into their classroom. These investigations are also intended to provide teachers with a way to help children develop fundamental literacy and mathematics skills in the context of science, because students read, write, and talk in the service of sense-making during each investigation found in this book and use mathematical concepts and practices to make sense of the data they collect during most of them. Finally, teachers can use these investigations to turn the classroom into a rich language-learning environment that provides emerging bilingual students with the opportunities they need to use receptive (listening and reading) and productive (speaking and writing) language. These investigations can therefore provide a context for emerging bilingual students to learn English and science at the same time as they interact with other people, the available materials, and ideas to develop a new understanding about how the world works.

We do not expect you to use every investigation included in this book. We do, however, recommend that you try to incorporate as many of these investigations into your fifth-grade science curriculum as possible to give students more opportunities to learn how to use disciplinary core ideas (DCIs), crosscutting concepts (CCs), and scientific and engineering practices (SEPs) to figure things out (NGSS Lead States 2013; NRC 2012). The more ADI investigations that students complete, the more progress that they will make on each aspect of science proficiency (Grooms, Enderle, and Sampson 2015; Sampson et al. 2013; Strimaitis et al. 2017).

These investigations are designed to function as stand-alone lessons, which gives you the flexibility needed to decide which ones to use and when to use them during the academic year. The investigations are organized by topic into different sections in the book. Sections 2–6 can be taught in any order. The investigations within a section, however, should be taught in order because many of the investigations within a section use similar DCIs and are designed to build on what students figured out during a previous investigation. For example, Investigations 4, 5, and 6 require students to use the same DCI (Matter and Its Interactions) to make sense of the world around them, but students can also use what they figured out during Investigation 4 (how to figure out if mixing two substances resulted in the creation of a new substance) as they attempt to figure out if the weight of a closed system changes when a chemical reaction takes place within that system during

Chapter 2

Investigation 5 or to identify unknown substances based on the chemical properties of those substances during Investigation 6.

We have aligned the investigations with the following sources to facilitate curriculum and lesson planning:

- *A Framework for K–12 Science Education* (NRC 2012; see Standards Matrix A in Appendix 1)
- The *Next Generation Science Standards,* or *NGSS* (NGSS Lead States 2013; see Standards Matrix B in Appendix 1)
- Aspects of the nature of scientific knowledge (NOSK) and the nature of scientific inquiry (NOSI) (see Standards Matrix C in Appendix 1; see also the discussion of NOSK and NOSI concepts in Chapter 1)

We wrote all the investigations included in this book to align with a specific performance expectation in the *NGSS* for fifth grade. Teachers who teach in states that use science standards other than the *NGSS* will need to determine how well each investigation aligns with the specific state standards for fifth grade. In states that have adopted standards that are based on the *Framework*, there is likely a great deal of overlap because the state standards will include many, if not all, of the DCIs, CCS, and SEPs that were used to write the *NGSS*; the standards will just be worded differently. In states that have standards that are not based on the *Framework*, there might be less overlap. However, given the fact that all 16 investigations are designed to give children an opportunity to learn science by doing science, it is likely that the investigations will still be useful for helping students learn the concepts and inquiry or process skills found in state standards that are based on something other than the *Framework* (such as the *Benchmarks for Science Literacy* [AAAS 1993] or the *National Science Education Standards* [NRC 1996]). It is important to note, however, that some of these investigations might be better aligned with content and inquiry or process skills standards that are addressed in third or fourth grade in states that have standards that are not the *NGSS* or based on the *Framework*. If these investigations align better with the content and inquiry or process skills for third or fourth grade, they can still be used in those grade levels as well as in fifth grade.

The investigations in this book, as noted earlier, create a context where students read, write, talk, and use mathematics to figure out how something works or why something happens. You can therefore use these investigations to help students develop important literacy skills, understand and use mathematical concepts and practices, or acquire a second language. With this in mind, we have also aligned each investigation with the following sources:

- The *Common Core State Standards* in English language arts, or *CCSS ELA* (NGAC and CCSSO 2010; see Standards Matrix D in Appendix 1)

- The *Common Core State Standards* in mathematics, or *CCSS Mathematics* (NGAC and CCSSO 2010; see Standards Matrix E in Appendix 1)
- The *English Language Proficiency (ELP) Standards* (CCSSO 2014; see Standards Matrix F in Appendix 1)

Teacher Notes

We have included Teacher Notes for each investigation to help you decide when to use each investigation and how to help guide students through each stage of an ADI. These notes include information about the purpose of the investigation, information about the DCI(s) and the CC(s) that students use during the investigation, and what students figure out by the end of it. The notes also include information about the time needed to implement each stage of the model, the materials that students need, safety precautions, and a detailed lesson plan by stage.

The Teacher Notes also include a "Alignment With Standards" section showing how each investigation is aligned with the *NGSS* performance expectations, *CCSS ELA*, *CCSS Mathematics*, and *ELP Standards*. In the following subsections, we will describe the information provided in each section of the Teacher Notes.

Purpose

This section describes what the students will do during the investigation. It also identifies the NOSK or NOSI concept that will be highlighted during the reflective discussion.

The DCI(s), CC(s), and SEPs That Students Use During This Investigation to Figure Things Out

This section of the Teacher Notes provides a basic overview of the DCI(s), CC(s), and SEPs that students will use during the investigation. The overview is based on the *Framework* and describes what students should know about the DCI(s) in grades 3–5. Please note that because of the nature of the ADI approach, you do not need to "teach the vocabulary first" or make sure that your students "know the content" before the investigation begins. Students will learn more about the DCI(s) and the CC(s) along with the SEPs and any important vocabulary as they work through each stage of ADI to *investigate* the phenomenon; *make sense* of the phenomenon; and *evaluate and refine* their ideas, explanations, and arguments.

Other Concepts That Students May Use During This Investigation

This section of the Teacher Notes provides a list of some other concepts that students may use during the investigation. These concepts are included in the "Introduction" section of the student handout. Again, because of the nature of the ADI approach, you do not need to "teach these ideas first" or make sure that your students "know the content" before the investigation begins. Students will learn more about these ideas during the investigation. This list is intended to help you get a better sense of what students will learn and talk about as they *investigate* the phenomenon, *make sense* of the phenomenon, and *evaluate and refine* their ideas, explanations, and arguments.

What Students Figure Out

This section of the Teacher Notes describes what new understanding of the natural world the students are likely to develop as they work through the investigation.

Background Information About This Investigation for the Teacher

This section provides an explanation of the phenomenon that the students will investigate.

Timeline

This section of the Teacher Notes provides information about how much time each stage of the investigation should take. Unlike typical science lessons, ADI investigations typically take between three and five hours to complete. An investigation can be completed over eight days (one day for each stage) during the designated science time in the daily schedule, or it can be completed in one day. The amount of time it will take to complete each investigation will vary depending on how long it takes to collect data and how familiar students are with the stages of ADI. The time needed to complete each stage of ADI will take longer the first few times that students work through the process, but the time the students need will be reduced as they become familiar with using DCIs, CCs, and SEPs to figure things out.

Materials and Preparation

This section describes the consumables and equipment that the students will need to complete the investigation. The quantities needed for each item are listed by group, by individual student, or by class. We have also included specific suggestions for some lab supplies, based on our findings that these supplies worked best during the field tests. However, if needed, substitutions can be made. Always be sure to test all materials before starting an investigation.

This section also describes any preparation that needs to be done *before* students can do the investigation. Please note that the preparation for some investigations may need to be done several days in advance.

Safety Precautions

This section provides an overview of potential safety hazards and summarizes safety protocols that should be followed to make the investigation safer for students. These are based on legal safety standards and current better professional safety practices. You should also review and follow all local polices and protocols used within your school district and/or school (e.g., the district chemical hygiene plan, Board of Education safety policies).

Lesson Plan by Stage

This section provides a detailed lesson plan for each stage that includes information about what you can do as you guide students through the investigation. The plan includes directions to give students and sample questions to ask students at different points in the lesson, but these directions and questions should not be followed like a script. As professionals, we believe that teachers know the unique needs of their students and how to best support them. The detailed lesson plans should be viewed as a starting point and should be modified and revised based on the unique needs of the students in a particular classroom and your professional judgment.

It is important to keep in mind that the goal of ADI is give students more opportunities to learn how to use DCIs, CCs, and SEPs to figure out how or why something happens, so you should resist the urge to tell students exactly what to do or how to complete a task before they start. It is also important not to reduce the complexity of the tasks by providing too much scaffolding. We believe that it is far better to keep the complexity of the task high by giving students more voice and choice, allowing for productive struggle during the investigation, and then providing scaffolding only when it is needed.

Many different teachers have tested these investigations and have helped us refine them and make them run smoother. As a result, we have collected a lot of advice about how to support children as they investigate a phenomenon, make sense of that phenomenon, and evaluate and refine their ideas, explanations, and arguments. We have organized this advice around some common questions that teachers have as they are guiding students through each stage, such as

- What are some things I can do to encourage productive talk among my students?
- What can I do to help my students comprehend more of what they read?
- What are some things I can do to support my emerging bilingual students?
- What are some things I need to remember during each stage of ADI?
- What should a student-designed investigation look like?
- What should a table or graph look like for this investigation?
- What are some things I can do if my students get stuck?

The "Lesson Plan by Stage" section also includes advice for supporting students as they work through each stage. Some of this advice can be found in "hint boxes" that describe, for example, what a student-designed investigation should look like and what a table or graph for this investigation should look like. In addition, teachers should refer to the hints for each stage of ADI found in Chapter 1 and to the frequently asked questions found in Appendix 3.

How to Use the Checkout Questions

This section describes how to use an optional assessment that we call Checkout Questions. It includes information about when to use this assessment and what types of answers to look for to determine if students understand the DCI(s) and the CC(s) from the investigation.

Alignment With Standards

This section is designed to inform curriculum and lesson planning by highlighting how the investigation can be used to address specific performance expectations from the *NGSS*, *CCSS ELA*, *CCSS Mathematics*, and *ELP Standards*.

Instructional Materials

The instructional materials included in this book are reproducible copy masters that are designed to support students as they participate in an ADI investigation. The materials needed for each investigation include an Investigation Handout, the peer-review guide and teacher scoring rubric (PRG/TSR), and a set of Checkout Questions. Some investigations also require supplementary materials.

Some schools will also purchase the student workbook that goes with this book. The instructional materials included in the workbook are similar to those included in this book, but instead of a handout there is an Investigation Log that includes, in addition to what is included in the handout, a peer-review guide, a place to write the final report, and a grading rubric.

Investigation Handout

At the beginning of each ADI investigation, each student should be given a copy of the Investigation Handout. This handout provides information about the phenomenon that they will investigate and a guiding question for the students to answer. In addition, the handout provides space for students to design their investigation; record their observations and measurements; and create tables, graphs, or pictures to analyze the data they collect during the investigation. The handout also has space for the students to keep track of critiques, suggestions for improvement, and good ideas that arise during the argumentation session. To help support students as they learn how to write in this context, the handout includes sentence starters that they can use when writing the draft report.

Peer-Review Guide and Teacher Scoring Rubric

The PRG/TSR is designed to make the criteria that are used to judge the quality of an investigation report explicit. The PRG/TSR can be found in Appendix 4. We recommend that you make one copy of the PRG/TSR for each student. Then during the peer-review stage, have each pair of reviewers fill out one PRG/TSR for each report that they review. The reviewers should rate the report on each criterion and then provide advice to the author about ways to improve it based on their rating. Once the review is complete, the author needs to revise his or her report and respond to the reviewers' rating and comments in the appropriate sections in the PRG/TSR. You should collect the PRG/TSR together with the first draft and the final report for a final evaluation.

To score the report, simply fill out the "Teacher Score" column of the PRG/TSR and then total the scores. There is also space at the bottom of the PRG/TSR for teacher feedback.

Checkout Questions

To facilitate formative assessment inside the classroom, we have included a set of Checkout Questions for each investigation. The questions target the DCI(s) and the CC(s) that the students used during the investigation. Students should complete the Checkout Questions on the same day they turn in their final report. One handout is needed for each student. The students should complete these questions on their own. You can use the students' responses to the Checkout Questions, along with what they write in their report, to determine if the students learned what they needed to learn during the lab, and then reteach as needed.

Supplementary Materials

Some investigations include supplementary materials such as data files, videos, or images. Students will need to be able to use these materials during their investigation. These materials can be downloaded from the book's Extras page at *www.nsta.org/adi-5th*.

References

American Association for the Advancement of Science (AAAS). 1993. *Benchmarks for science literacy*. Project 2061. New York: Oxford University Press.

Council of Chief State School Officers (CCSSO). 2014. *English language proficiency (ELP) standards*. Washington, DC: NGAC and CCSSO. *https://elpa21.org/wp-content/uploads/2019/03/Final-4_30-ELPA21-Standards_1.pdf*.

Grooms, J., P. Enderle, and V. Sampson. 2015. Coordinating scientific argumentation and the *Next Generation Science Standards* through argument driven inquiry. *Science Educator* 24 (1): 45–50.

National Governors Association Center for Best Practices and Council of Chief State School Officers (NGAC and CCSSO). 2010. *Common core state standards*. Washington, DC: NGAC and CCSSO.

National Research Council (NRC). 1996. *National Science Education Standards*. Washington, DC: National Academies Press.

National Research Council (NRC). 2012. *A framework for K–12 science education: Practices, crosscutting concepts, and core ideas*. Washington, DC: National Academies Press.

NGSS Lead States. 2013. *Next Generation Science Standards: For states, by states*. Washington, DC: National Academies Press. *www.nextgenscience.org/next-generation-science-standards*.

Sampson, V., P. Enderle, J. Grooms, and S. Witte. 2013. Writing to learn and learning to write during the school science laboratory: Helping middle and high school students develop argumentative writing skills as they learn core ideas. *Science Education* 97 (5): 643–670.

Strimaitis, A., S. Southerland, V. Sampson, P. Enderle, and J. Grooms. 2017. Promoting equitable biology lab instruction by engaging all students in a broad range of science practices: An exploratory study. *School Science and Mathematics* 117 (3–4): 92–103.

Section 2
Matter and Its Interactions

Teacher Notes

Investigation 1

Movement of Matter Into and Out of a System: What Causes the Water to Come Out of the Tube in the Balloon Water Dispenser?

Purpose

The purpose of this investigation is to give students an opportunity to use one disciplinary core idea (DCI), two crosscutting concepts (CCs), and eight scientific and engineering practices (SEPs) to figure out how a balloon water dispenser works. Students will also learn about the use of models as tools for reasoning about natural phenomena.

The DCI, CCs, and SEPs That Students Use During This Investigation to Figure Things Out

DCI

- *PS1.A: Structure and Properties of Matter:* Matter of any type can be subdivided into particles that are too small to see, but even then the matter still exists and can be detected by other means.

CCs

- *CC 2: Cause and Effect:* Cause-and-effect relationships are routinely identified, tested, and used to explain change. Events that occur together with regularity might or might not be a cause-and-effect relationship.
- *CC 4: Systems and System Models:* A system can be described in terms of its components and their interactions. A system is a group of related parts that make up a whole and can carry out functions its individual parts cannot.

SEPs

- *SEP 1: Asking Questions and Defining Problems:* Ask questions about what would happen if a variable is changed. Ask questions that can be investigated and predict reasonable outcomes based on patterns such as cause-and-effect relationships.
- *SEP 2: Developing and Using Models:* Develop and/or use models to describe and/or predict phenomena.
- *SEP 3: Planning and Carrying Out Investigations:* Plan and conduct an investigation collaboratively to produce data to serve as the basis for evidence, using fair tests in which variables are controlled and the number of trials considered. Evaluate appropriate methods and/or tools for collecting data.

Investigation 1. Movement of Matter Into and Out of a System: What Causes the Water to Come Out of the Tube in the Balloon Water Dispenser?

- *SEP 4: Analyzing and Interpreting Data:* Represent data in tables and/or various graphical displays (bar graphs, pictographs, and/or pie charts) to reveal patterns that indicate relationships. Analyze and interpret data to make sense of phenomena, using logical reasoning, mathematics, and/or computation. Compare and contrast data collected by different groups in order to discuss similarities and differences in their findings.
- *SEP 5: Using Mathematics and Computational Thinking:* Organize simple data sets to reveal patterns that suggest relationships. Describe, measure, estimate, and/or graph quantities (e.g., area, volume, weight, time) to address scientific and engineering questions and problems.
- *SEP 6: Constructing Explanations and Designing Solutions:* Construct an explanation of observed relationships. Use evidence to construct or support an explanation. Identify the evidence that supports particular points in an explanation.
- *SEP 7: Engaging in Argument From Evidence:* Compare and refine arguments based on an evaluation of the evidence presented. Distinguish among facts, reasoned judgment based on research findings, and speculation in an explanation. Respectfully provide and receive critiques from peers about a proposed procedure, explanation, or model by citing relevant evidence and posing specific questions.
- *SEP 8: Obtaining, Evaluating, and Communicating Information:* Read and comprehend grade-appropriate complex texts and/or other reliable media to summarize and obtain scientific and technical ideas. Combine information in written text with that contained in corresponding tables, diagrams, and/or charts to support the engagement in other scientific and/or engineering practices. Communicate scientific and/or technical information orally and/or in written formats, including various forms of media as well as tables, diagrams, and charts.

Other Concepts That Students May Use During This Investigation

Students might also use some of the following concepts:

- Substances like water and air have different properties because molecules of water and air are made up of different types of particles.
- Particles that make up substances are in constant motion.
- A solid has a fixed shape because the particles within a solid are held together by strong forces that prevent the particles from moving past each other.
- A liquid or a gas does not have a fixed shape because the forces holding the particles within a liquid or a gas together are not strong enough to lock them in place; therefore, the particles can move past each other.

Teacher Notes

What Students Figure Out

The balloon water dispenser works because the air in the balloon is a higher pressure than the air in the bottle. When the air in the balloon is released, some of the air moves into the bottle. The air then pushes on the water, and some of the water is pushed up through the tube and out into the cup. The more air that is in the balloon, the more water that will be pushed into the cup. Finally, if students remove the tape covering the holes in the container, no water will be pushed out of the tube because the air will escape through these openings.

Background Information About This Investigation for the Teacher

Matter, such as water or air, is made up of particles that are too small to be seen. These particles are called atoms. Matter that is made up of only one type of atom is called an element. Examples of elements include gold, aluminum, and oxygen. Individual atoms can combine with other atoms to form molecules. Matter that is made up of a specific type of molecule is called a compound. Water, for example, is made up of water molecules. A water molecule is made up of one atom of oxygen and two atoms of hydrogen. Air that we breathe is made up of several molecules, including oxygen (two oxygen atoms combined), nitrogen (two nitrogen atoms combined), and carbon dioxide (two oxygen atoms combined with one carbon atom).

A sample of matter can exist in different states such as solid, liquid, or gas. A sample of matter is a solid when the molecules that make up that matter have very little energy and are grouped closely together. Solids have a fixed shape because the molecules in a solid are locked in place and are unable to move past each other. Liquids, like water, do not have a fixed shape because the molecules in liquids are not locked in place. The molecules that make up a liquid are grouped more closely than in gases but not as closely as in solids. The molecules that make up a gas are the most loosely grouped and can move freely.

The molecules in different substances can exert forces on each other. This can be two solids coming into contact, like when you drop a rock onto a pile of sand. The molecules in the rock hit the molecules in the sand, and this results in a dimple in the sand. When dropping a solid into a liquid, like a rock into a pond of water, the force when the molecules collide results in ripples on the surface of the pond. Air molecules can also exert a force on liquids. When the wind blows across the surface of a pond, the air molecules will collide with the water molecules. The force from these collisions results in waves. In this investigation, students will need to use the idea that the particles that make up a gas can exert forces on a liquid or solid even when we cannot see these particles to be able to explain how the balloon water dispenser works.

Figure 1.1 is a picture of the balloon water dispenser. To make the balloon water dispenser, first cut three 3/8-inch holes in the sides of the bottle with a drill. The first two holes should be near the top of the bottle, and third hole should be on the side of the

Investigation 1. Movement of Matter Into and Out of a System:
What Causes the Water to Come Out of the Tube in the Balloon Water Dispenser?

bottle about 1 inch from the top. Next, place an 8-inch piece of 3/8-inch-wide vinyl tubing through the hole near the bottom of the bottle. The end of the tubing inside the bottle should be touching the bottom of the bottle. Seal the opening where the tubing enters the bottle with clear silicone waterproof sealant so no air can escape through the tiny gaps between the edge of the hole and the tubing. Use painter's tape to cover the two holes near the top of the bottle. The tape should not let air escape from the bottle but should be easy to remove. Place about 4 inches of water in the bottle. Blow up the balloon and pinch the end so no air can escape. Stretch the end of the balloon over the mouth of the bottle while still pinching the balloon closed. When you let go of the balloon, water will flow out of the bottle and into the cup.

FIGURE 1.1
The balloon water dispenser

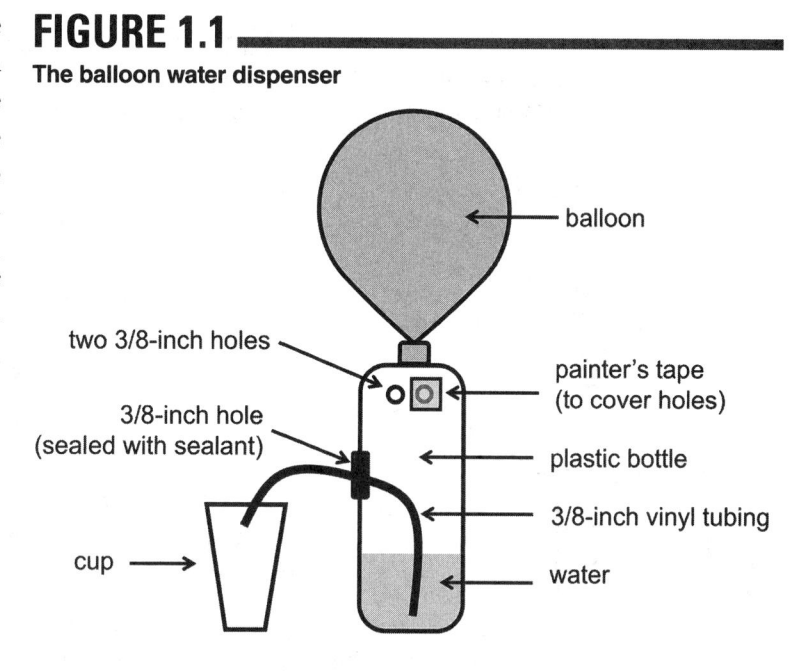

Figure 1.2 shows a model of air particles in the balloon and water dispenser *before* someone releases the air from the inflated balloon. At this point in time there are more air particles in the balloon than there are inside the bottle. The air inside the balloon is at a greater pressure than the air inside the bottle.

Once the air is released from the balloon, the balloon will start to deflate. When this happens, some of the air particles in the balloon will move into the bottle. In turn, this will cause the air particles already in the bottle to push down on the water molecules at the bottom of the bottle. The water molecules will be pushed up the tube and out into the cup. Figure 1.3 (p. 46) shows a model of the air particles in the balloon and the bottle as the balloon deflates. Notice that there are fewer air molecules in the balloon than at

FIGURE 1.2
A model of the air inside the balloon and the bottle before releasing the air out of the balloon

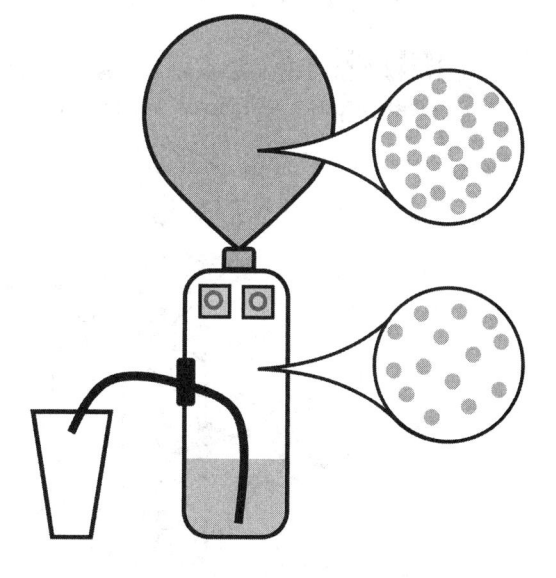

Argument-Driven Inquiry in **Fifth-Grade Science**: Three-Dimensional Investigations

Teacher Notes

FIGURE 1.3
A model of the air inside the balloon and the bottle after releasing air out of the balloon

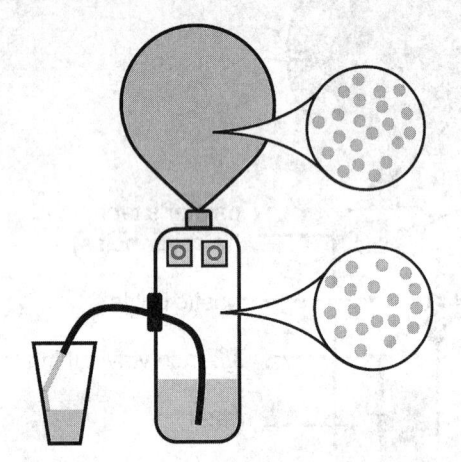

FIGURE 1.4
A model of the air inside the balloon and the bottle once the balloon completely deflates

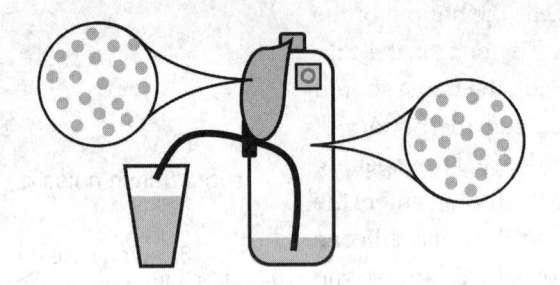

the beginning, more air molecules in the bottle than at the beginning, and more water in the cup than at the beginning.

Eventually, the balloon will fully deflate, and no more water will flow out of the dispenser because there are no longer any additional molecules from the balloon being pushed into the bottle. Figure 1.4 shows a model of the air particles after the balloon is fully deflated. Notice that there are still some air molecules in the balloon.

Finally, if students remove the tape covering the holes at the top of the bottle, as shown in Figure 1.5, no water will come out of the tube. This is because when air enters the bottle from the balloon, instead of pushing the water out, it will push the air already into the bottle out of the new opening. It takes a smaller force to push air through the opening than it does to push the water up the tube.

FIGURE 1.5
The balloon with a piece of tape removed

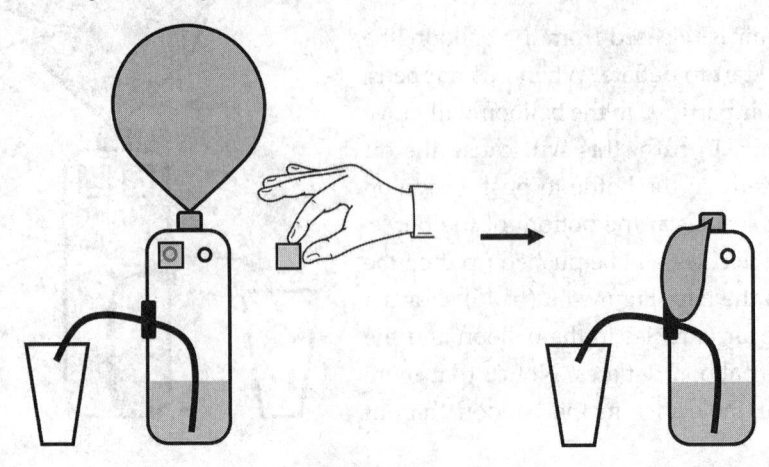

Investigation 1. Movement of Matter Into and Out of a System:
What Causes the Water to Come Out of the Tube in the Balloon Water Dispenser?

Timeline

The time needed to complete this investigation is 270 minutes (4 hours and 30 minutes). The amount of instructional time needed for each stage of the investigation is as follows:

- *Stage 1.* Introduce the task and the guiding question: 35 minutes
- *Stage 2.* Design a method and collect data: 50 minutes
- *Stage 3.* Create a draft argument: 45 minutes
- *Stage 4.* Argumentation session: 30 minutes
- *Stage 5.* Reflective discussion: 15 minutes
- *Stage 6.* Write a draft report: 30 minutes
- *Stage 7.* Peer review: 35 minutes
- *Stage 8.* Revise the report: 30 minutes

Materials and Preparation

The materials needed for this investigation are listed in Table 1.1 (p. 48). Except for the balloon water dispenser, all the items can be purchased from a big-box retail store such as Walmart or Target or through an online retailer such as Amazon. The materials for this investigation (including predrilled bottles and other materials to make the balloon water dispensers) can also be purchased as a complete kit (which includes enough materials for 24 students, or six groups of four students) at *www.argumentdriveninquiry.com*.

If you are not using the balloon water dispensers included in the kit, you will need to make the dispensers before starting the investigation; see the instructions and Figure 1.1 in the "Background Information About This Investigation for the Teacher" section. Table 1.1 (p. 48) lists the equipment and materials needed to make six dispensers as well as the rest of the materials needed for the investigation.

Teacher Notes

TABLE 1.1
Materials for Investigation 1

Item	Quantity
Safety goggles	1 per student
Balloon water dispenser	1 per group
Measuring cups	2 per group
Clothespins	2 per group
Cloth measuring tape	1 per group
Stopwatch	1 per group
Whiteboard, 2' × 3'*	1 per group
Investigation Handout	1 per student
Peer-review guide and teacher scoring rubric	1 per student
Checkout Questions (optional)	1 per student
Equipment and materials to make the balloon water dispensers	**Quantity**
3/8" drill bit	1
Electric drill	1
Scissors	1
Plastic bottles, 28 oz or 1 liter†	6
Latex balloons, 25-count bag‡	1
Roll of blue painter's tape	1
PVC clear vinyl tubing, 3/8" × 10"	1
Clear silicone waterproof sealant	1

*As an alternative, students can use computer and presentation software such as Microsoft PowerPoint or Apple Keynote to create their arguments.

†Although either size can be used, 28-oz plastic spray bottles work best because they are made of thicker plastic.

‡*Caution:* Some students may be allergic to latex and potentially have a serious physical reaction upon contact. Always check with the school nurse to see if any students are known to have this allergy; if any do, either find an alternative nonlatex balloon or have the students observe but not touch the latex material from a safe distance.

Safety Precautions

Remind students to follow all normal safety rules. In addition, tell the students to take the following safety precautions:

- Wear sanitized safety goggles during setup, investigation activity, and cleanup.
- Be sure to immediately clean up any spills.
- Make sure all materials are put away after completing the activity.

Investigation 1. Movement of Matter Into and Out of a System: What Causes the Water to Come Out of the Tube in the Balloon Water Dispenser?

- Wash their hands with soap and water when they are done cleaning up.

Lesson Plan by Stage

This lesson plan is only a suggestion. It is included here to illustrate what you can say and do during each stage of ADI for this specific investigation. We encourage you to modify this lesson plan by asking different questions, using different examples, and providing different scaffolds to better meet the needs of students in your class.

Stage 1: Introduce the Task and the Guiding Question (35 minutes)

1. Ask the students to sit in six groups, with three or four students in each group.
2. Ask the students to clear off their desks except for a pencil (and their *Student Workbook for Argument-Driven Inquiry in Fifth-Grade Science* if they have one).
3. Pass out an Investigation Handout to each student (or ask students to turn to the Investigation Log for Investigation 1 in their workbook).
4. Read the first paragraph of the "Introduction" aloud to the class. Ask the students to follow along as you read.
5. Demonstrate the balloon water dispenser for the class by blowing up a balloon and attaching it to the water dispenser.
6. Ask the students to record their observations and questions on the "NOTICED/WONDER" chart in the "Introduction."
7. After the students have recorded their observations and questions, ask them to share what they observed after you placed the balloon on the bottle.
8. Ask the students to share what questions they have about what happened when you put the balloon on the bottle.
9. Tell the students, "Some of your questions might be answered by reading the rest of the 'Introduction.'"
10. Ask the students to read the rest of the "Introduction" on their own *or* ask them to follow along as you read it aloud.
11. Once the students have read the rest of the "Introduction," ask them to fill out the "Things we KNOW" chart on their Investigation Handout (or in their Investigation Log) as a group.
12. Ask the students to share what they learned from the reading. Add these ideas to a class "Things we KNOW" chart.
13. Tell the students, "Let's see what we will need to figure out during our investigation."
14. Read the task and the guiding question aloud.
15. Tell the students, "I have lots of materials here that you can use."

Teacher Notes

16. Introduce the students to the materials available for them to use during the investigation by either (a) holding each one up and then asking what it might be used for or (b) giving them a kit with all the materials in it and giving them three to four minutes to play with them. If you give the students an opportunity to play with the materials, collect them from each group before moving on to stage 2.

Stage 2: Design a Method and Collect Data (50 minutes)

1. Tell the students, "I am now going to give you and the other members of your group about 15 minutes to plan your investigation. Before you begin, I want you all to take a couple of minutes to discuss the following questions with the rest of your group."

2. Show the following questions on the screen or board:
   - What might be the *cause* of the water coming out the tube?
   - What are the components of the *system* we are studying?

3. Tell the students, "Please take a few minutes to come up with an answer to these questions."

4. Give the students two to three minutes to discuss these two questions.

5. Ask two or three different groups to share their answers. Highlight or write down any important ideas on the board so students can refer to them later.

6. If possible, use a document camera to project an image of the graphic organizer for this investigation on a screen or board (or take a picture of it and project the picture on a screen or board). Tell the students, "I now want you all to plan out your investigation. To do that, you will need to fill out this investigation proposal."

7. Point to the box labeled "Our guiding question:" and tell the students, "You can put the question we are trying to answer in this box." Then ask, "Where can we find the guiding question?"

8. Wait for a student to answer.

9. Point to the box labeled "This is a picture of how we will set up the equipment:" and tell the students, "You can draw a picture in this box of how you will set up the equipment in order to carry out this investigation."

10. Point to the box labeled "We will collect the following data:" and tell the students, "You can list the measurements or observations that you will need to collect during the investigation in this box."

11. Point to the box labeled "These are the steps we will follow to collect data:" and tell the students, "You can list what you are going to do to collect the data you need and what you will do with your data once you have it. Be sure to give enough detail that I could do your investigation for you."

12. Ask the students, "Do you have any questions about what you need to do?"

Investigation 1. Movement of Matter Into and Out of a System: What Causes the Water to Come Out of the Tube in the Balloon Water Dispenser?

13. Answer any questions that come up.
14. Tell the students, "Once you are done, raise your hand and let me know. I'll then come by and look over your proposal and give you some feedback. You may not begin collecting data until I have approved your proposal by signing it. You need to have your proposal done in the next 15 minutes."
15. Give the students 15 minutes to work in their groups on their investigation proposal. As they work, move from group to group to check in, ask probing questions, and offer a suggestion if a group gets stuck.
16. As each group finishes its investigation proposal, read it over and determine if it will be productive or not. If you feel the investigation will be productive (not necessarily what you would do or what the other groups are doing), sign your name on the proposal and let the group start collecting data. If the plan needs to be changed, offer some suggestions or ask some probing questions, and have the group make the changes before you approve it.
17. Pass out the materials or have one student from each group collect the materials they need from a central supply table or cart for the groups that have an approved proposal.
18. Remind students of the safety rules and precautions for this investigation.
19. Tell the students to collect their data and record their observations or measurements in the "Collect Your Data" box in their Investigation Handout (or the Investigation Log in their workbook).
20. Give the students 30 minutes to collect their data (see Figure 1.6 for an example of students collecting data during this investigation). Collect the materials from each group before asking them to analyze their data.

FIGURE 1.6
Students collecting data

Argument-Driven Inquiry in **Fifth-Grade Science**: Three-Dimensional Investigations

Teacher Notes

What should a student-designed investigation look like?

There are a number of different investigations that students can design to answer the question "What causes the water to come out of the tube in the balloon water dispenser?" For example, one method might include the following steps:

1. Put 500 ml of water into the bottle.
2. Place a clothespin on the tube and cover the two holes in the bottle with tape.
3. Blow into the balloon once and put the balloon on the bottle.
4. Release the clothespin on the tube and let the balloon completely deflate.
5. Measure how much water is added to the cup.
6. Replace the water in the bottle.
7. Repeat steps 2–6 for blowing into the balloon twice and three times.

If students use this method, they will need to collect data on (1) how many times they blow into the balloon and (2) how much water comes out of the tube.

Another method might include the following steps:

1. Put 500 ml of water into the bottle.
2. Place a clothespin on the tube and cover the two holes in the bottle with tape.
3. Blow into the balloon once and put the balloon on the bottle.
4. Release the clothespin on the tube and let the balloon completely deflate.
5. Measure how much lower the water line in the bottle is.
6. Replace the water in the bottle.
7. Repeat steps 2–6 for blowing into the bottle twice and three times.
8. Remove the tape from the two holes and repeat steps 1–7.

If students use this method, they will need to collect data on (1) how many times they blow into the balloon, (2) how much lower the water level is in the bottle, and (3) if the tape was covering the holes in the bottle.

Investigation 1. Movement of Matter Into and Out of a System: What Causes the Water to Come Out of the Tube in the Balloon Water Dispenser?

Stage 3: Create a Draft Argument (45 minutes)

1. Tell the students, "Now that we have all this data, we need to analyze the data so we can figure out an answer to the guiding question."

2. If possible, project an image of the "Analyze Your Data" section for this investigation on a screen or board using a document camera (or take a picture of it and project the picture on a screen or board). Point to the section and tell the students, "You can create a graph as a way to analyze your data. You can make your graph in this section."

3. Ask the students, "What information do we need to include in a graph?"

4. Tell the students, "Please take a few minutes to discuss this question with your group, and be ready to share."

5. Give the students five minutes to discuss.

6. Ask two or three different groups to share their answers. Highlight or write down any important ideas on the board so students can refer to them later.

7. Tell the students, "I am now going to give you and the other members of your group about 10 minutes to create your graph." The graph they create should include the number of breaths into the balloon and how much water entered the cup or left the bottle. If the students are having trouble making a graph, you can take a few minutes to provide a mini-lesson about how to create a graph from a bunch of measurements (this strategy is called just-in-time instruction because it is offered only when students get stuck).

8. Give the students 10 minutes to analyze their data by creating a graph. As they work, move from group to group to check in, ask probing questions, and offer suggestions.

What should the graph for this investigation look like?

There are a number of different ways that students can analyze the measurements they collect during this investigation. One of the most straightforward ways is to create a bar graph with the number of breaths into the balloon on the horizontal axis, or *x*-axis, and how much water entered the cup or left the bottle on the vertical axis, or *y*-axis. There are other options for analyzing the collected data. Students often come up with some unique ways of analyzing their data, so be sure to give them some voice and choice during this stage.

9. Tell the students, "I am now going to give you and the other members of your group about 15 minutes to create an argument to share what you have learned

Teacher Notes

and convince others that they should believe you. Before you do that, we need to take a few minutes to discuss what you need to include in your argument."

10. If possible, use a document camera to project the "Argument Presentation on a Whiteboard" image from the "Draft Argument" section of the Investigation Handout (or the Investigation Log in their workbook) on a screen or board (or take a picture of it and project the picture on a screen or board).

11. Point to the box labeled "The Guiding Question:" and tell the students, "You can put the question we are trying to answer here on your whiteboard."

12. Point to the box labeled "Our Claim:" and tell the students, "You can put your claim here on your whiteboard. The claim is your answer to the guiding question."

13. Point to the box labeled "Our Evidence:" and tell the students, "You can put the evidence that you are using to support your claim here on your whiteboard. Your evidence will need to include the analysis you just did and an explanation of what your analysis means or shows. Scientists always need to support their claims with evidence."

14. Point to the box labeled "Our Justification of the Evidence:" and tell the students, "You can put your justification of your evidence here on your whiteboard. Your justification needs to explain why your evidence is important. Scientists often use core ideas to explain why the evidence they are using matters. Core ideas are important concepts that scientists use to help them make sense of what happens during an investigation."

15. Ask the students, "What are some core ideas that we read about earlier that might help us explain why the evidence we are using is important?"

16. Ask the students to share some of the core ideas from the "Introduction" section of the Investigation Handout (or the Investigation Log in the workbook). List these core ideas on the board.

17. Tell the students, "That is great. I would like to see everyone try to include these core ideas in your justification of the evidence. Your goal is to use these core ideas to help explain why your evidence matters and why the rest of us should pay attention to it."

18. Ask the students, "Do you have any questions about what you need to do?"

19. Answer any questions that come up.

20. Tell the students, "Okay, go ahead and start working on your arguments. You need to have your argument done in the next 15 minutes. It doesn't need to be perfect. We just need something down on the whiteboards so we can share our ideas."

21. Give the students 15 minutes to work in their groups on their arguments. As they work, move from group to group to check in, ask probing questions, and

Investigation 1. Movement of Matter Into and Out of a System:
What Causes the Water to Come Out of the Tube in the Balloon Water Dispenser?

offer a suggestion if a group gets stuck. Figure 1.7 shows an example of an argument created by students for this investigation.

FIGURE 1.7

Example of an argument

[Handwritten student whiteboard showing:]

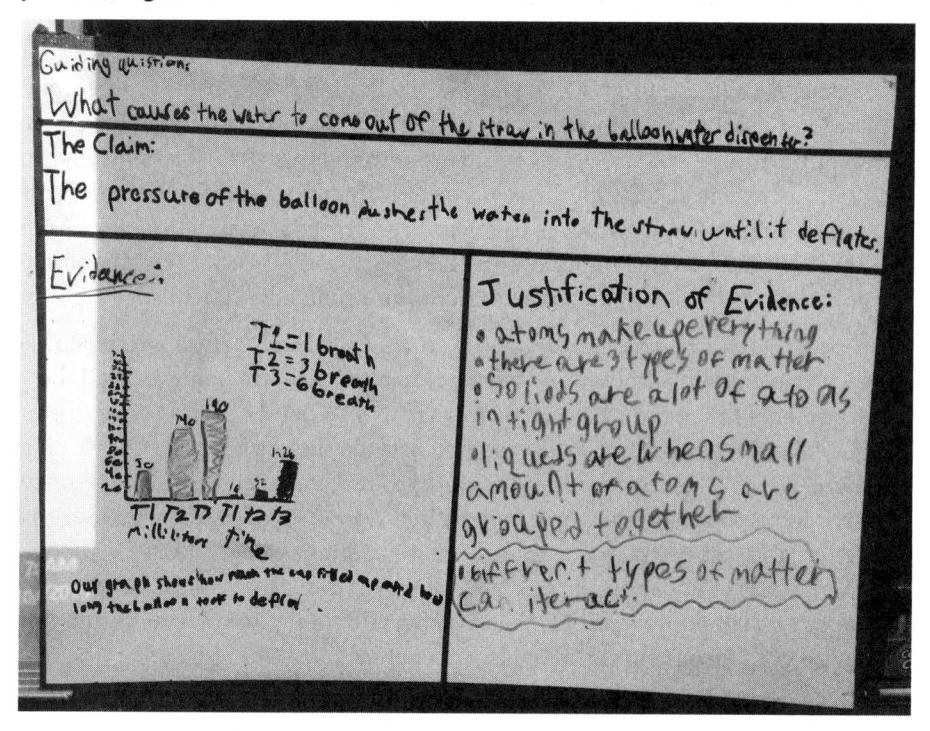

Guiding questions: What causes the water to come out of the straw in the balloon water dispenser?

The Claim: The pressure of the balloon pushes the water into the straw until it deflates.

Evidence: [Bar graph with T1=1 breath, T2=3 breaths, T3=6 breaths] Our graph shows how much the cup filled up and how long the balloon took to deflate.

Justification of Evidence:
- atoms make up everything
- there are 3 types of matter
- solids are a lot of atoms in tight group
- liquids are when small amount of atoms are grouped together
- different types of matter can interact

Stage 4: Argumentation Session (30 minutes)

The argumentation session can be conducted in a whole-class presentation format, a gallery walk format, or a modified gallery walk format. We recommend using a whole-class presentation format for the first investigation, but try to transition to either the gallery walk or modified gallery walk format as soon as possible because that will maximize student voice and choice inside the classroom. The following list shows the steps for the three formats; unless otherwise noted, the steps are the same for all three formats.

1. Begin by introducing the use of the whiteboard.
   - *If using the whole-class presentation format,* tell the students, "We are now going to share our arguments. Please set up your whiteboards so everyone can see them."
   - *If using the gallery walk or modified gallery walk format,* tell the students, "We are now going to share our arguments. Please set up your whiteboards so they are facing the walls."

Teacher Notes

2. Allow the students to set up their whiteboards.
   - *If using the whole-class presentation format*, the whiteboards should be set up on stands or chairs so they are facing toward the center of the room.
   - *If using the gallery walk or modified gallery walk format*, the whiteboards should be set up on stands or chairs so they are facing toward the outside of the room.

3. Give the following instructions to the students:
   - *If using the whole-class presentation format*, tell the students, "Okay, before we get started I want to explain what we are going to do next. Your group will have an opportunity to share your argument with the rest of the class. After you are done, everyone else in the class will have a chance to ask questions and offer some suggestions about ways to make your group's argument better. After we have a chance to listen to each other and learn something new, I'm going to give you some time to revise your arguments and make them better."
   - *If using the gallery walk format*, tell the students, "Okay, before we get started I want to explain what we are going to do next. You are going to read the arguments that were created by other groups. When I say 'go,' your group will go to a different group's station so you can see their argument. Once you are there, I'll give your group a few minutes to read and review their argument. Your job is to offer them some suggestions about ways to make their argument better. You can use sticky notes to give them suggestions. Please be specific about what you want to change and how you think they should change it. After we have a chance to learn from each other, I'm going to give you some time to revise your arguments and make them better."
   - *If using the modified gallery walk format*, tell the students, "Okay, before we get started I want to explain what we are going to do next. I'm going to ask some of you to present your arguments to your classmates. If you are presenting your argument, your job is to share your group's claim, evidence, and justification of the evidence. The rest of you will be travelers. If you are a traveler, your job is to listen to the presenters, ask the presenters questions if you do not understand something, and then offer them some suggestions about ways to make their argument better. After we have a chance to learn from each other, I'm going to give you some time to revise your arguments and make them better."

4. Use a document camera to project the "Ways to IMPROVE our argument …" box from the Investigation Handout (or the Investigation Log in their workbook) on a screen or board (or take a picture of it and project the picture on a screen or board).
   - *If using the whole-class presentation format*, point to the box and tell the students, "After your group presents your argument, you can write down the suggestions you get from your classmates here. If you are listening to a presentation and you see a good idea from another group, you can write down that idea here as

well. Once we are done with the presentations, I will give you a chance to use these suggestions or ideas to improve your arguments."

- *If using the gallery walk format,* point to the box and tell the students, "If you see a good idea from another group, you can write it down here. Once we are done reviewing the different arguments, I will give you a chance to use these ideas to improve your own arguments. It is important to share ideas like this."

- *If using the modified gallery walk format,* point to the box and tell the students, "If you are a presenter, you can write down the suggestions you get from the travelers here. If you are a traveler and you see a good idea from another group, you can write down that idea here. Once we are done with the presentations, I will give you a chance to use these suggestions or ideas to improve your arguments."

5. Ask the students, "Do you have any questions about what you need to do?"

6. Answer any questions that come up.

7. Give the following instructions:

    - *If using the whole-class presentation format,* tell the students, "Okay. Let's get started."

    - *If using the gallery walk format,* tell the students, "Okay, I'm now going to tell you which argument to go to and review."

    - *If using the modified gallery walk format,* tell the students, "Okay, I'm now going to assign you to be a presenter or a traveler." Assign one or two students from each group to be presenters and one or two students from each group to be travelers.

8. Give the students an opportunity to review the arguments.

    - *If using the whole-class presentation format,* have each group present their argument one at a time. Give each group only two to three minutes to present their argument. Then give the class two to three minutes to ask them questions and offer suggestions. Encourage as much participation from the students as possible.

    - *If using the gallery walk format,* tell the students, "Okay. Let's get started. Each group, move one argument to the left. Don't move to the next argument until I tell you to move. Once you get there, read the argument and then offer suggestions about how to make it better. I will put some sticky notes next to each argument. You can use the sticky notes to leave your suggestions." Give each group about three to four minutes to read the arguments, talk, and offer suggestions.

        a. After three to four minutes, tell the students, "Okay. Let's move on to the next argument. Please move one group to the left."

        b. Again, give each group three to four minutes to read, talk, and offer suggestions.

Teacher Notes

c. Repeat this process until each group has had their argument read and critiqued three times.

- *If using the modified gallery walk format,* tell the students, "Okay. Let's get started. Reviewers, move one group to the left. Don't move to the next group until I tell you to move. Presenters, go ahead and share your argument with the travelers when they get there." Give each group of presenters and travelers about three to four minutes to talk.

   a. Tell the students, "Okay. Let's move on to the next argument. Travelers, move one group to the left."

   b. Again, give each group of presenters and travelers about three to four minutes to talk.

   c. Repeat this process until each group has had their argument read and critiqued three times.

9. Tell the students to return to their workstations.

10. Give the following instructions about revising the argument:

    - *If using the whole-class presentation format,* tell the students, "I'm now going to give you all about 10 minutes to revise your argument. Take a few minutes to talk in your groups and determine what you want to change to make your argument better. Once you have decided what to change, go ahead and make the changes to your whiteboard."

    - *If using the gallery walk format,* tell the students, "I'm now going to give you all about 10 minutes to revise your argument. Take a few minutes to read the suggestions that were left at your argument. Then talk in your groups and determine what you want to change to make your argument better. Once you have decided what to change, go ahead and make the changes to your whiteboard."

    - *If using the modified gallery walk format,* tell the students, "I'm now going to give you all about 10 minutes to revise your argument. Please return to your original groups." Wait for the students to move back into their original groups and then tell the students, "Okay, take a few minutes to talk in your groups and determine what you want to change to make your argument better. Once you have decided what to change, go ahead and make the changes to your whiteboard."

11. Ask the students, "Do you have any questions about what you need to do?"

12. Answer any questions that come up.

13. Tell the students, "Okay. Let's get started."

14. Give the students 10 minutes to work in their groups on their arguments. As they work, move from group to group to check in, ask probing questions, and offer a suggestion if a group gets stuck.

Investigation 1. Movement of Matter Into and Out of a System:
What Causes the Water to Come Out of the Tube in the Balloon Water Dispenser?

Stage 5: Reflective Discussion (15 minutes)

1. Tell the students, "We are now going to take a minute to talk about some of the core ideas and crosscutting concepts that we have used during our investigation."

2. Show the students a video of a kite flying. You can use the video at *www.youtube.com/watch?v=rCOZTn74n1I* or a video of your choice. You only need to show about 30 seconds of a kite flying.

3. Ask the students, "What do you all see going on here?"

4. Allow the students to share their ideas.

5. Now show students this video of the wind pushing water back up a waterfall: *www.youtube.com/watch?v=rS-Ljbn9fvc*.

6. Ask the students, "What do you all see going on here?"

7. Allow the students to share their ideas.

8. Ask the students, "How can we explain what we saw in these two videos using what we know about matter and its interactions?"

9. Allow the students to share their ideas. As they share their ideas, ask different questions to encourage them to expand on their thinking (e.g., "Can you tell me more about that?"), clarify a contribution (e.g., "Can you say that in another way?"), support an idea (e.g., "Why do you think that?"), add to an idea mentioned by a classmate (e.g., "Would anyone like to add to the idea?"), re-voice an idea offered by a classmate (e.g., "Who can explain that to me in another way?"), or critique an idea during the discussion (e.g., "Do you agree or disagree with that idea and why?") until students are able to generate an adequate explanation.

10. Ask the students, "What other questions do you have about this?"

11. Answer any questions that come up.

12. Tell the students, "We also looked for cause-and-effect relationships during our investigation." Then ask, "Can anyone tell me why it is useful to look for cause-and-effect relationships?"

13. Allow the students to share their ideas.

14. Tell the students, "We also defined the system under study and made a model of it." Then ask, "Can anyone tell me why it is useful to describe and model systems?" and "What was the system we studied during this investigation?"

15. Allow the students to share their ideas.

16. Show an image of the question "What do you think are the most important core ideas or crosscutting concepts that we used during this investigation to help us make sense of what we observed?" Tell the students, "Okay, let's make sure we

Teacher Notes

FIGURE 1.8
Model of the balloon water dispenser

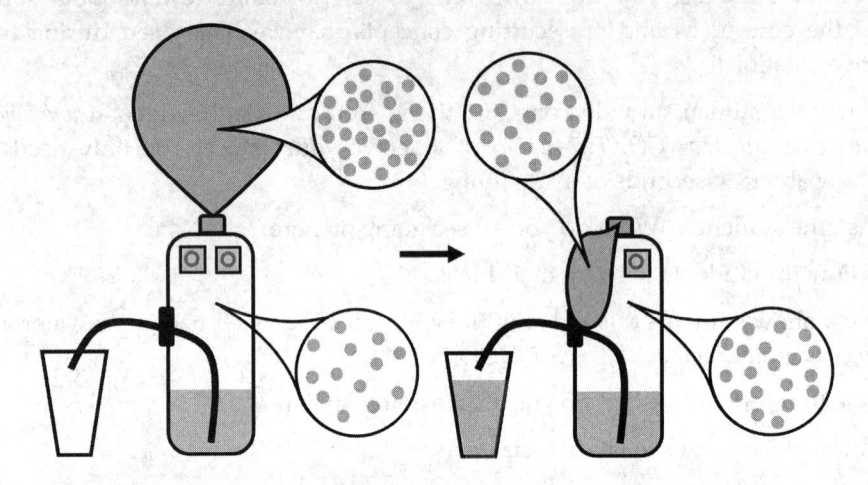

are all on the same page. Please take a moment to discuss this question with the other people in your group." Give them a few minutes to discuss the question.

17. Ask the students, "What do you all think? Who would like to share?"

18. Allow the students to share their ideas.

19. Tell the students, "We are now going to take a minute to talk about the use of models as tools for reasoning about natural phenomena."

20. Show an image of the question "What is a model?" on the screen. Tell the students, "Take a few minutes to talk about how you would answer this question with the other people in your group. Be ready to share with the rest of the class." Give the students two to three minutes to talk in their group.

21. Ask the students, "What do you all think? Who would like to share an idea?"

22. Allow the students to share their ideas.

23. Tell the students, "Okay, let's make sure we are all using the same definition. I think a model is something scientists use to help them understand how systems work. Models can be pictures. Or they can be graphs. Or they can even be physical models of things like a skeleton."

24. Show Figure 1.8 (without the caption) on a screen with the question "Is this a model?" on the screen.

25. Ask the students, "Do you think this is a model? Why or why not?"

26. Allow the students to share their ideas.

27. Tell the students, "I think this is a model because it allows us to understand what happened during the investigation."

Investigation 1. Movement of Matter Into and Out of a System: What Causes the Water to Come Out of the Tube in the Balloon Water Dispenser?

28. Ask the students, "What is the system being modeled?"
29. Allow the students to share their ideas.
30. Ask the students, "What questions do you have about how scientists use models?"
31. Answer any questions that come up.
32. Tell the students, "We are now going to take a minute to talk about what went well and what didn't go so well during our investigation. We need to talk about this because you all are going to be planning and carrying out your own investigations like this a lot this year, and I want to help you all get better at it."
33. Show an image of the question "What made your investigation scientific?" on the screen. Tell the students, "Take a few minutes to talk about how you would answer this question with the other people in your group. Be ready to share with the rest of the class." Give the students two to three minutes to talk in their group.
34. Ask the students, "What do you all think? Who would like to share an idea?"
35. Allow the students to share their ideas. Be sure to expand on their ideas about what makes an investigation scientific.
36. Show an image of the question "What made your investigation not so scientific?" on the screen. Tell the students, "Take a few minutes to talk about how you would answer this question with the other people in your group. Be ready to share with the rest of the class." Give the students two to three minutes to talk in their group.
37. Ask the students, "What do you all think? Who would like to share an idea?"
38. Allow the students to share their ideas. Be sure to expand on their ideas about what makes an investigation less scientific.
39. Show an image of the question "What rules can we put into place to help us make sure our next investigation is more scientific?" on the screen. Tell the students, "Take a few minutes to talk about how you would answer this question with the other people in your group. Be ready to share with the rest of the class." Give the students two to three minutes to talk in their group.
40. Ask the students, "What do you all think? Who would like to share an idea?"
41. Allow the students to share their ideas. Once they have shared their ideas, offer a suggestion for a possible class rule.
42. Ask the students, "What do you all think? Should we make this a rule?"
43. If the students agree, write the rule on the board or on a class "Rules for Scientific Investigation" chart so you can refer to it during the next investigation.

Teacher Notes

Stage 6: Write a Draft Report (30 minutes)

Your students will use either the Investigation Handout or the Investigation Log in the student workbook when writing the draft report. When you give the directions shown in quotes in the following steps, substitute "Investigation Log in your workbook" or just "Investigation Log" (as shown in brackets) for "handout" if they are using the workbook.

1. Tell the students, "You are now going to write an investigation report to share what you have learned. Please take out a pencil and turn to the 'Draft Report' section of your handout [Investigation Log in your workbook]."

2. If possible, use a document camera to project the "Introduction" section of the draft report from the Investigation Handout (or the Investigation Log in their workbook) on a screen or board (or take a picture of it and project the picture on a screen or board).

3. Tell the students, "The first part of the report is called the 'Introduction.' In this section of the report you want to explain to the reader what you were investigating, why you were investigating it, and what question you were trying to answer. All this information can be found in the text at the beginning of your handout [Investigation Log]." Point to the image. "Here are some sentence starters to help you begin writing."

4. Ask the students, "Do you have any questions about what you need to do?"

5. Answer any questions that come up.

6. Tell the students, "Okay, let's write."

7. Give the students 10 minutes to write the "Introduction" section of the report. As they work, move from student to student to check in, ask probing questions, and offer a suggestion if a student gets stuck.

8. If possible, use a document camera to project the "Method" section of the draft report from the Investigation Handout (or the Investigation Log in their workbook) on a screen or board (or take a picture of it and project the picture on a screen or board).

9. Tell the students, "The second part of the report is called the 'Method.' In this section of the report you want to explain to the reader what you did during the investigation, what data you collected and why, and how you went about analyzing your data. All this information can be found in the 'Plan Your Investigation' section of the handout [Investigation Log]. Remember that you all planned and carried out different investigations, so do not assume that the reader will know what you did." Point to the image. "Here are some sentence starters to help you begin writing."

10. Ask the students, "Do you have any questions about what you need to do?"

11. Answer any questions that come up.

Investigation 1. Movement of Matter Into and Out of a System: What Causes the Water to Come Out of the Tube in the Balloon Water Dispenser?

12. Tell the students, "Okay, let's write."
13. Give the students 10 minutes to write the "Method" section of the report. As they work, move from student to student to check in, ask probing questions, and offer a suggestion if a student gets stuck.
14. If possible, use a document camera to project the "Argument" section of the draft report from the Investigation Handout (or the Investigation Log in their workbook) on a screen or board (or take a picture of it and project the picture on a screen or board).
15. Tell the students, "The last part of the report is called the 'Argument.' In this section of the report you want to share your claim, evidence, and justification of the evidence with the reader. All this information can be found on your whiteboard." Point to the image. "Here are some sentence starters to help you begin writing."
16. Ask the students, "Do you have any questions about what you need to do?"
17. Answer any questions that come up.
18. Tell the students, "Okay, let's write."
19. Give the students 10 minutes to write the "Argument" section of the report. As they work, move from student to student to check in, ask probing questions, and offer a suggestion if a student gets stuck.

Stage 7: Peer Review (35 minutes)

Your students will use either the Investigation Handout or their workbook when doing the peer review. Except where noted below, the directions are the same whether using the handout or the workbook.

1. Tell the students, "We are now going to review our reports to find ways to make them better. I'm going to come around and collect your draft reports. While I do that, please take out a pencil."
2. Collect the handouts or the workbooks with the draft reports from the students.
3. If possible, use a document camera to project the peer-review guide (see Appendix 4) on a screen or board (or take a picture of it and project the picture on a screen or board).
4. Tell the students, "We are going to use this peer-review guide to give each other feedback." Point to the image.
5. Tell the students, "I'm going to ask you to work with a partner to do this. I'm going to give you and your partner a draft report to read. You two will then read the report together. Once you are done reading the report, I want you to answer each of the questions on the peer-review guide." Point to the review questions on the image of the peer-review guide.

Teacher Notes

6. Tell the students, "You can check 'no,' 'almost,' or 'yes' after each question." Point to the checkboxes on the image of the peer-review guide.

7. Tell the students, "This will be your rating for this part of the report. Make sure you agree on the rating you give the author. If you mark 'no' or 'almost,' then you need to tell the author what he or she needs to do to get a 'yes.'" Point to the space for the reviewer feedback on the image of the peer-review guide.

8. Tell the students, "It is really important for you to give the authors feedback that is helpful. That means you need to tell them exactly what they need to do to make their report better."

9. Ask the students, "Do you have any questions about what you need to do?"

10. Answer any questions that come up.

11. Tell the students, "Please sit with a partner who is not in your current group." Allow the students time to sit with a partner.

12. Tell the students, "Okay, I'm now going to give you one report to read." Pass out one Investigation Handout with a draft report or one workbook to each pair. Make sure that the report you give a pair was not written by one of the students in that pair. Give each pair one peer-review guide to fill out. If the students are using workbooks, the peer-review guide is included right after the draft report so you do not need to pass out copies of the peer-review guide.

13. Tell the students, "Okay, I'm going to give you 15 minutes to read the report I gave you and to fill out the peer-review guide. Go ahead and get started."

14. Give the students 15 minutes to work. As they work, move around from pair to pair to check in and see how things are going, answer questions, and offer advice.

15. After 15 minutes pass, tell the students, "Okay, time is up. Please give me the report and the peer-review guide that you filled out."

16. Collect the Investigation Handouts and the peer-review guides, or collect the workbooks if students are using them. If the students are using the Investigation Handouts and separate peer-review guides, be sure you keep each handout with its corresponding peer-review guide.

17. Tell the students, "Okay, I am now going to give you a different report to read and a new peer-review guide to fill out." Pass out one more report to each pair. Make sure that the report you give a pair was not written by one of the students in that pair. Give each pair a new peer-review guide to fill out as a group.

18. Tell the students, "Okay, I'm going to give you 15 minutes to read this new report and to fill out the peer-review guide. Go ahead and get started."

19. Give the students 15 minutes to work. As they work, move around from pair to pair to check in and see how things are going, answer questions, and offer advice.

Investigation 1. Movement of Matter Into and Out of a System:
What Causes the Water to Come Out of the Tube in the Balloon Water Dispenser?

20. After 15 minutes pass, tell the students, "Okay, time is up. Please give me the report and the peer-review guide that you filled out."

21. Collect the Investigation Handouts and the peer-review guides, or collect the workbooks if students are using them. If the students are using the Investigation Handouts and separate peer-review guides, be sure you keep each handout with its corresponding peer-review guide.

Stage 8: Revise the Report (30 minutes)

Your students will use either the Investigation Handout or their workbook when revising the report. Except where noted below, the directions are the same whether using the handout or the workbook.

1. Tell the students, "You are now going to revise your draft report based on the feedback you get from your classmates. Please take out a pencil."

2. Return the reports to the students.
   - *If the students used the Investigation Handout and a copy of the peer-review guide,* pass back the handout and the peer-review guide to each student.
   - *If the students used the workbook,* pass that back to each student.

3. Tell the students, "Please take a few minutes to read over the peer-review guide. You should use it to figure out what you need to change in your report and how you will change it."

4. Allow the students to read the peer-review guide.

5. *If the students used the workbook,* if possible use a document camera to project the "Write Your Final Report" section from the Investigation Log on a screen or board (or take a picture of it and project the picture on a screen or board).

6. Give the following directions about how to revise their reports:
   - *If the students used the Investigation Handout and a copy of the peer-review guide,* tell them, "Okay, let's revise our reports. Please take out a piece of paper. I would like you to rewrite your report. You can use your draft report as a starting point, but you also need to change it to make it better. Use the feedback on the peer-review guide to make it better."
   - *If the students used the workbook,* tell them, "Okay, let's revise our reports. I would like you to rewrite your report in the section of the Investigation Log called "Write Your Final Report." You can use your draft report as a starting point, but you also need to change it to make it better. Use the feedback on the peer-review guide to make it better."

7. Ask the students, "Do you have any questions about what you need to do?"

8. Answer any questions that come up.

Argument-Driven Inquiry in **Fifth-Grade Science:** Three-Dimensional Investigations

Teacher Notes

9. Tell the students, "Okay, let's write." Allow about 20 minutes for the students to revise their reports.

10. After about 20 minutes, give the following directions:
    - *If the students used the Investigation Handout,* tell them, "Okay, time's up. I will now come around and collect your Investigation Handout, the peer-review guide, and your final report."
    - *If the students used the workbook,* tell them, "Okay, time's up. I will now come around and collect your workbooks."

11. *If the students used the Investigation Handout,* collect all the Investigation Handouts, peer-review guides, and final reports. *If the students used the workbook,* collect all the workbooks.

12. *If the students used the Investigation Handout,* use the "Teacher Score" column in the peer-review guide to grade the final report. *If the students used the workbook,* use the "Investigation Report Grading Rubric" in the Investigation Log to grade the final report. Whether you are using the handout or the log, you can give the students feedback about their writing in the "Teacher Comments" section.

How to Use the Checkout Questions

The Checkout Questions are an optional assessment. We recommend giving them to students at the start of the next class period after the students finish stage 8 of the investigation. You can then look over the student answers to determine if you need to reteach the core idea from the investigation. Appendix 6 gives the answers to the Checkout Questions that should be given by a student who can apply the core idea correctly and can (1) explain the cause-and-effect relationship and (2) define the system.

Alignment With Standards

Table 1.2 highlights how the investigation can be used to address specific performance expectations from the *Next Generation Science Standards, Common Core State Standards for English Language Arts (CCSS ELA)* and *Common Core State Standards for Mathematics (CCSS Mathematics),* and *English Language Proficiency (ELP) Standards.*

TABLE 1.2

Investigation 1 alignment with standards

NGSS performance expectation	5-PS1-1: Develop a model to describe that matter is made up of particles too small to be seen.

Continued

Investigation 1. Movement of Matter Into and Out of a System: What Causes the Water to Come Out of the Tube in the Balloon Water Dispenser?

Table 1.2 (*continued*)

***CCSS ELA**—Reading: Informational Text*	Key ideas and details • CCSS.ELA-LITERACY.RI.5.1: Quote accurately from a text when explaining what the text says explicitly and when drawing inferences from the text. • CCSS.ELA-LITERACY.RI.5.2: Determine two or more main ideas of a text and explain how they are supported by key details; summarize the text. • CCSS.ELA-LITERACY.RI.5.3: Explain the relationships or interactions between two or more individuals, events, ideas, or concepts in a historical, scientific, or technical text based on specific information in the text. Craft and structure • CCSS.ELA-LITERACY.RI.5.4: Determine the meaning of general academic and domain-specific words and phrases in a text relevant to a *grade 5 topic or subject area*. • CCSS.ELA-LITERACY.RI.5.5: Compare and contrast the overall structure (e.g., chronology, comparison, cause/effect, problem/solution) of events, ideas, concepts, or information in two or more texts. • CCSS.ELA-LITERACY.RI.5.6: Analyze multiple accounts of the same event or topic, noting important similarities and differences in the point of view they represent. Integration of knowledge and ideas • CCSS.ELA-LITERACY.RI.5.7: Draw on information from multiple print or digital sources, demonstrating the ability to locate an answer to a question quickly or to solve a problem efficiently. • CCSS.ELA-LITERACY.RI.5.8: Explain how an author uses reasons and evidence to support particular points in a text, identifying which reasons and evidence support which point(s). Range of reading and level of text complexity • CCSS.ELA-LITERACY.RI.5.10: By the end of the year, read and comprehend informational texts, including history/social studies, science, and technical texts, at the high end of the grades 4–5 text complexity band independently and proficiently.

***CCSS ELA**—Writing*	Text types and purposes
	• CCSS.ELA-LITERACY.W.5.1: Write opinion pieces on topics or texts, supporting a point of view with reasons and information.
	   o CCSS.ELA-LITERACY.W.5.1.A: Introduce a topic or text clearly, state an opinion, and create an organizational structure in which ideas are logically grouped to support the writer's purpose.
	   o CCSS.ELA-LITERACY.W.5.1.B: Provide logically ordered reasons that are supported by facts and details.
	   o CCSS.ELA-LITERACY.W.5.1.C: Link opinion and reasons using words, phrases, and clauses (e.g., *consequently*, *specifically*).
	   o CCSS.ELA-LITERACY.W.5.1.D: Provide a concluding statement or section related to the opinion presented.

Continued

Table 1.2 (*continued*)

CCSS ELA—**Writing** (*continued*)	• CCSS.ELA-LITERACY.W.5.2: Write informative/explanatory texts to examine a topic and convey ideas and information clearly.
	    ○ CCSS.ELA-LITERACY.W.5.2.A: Introduce a topic clearly, provide a general observation and focus, and group related information logically; include formatting (e.g., headings), illustrations, and multimedia when useful to aiding comprehension.
	    ○ CCSS.ELA-LITERACY.W.5.2.B: Develop the topic with facts, definitions, concrete details, quotations, or other information and examples related to the topic.
	    ○ CCSS.ELA-LITERACY.W.5.2.C: Link ideas within and across categories of information using words, phrases, and clauses (e.g., *in contrast*, *especially*).
	    ○ CCSS.ELA-LITERACY.W.5.2.D: Use precise language and domain-specific vocabulary to inform about or explain the topic.
	    ○ CCSS.ELA-LITERACY.W.5.2.E: Provide a concluding statement or section related to the information or explanation presented.
	Production and distribution of writing
	• CCSS.ELA-LITERACY.W.5.4: Produce clear and coherent writing in which the development and organization are appropriate to task, purpose, and audience.
	• CCSS.ELA-LITERACY.W.5.5: With guidance and support from peers and adults, develop and strengthen writing as needed by planning, revising, editing, rewriting, or trying a new approach.
	• CCSS.ELA-LITERACY.W.5.6: With some guidance and support from adults, use technology, including the internet, to produce and publish writing as well as to interact and collaborate with others; demonstrate sufficient command of keyboarding skills to type a minimum of two pages in a single sitting.
	Research to build and present knowledge
	• CCSS.ELA-LITERACY.W.5.8: Recall relevant information from experiences or gather relevant information from print and digital sources; summarize or paraphrase information in notes and finished work, and provide a list of sources.
	• CCSS.ELA-LITERACY.W.5.9: Draw evidence from literary or informational texts to support analysis, reflection, and research.
	Range of writing
	• CCSS.ELA-LITERACY.W.5.10: Write routinely over extended time frames (time for research, reflection, and revision) and shorter time frames (a single sitting or a day or two) for a range of discipline-specific tasks, purposes, and audiences.

Continued

Investigation 1. Movement of Matter Into and Out of a System: What Causes the Water to Come Out of the Tube in the Balloon Water Dispenser?

Table 1.2 (*continued*)

CCSS ELA—Speaking and Listening	Comprehension and collaboration
	
	• CCSS.ELA-LITERACY.SL.5.1: Engage effectively in a range of collaborative discussions (one-on-one, in groups, and teacher-led) with diverse partners on *grade 5 topics and texts*, building on others' ideas and expressing their own clearly.
	    o CCSS.ELA-LITERACY.SL.5.1.A: Come to discussions prepared, having read or studied required material; explicitly draw on that preparation and other information known about the topic to explore ideas under discussion.
	    o CCSS.ELA-LITERACY.SL.5.1.B: Follow agreed-upon rules for discussions and carry out assigned roles.
	    o CCSS.ELA-LITERACY.SL.5.1.C: Pose and respond to specific questions by making comments that contribute to the discussion and elaborate on the remarks of others.
	    o CCSS.ELA-LITERACY.SL.5.1.D: Review the key ideas expressed and draw conclusions in light of information and knowledge gained from the discussions.
	• CCSS.ELA-LITERACY.SL.5.2: Summarize a written text read aloud or information presented in diverse media and formats, including visually, quantitatively, and orally.
	• CCSS.ELA-LITERACY.SL.5.3: Summarize the points a speaker makes and explain how each claim is supported by reasons and evidence.
	
	Presentation of knowledge and ideas
	
	• CCSS.ELA-LITERACY.SL.5.4: Report on a topic or text or present an opinion, sequencing ideas logically and using appropriate facts and relevant, descriptive details to support main ideas or themes; speak clearly at an understandable pace.
	• CCSS.ELA-LITERACY.SL.5.5: Include multimedia components (e.g., graphics, sound) and visual displays in presentations when appropriate to enhance the development of main ideas or themes.
	• CCSS.ELA-LITERACY.SL.5.6: Adapt speech to a variety of contexts and tasks, using formal English when appropriate to task and situation.

CCSS Mathematics—Operations and Algebraic Thinking	Write and interpret numerical expressions. • CCSS.MATH.CONTENT.5.OA.A.2: Write simple expressions that record calculations with numbers, and interpret numerical expressions without evaluating them.
CCSS Mathematics—Numbers and Operations in Base Ten	Perform operations with multi-digit whole numbers and with decimals to hundredths. • CCSS.MATH.CONTENT.5.NBT.B.5: Fluently multiply multi-digit whole numbers using the standard algorithm. • CCSS.MATH.CONTENT.5.NBT.B.7: Add, subtract, multiply, and divide decimals to hundredths.

Continued

Teacher Notes

Table 1.2 (*continued*)

CCSS Mathematics—Measurement and Data	Convert like measurement units within a given measurement system. • CCSS.MATH.CONTENT.5.MD.A.1: Convert among different-sized standard measurement units within a given measurement system (e.g., convert 5 cm to 0.05 m), and use these conversions in solving multi-step, real-world problems. Represent and interpret data. • CCSS.MATH.CONTENT.5.MD.B.2: Make a line plot to display a data set of measurements in fractions of a unit (1/2, 1/4, 1/8). Use operations on fractions for this grade to solve problems involving information presented in line plots. Geometric measurement: understand concepts of volume. • CCSS.MATH.CONTENT.5.MD.C.3: Recognize volume as an attribute of solid figures and understand concepts of volume measurement.
ELP Standards	Receptive modalities • ELP 1: Construct meaning from oral presentations and literary and informational text through grade-appropriate listening, reading, and viewing. • ELP 8: Determine the meaning of words and phrases in oral presentations and literary and informational text. Productive modalities • ELP 3: Speak and write about grade-appropriate complex literary and informational texts and topics. • ELP 4: Construct grade-appropriate oral and written claims and support them with reasoning and evidence. • ELP 7: Adapt language choices to purpose, task, and audience when speaking and writing. Interactive modalities • ELP 2: Participate in grade-appropriate oral and written exchanges of information, ideas, and analyses, responding to peer, audience, or reader comments and questions. • ELP 5: Conduct research and evaluate and communicate findings to answer questions or solve problems. • ELP 6: Analyze and critique the arguments of others orally and in writing. Linguistic structures of English • ELP 9: Create clear and coherent grade-appropriate speech and text. • ELP 10: Make accurate use of standard English to communicate in grade-appropriate speech and writing.

Investigation Handout

Investigation 1

Movement of Matter Into and Out of a System: What Causes the Water to Come Out of the Tube in the Balloon Water Dispenser?

Introduction

People use water dispensers to store drinking water. Your teacher will show you an example of a water dispenser and then demonstrate how to get water out of it. As you watch this demonstration, keep track of things you notice and things you wonder about in the boxes below.

Things I NOTICED …	Things I WONDER about …

The picture to the right shows a diagram of the balloon water dispenser that your teacher used a few minutes ago. The balloon water dispenser is a system. A *system* is a group of interacting parts. The balloon water dispenser system is made up of six different parts: (1) a bottle, (2) a tube, (3) a balloon, (4) tape, (5) air, and (6) water. Each part of the balloon water dispenser is made up of a different type of matter.

All matter is made up of tiny particles that are too small to be seen. These particles are in constant motion but are held together by attractive forces between them. Matter exists in different forms or *states* depending on how close the particles that make up a sample of matter are held together.

A sample of matter that is in a *solid* state is made up of particles that are held close together and cannot move past each other. When the particles

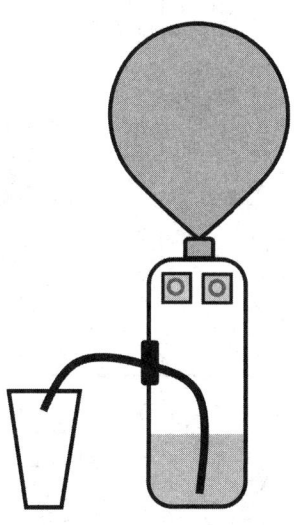

Argument-Driven Inquiry in **Fifth-Grade Science**: Three-Dimensional Investigations

that make up a sample of matter cannot move past each other, the sample of matter does not change shape without something bending, crushing, or stretching it. Solids therefore have a fixed shape. The straw and the bottle are examples of matter that is in a solid state.

A sample of matter that is in a *liquid* state or a *gas* state is made up of particles that are not held close together. The particles that make up a liquid or a gas, as a result, can move past each other. When the particles that make up a sample of matter can move past each other, the sample of matter will change shape when placed in a container. Liquids and gases therefore do not have a fixed shape. The water in the bottle is an example of a liquid, and the air in the balloon is an example of a gas.

When a solid, liquid, or gas that is moving *collides* with a solid, liquid, or gas that is not moving, the motion of all the matter that is involved in the collision can change. For example, when you turn a faucet on, liquid water that is moving can collide with pieces of solid food that are sitting in the sink. The moving water will cause those small solids to start moving toward the drain. The collision with the small solids will also cause the moving water to change the direction it is moving. A gas colliding with a liquid can also change the motion of the gas and the liquid. For example, if you blow air through a straw at a small pool of water on a table, the air will cause some of the water to move away from you. The air will also change direction as it collides with the pool of water on the table.

Things we KNOW from what we read …

Investigation 1. Movement of Matter Into and Out of a System:
What Causes the Water to Come Out of the Tube in the Balloon Water Dispenser?

Your Task

Use what you know about the properties of solids, liquids, and gases; cause-and-effect relationships; and systems to design and carry out an investigation to figure out how the balloon water dispenser works.

The *guiding question* of this investigation is, **What causes the water to come out of the tube in the balloon water dispenser?**

Materials

You may use any of the following materials during your investigation:

- Safety goggles (required)
- Balloon water dispenser
- 2 measuring cups
- Stopwatch
- Clothespin
- Cloth measuring tape

Safety Rules

Follow all normal safety rules. In addition, be sure to follow these rules:

- Wear sanitized safety goggles during setup, investigation activity, and cleanup.
- Immediately clean up any spills to avoid a slip or fall hazard.
- Wash your hands with soap and water when you are done cleaning up.

Plan Your Investigation

Prepare a plan for your investigation by filling out the chart on the next page; this plan is called an *investigation proposal*. Before you start developing your plan, be sure to discuss the following questions with the other members of your group:

- What might be the **cause** of the water coming out the tube?
- What are the components of the **system** that we are studying?

Investigation Handout

Our guiding question:

This is a picture of how we will set up the equipment:

We will collect the following data:

These are the steps we will follow to collect data:

I approve of this investigation proposal.

_____   _____
Teacher's signature   Date

Investigation 1. Movement of Matter Into and Out of a System: What Causes the Water to Come Out of the Tube in the Balloon Water Dispenser?

Collect Your Data

Keep a record of what you measure or observe during your investigation in the space below.

Analyze Your Data

You will need to analyze the data you collected before you can develop an answer to the guiding question. To analyze the data you collected, create a graph that shows the relationship between what you changed and what you measured or observed as a result of what you changed.

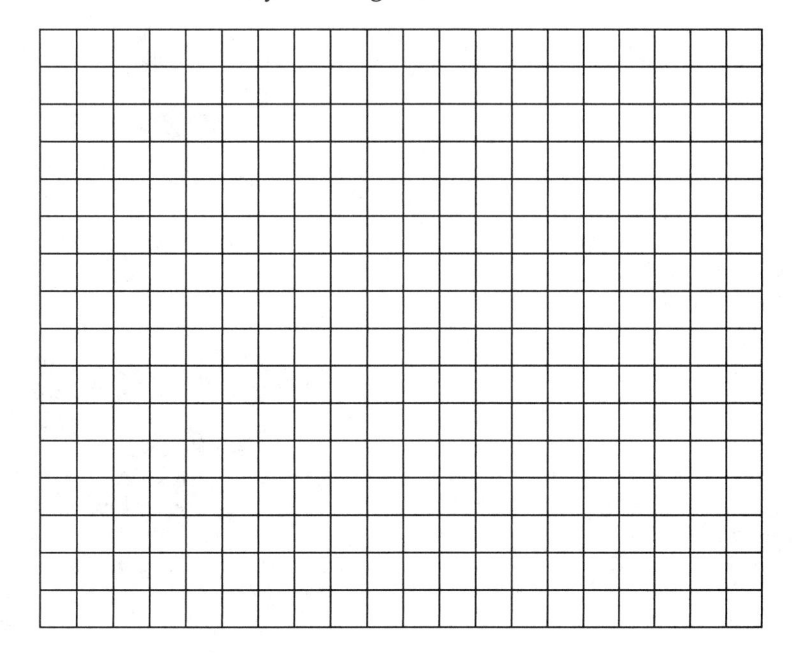

Investigation Handout

Draft Argument

Develop an argument on a whiteboard. It should include the following:

1. A *claim*: Your answer to the guiding question.
2. *Evidence*: An analysis of the data and an explanation of what the analysis means.
3. A *justification of the evidence*: Why your group thinks the evidence is important.

The Guiding Question:	
Our Claim:	
Our Evidence:	Our Justification of the Evidence:

Argumentation Session

Share your argument with your classmates. Be sure to ask them how to make your draft argument better. Keep track of their suggestions in the space below.

Ways to IMPROVE our argument …

Investigation 1. Movement of Matter Into and Out of a System: What Causes the Water to Come Out of the Tube in the Balloon Water Dispenser?

Draft Report

Prepare an *investigation report* to share what you have learned. Use the information in this handout and your group's final argument to write a *draft* of your investigation report.

Introduction

We have been studying _____ in class.

Before we started this investigation, we explored _____

_____

_____

We noticed _____

_____

_____

My goal for this investigation was to figure out _____

_____

_____

The guiding question was _____

_____

_____

Method

To gather the data I needed to answer this question, I _____

_____

_____

_____

_____

Investigation Handout

I then analyzed the data I collected by _____

_____

_____

_____

Argument

My claim is _____

_____

_____

_____

The graph below shows _____

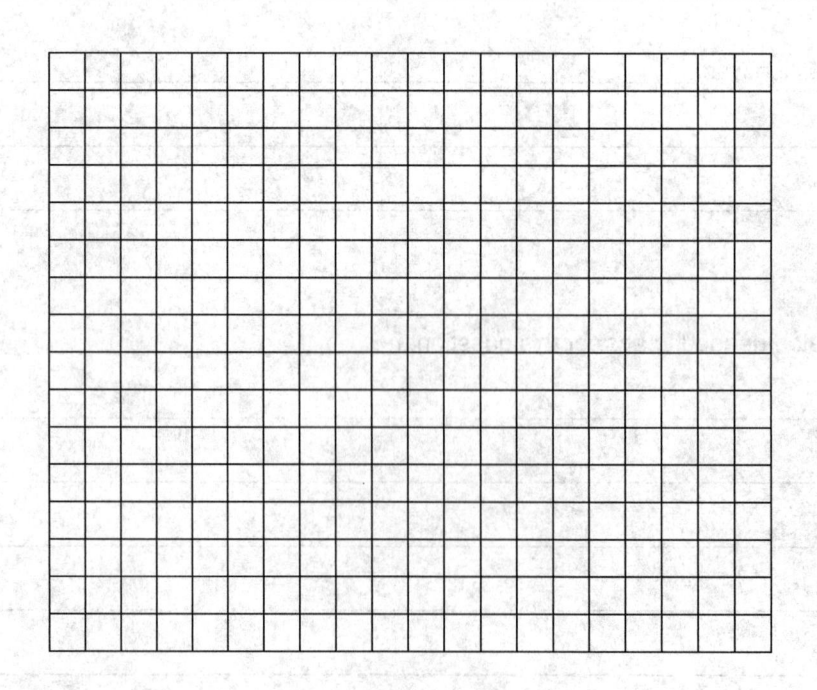

_____

_____

_____

_____

Investigation 1. Movement of Matter Into and Out of a System: What Causes the Water to Come Out of the Tube in the Balloon Water Dispenser?

This analysis of the data I collected suggests _____

_____

_____

_____

_____

This evidence is based on several important scientific concepts. The first one is _____

_____

_____

_____

_____

_____

Review

Your classmates need your help! Review the draft of their investigation reports and give them ideas about how to improve. Use the *peer-review guide* when doing your review.

Submit Your Final Report

Once you have received feedback from your classmates about your draft report, create your final investigation report and hand it in to your teacher.

Checkout Questions

Investigation 1. Movement of Matter Into and Out of a System

Use the following information to answer questions 1 and 2. The picture below shows a balloon water dispenser after it deflated. The student used six breaths to inflate the balloon, and 75 ml of water was in the cup after the balloon completely deflated.

1. If the student used eight breaths to inflate the balloon, how much water would be in the cup when it deflates?

    a. 75 ml

    b. More than 75 ml

    c. Less than 75 ml

2. How do you know? Use what you know about cause and effect as part of your answer.

    _____
    _____
    _____
    _____
    _____
    _____

Teacher Scoring Rubric for Checkout Questions 1 and 2

Level	Description
3	The student can apply the core idea correctly and can explain the cause-and-effect relationship.
2	The student can apply the core idea correctly but cannot explain the cause-and-effect relationship.
1	The student cannot apply the core idea correctly but can explain the cause-and-effect relationship.
0	The student cannot apply the core idea correctly and cannot explain the cause-and-effect relationship.

Investigation 1. Movement of Matter Into and Out of a System: What Causes the Water to Come Out of the Tube in the Balloon Water Dispenser?

Use the following information to answer questions 3 and 4. The picture below shows a balloon water dispenser before and after the balloon deflated. The student used only used three breaths to inflate the balloon.

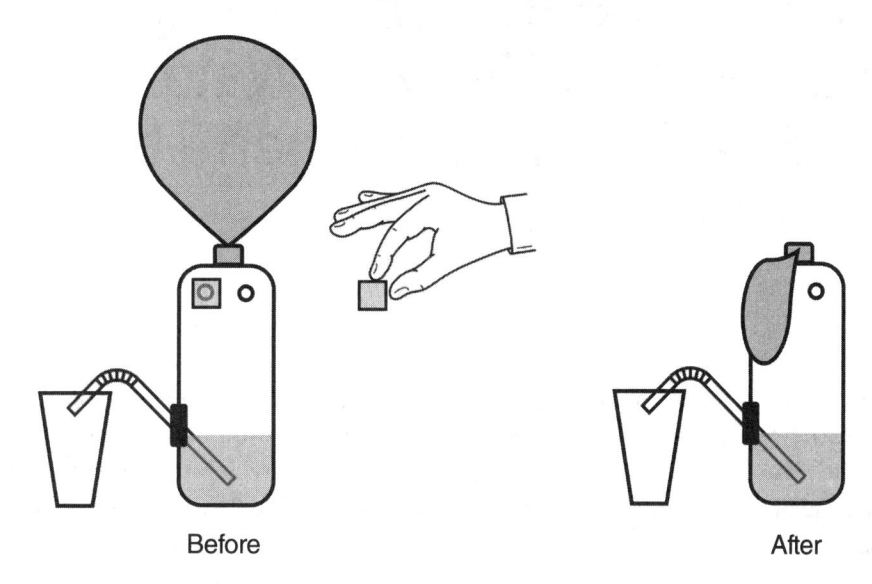

Before   After

3. Why is there no water in the cup?

   a. Three breaths are not enough to push the water up the straw.

   b. The tape covering a hole was removed.

   c. There was not enough water in the bottle to begin with.

   d. The balloon was stretched out.

   e. Unable to tell

4. Explain your thinking. Use what you know about systems as part of your answer.

_____

_____

_____

_____

_____

Teacher Scoring Rubric for Checkout Questions 3 and 4

Level	Description
3	The student can apply the core idea correctly and can define the system.
2	The student can apply the core idea correctly but cannot define the system.
1	The student cannot apply the core idea correctly but can define the system.
0	The student cannot apply the core idea correctly and cannot define the system.

Argument-Driven Inquiry in **Fifth-Grade Science**: Three-Dimensional Investigations

Teacher Notes

Investigation 2

Movement of Particles in a Liquid: Why Do People Use Hot Water Instead of Cold Water When They Make Tea?

Purpose

The purpose of this investigation is to give students an opportunity to use one disciplinary core idea (DCI), two crosscutting concepts (CCs), and eight scientific and engineering practices (SEPs) to figure out how temperature affects the rate at which tea spreads through a cup of water. Students will also learn about the difference between observations and inferences.

The DCI, CCs, and SEPs That Students Use During This Investigation to Figure Things Out

DCI

- *PS1.A: Structure and Properties of Matter:* Matter of any type can be subdivided into particles that are too small to see, but even then, the matter still exists and can be detected by other means.

CCs

- *CC 2: Cause and Effect:* Cause-and-effect relationships are routinely identified, tested, and used to explain change. Events that occur together with regularity might or might not be a cause-and-effect relationship.
- *CC 7: Stability and Change:* Change is measured in terms of differences over time and may occur at different rates.

SEPs

- *SEP 1: Asking Questions and Defining Problems:* Ask questions about what would happen if a variable is changed. Ask questions that can be investigated and predict reasonable outcomes based on patterns such as cause-and-effect relationships.
- *SEP 2: Developing and Using Models:* Develop and/or use models to describe and/or predict phenomena.
- *SEP 3: Planning and Carrying Out Investigations:* Plan and conduct an investigation collaboratively to produce data to serve as the basis for evidence, using fair tests in which variables are controlled and the number of trials considered. Evaluate appropriate methods and/or tools for collecting data.

Investigation 2. Movement of Particles in a Liquid:
Why Do People Use Hot Water Instead of Cold Water When They Make Tea?

- *SEP 4: Analyzing and Interpreting Data:* Represent data in tables and/or various graphical displays (bar graphs, pictographs, and/or pie charts) to reveal patterns that indicate relationships. Analyze and interpret data to make sense of phenomena, using logical reasoning, mathematics, and/or computation. Compare and contrast data collected by different groups in order to discuss similarities and differences in their findings.

- *SEP 5: Using Mathematics and Computational Thinking:* Organize simple data sets to reveal patterns that suggest relationships. Describe, measure, estimate, and/or graph quantities (e.g., area, volume, weight, time) to address scientific and engineering questions and problems.

- *SEP 6: Constructing Explanations and Designing Solutions:* Construct an explanation of observed relationships. Use evidence to construct or support an explanation. Identify the evidence that supports particular points in an explanation.

- *SEP 7: Engaging in Argument From Evidence:* Compare and refine arguments based on an evaluation of the evidence presented. Distinguish among facts, reasoned judgment based on research findings, and speculation in an explanation. Respectfully provide and receive critiques from peers about a proposed procedure, explanation, or model by citing relevant evidence and posing specific questions.

- *SEP 8: Obtaining, Evaluating, and Communicating Information:* Read and comprehend grade-appropriate complex texts and/or other reliable media to summarize and obtain scientific and technical ideas. Combine information in written text with that contained in corresponding tables, diagrams, and/or charts to support the engagement in other scientific and/or engineering practices. Communicate scientific and/or technical information orally and/or in written formats, including various forms of media as well as tables, diagrams, and charts.

Other Concepts That Students May Use During This Investigation

Students might also use some of the following concepts:

- Particles have energy because they are constantly moving.
- The temperature of an object or substance is a measure of the average energy of these particles.
- The particles that make up an object or substance stay together even though they are moving, because attractive forces pull them together.
- The attractive force between two or more individual particles is strongest when the particles are close together.

Teacher Notes

- An object or a substance is a solid when the particles that make up that object or substance have less energy and are held close to each other by the attractive force that exists between particles.
- A solid has a fixed shape because the particles within a solid are locked in place and are unable to move past each other.
- A liquid does not have a fixed shape because the particles within a liquid are not locked in place and can move past each other.

What Students Figure Out

Tea mixes with water faster at higher temperatures and slower at lower temperatures because the particles that make up tea and water move faster at higher temperatures and slower at lower temperatures.

Background Information About This Investigation for the Teacher

Matter, such as water or tea, is made up of particles that are too small to be seen. These particles are called atoms. Matter that is made up of only one type of atom is called an element. Examples of elements include gold, aluminum, and oxygen. Individual atoms can combine with other atoms to form molecules. Matter that is made up of a specific type of molecule is called a compound. Water, for example, is made up of water molecules. A water molecule is made up of one atom of oxygen and two atoms of hydrogen. Tea is made up of several different types of molecules, including catechin and theaflavin. These various molecules are made up of different combinations of carbon, oxygen, and hydrogen atoms. Different types of matter like water and tea have different properties because the molecules that make up these different types of matter have different combinations of atoms. Note that in the Investigation Handout for this investigation we use the term *water particles* in place of *water molecules* to describe the molecules found in water and the term *tea particles* to describe the many different types of molecules found in tea, to help make the ideas more accessible to children.

The molecules that make up water and tea, like all molecules, are in constant motion. Molecules in motion have kinetic energy. The more energy molecules absorb, the faster they move. The molecules that make up water or tea stay together even though they are moving because molecules attract each other. The attractive force between two or more individual molecules is strongest when molecules are close together.

An object (such as a tea leaf) or a substance (such as water) is a solid when the molecules that make up that object or a substance have little energy and are held close to each other by the attractive force that exists between molecules. Solids have a fixed shape because the molecules in a solid are locked in place and are unable to move past each other. If energy is transferred into a solid, the molecules that make up that object or substance will absorb that energy and start to move more and spread out. A solid will turn into a liquid (what

Investigation 2. Movement of Particles in a Liquid:
Why Do People Use Hot Water Instead of Cold Water When They Make Tea?

we often call melting) when enough molecules in the matter are able to move past each other. Liquids do not have a fixed shape because the molecules in liquids are not locked in place.

These ideas about the behavior of atoms or molecules in matter that is a solid or a liquid can help us understand why adding a tea bag to a cup of water will make the water change color over time. The tea and water, as noted earlier, are both made of molecules that are in constant motion. There is also empty space between these molecules (see Figure 2.1). When the tea is added to the water (see Figure 2.2), individual tea molecules will separate from the other tea molecules because the attractive forces from the water molecules pull them away from the other tea molecules. The individual tea molecules will then spread through the water because they are no longer fixed in place and can move around freely. This process is called diffusion.

FIGURE 2.1

A model of a bag of tea leaves and water before they are mixed

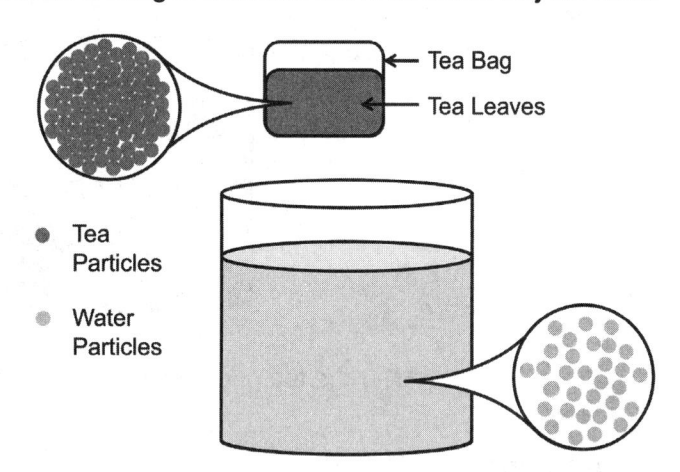

- Tea Particles
- Water Particles

Before a tea bag is added to a cup of water

FIGURE 2.2

A model of a bag of tea leaves and water immediately after the bag is added to the water

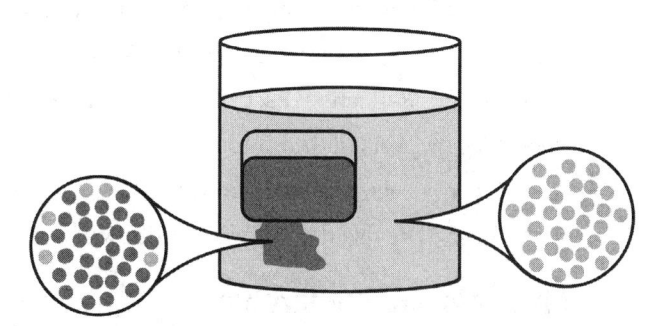

Right after a tea bag is added to a cup of water

The tea molecules will continue to spread through the container over time and mix with the water molecules. The diffusion of tea molecules will eventually result in an equal distribution of tea molecules throughout the water molecules inside the container (see Figure 2.3, p 86). The water in the cup changes color because the water becomes a mixture of tea molecules and water molecules. The water will turn a darker color over time as more and more individual tea molecules mix with the water molecules. The tea molecules will diffuse through the water faster at higher temperatures and slower at lower temperatures because the molecules that make up the tea and the water move faster at higher temperatures and slower at lower temperatures.

Teacher Notes

FIGURE 2.3
A model of tea molecules completely mixed with water molecules

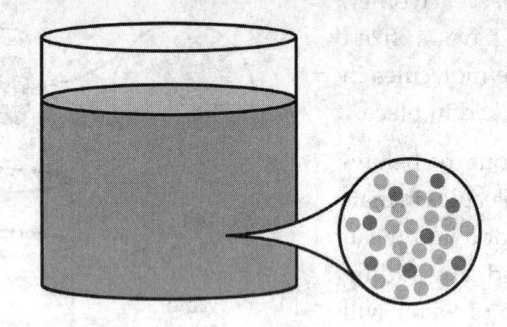

After the tea bag is removed from the cup of water

Timeline

The time needed to complete this investigation is 270 minutes (4 hours and 30 minutes). The amount of instructional time needed for each stage of the investigation is as follows:

- *Stage 1.* Introduce the task and the guiding question: 35 minutes
- *Stage 2.* Design a method and collect data: 50 minutes
- *Stage 3.* Create a draft argument: 45 minutes
- *Stage 4.* Argumentation session: 30 minutes
- *Stage 5.* Reflective discussion: 15 minutes
- *Stage 6.* Write a draft report: 30 minutes
- *Stage 7.* Peer review: 35 minutes
- *Stage 8.* Revise the report: 30 minutes

Materials and Preparation

The materials needed for this investigation are listed in Table 2.1. Most of the items can be purchased from a big-box retail store such as Walmart or Target or through an online retailer such as Amazon, but we suggest purchasing the clear plastic retail tubes from the supply store ULINE. The materials for this investigation can also be purchased as a complete kit (which includes enough materials for 24 students, or six groups of four students) at *www.argumentdriveninquiry.com*.

Investigation 2. Movement of Particles in a Liquid:
Why Do People Use Hot Water Instead of Cold Water When They Make Tea?

TABLE 2.1

Materials for Investigation 2

Item	Quantity
Safety goggles	1 per student
Chai or other black tea bags	4 per group
Thermometer	1 per group
Stopwatch	1 per group
Beaker, 250 ml	3 per group
Clear plastic retail tube, 1.5" × 6"	1 per group
Plastic ruler, 6"	1 per group
Secchi disk attached to the ruler	1 per group
Plastic ramekin	1 per group
Electric hot water kettle	1 per class
Pitcher (for hot and cold water)	3 per class
Whiteboard, 2' × 3'*	1 per group
Investigation Handout	1 per student
Peer-review guide and teacher scoring rubric	1 per student
Checkout Questions (optional)	1 per student

*As an alternative, students can use computer and presentation software such as Microsoft PowerPoint or Apple Keynote to create their arguments.

Students will need hot and cold water for their investigations. We recommend using an electric hot water kettle to heat water (to about 60°C) so students can use it as needed. You will also need three pitchers of ice water available for the students to use.

Students can use a clear plastic retail tube, a ruler, and a Secchi disk attached to the ruler to measure how much tea has spread through the water. A Secchi disk is simply a white and black disk that you can purchase or make yourself. The Secchi disk needs to be small so it can fit in a retail tube. You can make this disk by cutting a 1-inch circle out of a piece of paper and laminating it; then you can attach the laminated circle to the end of a ruler with tape. The Secchi disk is used to gauge the transparency of water by measuring the depth at which the disk ceases to be visible from the surface. This depth is called the Secchi depth.

Figure 2.4 (p. 88) shows how to measure the "darkness" or "cloudiness" of tea by finding the Secchi depth of the tea. First add the tea to the plastic retail tube so it is about 1 inch from the top of the tube, then submerge the ruler with the Secchi disk attachment until the disk is no longer visible. Record the depth of the disk using the units on the ruler—this depth is the Secchi depth of the tea. A higher Secchi depth indicates more transparent or "clear" water, and a lower Secchi depth indicates "darker" or "cloudier" water. In this investigation, tea that has more tea molecules (or particles) in it will be darker and have a lower Secchi depth than tea with fewer tea molecules (or particles) in it.

Teacher Notes

FIGURE 2.4
How to measure the cloudiness or darkness of tea using a Secchi disk and a ruler

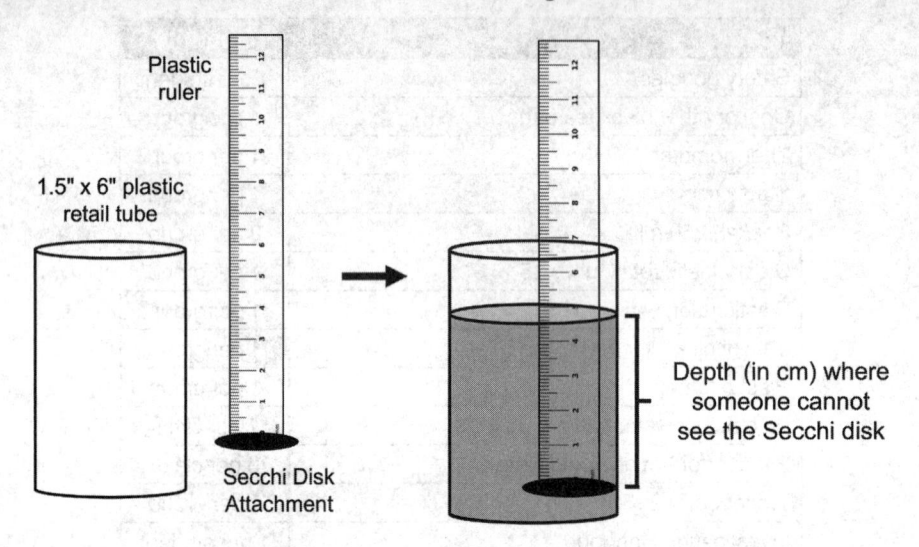

Plastic ruler

1.5" x 6" plastic retail tube

Secchi Disk Attachment

Depth (in cm) where someone cannot see the Secchi disk

Safety Precautions

Remind students to follow all normal safety rules. In addition, tell the students to take the following safety precautions:

- Wear sanitized safety goggles during setup, investigation activity, and cleanup.
- Be careful when working with hot water and use only under direct teacher supervision, because it can burn skin.
- Immediately clean up any spills to avoid a slip or fall hazard.
- Keep water sources away from electrical receptacles to prevent shock.
- Be careful when handling glassware, because it can shatter and cut skin.
- Make sure all materials are put away after completing the activity.
- Wash their hands with soap and water when they are done cleaning up.

Lesson Plan by Stage

This lesson plan is only a suggestion. It is included here to illustrate what you can say and do during each stage of ADI for this specific investigation. We encourage you to modify this lesson plan by asking different questions, using different examples, and providing different scaffolds as needed to better meet the needs of students in your class.

Investigation 2. Movement of Particles in a Liquid:
Why Do People Use Hot Water Instead of Cold Water When They Make Tea?

Stage 1: Introduce the Task and the Guiding Question (35 minutes)

1. Ask the students to sit in six groups, with three or four students in each group.
2. Ask the students to clear off their desks except for a pencil (and their *Student Workbook for Argument-Driven Inquiry in Fifth-Grade Science* if they have one).
3. Pass out an Investigation Handout to each student (or ask students to turn to the Investigation Log for Investigation 2 in their workbook).
4. Read the first paragraph of the "Introduction" aloud to the class. Ask the students to follow along as you read.
5. Give each group some tea leaves in a small plastic ramekin. Let the students look at and touch the tea leaves.
6. Pour some hot water into a beaker and add a tea bag. Be sure to use chai tea or another black tea, because it turns the water darker than green, white, yellow, or herbal teas. Encourage the students to watch what happens over time.
7. Ask the students to record their observations and questions about the tea leaves and the tea bag sitting in the hot water on the "NOTICED/WONDER" chart in the "Introduction."
8. After the students have recorded their observations and questions, ask them to share what they observed about the tea leaves or the tea bag in the hot water.
9. Ask the students to share what questions they have about the tea leaves or the tea bag in the hot water.
10. Tell the students, "Some of your questions might be answered by reading the rest of the 'Introduction.'"
11. Ask the students to read the rest of the "Introduction" on their own *or* ask them to follow along as you read it aloud.
12. Once the students have read the rest of the "Introduction," ask them to fill out the "Things we KNOW" chart on their Investigation Handout (or in their Investigation Log) as a group.
13. Ask the students to share what they learned from the reading. Add these ideas to a class "Things we KNOW" chart.
14. Tell the students, "Let's see what we will need to figure out during our investigation."
15. Read the task and the guiding question aloud.
16. Tell the students, "I have lots of materials here that you can use."
17. Introduce the students to the materials available for them to use during the investigation by either (a) holding each one up and then asking what it might be used for or (b) giving them a kit with all the materials in it and giving them three to four minutes to play with them. Once the students are familiar with the

Teacher Notes

materials, show them how to use a plastic retail tube, ruler, and Secchi disk to measure the "darkness" or cloudiness of tea. If you give the students an opportunity to play with the materials, collect them from each group before moving on to stage 2.

Stage 2: Design a Method and Collect Data (50 minutes)

1. Tell the students, "I am now going to give you and the other members of your group about 15 minutes to plan your investigation. Before you begin, I want you all to take a couple of minutes to discuss the following questions with the rest of your group."

2. Show the following questions on the screen or board:
   - How can we track or describe a *change* in the water *over time*?
   - Which standard *units* should we use to describe the darkness or cloudiness of the water?

3. Tell the students, "Please take a few minutes to come up with an answer to these questions."

4. Give the students two to three minutes to discuss these two questions.

5. Ask two or three different groups to share their answers. Highlight or write down any important ideas on the board so students can refer to them later.

6. If possible, use a document camera to project an image of the graphic organizer for this investigation on a screen or board (or take a picture of it and project the picture on a screen or board). Tell the students, "I now want you all to plan out your investigation. To do that, you will need to fill out this investigation proposal."

7. Point to the box labeled "Our guiding question:" and tell the students, "You can put the question we are trying to answer in this box." Then ask, "Where can we find the guiding question?"

8. Wait for a student to answer.

9. Point to the box labeled "This is a picture of how we will set up the equipment:" and tell the students, "You can draw a picture in this box of how you will set up the equipment in order to carry out this investigation."

10. Point to the box labeled "We will collect the following data:" and tell the students, "You can list the measurements or observations that you will need to collect during the investigation in this box."

11. Point to the box labeled "These are the steps we will follow to collect data:" and tell the students, "You can list what you are going to do to collect the data you need and what you will do with your data once you have it. Be sure to give enough detail that I could do your investigation for you."

Investigation 2. Movement of Particles in a Liquid:
Why Do People Use Hot Water Instead of Cold Water When They Make Tea?

12. Ask the students, "Do you have any questions about what you need to do?"

13. Answer any questions that come up.

14. Tell the students, "Once you are done, raise your hand and let me know. I'll then come by and look over your proposal and give you some feedback. You may not begin collecting data until I have approved your proposal by signing it. You need to have your proposal done in the next 15 minutes."

15. Give the students 15 minutes to work in their groups on their investigation proposal. As they work, move from group to group to check in, ask probing questions, and offer a suggestion if a group gets stuck.

16. As each group finishes its investigation proposal, read it over and determine if it will be productive or not. If you feel the investigation will be productive (not necessarily what you would do or what the other groups are doing), sign your name on the proposal and let the group start collecting data. If the plan needs to be changed, offer some suggestions or ask some probing questions, and have the group make the changes before you approve it.

17. Pass out the materials or have one student from each group collect the materials they need from a central supply table or cart for the groups that have an approved proposal.

18. Remind students of the safety rules and precautions for this investigation.

19. Tell the students to collect their data and record their observations or measurements in the "Collect Your Data" box in their Investigation Handout (or the Investigation Log in their workbook).

20. Give the students 30 minutes to collect their data. Collect the materials from each group before asking them to analyze their data.

Teacher Notes

> ### What should a student-designed investigation look like?
>
> There are a number of different investigations that students can design to answer the question "Why do people use hot water instead of cold water when they make tea?" For example, one method might include the following steps:
>
> 1. Put 200 ml of cold water in a 250-ml beaker.
> 2. Record the temperature of the water.
> 3. Add a tea bag.
> 4. Wait eight minutes (this is the minimum amount of time the students should let the tea diffuse through the water because this is how long it takes for tea to get dark enough to measure its "cloudiness" using a Secchi disk).
> 5. Remove the tea bag.
> 6. Transfer the tea to a plastic retail tube.
> 7. Measure the depth where a Secchi disk is still visible in the tea.
> 8. Repeat steps 1–7 for water of at least two different temperatures (such as 40°C and 60°C).
>
> If students use this method, they will need to collect data on (1) the temperature of water and (2) the depth at which the Secchi disk is still visible in the tea (the Secchi depth).

Stage 3: Create a Draft Argument (45 minutes)

1. Tell the students, "Now that we have all this data, we need to analyze the data so we can figure out an answer to the guiding question."
2. If possible, project an image of the "Analyze Your Data" section for this investigation on a screen or board using a document camera (or take a picture of it and project the picture on a screen or board). Point to the section and tell the students, "You can create a graph as a way to analyze your data. You can make your graph in this section."
3. Ask the students, "What information do we need to include in a graph?"
4. Tell the students, "Please take a few minutes to discuss this question with your group, and be ready to share."
5. Give the students five minutes to discuss.
6. Ask two or three different groups to share their answers. Highlight or write down any important ideas on the board so students can refer to them later.

7. Tell the students, "I am now going to give you and the other members of your group about 10 minutes to create your graph." The graph they create should include the different temperatures of water and the "darkness" or cloudiness of the tea. If the students are having trouble making a graph, you can take a few minutes to provide a mini-lesson about how to create a graph from a bunch of measurements (this strategy is called just-in-time instruction because it is offered only when students get stuck).

8. Give the students 10 minutes to analyze their data by creating a graph. As they work, move from group to group to check in, ask probing questions, and offer suggestions.

What should the graph for this investigation look like?

There are a number of different ways that students can analyze the measurements they collect during this investigation. One of the most straightforward ways is to create a line graph with the temperature of the water on the horizontal axis, or x-axis, and the Secchi depth (how deep a Secchi disk can still be seen) on the vertical axis, or y-axis. An example of this type of graph can be seen in Figure 2.5 (p. 95). There are other options for analyzing the collected data. Students often come up with some unique ways of analyzing their data, so be sure to give them some voice and choice during this stage.

9. Tell the students, "I am now going to give you and the other members of your group about 15 minutes to create an argument to share what you have learned and convince others that they should believe you. Before you do that, we need to take a few minutes to discuss what you need to include in your argument."

10. If possible, use a document camera to project the "Argument Presentation on a Whiteboard" image from the "Draft Argument" section of the Investigation Handout (or the Investigation Log in their workbook) on a screen or board (or take a picture of it and project the picture on a screen or board).

11. Point to the box labeled "The Guiding Question:" and tell the students, "You can put the question we are trying to answer here on your whiteboard."

12. Point to the box labeled "Our Claim:" and tell the students, "You can put your claim here on your whiteboard. The claim is your answer to the guiding question."

13. Point to the box labeled "Our Evidence:" and tell the students, "You can put the evidence that you are using to support your claim here on your whiteboard. Your evidence will need to include the analysis you just did and an explanation

Teacher Notes

of what your analysis means or shows. Scientists always need to support their claims with evidence."

14. Point to the box labeled "Our Justification of the Evidence:" and tell the students, "You can put your justification of your evidence here on your whiteboard. Your justification needs to explain why your evidence is important. Scientists often use core ideas to explain why the evidence they are using matters. Core ideas are important concepts that scientists use to help them make sense of what happens during an investigation."

15. Ask the students, "What are some core ideas that we read about earlier that might help us explain why the evidence we are using is important?"

16. Ask the students to share some of the core ideas from the "Introduction" section of the Investigation Handout (or the Investigation Log in the workbook). List these core ideas on the board.

17. Tell the students, "That is great. I would like to see everyone try to include these core ideas in your justification of the evidence. Your goal is to use these core ideas to help explain why your evidence matters and why the rest of us should pay attention to it."

18. Ask the students, "Do you have any questions about what you need to do?"

19. Answer any questions that come up.

20. Tell the students, "Okay, go ahead and start working on your arguments. You need to have your argument done in the next 15 minutes. It doesn't need to be perfect. We just need something down on the whiteboards so we can share our ideas."

21. Give the students 15 minutes to work in their groups on their arguments. As they work, move from group to group to check in, ask probing questions, and offer a suggestion if a group gets stuck. Figure 2.5 shows an example of an argument created by students for this investigation.

Stage 4: Argumentation Session (30 minutes)

The argumentation session can be conducted in a whole-class presentation format, a gallery walk format, or a modified gallery walk format. We recommend using a whole-class presentation format for the first investigation, but try to transition to either the gallery walk or modified gallery walk format as soon as possible because that will maximize student voice and choice inside the classroom. The following list shows the steps for the three formats; unless otherwise noted, the steps are the same for all three formats.

1. Begin by introducing the use of the whiteboard.
   - *If using the whole-class presentation format*, tell the students, "We are now going to share our arguments. Please set up your whiteboards so everyone can see them."

Investigation 2. Movement of Particles in a Liquid:
Why Do People Use Hot Water Instead of Cold Water When They Make Tea?

FIGURE 2.5

Example of an argument

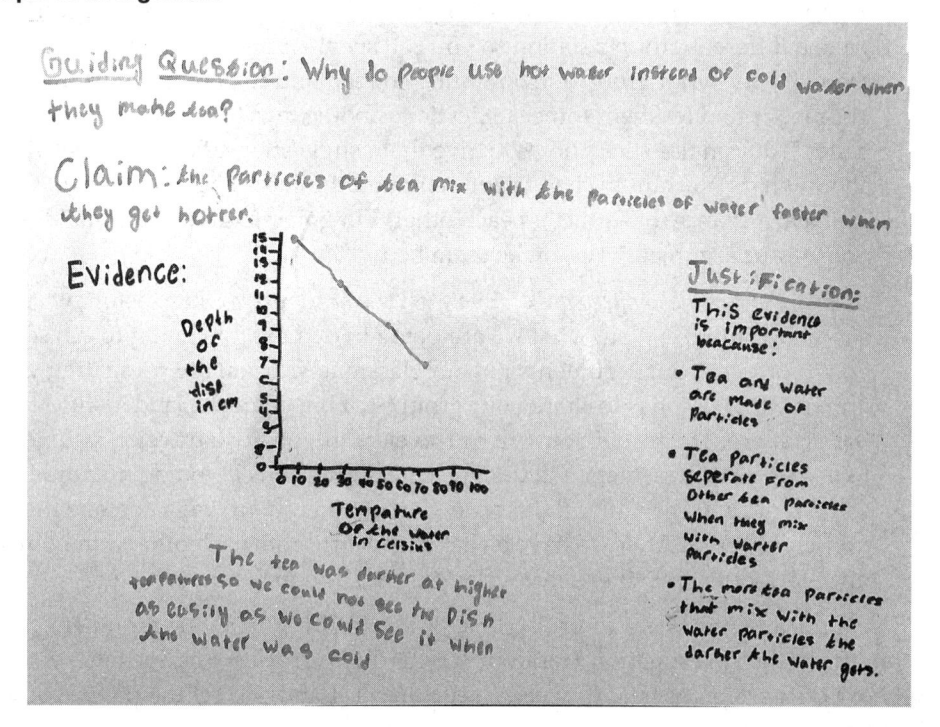

> **Guiding Question:** Why do people use hot water instead of cold water when they make tea?
>
> **Claim:** The particles of tea mix with the particles of water faster when they get hotter.
>
> **Evidence:** [graph with y-axis "Depth of the dish in cm" and x-axis "Tempature of the water in celsius" showing a decreasing trend]
>
> The tea was darker at higher temperatures so we could not see the dish as easily as we could see it when the water was cold.
>
> **Justification:** This evidence is important because:
> - Tea and water are made of particles
> - Tea particles seperate from other tea particles when they mix with water particles
> - The more tea particles that mix with the water particles the darker the water gets.

- *If using the gallery walk or modified gallery walk format,* tell the students, "We are now going to share our arguments. Please set up your whiteboards so they are facing the walls."

2. Allow the students to set up their whiteboards.
   - *If using the whole-class presentation format,* the whiteboards should be set up on stands or chairs so they are facing toward the center of the room.
   - *If using the gallery walk or modified gallery walk format,* the whiteboards should be set up on stands or chairs so they are facing toward the outside of the room.

3. Give the following instructions to the students:
   - *If using the whole-class presentation format,* tell the students, "Okay, before we get started I want to explain what we are going to do next. Your group will have an opportunity to share your argument with the rest of the class. After you are done, everyone else in the class will have a chance to ask questions and offer some suggestions about ways to make your group's argument better. After we have a chance to listen to each other and learn something new, I'm going to give you some time to revise your arguments and make them better."

Teacher Notes

- *If using the gallery walk format,* tell the students, "Okay, before we get started I want to explain what we are going to do next. You are going to read the arguments that were created by other groups. When I say 'go,' your group will go to a different group's station so you can see their argument. Once you are there, I'll give your group a few minutes to read and review their argument. Your job is to offer them some suggestions about ways to make their argument better. You can use sticky notes to give them suggestions. Please be specific about what you want to change and how you think they should change it. After we have a chance to learn from each other, I'm going to give you some time to revise your arguments and make them better."

- *If using the modified gallery walk format,* tell the students, "Okay, before we get started I want to explain what we are going to do next. I'm going to ask some of you to present your arguments to your classmates. If you are presenting your argument, your job is to share your group's claim, evidence, and justification of the evidence. The rest of you will be travelers. If you are a traveler, your job is to listen to the presenters, ask the presenters questions if you do not understand something, and then offer them some suggestions about ways to make their argument better. After we have a chance to learn from each other, I'm going to give you some time to revise your arguments and make them better."

4. Use a document camera to project the "Ways to IMPROVE our argument …" box from the Investigation Handout (or the Investigation Log in their workbook) on a screen or board (or take a picture of it and project the picture on a screen or board).

    - *If using the whole-class presentation format,* point to the box and tell the students, "After your group presents your argument, you can write down the suggestions you get from your classmates here. If you are listening to a presentation and you see a good idea from another group, you can write down that idea here as well. Once we are done with the presentations, I will give you a chance to use these suggestions or ideas to improve your arguments."

    - *If using the gallery walk format,* point to the box and tell the students, "If you see a good idea from another group, you can write it down here. Once we are done reviewing the different arguments, I will give you a chance to use these ideas to improve your own arguments. It is important to share ideas like this."

    - *If using the modified gallery walk format,* point to the box and tell the students, "If you are a presenter, you can write down the suggestions you get from the travelers here. If you are a traveler and you see a good idea from another group, you can write down that idea here. Once we are done with the presentations, I will give you a chance to use these suggestions or ideas to improve your arguments."

5. Ask the students, "Do you have any questions about what you need to do?"

6. Answer any questions that come up.

Investigation 2. Movement of Particles in a Liquid:
Why Do People Use Hot Water Instead of Cold Water When They Make Tea?

7. Give the following instructions:

    - *If using the whole-class presentation format,* tell the students, "Okay. Let's get started."

    - *If using the gallery walk format,* tell the students, "Okay, I'm now going to tell you which argument to go to and review."

    - *If using the modified gallery walk format,* tell the students, "Okay, I'm now going to assign you to be a presenter or a traveler." Assign one or two students from each group to be presenters and one or two students from each group to be travelers.

8. Give the students an opportunity to review the arguments.

    - *If using the whole-class presentation format,* have each group present their argument one at a time. Give each group only two to three minutes to present their argument. Then give the class two to three minutes to ask them questions and offer suggestions. Encourage as much participation from the students as possible.

    - *If using the gallery walk format,* tell the students, "Okay. Let's get started. Each group, move one argument to the left. Don't move to the next argument until I tell you to move. Once you get there, read the argument and then offer suggestions about how to make it better. I will put some sticky notes next to each argument. You can use the sticky notes to leave your suggestions." Give each group about three to four minutes to read the arguments, talk, and offer suggestions.

        a. After three to four minutes, tell the students, "Okay. Let's move on to the next argument. Please move one group to the left."

        b. Again, give each group three to four minutes to read, talk, and offer suggestions.

        c. Repeat this process until each group has had their argument read and critiqued three times.

    - *If using the modified gallery walk format,* tell the students, "Okay. Let's get started. Reviewers, move one group to the left. Don't move to the next group until I tell you to move. Presenters, go ahead and share your argument with the travelers when they get there." Give each group of presenters and travelers about three to four minutes to talk.

        a. Tell the students, "Okay. Let's move on to the next argument. Travelers, move one group to the left."

        b. Again, give each group of presenters and travelers about three to four minutes to talk.

        c. Repeat this process until each group has had their argument read and critiqued three times.

9. Tell the students to return to their workstations.

Teacher Notes

10. Give the following instructions about revising the argument:
    - *If using the whole-class presentation format,* tell the students, "I'm now going to give you all about 10 minutes to revise your argument. Take a few minutes to talk in your groups and determine what you want to change to make your argument better. Once you have decided what to change, go ahead and make the changes to your whiteboard."
    - *If using the gallery walk format,* tell the students, "I'm now going to give you all about 10 minutes to revise your argument. Take a few minutes to read the suggestions that were left at your argument. Then talk in your groups and determine what you want to change to make your argument better. Once you have decided what to change, go ahead and make the changes to your whiteboard."
    - *If using the modified gallery walk format,* tell the students, "I'm now going to give you all about 10 minutes to revise your argument. Please return to your original groups." Wait for the students to move back into their original groups and then tell the students, "Okay, take a few minutes to talk in your groups and determine what you want to change to make your argument better. Once you have decided what to change, go ahead and make the changes to your whiteboard."
11. Ask the students, "Do you have any questions about what you need to do?"
12. Answer any questions that come up.
13. Tell the students, "Okay. Let's get started."
14. Give the students 10 minutes to work in their groups on their arguments. As they work, move from group to group to check in, ask probing questions, and offer a suggestion if a group gets stuck.

Stage 5: Reflective Discussion (15 minutes)

1. Tell the students, "We are now going to take a minute to talk about some of the core ideas and crosscutting concepts that we have used during our investigation."
2. Go to *https://lab.concord.org/embeddable.html#interactives/sam/diffusion/1-dropping-dye-on-click.json* and show the simulation at this URL on the screen. This simulation was developed by the Concord Consortium and can be used to help students think about the behavior of water molecules.
3. Tell the students that this simulation is designed to model the behavior of water molecules. Click on the right-facing arrow at the bottom of the screen to start the simulation. Ask the students, "What do you all see going on here?"
4. Allow the students to share their ideas.

Investigation 2. Movement of Particles in a Liquid:
Why Do People Use Hot Water Instead of Cold Water When They Make Tea?

5. Ask the students, "What do you think will happen if I add some dye to the water?"
6. Allow the students to share their ideas.
7. Click anywhere in the simulation widow to add dye molecules. Ask the students, "What do you all see going on here?"

FIGURE 2.6
Food dye being added to water

Note: A full-color version of this figure is available on the book's Extras page at *www.nsta.org/adi-5th*.

8. Allow the students to share their ideas.
9. Show Figure 2.6 on the screen.
10. Ask the students, "What do you all see going on here?"
11. Allow the students to share their ideas.
12. Ask the students, "How can we explain this in terms of what we know about matter and its interactions?"
13. Allow the students to share their ideas. As they share their ideas, ask different questions to encourage them to expand on their thinking (e.g., "Can you tell me more about that?"), clarify a contribution (e.g., "Can you say that in another way?"), support an idea (e.g., "Why do you think that?"), add to an idea mentioned by a classmate (e.g., "Would anyone like to add to the idea?"), re-voice an idea offered by a classmate (e.g., "Who can explain that to me in another way?"), or critique an idea during the discussion (e.g., "Do you agree or disagree with that idea and why?") until students are able to generate an adequate explanation.
14. Ask the students, "What other questions do you have about this?"
15. Answer any questions that come up.

Argument-Driven Inquiry in **Fifth-Grade Science:** Three-Dimensional Investigations

Teacher Notes

16. Tell the students, "We also looked for cause-and-effect relationships during our investigation." Then ask, "Can anyone tell me why it is useful to look for cause-and-effect relationships?"

17. Allow the students to share their ideas.

18. Tell the students, "We also tracked change over time during our investigation." Then ask, "Can anyone tell me why it is useful to track change over time?"

19. Allow the students to share their ideas.

20. Show an image of the question "What do you think are the most important core ideas or crosscutting concepts that we used during this investigation to help us make sense of what we observed?" Tell the students, "Okay, let's make sure we are all on the same page. Please take a moment to discuss this question with the other people in your group." Give them a few minutes to discuss the question.

21. Ask the students, "What do you all think? Who would like to share?"

22. Allow the students to share their ideas.

23. Tell the students, "We are now going to take a minute to talk about different types of information in science."

24. Show an image of the question "What is the difference between an observation and an inference?" on the screen. Tell the students, "Take a few minutes to talk about how you would answer this question with the other people in your group. Be ready to share with the rest of the class." Give the students two to three minutes to talk in their group.

25. Ask the students, "What do you all think? Who would like to share an idea?"

26. Allow the students to share their ideas.

27. Tell the students, "Okay, let's make sure we are all using the same definition. I think an observation is a descriptive statement about something. An inference is an interpretation of an observation."

28. Show Figure 2.6 (p. 99) on a screen with the statement "The molecules of food dye spread out in between the water molecules over time" on the screen.

29. Ask the students, "Do you think this statement is an observation or an inference, and why do you think that?"

30. Allow the students to share their ideas.

31. Tell the students, "I think the statement is an inference because it is an interpretation of some observations."

32. Ask the students, "What are some observations we can make about what you see?"

33. Allow the students to share their ideas.

Investigation 2. Movement of Particles in a Liquid:
Why Do People Use Hot Water Instead of Cold Water When They Make Tea?

34. Ask the students, "What questions do you have about the difference between an observation and an inference?"

35. Answer any questions that come up.

36. Tell the students, "We are now going to take a minute to talk about what went well and what didn't go so well during our investigation. We need to talk about this because you all are going to be planning and carrying out your own investigations like this a lot this year, and I want to help you all get better at it."

37. Show an image of the question "What made your investigation scientific?" on the screen. Tell the students, "Take a few minutes to talk about how you would answer this question with the other people in your group. Be ready to share with the rest of the class." Give the students two to three minutes to talk in their group.

38. Ask the students, "What do you all think? Who would like to share an idea?"

39. Allow the students to share their ideas. Be sure to expand on their ideas about what makes an investigation scientific.

40. Show an image of the question "What made your investigation not so scientific?" on the screen. Tell the students, "Take a few minutes to talk about how you would answer this question with the other people in your group. Be ready to share with the rest of the class." Give the students two to three minutes to talk in their group.

41. Ask the students, "What do you all think? Who would like to share an idea?"

42. Allow the students to share their ideas. Be sure to expand on their ideas about what makes an investigation less scientific.

43. Show an image of the question "What rules can we put into place to help us make sure our next investigation is more scientific?" on the screen. Tell the students, "Take a few minutes to talk about how you would answer this question with the other people in your group. Be ready to share with the rest of the class." Give the students two to three minutes to talk in their group.

44. Ask the students, "What do you all think? Who would like to share an idea?"

45. Allow the students to share their ideas. Once they have shared their ideas, offer a suggestion for a possible class rule.

46. Ask the students, "What do you all think? Should we make this a rule?"

47. If the students agree, write the rule on the board or on a class "Rules for Scientific Investigation" chart so you can refer to it during the next investigation.

Stage 6: Write a Draft Report (30 minutes)

Your students will use either the Investigation Handout or the Investigation Log in the student workbook when writing the draft report. When you give the directions shown

Teacher Notes

in quotes in the following steps, substitute "Investigation Log in your workbook" or just "Investigation Log" (as shown in brackets) for "handout" if they are using the workbook.

1. Tell the students, "You are now going to write an investigation report to share what you have learned. Please take out a pencil and turn to the 'Draft Report' section of your handout [Investigation Log in your workbook]."

2. If possible, use a document camera to project the "Introduction" section of the draft report from the Investigation Handout (or the Investigation Log in their workbook) on a screen or board (or take a picture of it and project the picture on a screen or board).

3. Tell the students, "The first part of the report is called the 'Introduction.' In this section of the report you want to explain to the reader what you were investigating, why you were investigating it, and what question you were trying to answer. All this information can be found in the text at the beginning of your handout [Investigation Log]." Point to the image. "Here are some sentence starters to help you begin writing."

4. Ask the students, "Do you have any questions about what you need to do?"

5. Answer any questions that come up.

6. Tell the students, "Okay, let's write."

7. Give the students 10 minutes to write the "Introduction" section of the report. As they work, move from student to student to check in, ask probing questions, and offer a suggestion if a student gets stuck.

8. If possible, use a document camera to project the "Method" section of the draft report from the Investigation Handout (or the Investigation Log in their workbook) on a screen or board (or take a picture of it and project the picture on a screen or board).

9. Tell the students, "The second part of the report is called the 'Method.' In this section of the report you want to explain to the reader what you did during the investigation, what data you collected and why, and how you went about analyzing your data. All this information can be found in the 'Plan Your Investigation' section of the handout [Investigation Log]. Remember that you all planned and carried out different investigations, so do not assume that the reader will know what you did." Point to the image. "Here are some sentence starters to help you begin writing."

10. Ask the students, "Do you have any questions about what you need to do?"

11. Answer any questions that come up.

12. Tell the students, "Okay, let's write."

Investigation 2. Movement of Particles in a Liquid:
Why Do People Use Hot Water Instead of Cold Water When They Make Tea?

13. Give the students 10 minutes to write the "Method" section of the report. As they work, move from student to student to check in, ask probing questions, and offer a suggestion if a student gets stuck.

14. If possible, use a document camera to project the "Argument" section of the draft report from the Investigation Handout (or the Investigation Log in their workbook) on a screen or board (or take a picture of it and project the picture on a screen or board).

15. Tell the students, "The last part of the report is called the 'Argument.' In this section of the report you want to share your claim, evidence, and justification of the evidence with the reader. All this information can be found on your whiteboard." Point to the image. "Here are some sentence starters to help you begin writing."

16. Ask the students, "Do you have any questions about what you need to do?"

17. Answer any questions that come up.

18. Tell the students, "Okay, let's write."

19. Give the students 10 minutes to write the "Argument" section of the report. As they work, move from student to student to check in, ask probing questions, and offer a suggestion if a student gets stuck.

Stage 7: Peer Review (35 minutes)

Your students will use either the Investigation Handout or their workbook when doing the peer review. Except where noted below, the directions are the same whether using the handout or the workbook.

1. Tell the students, "We are now going to review our reports to find ways to make them better. I'm going to come around and collect your draft reports. While I do that, please take out a pencil."

2. Collect the handouts or the workbooks with the draft reports from the students.

3. If possible, use a document camera to project the peer-review guide (see Appendix 4) on a screen or board (or take a picture of it and project the picture on a screen or board).

4. Tell the students, "We are going to use this peer-review guide to give each other feedback." Point to the image.

5. Tell the students, "I'm going to ask you to work with a partner to do this. I'm going to give you and your partner a draft report to read. You two will then read the report together. Once you are done reading the report, I want you to answer each of the questions on the peer-review guide." Point to the review questions on the image of the peer-review guide.

Teacher Notes

6. Tell the students, "You can check 'no,' 'almost,' or 'yes' after each question." Point to the checkboxes on the image of the peer-review guide.

7. Tell the students, "This will be your rating for this part of the report. Make sure you agree on the rating you give the author. If you mark 'no' or 'almost,' then you need to tell the author what he or she needs to do to get a 'yes.'" Point to the space for the reviewer feedback on the image of the peer-review guide.

8. Tell the students, "It is really important for you to give the authors feedback that is helpful. That means you need to tell them exactly what they need to do to make their report better."

9. Ask the students, "Do you have any questions about what you need to do?"

10. Answer any questions that come up.

11. Tell the students, "Please sit with a partner who is not in your current group." Allow the students time to sit with a partner.

12. Tell the students, "Okay, I'm now going to give you one report to read." Pass out one Investigation Handout with a draft report or one workbook to each pair. Make sure that the report you give a pair was not written by one of the students in that pair. Give each pair one peer-review guide to fill out. If the students are using workbooks, the peer-review guide is included right after the draft report so you do not need to pass out copies of the peer-review guide.

13. Tell the students, "Okay, I'm going to give you 15 minutes to read the report I gave you and to fill out the peer-review guide. Go ahead and get started."

14. Give the students 15 minutes to work. As they work, move around from pair to pair to check in and see how things are going, answer questions, and offer advice.

15. After 15 minutes pass, tell the students, "Okay, time is up. Please give me the report and the peer-review guide that you filled out."

16. Collect the Investigation Handouts and the peer-review guides, or collect the workbooks if students are using them. If the students are using the Investigation Handouts and separate peer-review guides, be sure you keep each handout with its corresponding peer-review guide.

17. Tell the students, "Okay, I am now going to give you a different report to read and a new peer-review guide to fill out." Pass out one more report to each pair. Make sure that the report you give a pair was not written by one of the students in that pair. Give each pair a new peer-review guide to fill out as a group.

18. Tell the students, "Okay, I'm going to give you 15 minutes to read this new report and to fill out the peer-review guide. Go ahead and get started."

19. Give the students 15 minutes to work. As they work, move around from pair to pair to check in and see how things are going, answer questions, and offer advice.

Investigation 2. Movement of Particles in a Liquid:
Why Do People Use Hot Water Instead of Cold Water When They Make Tea?

20. After 15 minutes pass, tell the students, "Okay, time is up. Please give me the report and the peer-review guide that you filled out."

21. Collect the Investigation Handouts and the peer-review guides, or collect the workbooks if students are using them. If the students are using the Investigation Handouts and separate peer-review guides, be sure you keep each handout with its corresponding peer-review guide.

Stage 8: Revise the Report (30 minutes)

Your students will use either the Investigation Handout or their workbook when revising the report. Except where noted below, the directions are the same whether using the handout or the workbook.

1. Tell the students, "You are now going to revise your draft report based on the feedback you get from your classmates. Please take out a pencil."

2. Return the reports to the students.
   - *If the students used the Investigation Handout and a copy of the peer-review guide,* pass back the handout and the peer-review guide to each student.
   - *If the students used the workbook,* pass that back to each student.

3. Tell the students, "Please take a few minutes to read over the peer-review guide. You should use it to figure out what you need to change in your report and how you will change it."

4. Allow the students to read the peer-review guide.

5. *If the students used the workbook,* if possible use a document camera to project the "Write Your Final Report" section from the Investigation Log on a screen or board (or take a picture of it and project the picture on a screen or board).

6. Give the following directions about how to revise their reports:
   - *If the students used the Investigation Handout and a copy of the peer-review guide,* tell them, "Okay, let's revise our reports. Please take out a piece of paper. I would like you to rewrite your report. You can use your draft report as a starting point, but you also need to change it to make it better. Use the feedback on the peer-review guide to make it better."
   - *If the students used the workbook,* tell them, "Okay, let's revise our reports. I would like you to rewrite your report in the section of the Investigation Log called "Write Your Final Report." You can use your draft report as a starting point, but you also need to change it to make it better. Use the feedback on the peer-review guide to make it better."

7. Ask the students, "Do you have any questions about what you need to do?"

8. Answer any questions that come up.

Teacher Notes

9. Tell the students, "Okay, let's write." Allow about 20 minutes for the students to revise their reports.

10. After about 20 minutes, give the following directions:
    - *If the students used the Investigation Handout,* tell them, "Okay, time's up. I will now come around and collect your Investigation Handout, the peer-review guide, and your final report."
    - *If the students used the workbook,* tell them, "Okay, time's up. I will now come around and collect your workbooks."

11. *If the students used the Investigation Handout,* collect all the Investigation Handouts, peer-review guides, and final reports. *If the students used the workbook,* collect all the workbooks.

12. *If the students used the Investigation Handout,* use the "Teacher Score" column in the peer-review guide to grade the final report. *If the students used the workbook,* use the "Investigation Report Grading Rubric" in the Investigation Log to grade the final report. Whether you are using the handout or the log, you can give the students feedback about their writing in the "Teacher Comments" section.

How to Use the Checkout Questions

The Checkout Questions are an optional assessment. We recommend giving them to students at the start of the next class period after the students finish stage 8 of the investigation. You can then look over the student answers to determine if you need to reteach the core idea from the investigation. Appendix 6 gives the answers to the Checkout Questions that should be given by a student who can (1) apply the core idea correctly in all cases and explain the cause-and-effect relationship and (2) create an accurate particulate model and explain the change in the cup over time.

Alignment With Standards

Table 2.2 highlights how the investigation can be used to address specific performance expectations from the *Next Generation Science Standards, Common Core State Standards for English Language Arts (CCSS ELA)* and *Common Core State Standards for Mathematics (CCSS Mathematics),* and *English Language Proficiency (ELP) Standards.*

TABLE 2.2

Investigation 2 alignment with standards

NGSS performance expectation	5-PS1-1: Develop a model to describe that matter is made up of particles too small to be seen.

Continued

Table 2.2 (*continued*)

***CCSS ELA*—Reading: Informational Text**	Key ideas and details • CCSS.ELA-LITERACY.RI.5.1: Quote accurately from a text when explaining what the text says explicitly and when drawing inferences from the text. • CCSS.ELA-LITERACY.RI.5.2: Determine two or more main ideas of a text and explain how they are supported by key details; summarize the text. • CCSS.ELA-LITERACY.RI.5.3: Explain the relationships or interactions between two or more individuals, events, ideas, or concepts in a historical, scientific, or technical text based on specific information in the text. Craft and structure • CCSS.ELA-LITERACY.RI.5.4: Determine the meaning of general academic and domain-specific words and phrases in a text relevant to a *grade 5 topic or subject area*. • CCSS.ELA-LITERACY.RI.5.5: Compare and contrast the overall structure (e.g., chronology, comparison, cause/effect, problem/solution) of events, ideas, concepts, or information in two or more texts. • CCSS.ELA-LITERACY.RI.5.6: Analyze multiple accounts of the same event or topic, noting important similarities and differences in the point of view they represent. Integration of knowledge and ideas • CCSS.ELA-LITERACY.RI.5.7: Draw on information from multiple print or digital sources, demonstrating the ability to locate an answer to a question quickly or to solve a problem efficiently. • CCSS.ELA-LITERACY.RI.5.8: Explain how an author uses reasons and evidence to support particular points in a text, identifying which reasons and evidence support which point(s). Range of reading and level of text complexity • CCSS.ELA-LITERACY.RI.5.10: By the end of the year, read and comprehend informational texts, including history/social studies, science, and technical texts, at the high end of the grades 4–5 text complexity band independently and proficiently.

***CCSS ELA*—Writing**	Text types and purposes
	• CCSS.ELA-LITERACY.W.5.1: Write opinion pieces on topics or texts, supporting a point of view with reasons.
	   ○ CCSS.ELA-LITERACY.W.5.1.A: Introduce a topic or text clearly, state an opinion, and create an organizational structure in which ideas are logically grouped to support the writer's purpose.
	   ○ CCSS.ELA-LITERACY.W.5.1.B: Provide logically ordered reasons that are supported by facts and details.
	   ○ CCSS.ELA-LITERACY.W.5.1.C: Link opinion and reasons using words, phrases, and clauses (e.g., *consequently*, *specifically*).
	   ○ CCSS.ELA-LITERACY.W.5.1.D: Provide a concluding statement or section related to the opinion presented.

Continued

Table 2.2 (continued)

CCSS ELA—Writing (continued)	Text types and purposes (continued)
	• CCSS.ELA-LITERACY.W.5.2: Write informative/explanatory texts to examine a topic and convey ideas and information clearly.
	    ○ CCSS.ELA-LITERACY.W.5.2.A: Introduce a topic clearly, provide a general observation and focus, and group related information logically; include formatting (e.g., headings), illustrations, and multimedia when useful to aiding comprehension.
	    ○ CCSS.ELA-LITERACY.W.5.2.B: Develop the topic with facts, definitions, concrete details, quotations, or other information and examples related to the topic.
	    ○ CCSS.ELA-LITERACY.W.5.2.C: Link ideas within and across categories of information using words, phrases, and clauses (e.g., *in contrast*, *especially*).
	    ○ CCSS.ELA-LITERACY.W.5.2.D: Use precise language and domain-specific vocabulary to inform about or explain the topic.
	    ○ CCSS.ELA-LITERACY.W.5.2.E: Provide a concluding statement or section related to the information or explanation presented. Production and distribution of writing
	• CCSS.ELA-LITERACY.W.5.4: Produce clear and coherent writing in which the development and organization are appropriate to task, purpose, and audience.
	• CCSS.ELA-LITERACY.W.5.5: With guidance and support from peers and adults, develop and strengthen writing as needed by planning, revising, editing, rewriting, or trying a new approach.
	• CCSS.ELA-LITERACY.W.5.6: With some guidance and support from adults, use technology, including the internet, to produce and publish writing as well as to interact and collaborate with others; demonstrate sufficient command of keyboarding skills to type a minimum of two pages in a single sitting.
	Research to build and present knowledge
	• CCSS.ELA-LITERACY.W.5.8: Recall relevant information from experiences or gather relevant information from print and digital sources; summarize or paraphrase information in notes and finished work, and provide a list of sources.
	• CCSS.ELA-LITERACY.W.5.9: Draw evidence from literary or informational texts to support analysis, reflection, and research.
	Range of writing
	• CCSS.ELA-LITERACY.W.5.10: Write routinely over extended time frames (time for research, reflection, and revision) and shorter time frames (a single sitting or a day or two) for a range of discipline-specific tasks, purposes, and audiences.

Continued

Investigation 2. Movement of Particles in a Liquid:
Why Do People Use Hot Water Instead of Cold Water When They Make Tea?

Table 2.2 (*continued*)

CCSS ELA—Speaking and Listening	Comprehension and collaboration • CCSS.ELA-LITERACY.SL.5.1: Engage effectively in a range of collaborative discussions (one-on-one, in groups, and teacher-led) with diverse partners on *grade 5 topics and texts,* building on others' ideas and expressing their own clearly. • CCSS.ELA-LITERACY.SL.5.1.A: Come to discussions prepared, having read or studied required material; explicitly draw on that preparation and other information known about the topic to explore ideas under discussion. • CCSS.ELA-LITERACY.SL.5.1.B: Follow agreed-upon rules for discussions and carry out assigned roles. • CCSS.ELA-LITERACY.SL.5.1.C: Pose and respond to specific questions by making comments that contribute to the discussion and elaborate on the remarks of others. • CCSS.ELA-LITERACY.SL.5.1.D: Review the key ideas expressed and draw conclusions in light of information and knowledge gained from the discussions. • CCSS.ELA-LITERACY.SL.5.2: Summarize a written text read aloud or information presented in diverse media and formats, including visually, quantitatively, and orally. • CCSS.ELA-LITERACY.SL.5.3: Summarize the points a speaker makes and explain how each claim is supported by reasons and evidence. Presentation of knowledge and ideas • CCSS.ELA-LITERACY.SL.5.4: Report on a topic or text or present an opinion, sequencing ideas logically and using appropriate facts and relevant, descriptive details to support main ideas or themes; speak clearly at an understandable pace. • CCSS.ELA-LITERACY.SL.5.5: Include multimedia components (e.g., graphics, sound) and visual displays in presentations when appropriate to enhance the development of main ideas or themes. • CCSS.ELA-LITERACY.SL.5.6: Adapt speech to a variety of contexts and tasks, using formal English when appropriate to task and situation.
CCSS Mathematics—Numbers and Operations in Base Ten	Understand the place value system. • CCSS.MATH.CONTENT.5.NBT.A.4: Use place value understanding to round decimals to any place. Perform operations with multi-digit whole numbers and with decimals to hundredths. • CCSS.MATH.CONTENT.5.NBT.B.7: Add, subtract, multiply, and divide decimals to hundredths.

Continued

Teacher Notes

Table 2.2 (*continued*)

CCSS Mathematics—Measurement and Data	Convert like measurement units within a given measurement system. • CCSS.MATH.CONTENT.5.MD.A.1: Convert among different-sized standard measurement units within a given measurement system (e.g., convert 5 cm to 0.05 m), and use these conversions in solving multi-step, real-world problems. • Represent and interpret data. • CCSS.MATH.CONTENT.5.MD.B.2: Make a line plot to display a data set of measurements in fractions of a unit (1/2, 1/4, 1/8). Use operations on fractions for this grade to solve problems involving information presented in line plots. Geometric measurement: understand concepts of volume. • CCSS.MATH.CONTENT.5.MD.C.3: Recognize volume as an attribute of solid figures and understand concepts of volume measurement.
ELP Standards	Receptive modalities • ELP 1: Construct meaning from oral presentations and literary and informational text through grade-appropriate listening, reading, and viewing. • ELP 8: Determine the meaning of words and phrases in oral presentations and literary and informational text. Productive modalities • ELP 3: Speak and write about grade-appropriate complex literary and informational texts and topics. • ELP 4: Construct grade-appropriate oral and written claims and support them with reasoning and evidence. • ELP 7: Adapt language choices to purpose, task, and audience when speaking and writing. Interactive modalities • ELP 2: Participate in grade-appropriate oral and written exchanges of information, ideas, and analyses, responding to peer, audience, or reader comments and questions. • ELP 5: Conduct research and evaluate and communicate findings to answer questions or solve problems. • ELP 6: Analyze and critique the arguments of others orally and in writing. Linguistic structures of English • ELP 9: Create clear and coherent grade-appropriate speech and text. • ELP 10: Make accurate use of standard English to communicate in grade-appropriate speech and writing.

Investigation Handout

Investigation 2

Movement of Particles in a Liquid: Why Do People Use Hot Water Instead of Cold Water When They Make Tea?

Introduction

Tea is a pleasant-smelling drink that is usually made by mixing the leaves of certain types of plants with hot water. Many people drink tea with a meal, in the morning when they wake up, in the evening before they go to sleep, and when they are visiting with friends or family. Take a few minutes to look at some tea leaves. After you look at some tea leaves, your teacher will add a bag filled with tea leaves to a cup of hot water. After your teacher adds the bag of tea leaves to the cup of hot water, keep track of things you notice and things you wonder about in the boxes below.

Things I NOTICED ...	Things I WONDER about ...

Water and tea leaves are examples of different types of matter. All matter, including water and tea leaves, is made up of particles that are too small to be seen. These particles have energy because they

Investigation Handout

are constantly moving. The *temperature* of an object or substance is a measure of the average energy of these particles. The particles that make up an object (such as a tea leaf) or a substance (such as water) stay together even though they are moving because attractive forces pull them together. The *attractive force* between particles is strongest when these particles are close together and gets weaker the farther the particles are from each other.

When a tea bag is added to a cup of water, some particles of tea will start to move away from the other particles of tea in the tea leaves and mix with the particles of water. The tea particles separate from the other tea particles because the attractive forces from all the water particles can pull the tea particles apart. The individual tea particles then mix with the water particles. The water in the cup changes color when a tea bag is added to it because the water becomes a mixture of tea particles and water particles. The water will turn a darker color over time as more and more individual tea particles separate from the other tea particles and mix with the water particles.

These ideas can help us understand why adding a tea bag to a cup of water will make the water in the cup change color, but they do not tell us anything about how fast the tea particles will spread through the water or what might cause them to spread through the water faster or slower. These ideas also do not help us understand why people use hot water instead of cold water to make a cup of tea. Your goal in this investigation is to figure out what makes tea spread through water faster or slower and if the temperature of the water matters.

Things we KNOW from what we read ...

Investigation 2. Movement of Particles in a Liquid:
Why Do People Use Hot Water Instead of Cold Water When They Make Tea?

Your Task

Use what you know about matter and its interactions, cause and effect, and rates of change to design and carry out an experiment. You will need to figure out how a change in the temperature of water (a cause) affects how much tea will spread through a cup of water in eight minutes (the effect). Your teacher will show you a way to measure how much tea spreads through a cup of water.

The *guiding question* of this investigation is, **Why do people use hot water instead of cold water when they make tea?**

Materials

You may use any of the following materials during your investigation:

- Safety goggles (required)
- Chai or other black tea bags
- Thermometer
- 3 beakers (250 ml)
- Stopwatch
- Plastic ruler with Secchi disk attachment
- 1.5" × 6" clear plastic retail tube

Safety Rules

Follow all normal safety rules. In addition, be sure to follow these rules:

- Wear sanitized safety goggles during setup, investigation activity, and cleanup.
- Be careful when working with hot water and use only under direct teacher supervision, because it can burn skin.
- Immediately clean up any spills to avoid a slip or fall hazard.
- Keep water sources away from electrical receptacles to prevent shock.
- Be careful when handling glassware, because it can shatter and cut skin.
- Wash your hands with soap and water when you are done cleaning up.

Plan Your Investigation

Prepare a plan for your investigation by filling out the chart on the next page; this plan is called an *investigation proposal*. Before you start developing your plan, be sure to discuss the following questions with the other members of your group:

- How can we track or describe a **change** in the water **over time?**
- Which standard **units** should we use to describe the darkness or cloudiness of the water?

Investigation Handout

Our guiding question:

This is a picture of how we will set up the equipment:

We will collect the following data:

These are the steps we will follow to collect data:

I approve of this investigation proposal.

_____ _____
Teacher's signature   Date

Investigation 2. Movement of Particles in a Liquid: Why Do People Use Hot Water Instead of Cold Water When They Make Tea?

Collect Your Data

Keep a record of what you measure or observe during your investigation in the space below.

Analyze Your Data

You will need to analyze the data you collected before you can develop an answer to the guiding question. To analyze the data you collected, create a graph that shows the relationship between the cause and effect.

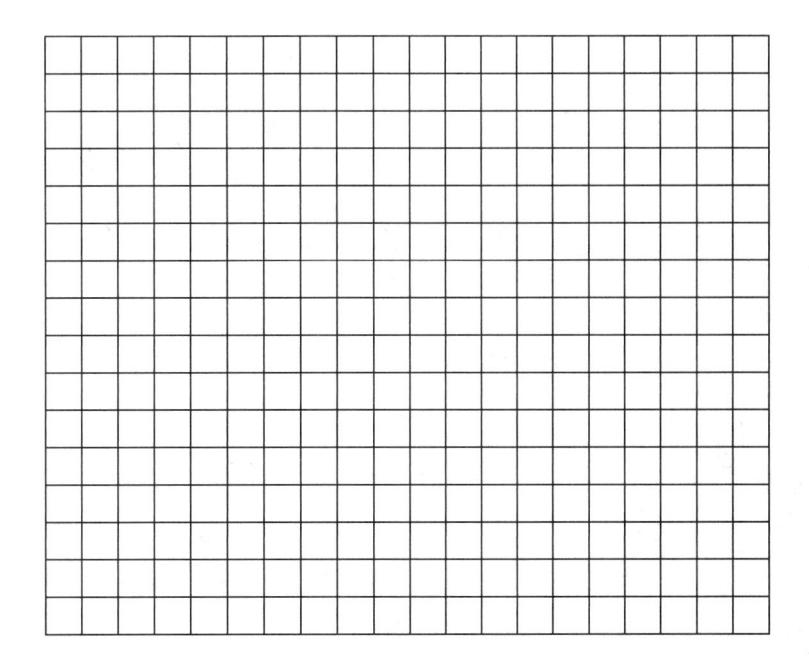

Argument-Driven Inquiry in **Fifth-Grade Science**: Three-Dimensional Investigations

Investigation Handout

Draft Argument

Develop an argument on a whiteboard. It should include the following:

1. A *claim*: Your answer to the guiding question.
2. *Evidence*: An analysis of the data and an explanation of what the analysis means.
3. A *justification of the evidence*: Why your group thinks the evidence is important.

The Guiding Question:	
Our Claim:	
Our Evidence:	Our Justification of the Evidence:

Argumentation Session

Share your argument with your classmates. Be sure to ask them how to make your draft argument better. Keep track of their suggestions in the space below.

Ways to IMPROVE our argument …

Investigation 2. Movement of Particles in a Liquid:
Why Do People Use Hot Water Instead of Cold Water When They Make Tea?

Draft Report

Prepare an *investigation* report to share what you have learned. Use the information in this handout and your group's final argument to write a draft of your investigation report.

Introduction

We have been studying _____ in class.

Before we started this investigation, we explored _____

_____

_____

We noticed _____

_____

_____

My goal for this investigation was to figure out _____

_____

_____

The guiding question was _____

_____

_____

Method

To gather the data I needed to answer this question, I _____

_____

_____

_____

_____

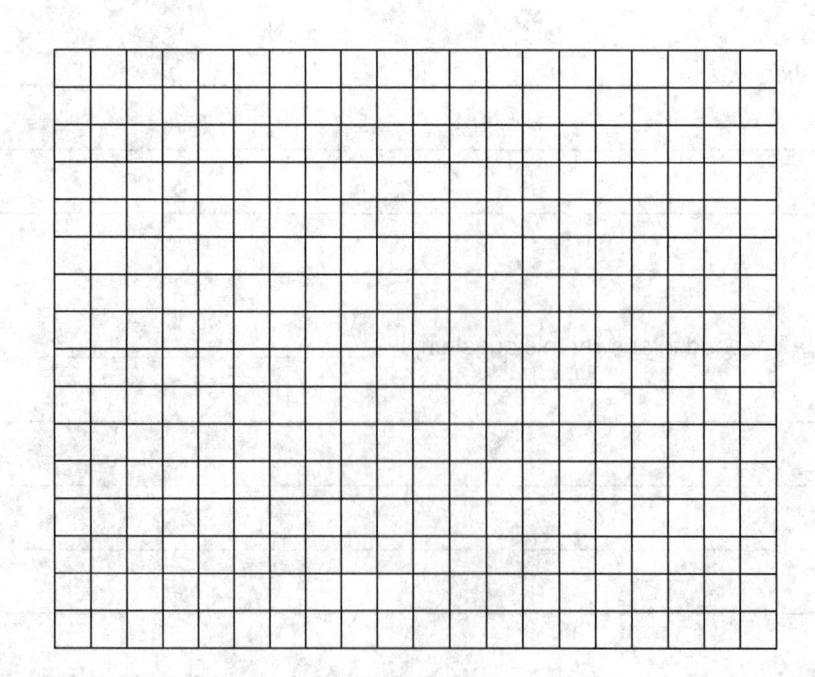

Investigation Handout

I then analyzed the data I collected by _____

_____

_____

_____

Argument

My claim is _____

_____

_____

_____

The graph below shows _____

_____

_____

_____

_____

Investigation 2. Movement of Particles in a Liquid:
Why Do People Use Hot Water Instead of Cold Water When They Make Tea?

This analysis of the data I collected suggests _____

_____

_____

_____

_____

This evidence is based on several important scientific concepts. The first one is _____

_____

_____

_____

_____

_____

Review

Your classmates need your help! Review the draft of their investigation reports and give them ideas about how to improve. Use the *peer-review guide* when doing your review.

Submit Your Final Report

Once you have received feedback from your classmates about your draft report, create your final investigation report and hand it in to your teacher.

Argument-Driven Inquiry in **Fifth-Grade Science:** Three-Dimensional Investigations

Checkout Questions

Investigation 2. Movement of Particles in a Liquid

Use the following information to answer questions 1–3. The picture below shows three different beakers of water. You add a tea bag to each one.

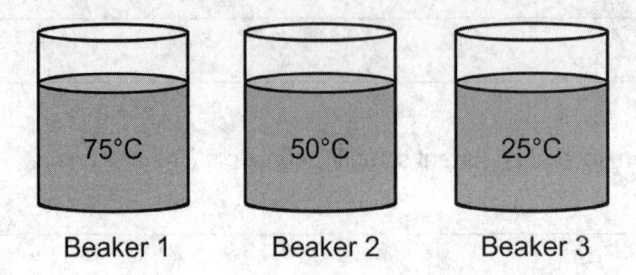

Beaker 1: 75°C   Beaker 2: 50°C   Beaker 3: 25°C

1. In which beaker will the tea spread through the water the fastest?

    a. Beaker 1

    b. Beaker 2

    c. Beaker 3

    d. It will take the same amount of time in each beaker.

2. In which beaker will the tea spread through the water the slowest?

    a. Beaker 1

    b. Beaker 2

    c. Beaker 3

    d. It will take the same amount of time in each beaker.

3. Explain your thinking. What cause-and-effect relationship did you use to answer questions 1 and 2?

_____
_____
_____
_____
_____
_____

Teacher Scoring Rubric for Checkout Questions 1–3

Level	Description
3	The student can apply the core idea correctly and can explain the cause-and-effect relationship.
2	The student can apply the core idea correctly but cannot explain the cause-and-effect relationship.
1	The student cannot apply the core idea correctly but can explain the cause-and-effect relationship.
0	The student cannot apply the core idea correctly and cannot explain the cause-and-effect relationship.

Investigation 2. Movement of Particles in a Liquid:
Why Do People Use Hot Water Instead of Cold Water When They Make Tea?

4. Create a model that shows how tea mixes with the water in a cup over time. Your model must include the molecules that make up both types of matter. This model must be able to explain how temperature affects the rate of this process.

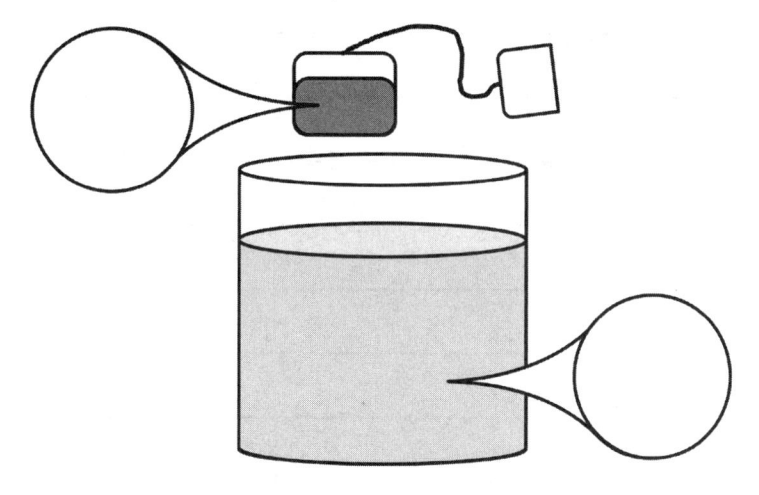

Right before a tea bag is added to the water

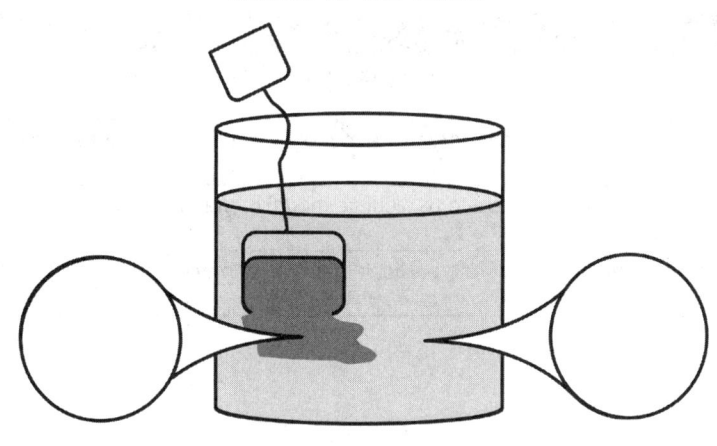

Right after a tea bag is added to the water

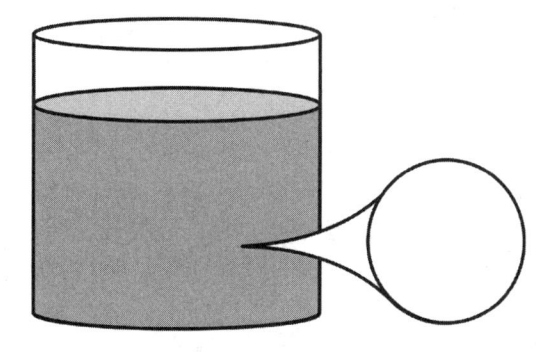

After the tea bag is removed from the water

Checkout Questions

5. Explain your thinking. How does thinking about change over time help you understand what happens as the tea spreads through the water?

_____

_____

_____

_____

_____

_____

_____

Teacher Scoring Rubric for Checkout Questions 4 and 5

Level	Description
3	The student can create an accurate particulate model of matter and can explain the change in the cup over time.
2	The student can create an accurate particulate model of matter but cannot explain the change in the cup over time.
1	The student cannot create an accurate particulate model of matter but can explain the change in the cup over time.
0	The student cannot create an accurate particulate model of matter and cannot explain the change in the cup over time.

Teacher Notes

Investigation 3

States of Matter and Weight: What Happens to the Weight of a Substance When It Changes From a Solid to a Liquid or a Liquid to a Solid?

Purpose

The purpose of this investigation is to give students an opportunity to use one disciplinary core idea (DCI), two crosscutting concepts (CCs), and seven scientific and engineering practices (SEPs) to figure out how the total weight of a sample of matter changes when it changes from a solid to a liquid or a liquid to a solid. Students will also learn about the assumptions that scientists make about order and consistency in nature.

The DCI, CCs, and SEPs That Students Use During This Investigation to Figure Things Out

DCI

- *PS1.A: Structure and Properties of Matter:* Different kinds of matter exist and many of them can be either solid or liquid, depending on temperature. Matter can be described and classified by its observable properties.

CCs

- *CC 3: Scale, Proportion, and Quantity:* Standard units are used to measure and describe physical quantities such as weight, time, temperature, and volume.
- *CC 4: Systems and System Models*: A system can be described in terms of its components and their interactions. A system is a group of related parts that make up a whole and can carry out functions its individual parts cannot.

SEPs

- *SEP 1: Asking Questions and Defining Problems:* Ask questions about what would happen if a variable is changed. Ask questions that can be investigated and predict reasonable outcomes based on patterns such as cause-and-effect relationships.
- *SEP 3: Planning and Carrying Out Investigations:* Plan and conduct an investigation collaboratively to produce data to serve as the basis for evidence, using fair tests in which variables are controlled and the number of trials considered. Evaluate appropriate methods and/or tools for collecting data.

Teacher Notes

- *SEP 4: Analyzing and Interpreting Data:* Represent data in tables and/or various graphical displays (bar graphs, pictographs, and/or pie charts) to reveal patterns that indicate relationships. Analyze and interpret data to make sense of phenomena, using logical reasoning, mathematics, and/or computation. Compare and contrast data collected by different groups in order to discuss similarities and differences in their findings.
- *SEP 5: Using Mathematics and Computational Thinking:* Organize simple data sets to reveal patterns that suggest relationships. Describe, measure, estimate, and/or graph quantities (e.g., area, volume, weight, time) to address scientific and engineering questions and problems.
- *SEP 6: Constructing Explanations and Designing Solutions:* Construct an explanation of observed relationships. Use evidence to construct or support an explanation. Identify the evidence that supports particular points in an explanation.
- *SEP 7: Engaging in Argument From Evidence:* Compare and refine arguments based on an evaluation of the evidence presented. Distinguish among facts, reasoned judgment based on research findings, and speculation in an explanation. Respectfully provide and receive critiques from peers about a proposed procedure, explanation, or model by citing relevant evidence and posing specific questions.
- *SEP 8: Obtaining, Evaluating, and Communicating Information:* Read and comprehend grade-appropriate complex texts and/or other reliable media to summarize and obtain scientific and technical ideas. Combine information in written text with that contained in corresponding tables, diagrams, and/or charts to support the engagement in other scientific and/or engineering practices. Communicate scientific and/or technical information orally and/or in written formats, including various forms of media as well as tables, diagrams, and charts.

Other Concepts That Students May Use During This Investigation

Students might also use some of the following concepts:

- A solid has a fixed shape and volume.
- A liquid has a fixed volume but does not have a fixed shape.
- A gas state does not have a fixed shape or volume.
- A sample of matter will change states depending on temperature.
- The temperature of matter changes when it gains or loses thermal energy.
- Temperature is a measure of the average kinetic energy of the atoms or molecules in a system (or a measure of the hotness or coldness of a substance).
- Melting point is the temperature at which matter changes from a solid to a liquid state.

Investigation 3. States of Matter and Weight: What Happens to the Weight of a Substance When It Changes From a Solid to a Liquid or a Liquid to a Solid?

- Freezing point is the temperature at which matter changes from a liquid to a solid state.

What Students Figure Out

The weight (amount) of matter does not change when it changes state.

Background Information About This Investigation for the Teacher

Matter can exist in different forms or states. The three states of matter that we most often see on Earth are called solid, liquid, and gas. A sample of matter that is in a solid state has a fixed shape and volume. Matter with a fixed shape and volume does not change shape and does not take up more or less space when it is placed in a container. A sample of matter that is in a liquid state has a fixed volume but does not have a fixed shape. A liquid will therefore change shape when it is placed in a container, but the total amount of space it takes up will stay the same. A sample of matter that is in a gas state does not have a fixed shape or volume. A gas will change shape and volume when it is placed in a container.

A sample of matter will change states depending on temperature. The temperature of matter changes when it gains or loses thermal energy. When a sample of matter gains thermal energy, it increases in temperature. When a sample of matter loses thermal energy, it decreases in temperature. A sample of matter will change from a solid into a liquid when it reaches a specific temperature; this temperature is called the melting point. A liquid will change into a solid when it cools to a specific temperature; this temperature is called the freezing point. The amount of matter does not change (it is conserved) when it changes states, even in transitions in which it seems to vanish. The weight of a sample of matter will not change when it changes from a solid to a liquid or a liquid to a solid.

Timeline

The time needed to complete this investigation is 270 minutes (4 hours and 30 minutes). The amount of instructional time needed for each stage of the investigation is as follows:

- *Stage 1.* Introduce the task and the guiding question: 35 minutes
- *Stage 2.* Design a method and collect data: 50 minutes
- *Stage 3.* Create a draft argument: 45 minutes
- *Stage 4.* Argumentation session: 30 minutes
- *Stage 5.* Reflective discussion: 15 minutes
- *Stage 6.* Write a draft report: 30 minutes
- *Stage 7.* Peer review: 35 minutes
- *Stage 8.* Revise the report: 30 minutes

Teacher Notes

Materials and Preparation

The materials needed for this investigation are listed in Table 3.1. The items can be purchased from a big-box retail store such as Walmart or Target or through an online retailer such as Amazon. The materials for this investigation can also be purchased as a complete kit (which includes enough materials for 24 students, or six groups of four students) at *www.argumentdriveninquiry.com*.

In addition to the items listed in Table 3.1, you will need access to a freezer so you can freeze the test tubes filled with water as described later in this section. Most schools have a freezer in the break room(s) and cafeteria.

TABLE 3.1
Materials for Investigation 3

Item	Quantity
Indirectly vented chemical-splash goggles	1 per student
Nonlatex apron	1 per student
Vinyl gloves	1 pair per student
Chocolate chips, 12-oz bag	1 per class
Pyrex 2-quart baking dish (for hot water bath)	1 per class
Pyrex measuring cup	1 per class
Test tube rack (for hot water bath)	1 per class
Hot plate (for hot water bath)	1 per class
Test tube with screw cap filled with water	1 per group
Test tube with screw cap filled with wax	1 per group
Test tube with screw cap filled with vegetable shortening	1 per group
Test tube clamp	1 per group
Electronic scale	1 per group
Whiteboard, 2' × 3'*	1 per group
Investigation Handout	1 per student
Peer-review guide and teacher scoring rubric	1 per student
Checkout Questions (optional)	1 per student

*As an alternative, students can use computer and presentation software such as Microsoft PowerPoint or Apple Keynote to create their arguments.

The day before you plan to start the investigation, add water to 6 test tubes, wax to 6 test tubes, and vegetable shortening to 6 test tubes. Seal all 18 test tubes using the screw cap and label them so students know what is inside each tube. Place the test tubes filled with water in a freezer overnight so the water turns from a liquid to a solid before the

Investigation 3. States of Matter and Weight: What Happens to the Weight of a Substance When It Changes From a Solid to a Liquid or a Liquid to a Solid?

investigation begins (be sure that there is room inside the tube for the water to expand when it freezes). The wax and vegetable shortening will be solid at room temperature.

You will need to create a hot water bath to heat the ice, wax, and vegetable shortening in order to turn them into liquids during the investigation. You can use a hot plate, a Pyrex baking dish and a test tube rack to create a hot water bath (see Figure 3.1), using the following procedure: Place the test tube rack in the baking dish, fill the baking dish with water, place the baking dish on the hot plate, and set the hot plate to low so the water heats up but does not boil. The water in the hot water bath only needs to be 100°F (38°C) to melt the wax and vegetable shortening.

FIGURE 3.1

Equipment setup for creating a hot water bath

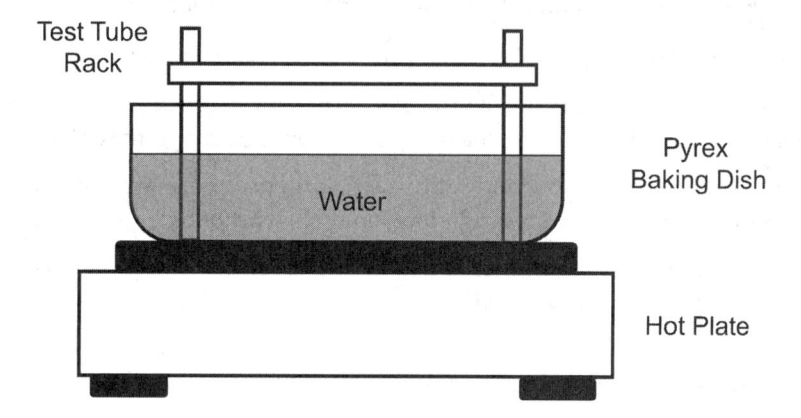

Safety Precautions

Remind students to follow all normal safety rules. In addition, tell the students to take the following safety precautions:

- Wear sanitized indirectly vented chemical-splash goggles, nonlatex aprons, and vinyl gloves during setup, investigation activity, and cleanup.
- Do not taste any food products used in this activity.
- Be careful when working with hot plates, because they can seriously burn skin and cause fires.
- Be careful when working with hot water and use only under direct teacher supervision, because it can burn skin.
- Immediately clean up any spills to avoid a slip or fall hazard.
- Keep water sources away from electrical receptacles to prevent shock.

Teacher Notes

- Be careful when handling glassware (test tubes, etc.), because it can shatter and cut skin.
- Make sure all materials are put away after completing the activity.
- Wash their hands with soap and water when done collecting the data and when done cleaning up.

Lesson Plan by Stage

This lesson plan is only a suggestion. It is included here to illustrate what you can say and do during each stage of ADI for this specific investigation. We encourage you to modify this lesson plan by asking different questions, using different examples, and providing different scaffolds as needed to better meet the needs of students in your class.

Stage 1: Introduce the Task and the Guiding Question (35 minutes)

1. Ask the students to sit in six groups, with three or four students in each group.
2. Ask the students to clear off their desks except for a pencil (and their *Student Workbook for Argument-Driven Inquiry in Fifth-Grade Science* if they have one).
3. Pass out an Investigation Handout to each student (or ask students to turn to the Investigation Log for Investigation 3 in their workbook).
4. Read the first paragraph of the "Introduction" aloud to the class. Ask the students to follow along as you read.
5. Heat up a few chocolate chips using the hot water bath (without the test tube rack) and a Pyrex measuring cup (see Figure 3.2).

FIGURE 3.2

Equipment setup for melting chocolate chips in a hot water bath

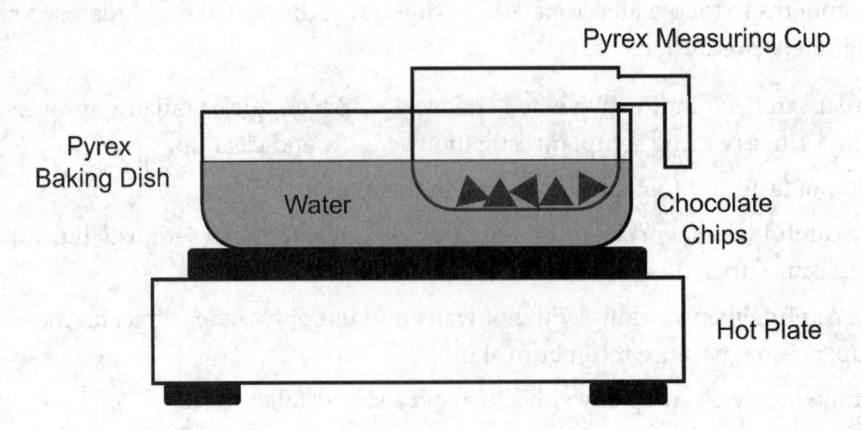

Investigation 3. States of Matter and Weight:
What Happens to the Weight of a Substance When It Changes
From a Solid to a Liquid or a Liquid to a Solid?

6. Ask the students to watch what happens to the chocolate chips as they get hotter and then record their observations and questions in the "NOTICED/ WONDER" chart in the "Introduction."

7. After the students have recorded their observations and questions, ask them to share what they observed.

8. Ask the students to share what questions they have about the chocolate chips.

9. Tell the students, "Some of your questions might be answered by reading the rest of the 'Introduction.'"

10. Ask the students to read the rest of the "Introduction" on their own *or* ask them to follow along as you read it aloud.

11. Once the students have read the rest of the "Introduction," ask them to fill out the "Things we KNOW" chart on their Investigation Handout (or in their Investigation Log) as a group.

12. Ask the students to share what they learned from the reading. Add these ideas to a class "Things we KNOW" chart.

13. Tell the students, "Let's see what we will need to figure out during our investigation."

14. Read the task and the guiding question aloud.

15. Tell the students, "I have lots of materials here that you can use."

16. Introduce the students to the materials available for them to use during the investigation by either (a) holding each one up and then asking what it might be used for or (b) giving them a kit with all the materials in it and giving them three to four minutes to play with them. If you give the students an opportunity to play with the materials, collect them from each group before moving on to stage 2.

Stage 2: Design a Method and Collect Data (50 minutes)

1. Tell the students, "I am now going to give you and the other members of your group about 15 minutes to plan your investigation. Before you begin, I want you all to take a couple of minutes to discuss the following questions with the rest of your group."

2. Show the following questions on the screen or board:
   - Which standard *units* should we use to describe the physical quantities of the matter?
   - What are the components of the *system* that we are studying?

3. Tell the students, "Please take a few minutes to come up with an answer to these questions."

Teacher Notes

4. Give the students two or three minutes to discuss these two questions.

5. Ask two or three different groups to share their answers. Highlight or write down any important ideas on the board so students can refer to them later.

6. If possible, use a document camera to project an image of the graphic organizer for this investigation on a screen or board (or take a picture of it and project the picture on a screen or board). Tell the students, "I now want you all to plan out your investigation. To do that, you will need to fill out this investigation proposal."

7. Point to the box labeled "Our guiding question:" and tell the students, "You can put the question we are trying to answer in this box." Then ask, "Where can we find the guiding question?"

8. Wait for a student to answer.

9. Point to the box labeled "This is a picture of how we will set up the equipment:" and tell the students, "You can draw a picture in this box of how you will set up the equipment in order to carry out this investigation."

10. Point to the box labeled "We will collect the following data:" and tell the students, "You can list the measurements or observations that you will need to collect during the investigation in this box."

11. Point to the box labeled "These are the steps we will follow to collect data:" and tell the students, "You can list what you are going to do to collect the data you need and what you will do with your data once you have it. Be sure to give enough detail that I could do your investigation for you."

12. Ask the students, "Do you have any questions about what you need to do?"

13. Answer any questions that come up.

14. Tell the students, "Once you are done, raise your hand and let me know. I'll then come by and look over your proposal and give you some feedback. You may not begin collecting data until I have approved your proposal by signing it. You need to have your proposal done in the next 15 minutes."

15. Give the students 15 minutes to work in their groups on their investigation proposal. As they work, move from group to group to check in, ask probing questions, and offer a suggestion if a group gets stuck.

16. As each group finishes its investigation proposal, read it over and determine if it will be productive or not. If you feel the investigation will be productive (not necessarily what you would do or what the other groups are doing), sign your name on the proposal and let the group start collecting data. If the plan needs to be changed, offer some suggestions or ask some probing questions, and have the group make the changes before you approve it.

Investigation 3. States of Matter and Weight:
What Happens to the Weight of a Substance When It Changes
From a Solid to a Liquid or a Liquid to a Solid?

17. Pass out the materials or have one student from each group collect the materials they need from a central supply table or cart for the groups that have an approved proposal.
18. Remind students of the safety rules and precautions for this investigation.
19. Tell the students to collect their data and record their observations or measurements in the "Collect Your Data" box in their Investigation Handout (or the Investigation Log in their workbook).
20. Give the students 30 minutes to collect their data. Collect the materials from each group before asking them to analyze their data.

What should a student-designed investigation look like?

There are a number of different investigations that students can design to answer the question "What happens to the weight of a substance when it changes from a solid to a liquid or a liquid to a solid?" For example, one method might include the following steps:

1. Weigh each test tube.
2. Place each test tube in a hot water bath until the substances change from solid to liquid.
3. Weigh each test tube.
4. Put each test tube in the freezer until the substances change from liquid to solid.
5. Weigh each test tube.

If students use this method, they will need to collect data on (1) the weight of each test tube and (2) the state (solid or liquid) of the substance inside each test tube.

Stage 3: Create a Draft Argument (45 minutes)

1. Tell the students, "Now that we have all this data, we need to analyze the data so we can figure out an answer to the guiding question."
2. If possible, project an image of the "Analyze Your Data" section for this investigation on a screen or board using a document camera (or take a picture of it and project the picture on a screen or board). Point to the section and tell the students, "You can create a graph as a way to analyze your data. You can make your graph in this section."
3. Ask the students, "What information do we need to include in a graph?"

Teacher Notes

4. Tell the students, "Please take a few minutes to discuss this question with your group and be ready to share."

5. Give the students five minutes to discuss.

6. Ask two or three different groups to share their answers. Highlight or write down any important ideas on the board so students can refer to them later.

7. Tell the students, "I am now going to give you and the other members of your group about 10 minutes to create your graph." The graph they create should include the names and the states of each substance and the weight of each test tube. If the students are having trouble making a graph, you can take a few minutes to provide a mini-lesson about how to create a graph from a bunch of measurements (this strategy is called just-in-time instruction because it is offered only when students get stuck).

8. Give the students 10 minutes to analyze their data by creating a graph. As they work, move from group to group to check in, ask probing questions, and offer suggestions.

9. Tell the students, "I am now going to give you and the other members of your group about 15 minutes to create an argument to share what you have learned and convince others that they should believe you. Before you do that, we need to take a few minutes to discuss what you need to include in your argument."

10. If possible, use a document camera to project the "Argument Presentation on a Whiteboard" image from the "Draft Argument" section of the Investigation Handout (or the Investigation Log in their workbook) on a screen or board (or take a picture of it and project the picture on a screen or board).

11. Point to the box labeled "The Guiding Question:" and tell the students, "You can put the question we are trying to answer here on your whiteboard."

12. Point to the box labeled "Our Claim:" and tell the students, "You can put your claim here on your whiteboard. The claim is your answer to the guiding question."

13. Point to the box labeled "Our Evidence:" and tell the students, "You can put the evidence that you are using to support your claim here on your whiteboard. Your evidence will need to include the analysis you just did and an explanation of what your analysis means or shows. Scientists always need to support their claims with evidence."

14. Point to the box labeled "Our Justification of the Evidence:" and tell the students, "You can put your justification of your evidence here on your whiteboard. Your justification needs to explain why your evidence is important. Scientists often use core ideas to explain why the evidence they are using matters. Core ideas are important concepts that scientists use to help them make sense of what happens during an investigation."

Investigation 3. States of Matter and Weight:
What Happens to the Weight of a Substance When It Changes
From a Solid to a Liquid or a Liquid to a Solid?

15. Ask the students, "What are some core ideas that we read about earlier that might help us explain why the evidence we are using is important?"

16. Ask the students to share some of the core ideas from the "Introduction" section of the Investigation Handout (or the Investigation Log in the workbook). List these core ideas on the board.

17. Tell the students, "That is great. I would like to see everyone try to include these core ideas in your justification of the evidence. Your goal is to use these core ideas to help explain why your evidence matters and why the rest of us should pay attention to it."

18. Ask the students, "Do you have any questions about what you need to do?"

19. Answer any questions that come up.

20. Tell the students, "Okay, go ahead and start working on your arguments. You need to have your argument done in the next 15 minutes. It doesn't need to be perfect. We just need something down on the whiteboards so we can share our ideas."

21. Give the students 15 minutes to work in their groups on their arguments. As they work, move from group to group to check in, ask probing questions, and offer a suggestion if a group gets stuck. Figure 3.3 (p. 134) shows an example of an argument created by students for this investigation.

What should the graph for this investigation look like?

There are a number of different ways that students can analyze the measurements they collect during this investigation. One of the most straightforward ways is to create a grouped bar graph with the substances (water, wax, and vegetable shortening) on the horizontal axis, or *x*-axis, and the weight on the vertical axis, or *y*-axis. The groups of bars can be labeled by state (solid or liquid). An example of a grouped bar graph can be seen in Figure 3.3. There are other options for analyzing the collected data. Students often come up with some unique ways of analyzing their data, so be sure to give them some voice and choice during this stage.

Teacher Notes

FIGURE 3.3
Example of an argument

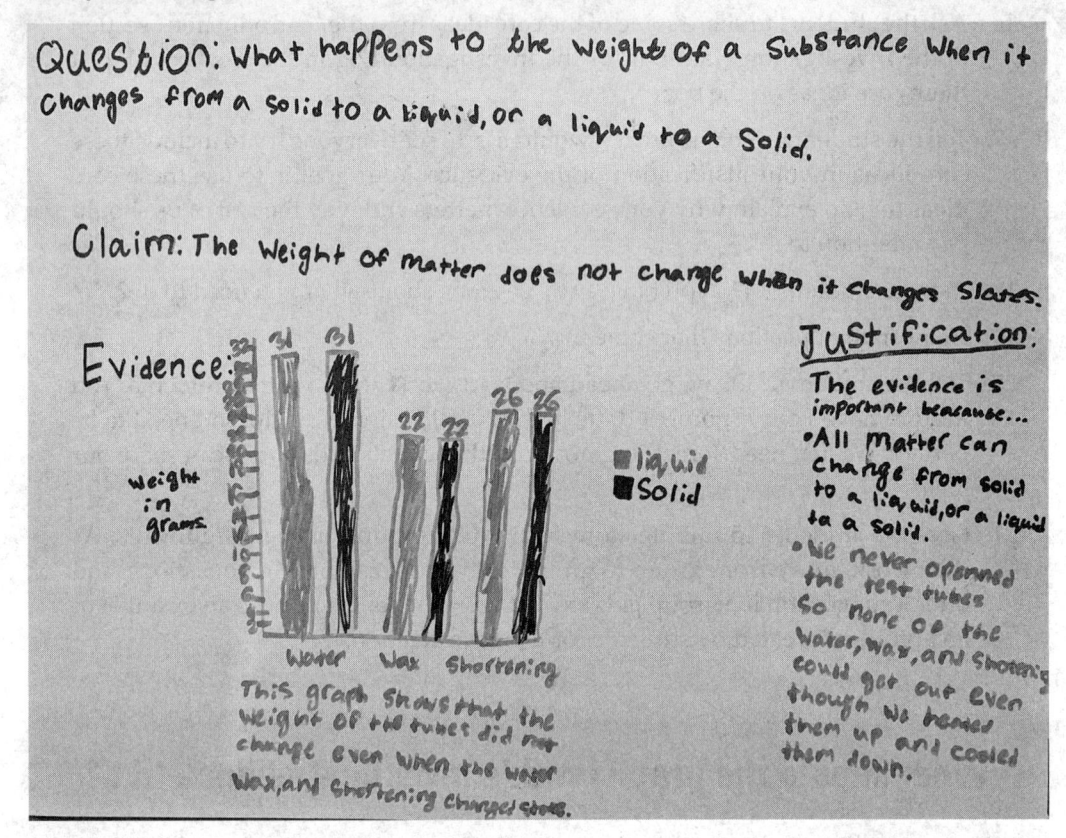

> **Question:** What happens to the weight of a substance when it changes from a solid to a liquid, or a liquid to a solid.
>
> **Claim:** The weight of matter does not change when it changes states.
>
> **Evidence:**
> [Bar graph showing weight in grams for Water (31, 31), Wax (22, 22), and Shortening (26, 26) comparing liquid and solid states]
>
> This graph shows that the weight of the tubes did not change even when the water, wax, and shortening changed states.
>
> **Justification:**
> The evidence is important because...
> - All matter can change from solid to a liquid, or a liquid to a solid.
> - We never opened the test tubes so none of the water, wax, and shortening could get out even though we heated them up and cooled them down.

Stage 4: Argumentation Session (30 minutes)

The argumentation session can be conducted in a whole-class presentation format, a gallery walk format, or a modified gallery walk format. We recommend using a whole-class presentation format for the first investigation, but try to transition to either the gallery walk or modified gallery walk format as soon as possible because that will maximize student voice and choice inside the classroom. The following list shows the steps for the three formats; unless otherwise noted, the steps are the same for all three formats.

1. Begin by introducing the use of the whiteboard.
   - *If using the whole-class presentation format,* tell the students, "We are now going to share our arguments. Please set up your whiteboards so everyone can see them."
   - *If using the gallery walk or modified gallery walk format,* tell the students, "We are now going to share our arguments. Please set up your whiteboards so they are facing the walls."

Investigation 3. States of Matter and Weight: What Happens to the Weight of a Substance When It Changes From a Solid to a Liquid or a Liquid to a Solid?

2. Allow the students to set up their whiteboards.
   - *If using the whole-class presentation format,* the whiteboards should be set up on stands or chairs so they are facing toward the center of the room.
   - *If using the gallery walk or modified gallery walk format,* the whiteboards should be set up on stands or chairs so they are facing toward the outside of the room.

3. Give the following instructions to the students:
   - *If using the whole-class presentation format,* tell the students, "Okay, before we get started I want to explain what we are going to do next. Your group will have an opportunity to share your argument with the rest of the class. After you are done, everyone else in the class will have a chance to ask questions and offer some suggestions about ways to make your group's argument better. After we have a chance to listen to each other and learn something new, I'm going to give you some time to revise your arguments and make them better."
   - *If using the gallery walk format,* tell the students, "Okay, before we get started I want to explain what we are going to do next. You are going to read the arguments that were created by other groups. When I say 'go,' your group will go to a different group's station so you can see their argument. Once you are there, I'll give your group a few minutes to read and review their argument. Your job is to offer them some suggestions about ways to make their argument better. You can use sticky notes to give them suggestions. Please be specific about what you want to change and how you think they should change it. After we have a chance to learn from each other, I'm going to give you some time to revise your arguments and make them better."
   - *If using the modified gallery walk format,* tell the students, "Okay, before we get started I want to explain what we are going to do next. I'm going to ask some of you to present your arguments to your classmates. If you are presenting your argument, your job is to share your group's claim, evidence, and justification of the evidence. The rest of you will be travelers. If you are a traveler, your job is to listen to the presenters, ask the presenters questions if you do not understand something, and then offer them some suggestions about ways to make their argument better. After we have a chance to learn from each other, I'm going to give you some time to revise your arguments and make them better."

4. Use a document camera to project the "Ways to IMPROVE our argument ..." box from the Investigation Handout (or the Investigation Log in their workbook) on a screen or board (or take a picture of it and project the picture on a screen or board).
   - *If using the whole-class presentation format,* point to the box and tell the students, "After your group presents your argument, you can write down the suggestions you get from your classmates here. If you are listening to a presentation and you see a good idea from another group, you can write down that idea here as

Teacher Notes

well. Once we are done with the presentations, I will give you a chance to use these suggestions or ideas to improve your arguments."

- *If using the gallery walk format,* point to the box and tell the students, "If you see a good idea from another group, you can write it down here. Once we are done reviewing the different arguments, I will give you a chance to use these ideas to improve your own arguments. It is important to share ideas like this."

- *If using the modified gallery walk format,* point to the box and tell the students, "If you are a presenter, you can write down the suggestions you get from the travelers here. If you are a traveler and you see a good idea from another group, you can write down that idea here. Once we are done with the presentations, I will give you a chance to use these suggestions or ideas to improve your arguments."

5. Ask the students, "Do you have any questions about what you need to do?"

6. Answer any questions that come up.

7. Give the following instructions:

    - *If using the whole-class presentation format,* tell the students, "Okay. Let's get started."

    - *If using the gallery walk format,* tell the students, "Okay, I'm now going to tell you which argument to go to and review."

    - *If using the modified gallery walk format,* tell the students, "Okay, I'm now going to assign you to be a presenter or a traveler." Assign one or two students from each group to be presenters and one or two students from each group to be travelers.

8. Give the students an opportunity to review the arguments.

    - *If using the whole-class presentation format,* have each group present their argument one at a time. Give each group only two to three minutes to present their argument. Then give the class two to three minutes to ask them questions and offer suggestions. Encourage as much participation from the students as possible.

    - *If using the gallery walk format,* tell the students, "Okay. Let's get started. Each group, move one argument to the left. Don't move to the next argument until I tell you to move. Once you get there, read the argument and then offer suggestions about how to make it better. I will put some sticky notes next to each argument. You can use the sticky notes to leave your suggestions." Give each group about three to four minutes to read the arguments, talk, and offer suggestions.

        a. After three to four minutes, tell the students, "Okay. Let's move on to the next argument. Please move one group to the left."

        b. Again, give each group three to four minutes to read, talk, and offer suggestions.

Investigation 3. States of Matter and Weight:
What Happens to the Weight of a Substance When It Changes
From a Solid to a Liquid or a Liquid to a Solid?

      c. Repeat this process until each group has had their argument read and critiqued three times.

- *If using the modified gallery walk format,* tell the students, "Okay. Let's get started. Reviewers, move one group to the left. Don't move to the next group until I tell you to move. Presenters, go ahead and share your argument with the travelers when they get there." Give each group of presenters and travelers about three to four minutes to talk.

      a. Tell the students, "Okay. Let's move on to the next argument. Travelers, move one group to the left."

      b. Again, give each group of presenters and travelers about three to four minutes to talk.

      c. Repeat this process until each group has had their argument read and critiqued three times.

9. Tell the students to return to their workstations.

10. Give the following instructions about revising the argument:

- *If using the whole-class presentation format,* tell the students, "I'm now going to give you all about 10 minutes to revise your argument. Take a few minutes to talk in your groups and determine what you want to change to make your argument better. Once you have decided what to change, go ahead and make the changes to your whiteboard."

- *If using the gallery walk format,* tell the students, "I'm now going to give you all about 10 minutes to revise your argument. Take a few minutes to read the suggestions that were left at your argument. Then talk in your groups and determine what you want to change to make your argument better. Once you have decided what to change, go ahead and make the changes to your whiteboard."

- *If using the modified gallery walk format,* tell the students, "I'm now going to give you all about 10 minutes to revise your argument. Please return to your original groups." Wait for the students to move back into their original groups and then tell the students, "Okay, take a few minutes to talk in your groups and determine what you want to change to make your argument better. Once you have decided what to change, go ahead and make the changes to your whiteboard."

11. Ask the students, "Do you have any questions about what you need to do?"

12. Answer any questions that come up.

13. Tell the students, "Okay. Let's get started."

14. Give the students 10 minutes to work in their groups on their arguments. As they work, move from group to group to check in, ask probing questions, and offer a suggestion if a group gets stuck.

Teacher Notes

Stage 5: Reflective Discussion (15 minutes)

1. Tell the students, "We are now going to take a minute to talk about some of the core ideas and crosscutting concepts that we have used during our investigation."
2. Show Figure 3.4 on the screen
3. Ask the students, "What do you all see going on here?"
4. Allow the students to share their ideas.
5. Show Figure 3.5 on the screen.
6. Ask the students, "What do you all see going on here?"
7. Allow the students to share their ideas.
8. Show Figures 3.4 and 3.5 on the screen at the same time and ask the students, "How can we explain these two phenomena in terms of the properties of matter?"
9. Allow the students to share their ideas. As they share their ideas, ask different questions to encourage them to expand on their thinking (e.g., "Can you tell me more about that?"), clarify a contribution (e.g., "Can you say that in another way?"), support an idea (e.g., "Why do you think that?"), add to an idea mentioned by a classmate (e.g., "Would anyone like to add to the idea?"), re-voice an idea offered by a classmate (e.g., "Who can explain that to me in another way?"), or critique an idea during the discussion (e.g., "Do you agree or disagree with that idea and why?") until students are able to generate an adequate explanation.
10. Tell the students, "We used standard units to measure weight during our investigation." Then ask, "Can anyone tell me why this was important?"
11. Allow the students to share their ideas.
12. Tell the students, "I think using standard units to measure and describe physical quantities such as weight is important because we need to have a common understanding of what our measurements mean. This allows us to share and critique each other's ideas and evidence."
13. Ask the students, "What units did you use today?"
14. Allow the students to share their ideas.
15. Show an image of the question "What do you think are the most important core ideas or crosscutting concepts that we used during this investigation to help us

FIGURE 3.4

Melting ice cream

FIGURE 3.5

A water bottle filled with water before and after being placed in a freezer

BEFORE being placed in a freezer   AFTER being placed in a freezer

make sense of what we observed?" Tell the students, "Okay, let's make sure we are all on the same page. Please take a moment to discuss this question with the other people in your group." Give them a few minutes to discuss the question.

16. Ask the students, "What do you all think? Who would like to share?"

17. Allow the students to share their ideas.

18. Tell the students, "We are now going to take a minute to talk about how scientists think about the world."

19. Show an image of the question "Are the laws of nature the same everywhere?" on the screen. Tell the students, "Take a few minutes to talk about how you would answer this question with the other people in your group. Be ready to share with the rest of the class." Give the students two to three minutes to talk in their group.

20. Ask the students, "What do you all think? Who would like to share an idea?"

21. Allow the students to share their ideas.

22. Ask the students, "Why would it be important for scientists to assume that the laws of nature are the same everywhere?"

23. Allow the students to share their ideas.

24. Tell the students, "I think it is important that scientists assume that the laws of nature are the same everywhere because it allows them to make predictions based on those laws no matter where they are located."

25. Ask the students, "Suppose I had some ice cream in Europe, in Africa, and in Australia: Would the ice cream change from a solid to a liquid in all three of these locations when it increases in temperature?"

Teacher Notes

26. Allow the students to share their ideas.

27. Ask the students, "Does anyone have any questions about why scientists assume that the laws of nature are the same everywhere?"

28. Answer any questions that come up.

29. Tell the students, "We are now going to take a minute to talk about what went well and what didn't go so well during our investigation. We need to talk about this because you all are going to be planning and carrying out your own investigations like this a lot this year, and I want to help you all get better at it."

30. Show an image of the question "What made your investigation scientific?" on the screen. Tell the students, "Take a few minutes to talk about how you would answer this question with the other people in your group. Be ready to share with the rest of the class." Give the students two to three minutes to talk in their group.

31. Ask the students, "What do you all think? Who would like to share an idea?"

32. Allow the students to share their ideas. Be sure to expand on their ideas about what makes an investigation scientific.

33. Show an image of the question "What made your investigation not so scientific?" on the screen. Tell the students, "Take a few minutes to talk about how you would answer this question with the other people in your group. Be ready to share with the rest of the class." Give the students two to three minutes to talk in their group.

34. Ask the students, "What do you all think? Who would like to share an idea?"

35. Allow the students to share their ideas. Be sure to expand on their ideas about what makes an investigation less scientific.

36. Show an image of the question "What rules can we put into place to help us make sure our next investigation is more scientific?" on the screen. Tell the students, "Take a few minutes to talk about how you would answer this question with the other people in your group. Be ready to share with the rest of the class." Give the students two to three minutes to talk in their group.

37. Ask the students, "What do you all think? Who would like to share an idea?"

38. Allow the students to share their ideas. Once they have shared their ideas, offer a suggestion for a possible class rule.

39. Ask the students, "What do you all think? Should we make this a rule?"

40. If the students agree, write the rule on the board or on a class "Rules for Scientific Investigation" chart so you can refer to it during the next investigation.

Investigation 3. States of Matter and Weight:
What Happens to the Weight of a Substance When It Changes
From a Solid to a Liquid or a Liquid to a Solid?

Stage 6: Write a Draft Report (30 minutes)

Your students will use either the Investigation Handout or the Investigation Log in the student workbook when writing the draft report. When you give the directions shown in quotes in the following steps, substitute "Investigation Log in your workbook" or just "Investigation Log" (as shown in brackets) for "handout" if they are using the workbook.

1. Tell the students, "You are now going to write an investigation report to share what you have learned. Please take out a pencil and turn to the 'Draft Report' section of your handout [Investigation Log in your workbook]."

2. If possible, use a document camera to project the "Introduction" section of the draft report from the Investigation Handout (or the Investigation Log in their workbook) on a screen or board (or take a picture of it and project the picture on a screen or board).

3. Tell the students, "The first part of the report is called the 'Introduction.' In this section of the report you want to explain to the reader what you were investigating, why you were investigating it, and what question you were trying to answer. All this information can be found in the text at the beginning of your handout [Investigation Log]." Point to the image. "Here are some sentence starters to help you begin writing."

4. Ask the students, "Do you have any questions about what you need to do?"

5. Answer any questions that come up.

6. Tell the students, "Okay, let's write."

7. Give the students 10 minutes to write the "Introduction" section of the report. As they work, move from student to student to check in, ask probing questions, and offer a suggestion if a student gets stuck.

8. If possible, use a document camera to project the "Method" section of the draft report from the Investigation Handout (or the Investigation Log in their workbook) on a screen or board (or take a picture of it and project the picture on a screen or board).

9. Tell the students, "The second part of the report is called the 'Method.' In this section of the report you want to explain to the reader what you did during the investigation, what data you collected and why, and how you went about analyzing your data. All this information can be found in the 'Plan Your Investigation' section of the handout [Investigation Log]. Remember that you all planned and carried out different investigations, so do not assume that the reader will know what you did." Point to the image. "Here are some sentence starters to help you begin writing."

10. Ask the students, "Do you have any questions about what you need to do?"

11. Answer any questions that come up.

Teacher Notes

12. Tell the students, "Okay, let's write."
13. Give the students 10 minutes to write the "Method" section of the report. As they work, move from student to student to check in, ask probing questions, and offer a suggestion if a student gets stuck.
14. If possible, use a document camera to project the "Argument" section of the draft report from the Investigation Handout (or the Investigation Log in their workbook) on a screen or board (or take a picture of it and project the picture on a screen or board).
15. Tell the students, "The last part of the report is called the 'Argument.' In this section of the report you want to share your claim, evidence, and justification of the evidence with the reader. All this information can be found on your whiteboard." Point to the image. "Here are some sentence starters to help you begin writing."
16. Ask the students, "Do you have any questions about what you need to do?"
17. Answer any questions that come up.
18. Tell the students, "Okay, let's write."
19. Give the students 10 minutes to write the "Argument" section of the report. As they work, move from student to student to check in, ask probing questions, and offer a suggestion if a student gets stuck.

Stage 7: Peer Review (35 minutes)

Your students will use either the Investigation Handout or their workbook when doing the peer review. Except where noted below, the directions are the same whether using the handout or the workbook.

1. Tell the students, "We are now going to review our reports to find ways to make them better. I'm going to come around and collect your draft reports. While I do that, please take out a pencil."
2. Collect the handouts or the workbooks with the draft reports from the students.
3. If possible, use a document camera to project the peer-review guide (see Appendix 4) on a screen or board (or take a picture of it and project the picture on a screen or board).
4. Tell the students, "We are going to use this peer-review guide to give each other feedback." Point to the image.
5. Tell the students, "I'm going to ask you to work with a partner to do this. I'm going to give you and your partner a draft report to read. You two will then read the report together. Once you are done reading the report, I want you to answer each of the questions on the peer-review guide." Point to the review questions on the image of the peer-review guide.

6. Tell the students, "You can check 'no,' 'almost,' or 'yes' after each question." Point to the checkboxes on the image of the peer-review guide.

7. Tell the students, "This will be your rating for this part of the report. Make sure you agree on the rating you give the author. If you mark 'no' or 'almost,' then you need to tell the author what he or she needs to do to get a 'yes.'" Point to the space for the reviewer feedback on the image of the peer-review guide.

8. Tell the students, "It is really important for you to give the authors feedback that is helpful. That means you need to tell them exactly what they need to do to make their report better."

9. Ask the students, "Do you have any questions about what you need to do?"

10. Answer any questions that come up.

11. Tell the students, "Please sit with a partner who is not in your current group." Allow the students time to sit with a partner.

12. Tell the students, "Okay, I'm now going to give you one report to read." Pass out one Investigation Handout with a draft report or one workbook to each pair. Make sure that the report you give a pair was not written by one of the students in that pair. Give each pair one peer-review guide to fill out. If the students are using workbooks, the peer-review guide is included right after the draft report so you do not need to pass out copies of the peer-review guide.

13. Tell the students, "Okay, I'm going to give you 15 minutes to read the report I gave you and to fill out the peer-review guide. Go ahead and get started."

14. Give the students 15 minutes to work. As they work, move around from pair to pair to check in and see how things are going, answer questions, and offer advice.

15. After 15 minutes pass, tell the students, "Okay, time is up. Please give me the report and the peer-review guide that you filled out."

16. Collect the Investigation Handouts and the peer-review guides, or collect the workbooks if students are using them. If the students are using the Investigation Handouts and separate peer-review guides, be sure you keep each handout with its corresponding peer-review guide.

17. Tell the students, "Okay, I am now going to give you a different report to read and a new peer-review guide to fill out." Pass out one more report to each pair. Make sure that the report you give a pair was not written by one of the students in that pair. Give each pair a new peer-review guide to fill out as a group.

18. Tell the students, "Okay, I'm going to give you 15 minutes to read this new report and to fill out the peer-review guide. Go ahead and get started."

19. Give the students 15 minutes to work. As they work, move around from pair to pair to check in and see how things are going, answer questions, and offer advice.

Teacher Notes

20. After 15 minutes pass, tell the students, "Okay, time is up. Please give me the report and the peer-review guide that you filled out."

21. Collect the Investigation Handouts and the peer-review guides, or collect the workbooks if students are using them. If the students are using the Investigation Handouts and separate peer-review guides, be sure you keep each handout with its corresponding peer-review guide.

Stage 8: Revise the Report (30 minutes)

Your students will use either the Investigation Handout or their workbook when revising the report. Except where noted below, the directions are the same whether using the handout or the workbook.

1. Tell the students, "You are now going to revise your draft report based on the feedback you get from your classmates. Please take out a pencil."

2. Return the reports to the students.
   - *If the students used the Investigation Handout and a copy of the peer-review guide,* pass back the handout and the peer-review guide to each student.
   - *If the students used the workbook,* pass that back to each student.

3. Tell the students, "Please take a few minutes to read over the peer-review guide. You should use it to figure out what you need to change in your report and how you will change it."

4. Allow the students to read the peer-review guide.

5. *If the students used the workbook,* if possible use a document camera to project the "Write Your Final Report" section from the Investigation Log on a screen or board (or take a picture of it and project the picture on a screen or board).

6. Give the following directions about how to revise their reports:
   - *If the students used the Investigation Handout and a copy of the peer-review guide,* tell them, "Okay, let's revise our reports. Please take out a piece of paper. I would like you to rewrite your report. You can use your draft report as a starting point, but you also need to change it to make it better. Use the feedback on the peer-review guide to make it better."
   - *If the students used the workbook,* tell them, "Okay, let's revise our reports. I would like you to rewrite your report in the section of the Investigation Log called "Write Your Final Report." You can use your draft report as a starting point, but you also need to change it to make it better. Use the feedback on the peer-review guide to make it better."

7. Ask the students, "Do you have any questions about what you need to do?"

8. Answer any questions that come up.

Investigation 3. States of Matter and Weight:
What Happens to the Weight of a Substance When It Changes
From a Solid to a Liquid or a Liquid to a Solid?

9. Tell the students, "Okay, let's write."
10. Give the following directions:
    - *If the students used the Investigation Handout,* tell them, "Okay, time's up. I will now come around and collect your Investigation Handout, the peer-review guide, and your final report."
    - *If the students used the workbook,* tell them, "Okay, time's up. I will now come around and collect your workbooks."
11. *If the students used the Investigation Handout,* collect all the Investigation Handouts, peer-review guides, and final reports. *If the students used the workbook,* collect all the workbooks.
12. *If the students used the Investigation Handout,* use the "Teacher Score" column in the peer-review guide to grade the final report. *If the students used the workbook,* use the "Investigation Report Grading Rubric" in the Investigation Log to grade the final report. Whether you are using the handout or the log, you can give the students feedback about their writing in the "Teacher Comments" section.

How to Use the Checkout Questions

The Checkout Questions are an optional assessment. We recommend giving them to students at the start of the next class period after the students finish stage 8 of the investigation. You can then look over the student answers to determine if you need to reteach the core idea from the investigation. Appendix 6 gives the answers to the Checkout Questions that should be given by a student who can apply the core idea correctly and can explain the importance of (1) thinking about systems and (2) using standard units to measure or describe the physical quantities of matter.

Alignment With Standards

Table 3.2 highlights how the investigation can be used to address specific performance expectations from the *Next Generation Science Standards, Common Core State Standards for English Language Arts (CCSS ELA)* and *Common Core State Standards for Mathematics (CCSS Mathematics),* and *English Language Proficiency (ELP) Standards.*

TABLE 3.2

Investigation 3 alignment with standards

NGSS performance expectation	5-PS1-2: Measure and graph quantities to provide evidence that regardless of the type of change that occurs when heating, cooling, or mixing substances, the total weight of matter is conserved.

Continued

Teacher Notes

Table 3.2 (*continued*)

***CCSS ELA*—Reading: Informational Text**	Key ideas and details • CCSS.ELA-LITERACY.RI.5.1: Quote accurately from a text when explaining what the text says explicitly and when drawing inferences from the text. • CCSS.ELA-LITERACY.RI.5.2: Determine two or more main ideas of a text and explain how they are supported by key details; summarize the text. • CCSS.ELA-LITERACY.RI.5.3: Explain the relationships or interactions between two or more individuals, events, ideas, or concepts in a historical, scientific, or technical text based on specific information in the text. Craft and structure • CCSS.ELA-LITERACY.RI.5.4: Determine the meaning of general academic and domain-specific words and phrases in a text relevant to a *grade 5 topic or subject area*. • CCSS.ELA-LITERACY.RI.5.5: Compare and contrast the overall structure (e.g., chronology, comparison, cause/effect, problem/solution) of events, ideas, concepts, or information in two or more texts. • CCSS.ELA-LITERACY.RI.5.6: Analyze multiple accounts of the same event or topic, noting important similarities and differences in the point of view they represent. Integration of knowledge and ideas • CCSS.ELA-LITERACY.RI.5.7: Draw on information from multiple print or digital sources, demonstrating the ability to locate an answer to a question quickly or to solve a problem efficiently. • CCSS.ELA-LITERACY.RI.5.8: Explain how an author uses reasons and evidence to support particular points in a text, identifying which reasons and evidence support which point(s). Range of reading and level of text complexity • CCSS.ELA-LITERACY.RI.5.10: By the end of the year, read and comprehend informational texts, including history/social studies, science, and technical texts, at the high end of the grades 4–5 text complexity band independently and proficiently.

***CCSS ELA*—Writing**	Text types and purposes
	• CCSS.ELA-LITERACY.W.5.1: Write opinion pieces on topics or texts, supporting a point of view with reasons.
	   ○ CCSS.ELA-LITERACY.W.5.1.A: Introduce a topic or text clearly, state an opinion, and create an organizational structure in which ideas are logically grouped to support the writer's purpose.
	   ○ CCSS.ELA-LITERACY.W.5.1.B: Provide logically ordered reasons that are supported by facts and details.
	   ○ CCSS.ELA-LITERACY.W.5.1.C: Link opinion and reasons using words, phrases, and clauses (e.g., *consequently*, *specifically*).
	   ○ CCSS.ELA-LITERACY.W.5.1.D: Provide a concluding statement or section related to the opinion presented.

Continued

Table 3.2 (*continued*)

CCSS ELA—Writing (*continued*)	• CCSS.ELA-LITERACY.W.5.2: Write informative/explanatory texts to examine a topic and convey ideas and information clearly.
	○ CCSS.ELA-LITERACY.W.5.2.A: Introduce a topic clearly, provide a general observation and focus, and group related information logically; include formatting (e.g., headings), illustrations, and multimedia when useful to aiding comprehension.
	○ CCSS.ELA-LITERACY.W.5.2.B: Develop the topic with facts, definitions, concrete details, quotations, or other information and examples related to the topic.
	○ CCSS.ELA-LITERACY.W.5.2.C: Link ideas within and across categories of information using words, phrases, and clauses (e.g., *in contrast*, *especially*).
	○ CCSS.ELA-LITERACY.W.5.2.D: Use precise language and domain-specific vocabulary to inform about or explain the topic.
	○ CCSS.ELA-LITERACY.W.5.2.E: Provide a concluding statement or section related to the information or explanation presented.
	Production and distribution of writing
	• CCSS.ELA-LITERACY.W.5.4: Produce clear and coherent writing in which the development and organization are appropriate to task, purpose, and audience.
	• CCSS.ELA-LITERACY.W.5.5: With guidance and support from peers and adults, develop and strengthen writing as needed by planning, revising, editing, rewriting, or trying a new approach.
	• CCSS.ELA-LITERACY.W.5.6: With some guidance and support from adults, use technology, including the internet, to produce and publish writing as well as to interact and collaborate with others; demonstrate sufficient command of keyboarding skills to type a minimum of two pages in a single sitting.
	Research to build and present knowledge
	• CCSS.ELA-LITERACY.W.5.8: Recall relevant information from experiences or gather relevant information from print and digital sources; summarize or paraphrase information in notes and finished work, and provide a list of sources.
	• CCSS.ELA-LITERACY.W.5.9: Draw evidence from literary or informational texts to support analysis, reflection, and research.
	Range of writing
	• CCSS.ELA-LITERACY.W.5.10: Write routinely over extended time frames (time for research, reflection, and revision) and shorter time frames (a single sitting or a day or two) for a range of discipline-specific tasks, purposes, and audiences.

Continued

Teacher Notes

Table 3.2 (*continued*)

CCSS ELA—Speaking and Listening	Comprehension and collaboration
	• CCSS.ELA-LITERACY.SL.5.1: Engage effectively in a range of collaborative discussions (one-on-one, in groups, and teacher-led) with diverse partners on *grade 5 topics and texts,* building on others' ideas and expressing their own clearly.
	    o CCSS.ELA-LITERACY.SL.5.1.A: Come to discussions prepared, having read or studied required material; explicitly draw on that preparation and other information known about the topic to explore ideas under discussion.
	    o CCSS.ELA-LITERACY.SL.5.1.B: Follow agreed-upon rules for discussions and carry out assigned roles.
	    o CCSS.ELA-LITERACY.SL.5.1.C: Pose and respond to specific questions by making comments that contribute to the discussion and elaborate on the remarks of others.
	    o CCSS.ELA-LITERACY.SL.5.1.D: Review the key ideas expressed and draw conclusions in light of information and knowledge gained from the discussions.
	• CCSS.ELA-LITERACY.SL.5.2: Summarize a written text read aloud or information presented in diverse media and formats, including visually, quantitatively, and orally.
	• CCSS.ELA-LITERACY.SL.5.3: Summarize the points a speaker makes and explain how each claim is supported by reasons and evidence.
	Presentation of knowledge and ideas
	• CCSS.ELA-LITERACY.SL.5.4: Report on a topic or text or present an opinion, sequencing ideas logically and using appropriate facts and relevant, descriptive details to support main ideas or themes; speak clearly at an understandable pace.
	• CCSS.ELA-LITERACY.SL.5.5: Include multimedia components (e.g., graphics, sound) and visual displays in presentations when appropriate to enhance the development of main ideas or themes.
	• CCSS.ELA-LITERACY.SL.5.6: Adapt speech to a variety of contexts and tasks, using formal English when appropriate to task and situation.

CCSS Mathematics—Numbers and Operations in Base Ten	Understand the place value system. • CCSS.MATH.CONTENT.5.NBT.A.1: Recognize that in a multi-digit number, a digit in one place represents 10 times as much as it represents in the place to its right and 1/10 of what it represents in the place to its left • CCSS.MATH.CONTENT.5.NBT.A.3: Read, write, and compare decimals to thousandths. • CCSS.MATH.CONTENT.5.NBT.A.4: Use place value understanding to round decimals to any place. Perform operations with multi-digit whole numbers and with decimals to hundredths. • CCSS.MATH.CONTENT.5.NBT.B.7: Add, subtract, multiply, and divide decimals to hundredths.

Continued

Investigation 3. States of Matter and Weight: What Happens to the Weight of a Substance When It Changes From a Solid to a Liquid or a Liquid to a Solid?

Table 3.2 (*continued*)

CCSS Mathematics—Measurement and Data	Convert like measurement units within a given measurement system. • CCSS.MATH.CONTENT.5.MD.A.1: Convert among different-sized standard measurement units within a given measurement system (e.g., convert 5 cm to 0.05 m), and use these conversions in solving multi-step, real-world problems. Represent and interpret data. • CCSS.MATH.CONTENT.5.MD.B.2: Make a line plot to display a data set of measurements in fractions of a unit (1/2, 1/4, 1/8). Use operations on fractions for this grade to solve problems involving information presented in line plots.
ELP Standards	Receptive modalities • ELP 1: Construct meaning from oral presentations and literary and informational text through grade-appropriate listening, reading, and viewing. • ELP 8: Determine the meaning of words and phrases in oral presentations and literary and informational text. Productive modalities • ELP 3: Speak and write about grade-appropriate complex literary and informational texts and topics. • ELP 4: Construct grade-appropriate oral and written claims and support them with reasoning and evidence. • ELP 7: Adapt language choices to purpose, task, and audience when speaking and writing. Interactive modalities • ELP 2: Participate in grade-appropriate oral and written exchanges of information, ideas, and analyses, responding to peer, audience, or reader comments and questions. • ELP 5: Conduct research and evaluate and communicate findings to answer questions or solve problems. • ELP 6: Analyze and critique the arguments of others orally and in writing. Linguistic structures of English • ELP 9: Create clear and coherent grade-appropriate speech and text. • ELP 10: Make accurate use of standard English to communicate in grade-appropriate speech and writing.

Investigation Handout

Investigation 3

States of Matter and Weight: What Happens to the Weight of a Substance When It Changes From a Solid to a Liquid or a Liquid to a Solid?

Introduction

Matter is anything that has mass and takes up space. There are many different types of matter. Water, aluminum, wax, and chocolate are all examples of different types of matter. Take a few minutes to watch what happens when your teacher changes the temperature of chocolate chips. As you watch this demonstration, keep track of things you notice and things you wonder about in the boxes below.

Things I NOTICED ...	Things I WONDER about ...

A sample of matter can exist in different forms or *states*. A sample of matter that is in a *solid* state has a fixed shape and volume. Matter with a fixed shape and volume does not change shape and does not take up more or less space when it is placed in a container. The chocolate chips you observed earlier are an

Investigation 3. States of Matter and Weight: What Happens to the Weight of a Substance When It Changes From a Solid to a Liquid or a Liquid to a Solid?

example of a solid. A sample of matter that is in a *liquid* state has a fixed volume but does not have a fixed shape. A liquid will therefore change shape when it is placed in a container, but the total amount of space it takes up will stay the same. The melted chocolate chips are an example of a liquid. The liquid chocolate changes shape, but it does not take up more or less space. A sample of matter that is in a gas state does not have a fixed shape or volume. A gas will change shape and volume when it is placed in a container. Therefore, a gas will spread out and fill its container, and it can also be squeezed into a small container.

A sample of matter will change states depending on temperature. The temperature of matter changes when it gains or loses *thermal energy*. When a sample of matter gains thermal energy, it increases in temperature. When a sample of matter loses thermal energy, it decreases in temperature. A sample of matter will change from a solid into a liquid when it reaches a specific temperature. Chocolate, for example, melts at 90°F (32°C). When you place a piece of chocolate in your mouth it will gain thermal energy from your body, which is about 98°F (37°C), and increase in temperature. The chocolate will melt and turn into a liquid when its temperature reaches 90°F (32°C). A liquid will change into a solid when it cools to a specific temperature. Water freezes at 32°F (0°C). Water is therefore a liquid at room temperature, but when it is placed in the freezer and cooled to a temperature of 32°F (0°C) it turns into a solid. Water that is in a solid state is called ice. When the temperature of ice increases above 32°F (0°C), it turns back into a liquid.

These ideas about how matter can be found in different states and how matter changes states depending on temperature can help us understand why objects melt when they warm up and freeze when they cool down. These ideas can also help us predict what will happen to the shape and volume of a sample of matter when it changes from a solid to a liquid or when it changes from a liquid to a solid. Although useful, these ideas do not tell us anything about what happens to the weight of different types of matter when they change state. Your goal in this investigation is to figure out if a sample of matter will become heavier, lighter, or stay the same weight when it changes from a solid to a liquid or a liquid to a solid.

Things we KNOW from what we read …

Investigation Handout

Your Task

Use what you know about the structure and properties of matter, tracking matter in a system, and cause-and-effect relationships to design and carry out an experiment to figure out how a change of state affects the weight of a sample of water, a sample of wax, and a sample of vegetable shortening.

The *guiding question* of this investigation is, **What happens to the weight of a substance when it changes from a solid to a liquid or a liquid to a solid?**

Materials

You may use any of the following materials during your investigation:

- Safety goggles, nonlatex apron, and vinyl gloves (required)
- Test tube with screw cap filled with water
- Test tube with screw cap filled with wax
- Test tube with screw cap filled with vegetable shortening
- Test tube clamp
- Electronic scale
- Hot water bath
- Freezer

Safety Rules

Follow all normal safety rules. In addition, be sure to follow these rules:

- Wear sanitized indirectly vented chemical-splash goggles, nonlatex aprons, and vinyl gloves during setup, investigation activity, and cleanup.
- Do not taste any food products used in this activity.
- Be careful when working with hot plates, because they can seriously burn skin and cause fires.
- Be careful when working with hot water and use only under direct teacher supervision, because it can burn skin.
- Immediately clean up any spills to avoid a slip or fall hazard.
- Keep water sources away from electrical receptacles to prevent shock.
- Be careful when handling glassware (test tubes, etc.), because it can shatter and cut skin.
- Wash your hands with soap and water when you are done cleaning up.

Plan Your Investigation

Prepare a plan for your investigation by filling out the chart on the next page; this plan is called an *investigation proposal*. Before you start developing your plan, be sure to discuss the following questions with the other members of your group:

Investigation 3. States of Matter and Weight:
What Happens to the Weight of a Substance When It Changes
From a Solid to a Liquid or a Liquid to a Solid?

- Which standard **units** should we use to describe the physical quantities of the matter?
- What are the components of the **system** that we are studying?

Our guiding question:

This is a picture of how we will set up the equipment:

We will collect the following data:

These are the steps we will follow to collect data:

I approve of this investigation proposal.

_____   _____
Teacher's signature   Date

Argument-Driven Inquiry in **Fifth-Grade Science:** Three-Dimensional Investigations

Investigation Handout

Collect Your Data

Keep a record of what you measure or observe during your investigation in the space below.

Analyze Your Data

You will need to analyze the data you collected before you can develop an answer to the guiding question. To analyze the data you collected, create a graph that shows the relationship between the cause and effect.

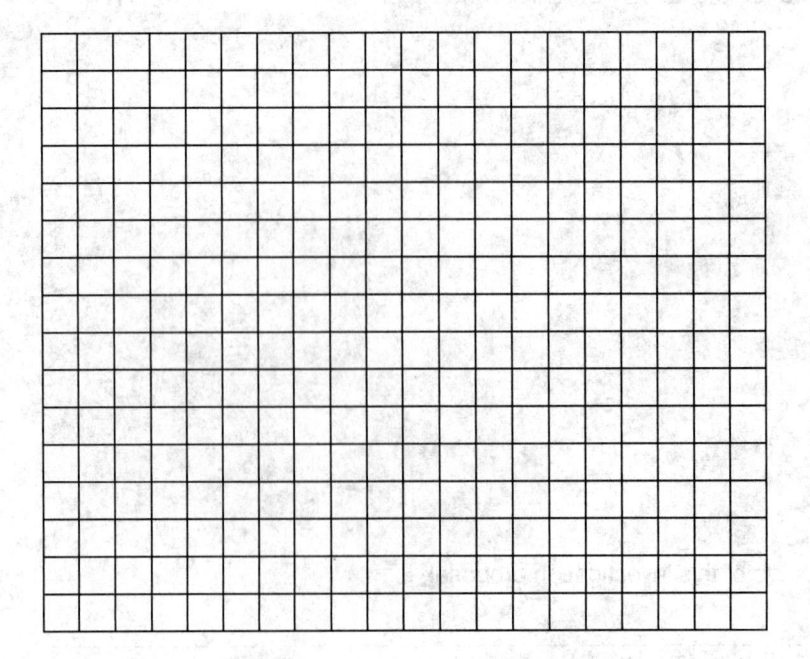

Investigation 3. States of Matter and Weight: What Happens to the Weight of a Substance When It Changes From a Solid to a Liquid or a Liquid to a Solid?

Draft Argument

Develop an argument on a whiteboard. It should include the following:

1. A *claim*: Your answer to the guiding question.
2. *Evidence*: An analysis of the data and an explanation of what the analysis means.
3. A *justification of the evidence*: Why your group thinks the evidence is important.

The Guiding Question:	
Our Claim:	
Our Evidence:	Our Justification of the Evidence:

Argumentation Session

Share your argument with your classmates. Be sure to ask them how to make your draft argument better. Keep track of their suggestions in the space below.

Ways to IMPROVE our argument …

Investigation Handout

Draft Report

Prepare an *investigation report* to share what you have learned. Use the information in this handout and your group's final argument to write a *draft* of your investigation report.

Introduction

We have been studying _____ in class.

Before we started this investigation, we explored _____
_____
_____

We noticed _____
_____
_____

My goal for this investigation was to figure out _____
_____
_____

The guiding question was _____
_____
_____

Method

To gather the data I needed to answer this question, I _____
_____
_____
_____
_____

Investigation 3. States of Matter and Weight:
What Happens to the Weight of a Substance When It Changes
From a Solid to a Liquid or a Liquid to a Solid?

I then analyzed the data I collected by _____

_____

_____

_____

Argument

My claim is _____

_____

_____

_____

The graph below shows _____

_____

_____

_____

_____

_____

Investigation Handout

This analysis of the data I collected suggests _____

_____

_____

_____

_____

This evidence is based on several important scientific concepts. The first one is _____

_____

_____

_____

_____

Review

Your classmates need your help! Review the draft of their investigation reports and give them ideas about how to improve. Use the peer-review guide when doing your review.

Submit Your Final Report

Once you have received feedback from your classmates about your draft report, create your final investigation report and hand it in to your teacher.

Checkout Questions

Investigation 3. States of Matter and Weight

Use the following information to answer questions 1 and 2. The picture below shows a container filled with water. The container is sealed. The container and the water together weigh 56 grams. The sealed container is placed in a freezer until the water inside the container turns from liquid to solid, with the container remaining sealed the entire time. Then the container is removed from the freezer.

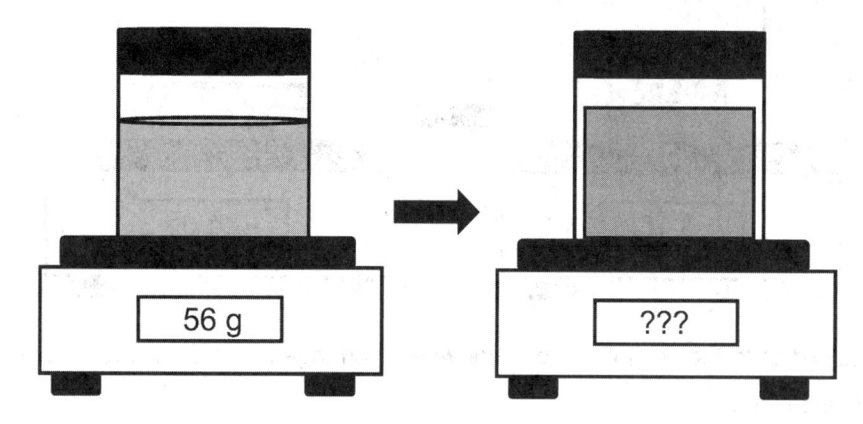

1. How much do you think the container filled with water weighs after it is removed from the freezer?

    a. 56 grams

    b. More than 56 grams

    c. Less than 56 grams

2. How do you know? Use what you know about systems to explain your answer.

_____

_____

_____

_____

_____

_____

Teacher Scoring Rubric for Checkout Questions 1 and 2

Level	Description
3	The student can apply the core idea correctly and can explain the importance of thinking about systems.
2	The student cannot apply the core idea correctly but can explain the importance of thinking about systems.
1	The student can apply the core idea correctly but cannot explain the importance of thinking about systems.
0	The student cannot apply the core idea correctly and cannot explain the importance of thinking about systems.

Argument-Driven Inquiry in **Fifth-Grade Science**: Three-Dimensional Investigations

Checkout Questions

Use the following information to answer questions 3 and 4. The picture below shows a container with an ice cube in it. The container is sealed. The container and the ice cube together weigh 35.15 grams. The sealed container is left on a table until the water inside the container turns from a solid to a liquid. The container and the water are then weighed for a second time. This time the weight of the container and the water in it is 1.24 ounces.

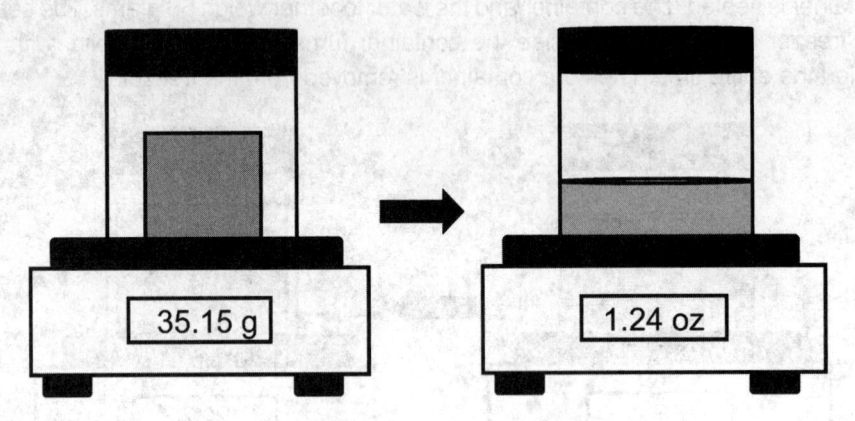

3. What happened to the weight of the water when it changed from a solid to a liquid?

    a. It decreased.

    b. It increased.

    c. It stayed the same.

    d. Unable to tell.

4. Explain your thinking. Use what you know about using standard units during an investigation as part of your answer.

_____

_____

_____

_____

_____

Teacher Scoring Rubric for Checkout Questions 3 and 4

Level	Description
3	The student can apply the core idea correctly and can explain the importance of using standard units to measure or describe the physical quantities of matter.
2	The student can apply the core idea correctly but cannot explain the importance of using standard units to measure or describe the physical quantities of matter.
1	The student cannot apply the core idea correctly but can explain the importance of using standard units to measure or describe the physical quantities of matter.
0	The student cannot apply the core idea correctly and cannot explain the importance of using standard units to measure or describe the physical quantities of matter.

Teacher Notes

Investigation 4

Chemical Reactions: Which Pairs of the Available Liquids Produce a New Substance When They Are Mixed Together?

Purpose

The purpose of this investigation is to give students an opportunity to use two disciplinary core ideas (DCIs), two crosscutting concepts (CCs), and six scientific and engineering practices (SEPs) to figure out whether the mixing of two substances results in a new substance. Students will also learn about the types of questions that scientists investigate.

The DCIs, CCs, and SEPs That Students Use During This Investigation to Figure Things Out

DCIs

- *PS1.A: Structure and Properties of Matter:* Measurements of a variety of properties can be used to identify materials.
- *PS1.B: Chemical Reactions:* When two or more different substances are mixed, a new substance with different properties may be formed.

CCs

- *CC 2: Cause and Effect:* Cause-and-effect relationships are routinely identified, tested, and used to explain change. Events that occur together with regularity might or might not be a cause-and-effect relationship.
- *CC 3: Scale, Proportion, and Quantity:* Standard units are used to measure and describe physical quantities such as weight, time, temperature, and volume.

SEPs

- *SEP 1: Asking Questions and Defining Problems:* Ask questions about what would happen if a variable is changed. Ask questions that can be investigated and predict reasonable outcomes based on patterns such as cause-and-effect relationships.
- *SEP 3: Planning and Carrying Out Investigations:* Plan and conduct an investigation collaboratively to produce data to serve as the basis for evidence, using fair tests in which variables are controlled and the number of trials considered. Evaluate appropriate methods and/or tools for collecting data.

Teacher Notes

- *SEP 4: Analyzing and Interpreting Data:* Represent data in tables and/or various graphical displays (bar graphs, pictographs, and/or pie charts) to reveal patterns that indicate relationships. Analyze and interpret data to make sense of phenomena, using logical reasoning, mathematics, and/or computation. Compare and contrast data collected by different groups in order to discuss similarities and differences in their findings.
- *SEP 6: Constructing Explanations and Designing Solutions:* Construct an explanation of observed relationships. Use evidence to construct or support an explanation. Identify the evidence that supports particular points in an explanation.
- *SEP 7: Engaging in Argument From Evidence:* Compare and refine arguments based on an evaluation of the evidence presented. Distinguish among facts, reasoned judgment based on research findings, and speculation in an explanation. Respectfully provide and receive critiques from peers about a proposed procedure, explanation, or model by citing relevant evidence and posing specific questions.
- *SEP 8: Obtaining, Evaluating, and Communicating Information:* Read and comprehend grade-appropriate complex texts and/or other reliable media to summarize and obtain scientific and technical ideas. Combine information in written text with that contained in corresponding tables, diagrams, and/or charts to support the engagement in other scientific and/or engineering practices. Communicate scientific and/or technical information orally and/or in written formats, including various forms of media as well as tables, diagrams, and charts.

Other Concepts That Students May Use During This Investigation

Students might also use some of the following concepts:

- Physical properties are characteristics of matter that can be observed without mixing it with another type of matter.
- Chemical properties are characteristics of matter that can only be observed when different types of matter are mixed.
- The physical and chemical properties of a specific type of matter do not change.
- Matter of any type is made up of particles that are too small to see.
- Different types of matter have different physical and chemical properties because they are made of different types of particles. These particles have a unique structure.

What Students Figure Out

Four pairs of liquids produce a new substance when mixed:

1. Sodium bicarbonate in water and calcium chloride in water
2. Sodium carbonate in water and calcium chloride in water
3. Sodium carbonate in water and magnesium sulfate in water
4. Calcium chloride in water and magnesium sulfate in water

Two pairs of liquids do not produce a new substance when mixed:

1. Sodium bicarbonate in water and magnesium sulfate in water
2. Sodium bicarbonate in water and sodium carbonate in water

Background Information About This Investigation for the Teacher

A chemical reaction occurs when the atoms that make up one or more different substances break apart from each other, rearrange, and then recombine in a different way. The rearrangement of atoms during a chemical reaction creates one or more different substances. To illustrate, consider the chemical reaction that takes place when sodium bicarbonate (baking soda), calcium chloride (pool salt), and water are mixed together. The atoms that make up the sodium bicarbonate and the atoms that make up the calcium chloride will separate from each other and then recombine in a different way to produce four new substances. Water molecules are not involved in this reaction; the water just provides a medium that allows the atoms that make up the sodium bicarbonate and calcium chloride to separate and then rearrange. The four new substances that are produced during this chemical reaction are water, sodium chloride (table salt), calcium carbonate, and carbon dioxide. This reaction also releases energy.

Different types of substances have different properties. A property is any measurable or observable characteristic. There are two kinds of properties that we use to describe a substance. *Physical properties* are characteristics of a substance that can be observed without mixing it with another type of matter. Examples of physical properties are color, texture, and state. Sodium bicarbonate, for example, is a white solid at room temperature. It has a fine texture. Calcium chloride is also a white solid at room temperature. This substance, unlike sodium bicarbonate, is a granular texture. *Chemical properties* are characteristics of substances that can only be observed when a substance is mixed with another substance. Examples of chemical properties are how something reacts with water or how something reacts with an acid (such as vinegar). The physical and chemical properties of a specific type of substance do not change. Scientists can therefore use physical and chemical properties to identify different substances. They can also use these properties to determine if a chemical reaction has occurred, because a chemical reaction will produce one or more new substances with physical or chemical properties that are different from the properties of the original substances.

Teacher Notes

TABLE 4.1
Pairs of liquids that can be mixed by the students during the investigation

Liquid 1	Liquid 2	Result
Sodium bicarbonate	Sodium carbonate	No new substance (no reaction)
Sodium bicarbonate	Calcium chloride	Produces a gas (carbon dioxide)
Sodium bicarbonate	Magnesium sulfate	No new substance (no reaction)
Sodium carbonate	Calcium chloride	Produces a white solid (calcium carbonate)
Sodium carbonate	Magnesium sulfate	Produces a white solid (magnesium carbonate)
Calcium chloride	Magnesium sulfate	Produces a white solid (calcium sulfate)

The students can test six different combinations of liquids during this investigation. Table 4.1 provides a list of each combination and the result of mixing the liquids together. To determine if a new substance is produced as a result of mixing two liquids together, the students will need to add about 10 drops of two different liquids to the same test tube (see Figure 4.1). They need to make observations of the properties of the liquids (color, texture, and state) before they mix them together. Then they can make observations about the properties of any substance that they can see in the test tube after they mix the liquids. The students can use any difference in physical properties before and after mixing as evidence of a new substance.

FIGURE 4.1
An example of how to mix the liquids

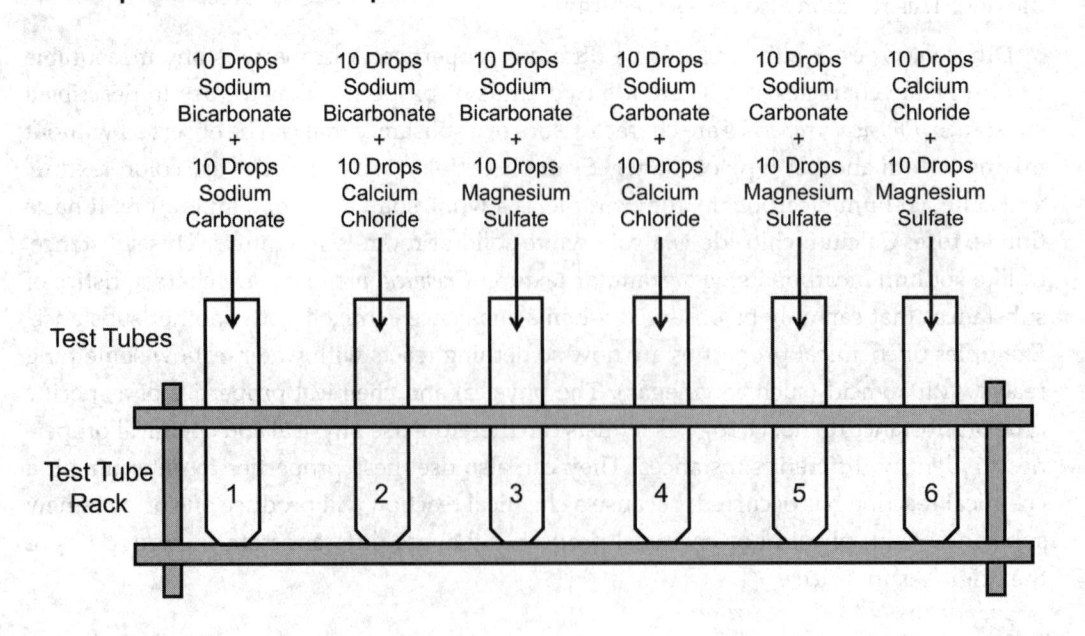

Investigation 4. Chemical Reactions: Which Pairs of the Available Liquids Produce a New Substance When They Are Mixed Together?

Timeline

The time needed to complete this investigation is 260 minutes (4 hours and 20 minutes). The amount of instructional time needed for each stage of the investigation is as follows:

- *Stage 1.* Introduce the task and the guiding question: 35 minutes
- *Stage 2.* Design a method and collect data: 40 minutes
- *Stage 3.* Create a draft argument: 45 minutes
- *Stage 4.* Argumentation session: 30 minutes
- *Stage 5.* Reflective discussion: 15 minutes
- *Stage 6.* Write a draft report: 30 minutes
- *Stage 7.* Peer review: 35 minutes
- *Stage 8.* Revise the report: 30 minutes

Materials and Preparation

The materials needed for this investigation are listed in Table 4.2. The items can be purchased from a science education supply company such as Ward's Science (*www.wardsci.com*) or Flinn Scientific (*www.flinnsci.com*); at a big-box retail store such as Walmart or Target; or through an online retailer such as Amazon. The materials for this investigation can also be purchased as a complete kit (which includes enough materials for 24 students, or six groups of four students) at *www.argumentdriveninquiry.com*.

TABLE 4.2
Materials for Investigation 4

Item	Quantity
Indirectly vented chemical-splash goggles	1 per student
Nonlatex apron	1 per student
Vinyl gloves	1 pair per student
Calcium chloride (pool salt) (for opening activity)	10 g per class
Sodium bicarbonate (baking soda) (for opening activity)	10 g per class
Water in a cup	100 ml per class
Plastic freezer bag, 1-gallon size	2 per class
Calcium chloride	50 g per class
Magnesium sulfate (Epsom salt)	60 g per class
Sodium carbonate (washing soda)	50 g per class
Sodium bicarbonate	40 g per class

Continued

Teacher Notes

Table 4.2. (*continued*)

Item	Quantity
Beaker, 1,000 ml (for mixing solutions)	1 per class
Calcium chloride solution in a 30-ml plastic dropper bottle with cap	1 bottle per group
Magnesium sulfate solution in a 30-ml plastic dropper bottle with cap	1 bottle per group
Sodium carbonate solution in a 30-ml plastic dropper bottle with cap	1 bottle per group
Sodium bicarbonate solution in a 30-ml plastic dropper bottle with cap	1 bottle per group
Test tubes	6 per group
Test tube rack	1 per group
Whiteboard, 2' × 3'*	1 per group
Investigation Handout	1 per student
Peer-review guide and teacher scoring rubric	1 per student
Checkout Questions (optional)	1 per student

*As an alternative, students can use computer and presentation software such as Microsoft PowerPoint or Apple Keynote to create their arguments.

The day before you plan to start the investigation, you will need to prepare the solutions for the students to use. Prepare the solutions as follows:

- *Calcium chloride solution*: Add 50 g of calcium chloride to 500 ml of water in a beaker. Mix thoroughly to dissolve. Fill a 30-ml plastic dropper bottle with the calcium chloride solution. Place the cap on the bottle and label it. Make one bottle for each group.

- *Magnesium sulfate solution*: Add 60 g of magnesium sulfate to 500 ml of water in a beaker. Mix thoroughly to dissolve. Fill a 30-ml plastic dropper bottle with the magnesium sulfate solution. Place the cap on the bottle and label it. Make one bottle for each group.

- *Sodium carbonate solution*: Add 50 g of sodium carbonate to 500 ml of water in a beaker. Mix thoroughly to dissolve. Fill a 30-ml plastic dropper bottle with the sodium carbonate solution. Place the cap on the bottle and label it. Make one bottle for each group.

- *Sodium bicarbonate solution*: Add 40 g of sodium bicarbonate to 500 ml of water in a beaker. Mix thoroughly to dissolve. Fill a 30-ml plastic dropper bottle with the sodium bicarbonate solution. Place the cap on the bottle and label it. Make one bottle for each group.

Investigation 4. Chemical Reactions:
Which Pairs of the Available Liquids Produce
a New Substance When They Are Mixed Together?

Excess solution may be flushed down the drain with a large quantity of water. The students may also dispose of any liquid waste from their investigations by flushing it down the drain with a large quantity of water.

Safety Precautions

Remind students to follow all normal safety rules. In addition, tell the students to take the following safety precautions:

- Wear sanitized indirectly vented chemical-splash goggles, nonlatex aprons, and vinyl gloves during setup, investigation activity, and cleanup.
- Do not put any liquids in their mouth.
- Immediately clean up any spills to avoid a slip or fall hazard.
- Keep water sources away from electrical receptacles to prevent shock.
- Be careful when handling glassware, because it can shatter and cut skin.
- Make sure all materials are put away after completing the activity.
- Wash their hands with soap and water when they are done cleaning up.

Lesson Plan by Stage

This lesson plan is only a suggestion. It is included here to illustrate what you can say and do during each stage of ADI for this specific investigation. We encourage you to modify this lesson plan by asking different questions, using different examples, and providing different scaffolds as needed to better meet the needs of students in your class.

Stage 1: Introduce the Task and the Guiding Question (35 minutes)

1. Ask the students to sit in six groups, with three or four students in each group.
2. Ask the students to clear off their desks except for a pencil (and their *Student Workbook for Argument-Driven Inquiry in Fifth-Grade Science* if they have one).
3. Pass out an Investigation Handout to each student (or ask students to turn to the Investigation Log for Investigation 4 in their workbook).
4. Read the first paragraph of the "Introduction" aloud to the class. Ask the students to follow along as you read.
5. Hold up a 1-gallon clear plastic freezer bag so everyone can see it.
6. Add 5 g of sodium bicarbonate to the bag. Wait 10 seconds. Add 5 g of calcium chloride to the bag. Wait another 10 seconds (at this point in time there will be nothing happening in the bag). Add about 10 ml of water to the bag and seal it (at this point in time a chemical reaction will begin).

Teacher Notes

7. Tell the students, "Pass this bag around the room so everyone can see what is happening inside it, but do not open the bag." If needed, you can use the remaining sodium bicarbonate, calcium chloride, and water to create a second "reaction in a bag" for the students to look at and pass around.

8. Ask the students to record their observations and questions on the "NOTICED/WONDER" chart in the "Introduction."

9. After the students have recorded their observations and questions, ask them to share what they observed about what happened inside the bag.

10. Ask the students to share what questions they have about what happened inside the bag.

11. Tell the students, "Some of your questions might be answered by reading the rest of the 'Introduction.'"

12. Ask the students to read the rest of the "Introduction" on their own *or* ask them to follow along as you read it aloud.

13. Once the students have read the rest of the "Introduction," ask them to fill out the "Things we KNOW" chart on their Investigation Handout (or in their Investigation Log) as a group.

14. Ask the students to share what they learned from the reading. Add these ideas to a class "Things we KNOW" chart.

15. Tell the students, "Let's see what we will need to figure out during our investigation."

16. Read the task and the guiding question aloud.

17. Tell the students, "I have lots of materials here that you can use."

18. Introduce the students to the materials available for them to use during the investigation by holding each one up and asking how it might be used.

Stage 2: Design a Method and Collect Data (40 minutes)

1. Tell the students, "I am now going to give you and the other members of your group about 15 minutes to plan your investigation. Before you begin, I want you all to take a couple of minutes to discuss the following questions with the rest of your group."

2. Show the following question on the screen or board:
   - What *causes* a chemical reaction and what is the *effect* of that reaction?
   - What information should we collect so we can *describe the properties* of a substance?

3. Tell the students, "Please take a few minutes to come up with an answer to these questions."

Investigation 4. Chemical Reactions: Which Pairs of the Available Liquids Produce a New Substance When They Are Mixed Together?

4. Give the students two to three minutes to discuss these two questions.

5. Ask two or three different groups to share their answers. Highlight or write down any important ideas on the board so students can refer to them later.

6. If possible, use a document camera to project an image of the graphic organizer for this investigation on a screen or board (or take a picture of it and project the picture on a screen or board). Tell the students, "I now want you all to plan out your investigation. To do that, you will need to fill out this investigation proposal."

7. Point to the box labeled "Our guiding question:" and tell the students, "You can put the question we are trying to answer in this box." Then ask, "Where can we find the guiding question?"

8. Wait for a student to answer.

9. Point to the box labeled "This is a picture of how we will set up the equipment:" and tell the students, "You can draw a picture in this box of how you will set up the equipment in order to carry out this investigation."

10. Point to the box labeled "We will collect the following data:" and tell the students, "You can list the measurements or observations that you will need to collect during the investigation in this box."

11. Point to the box labeled "These are the steps we will follow to collect data:" and tell the students, "You can list what you are going to do to collect the data you need and what you will do with your data once you have it. Be sure to give enough detail that I could do your investigation for you."

12. Ask the students, "Do you have any questions about what you need to do?"

13. Answer any questions that come up.

14. Tell the students, "Once you are done, raise your hand and let me know. I'll then come by and look over your proposal and give you some feedback. You may not begin collecting data until I have approved your proposal by signing it. You need to have your proposal done in the next 15 minutes."

15. Give the students 15 minutes to work in their groups on their investigation proposal. As they work, move from group to group to check in, ask probing questions, and offer a suggestion if a group gets stuck.

16. As each group finishes its investigation proposal, read it over and determine if it will be productive or not. If you feel the investigation will be productive (not necessarily what you would do or what the other groups are doing), sign your name on the proposal and let the group start collecting data. If the plan needs to be changed, offer some suggestions or ask some probing questions, and have the group make the changes before you approve it.

Teacher Notes

17. Pass out the materials or have one student from each group collect the materials they need from a central supply table or cart for the groups that have an approved proposal.
18. Remind students of the safety rules and precautions for this investigation.

What should a student-designed investigation look like?

There are a number of different investigations that students can design to answer the question "Which pairs of the available liquids produce a new substance when they are mixed together?" For example, one method might include the following steps:

1. Make observations about the properties of the four liquids (sodium bicarbonate solution, sodium carbonate solution, calcium chloride solution, and magnesium sulfate solution).
2. Mix 10 drops of sodium bicarbonate solution and 10 drops of sodium carbonate solution in a test tube.
3. Make observations about the properties of what is in the test tube.
4. Mix 10 drops of sodium bicarbonate solution and 10 drops of calcium chloride solution in a test tube.
5. Make observations about the properties of what is in the test tube.
6. Mix 10 drops of sodium bicarbonate solution and 10 drops of magnesium sulfate solution in a test tube.
7. Make observations about the properties of what is in the test tube.
8. Mix 10 drops of sodium carbonate solution and 10 drops of calcium chloride solution in a test tube.
9. Make observations about the properties of what is in the test tube.
10. Mix 10 drops of sodium carbonate and 10 drops of magnesium sulfate solution in a test tube.
11. Make observations about the properties of what is in the test tube.
12. Mix 10 drops of calcium chloride solution and 10 drops of magnesium sulfate in a test tube.
13. Make observations about the properties of what is in the test tube.

Continued

> If students use this method, they will need to collect the following data:
>
> 1. Color of each liquid (a physical property)
> 2. Texture of each liquid (a physical property)
> 3. Color of any substance found in the test tube after two liquids are mixed (a physical property)
> 4. Texture of any substance found in the test tube after two liquids are mixed (a physical property)
> 5. State of any substance found in the test tube after two liquids are mixed (a physical property)

19. Tell the students to collect their data and record their observations or measurements in the "Collect Your Data" box in their Investigation Handout (or the Investigation Log in their workbook).
20. Give the students 20 minutes to collect their data. Collect the materials from each group before asking them to analyze their data.

Stage 3: Create a Draft Argument (45 minutes)

1. Tell the students, "Now that we have all this data, we need to analyze the data so we can figure out an answer to the guiding question."
2. If possible, project an image of the "Analyze Your Data" section for this investigation on a screen or board using a document camera (or take a picture of it and project the picture on a screen or board). Point to the section and tell the students, "You can create a table, graph, or other representation as a way to analyze your data. You can make your table, graph, or other representation in this section."
3. Ask the students, "What information do we need to include in this analysis?"
4. Tell the students, "Please take a few minutes to discuss this question with your group, and be ready to share."
5. Give the students five minutes to discuss.
6. Ask two or three different groups to share their answers. Highlight or write down any important ideas on the board so students can refer to them later.
7. Tell the students, "I am now going to give you and the other members of your group about 10 minutes to analyze your data." If the students are having trouble analyzing their data, you can take a few minutes to provide a mini-lesson about possible ways to analyze the data they collected (this strategy is called just-in-time instruction because it is offered only when students get stuck).

Teacher Notes

8. Give the students 10 minutes to analyze their data. As they work, move from group to group to check in, ask probing questions, and offer suggestions.

9. Tell the students, "I am now going to give you and the other members of your group about 15 minutes to create an argument to share what you have learned and convince others that they should believe you. Before you do that, we need to take a few minutes to discuss what you need to include in your argument."

10. If possible, use a document camera to project the "Argument Presentation on a Whiteboard" image from the "Draft Argument" section of the Investigation Handout (or the Investigation Log in their workbook) on a screen or board (or take a picture of it and project the picture on a screen or board).

11. Point to the box labeled "The Guiding Question:" and tell the students, "You can put the question we are trying to answer here on your whiteboard."

12. Point to the box labeled "Our Claim:" and tell the students, "You can put your claim here on your whiteboard. The claim is your answer to the guiding question."

What should the table, graph, or other visualization for this investigation look like?

There are a number of different ways that students can analyze the qualitative data they collect during this investigation. One of the most straightforward ways is to create a four-column table. With this approach, the heading for column 1 could be "solutions mixed," the heading for column 2 could be "color change," the heading for column 3 could be "gas produced," and the heading for column 4 could be "solid produced" (see Figure 4.2 for an example of an argument that uses a four-column table but with different headings).

Another way to analyze the data is to create a reaction matrix, which is a type of data table that allows you to document what happens when you react a series of chemicals with each other. The reaction matrix accounts for all the different possible reactions within the set of substances you have available. With this approach, students create a 4 × 4 table—that is, four rows by four columns. Each row is labeled as a different solution (calcium chloride, magnesium sulfate, sodium bicarbonate, and sodium carbonate). Each column is also labeled as a different solution. Each cell in the table then contains what was observed after the liquids labeled in that column and that row were mixed together.

There are other options for analyzing the collected data. Students often come up with some unique ways of analyzing their data, so be sure to give them some voice and choice during this stage.

Investigation 4. Chemical Reactions: Which Pairs of the Available Liquids Produce a New Substance When They Are Mixed Together?

13. Point to the box labeled "Our Evidence:" and tell the students, "You can put the evidence that you are using to support your claim here on your whiteboard. Your evidence will need to include the analysis you just did and an explanation of what your analysis means or shows. Scientists always need to support their claims with evidence."

14. Point to the box labeled "Our Justification of the Evidence:" and tell the students, "You can put your justification of your evidence here on your whiteboard. Your justification needs to explain why your evidence is important. Scientists often use core ideas to explain why the evidence they are using matters. Core ideas are important concepts that scientists use to help them make sense of what happens during an investigation."

15. Ask the students, "What are some core ideas that we read about earlier that might help us explain why the evidence we are using is important?"

16. Ask the students to share some of the core ideas from the "Introduction" section of the Investigation Handout (or the Investigation Log in the workbook). List these core ideas on the board.

17. Tell the students, "That is great. I would like to see everyone try to include these core ideas in your justification of the evidence. Your goal is to use these core ideas to help explain why your evidence matters and why the rest of us should pay attention to it."

18. Ask the students, "Do you have any questions about what you need to do?"

19. Answer any questions that come up.

20. Tell the students, "Okay, go ahead and start working on your arguments. You need to have your argument done in the next 15 minutes. It doesn't need to be perfect. We just need something down on the whiteboards so we can share our ideas."

21. Give the students 15 minutes to work in their groups on their arguments. As they work, move from group to group to check in, ask probing questions, and offer a suggestion if a group gets stuck. Figure 4.2 shows an example

FIGURE 4.2
Example of an argument

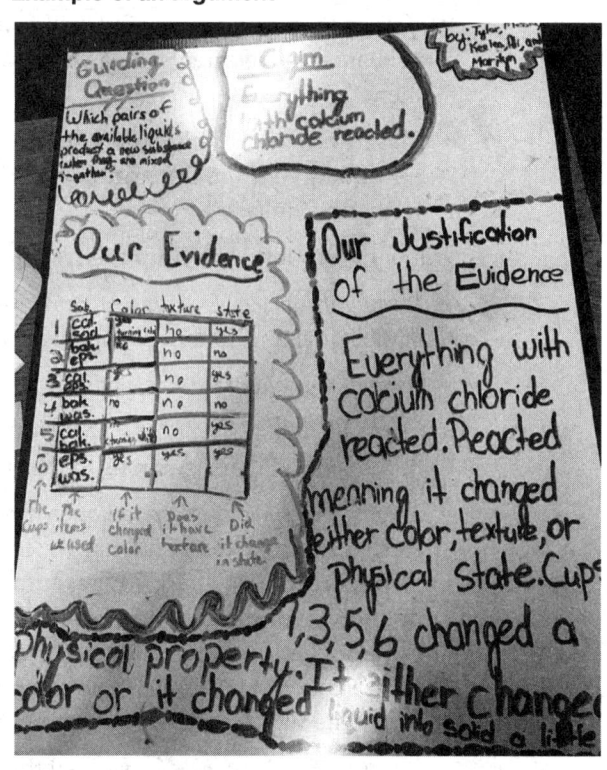

Teacher Notes

of an argument created by students for this investigation. Notice, however, that the information that is included in the "Our Justification of the Evidence" section of this argument is really the student's interpretation of their analysis of the data they collected (which is the table in the evidence section of the whiteboard). A more appropriate justification of the evidence would include ideas such as "a chemical reaction happens when two or more substances are mixed together and a new substance with different properties is formed" and "we can use physical properties to tell the difference between two or more substances."

Stage 4: Argumentation Session (30 minutes)

The argumentation session can be conducted in a whole-class presentation format, a gallery walk format, or a modified gallery walk format. We recommend using a whole-class presentation format for the first investigation, but try to transition to either the gallery walk or modified gallery walk format as soon as possible because that will maximize student voice and choice inside the classroom. The following list shows the steps for the three formats; unless otherwise noted, the steps are the same for all three formats.

1. Begin by introducing the use of the whiteboard.
   - *If using the whole-class presentation format,* tell the students, "We are now going to share our arguments. Please set up your whiteboards so everyone can see them."
   - *If using the gallery walk or modified gallery walk format,* tell the students, "We are now going to share our arguments. Please set up your whiteboards so they are facing the walls."

2. Allow the students to set up their whiteboards.
   - *If using the whole-class presentation format,* the whiteboards should be set up on stands or chairs so they are facing toward the center of the room.
   - *If using the gallery walk or modified gallery walk format,* the whiteboards should be set up on stands or chairs so they are facing toward the outside of the room.

3. Give the following instructions to the students:
   - *If using the whole-class presentation format,* tell the students, "Okay, before we get started I want to explain what we are going to do next. Your group will have an opportunity to share your argument with the rest of the class. After you are done, everyone else in the class will have a chance to ask questions and offer some suggestions about ways to make your group's argument better. After we have a chance to listen to each other and learn something new, I'm going to give you some time to revise your arguments and make them better."
   - *If using the gallery walk format,* tell the students, "Okay, before we get started I want to explain what we are going to do next. You are going to read the arguments that were created by other groups. When I say 'go,' your group will

go to a different group's station so you can see their argument. Once you are there, I'll give your group a few minutes to read and review their argument. Your job is to offer them some suggestions about ways to make their argument better. You can use sticky notes to give them suggestions. Please be specific about what you want to change and how you think they should change it. After we have a chance to learn from each other, I'm going to give you some time to revise your arguments and make them better."

- *If using the modified gallery walk format,* tell the students, "Okay, before we get started I want to explain what we are going to do next. I'm going to ask some of you to present your arguments to your classmates. If you are presenting your argument, your job is to share your group's claim, evidence, and justification of the evidence. The rest of you will be travelers. If you are a traveler, your job is to listen to the presenters, ask the presenters questions if you do not understand something, and then offer them some suggestions about ways to make their argument better. After we have a chance to learn from each other, I'm going to give you some time to revise your arguments and make them better."

4. Use a document camera to project the "Ways to IMPROVE our argument …" box from the Investigation Handout (or the Investigation Log in their workbook) on a screen or board (or take a picture of it and project the picture on a screen or board).

- *If using the whole-class presentation format,* point to the box and tell the students, "After your group presents your argument, you can write down the suggestions you get from your classmates here. If you are listening to a presentation and you see a good idea from another group, you can write down that idea here as well. Once we are done with the presentations, I will give you a chance to use these suggestions or ideas to improve your arguments."

- *If using the gallery walk format,* point to the box and tell the students, "If you see a good idea from another group, you can write it down here. Once we are done reviewing the different arguments, I will give you a chance to use these ideas to improve your own arguments. It is important to share ideas like this."

- *If using the modified gallery walk format,* point to the box and tell the students, "If you are a presenter, you can write down the suggestions you get from the travelers here. If you are a traveler and you see a good idea from another group, you can write down that idea here. Once we are done with the presentations, I will give you a chance to use these suggestions or ideas to improve your arguments."

5. Ask the students, "Do you have any questions about what you need to do?"

6. Answer any questions that come up.

7. Give the following instructions:

Teacher Notes

- *If using the whole-class presentation format,* tell the students, "Okay. Let's get started."
- *If using the gallery walk format,* tell the students, "Okay, I'm now going to tell you which argument to go to and review."
- *If using the modified gallery walk format,* tell the students, "Okay, I'm now going to assign you to be a presenter or a traveler." Assign one or two students from each group to be presenters and one or two students from each group to be travelers.

8. Give the students an opportunity to review the arguments.

- *If using the whole-class presentation format,* have each group present their argument one at a time. Give each group only two to three minutes to present their argument. Then give the class two to three minutes to ask them questions and offer suggestions. Encourage as much participation from the students as possible.
- *If using the gallery walk format,* tell the students, "Okay. Let's get started. Each group, move one argument to the left. Don't move to the next argument until I tell you to move. Once you get there, read the argument and then offer suggestions about how to make it better. I will put some sticky notes next to each argument. You can use the sticky notes to leave your suggestions." Give each group about three to four minutes to read the arguments, talk, and offer suggestions.

    a. After three to four minutes, tell the students, "Okay. Let's move on to the next argument. Please move one group to the left."

    b. Again, give each group three to four minutes to read, talk, and offer suggestions.

    c. Repeat this process until each group has had their argument read and critiqued three times.

- *If using the modified gallery walk format,* tell the students, "Okay. Let's get started. Reviewers, move one group to the left. Don't move to the next group until I tell you to move. Presenters, go ahead and share your argument with the travelers when they get there." Give each group of presenters and travelers about three to four minutes to talk.

    a. Tell the students, "Okay. Let's move on to the next argument. Travelers, move one group to the left."

    b. Again, give each group of presenters and travelers about three to four minutes to talk.

    c. Repeat this process until each group has had their argument read and critiqued three times.

9. Tell the students to return to their workstations.

10. Give the following instructions about revising the argument:

Investigation 4. Chemical Reactions: Which Pairs of the Available Liquids Produce a New Substance When They Are Mixed Together?

- *If using the whole-class presentation format,* tell the students, "I'm now going to give you all about 10 minutes to revise your argument. Take a few minutes to talk in your groups and determine what you want to change to make your argument better. Once you have decided what to change, go ahead and make the changes to your whiteboard."

- *If using the gallery walk format,* tell the students, "I'm now going to give you all about 10 minutes to revise your argument. Take a few minutes to read the suggestions that were left at your argument. Then talk in your groups and determine what you want to change to make your argument better. Once you have decided what to change, go ahead and make the changes to your whiteboard."

- *If using the modified gallery walk format,* tell the students, "I'm now going to give you all about 10 minutes to revise your argument. Please return to your original groups." Wait for the students to move back into their original groups and then tell the students, "Okay, take a few minutes to talk in your groups and determine what you want to change to make your argument better. Once you have decided what to change, go ahead and make the changes to your whiteboard."

11. Ask the students, "Do you have any questions about what you need to do?"

12. Answer any questions that come up.

13. Tell the students, "Okay. Let's get started."

14. Give the students 10 minutes to work in their groups on their arguments. As they work, move from group to group to check in, ask probing questions, and offer a suggestion if a group gets stuck.

Stage 5: Reflective Discussion (15 minutes)

1. Tell the students, "We are now going to take a minute to talk about some of the core ideas and crosscutting concepts that we have used during our investigation."

2. Show Figure 4.3 on the screen.

3. Ask the students, "What do you all see going on here?"

4. Allow the students to share their ideas.

5. Ask the students, "How can we use what we know about the chemical reactions to explain what is going on here?"

FIGURE 4.3

A precipitate reaction

Note: A full-color version of this figure is available on the book's Extras page at *www.nsta.org/adi-5th*.

Teacher Notes

6. Allow the students to share their ideas. As they share their ideas, ask different questions to encourage them to expand on their thinking (e.g., "Can you tell me more about that?"), clarify a contribution (e.g., "Can you say that in another way?"), support an idea (e.g., "Why do you think that?"), add to an idea mentioned by a classmate (e.g., "Would anyone like to add to the idea?"), re-voice an idea offered by a classmate (e.g., "Who can explain that to me in another way?"), or critique an idea during the discussion (e.g., "Do you agree or disagree with that idea and why?") until students are able to generate an adequate explanation.

7. Tell the students, "We had to think about the cause of chemical reactions and the effect of chemical reactions when designing our investigations." Then ask, "Can anyone tell me why this was important?"

8. Allow the students to share their ideas.

9. Tell the students, "I think using standard units to measure and describe physical properties such as color or texture is important because we need to have a common understanding of what our observations mean. This allows us to share and critique each other's ideas and evidence."

10. Ask the students, "How did you try to standardize your observations today?"

11. Tell the students, "Scientists often use standard units to describe the properties of matter." Then ask, "Can anyone tell me why this is important?"

12. Allow the students to share their ideas.

13. Show an image of the question "What do you think are the most important core ideas or crosscutting concepts that we used during this investigation to help us make sense of what we observed?" Tell the students, "Okay, let's make sure we are all on the same page. Please take a moment to discuss this question with the other people in your group." Give them a few minutes to discuss the question.

14. Ask the students, "What do you all think? Who would like to share?"

15. Allow the students to share their ideas.

16. Tell the students, "We are now going to take a minute to talk about the types of questions that scientists investigate."

17. Show Figure 4.3 on the screen along with the question "Is the result of this reaction cool?" on the screen.

18. Tell the students, "Take a few minutes to talk in your group about whether or not this is a scientific question and why. Be ready to share with the rest of the class." Give the students two to three minutes to talk in their group.

19. Ask the students, "What do you all think? Who would like to share an idea?"

20. Allow the students to share their ideas.

Investigation 4. Chemical Reactions: Which Pairs of the Available Liquids Produce a New Substance When They Are Mixed Together?

21. Tell the students, "I think this is not a scientific question because scientists do not attempt to answer questions about whether something is cool, fun, or exciting. What scientists do attempt to answer are questions about what can happen, why things happen, or how things happen in the natural or material world. They do not attempt to answer questions about what should happen, because that would require consideration of ethical values, moral values, and political issues. Not all questions can be answered by science. Science and technology may raise ethical issues for which science, by itself, does not provide answers and solutions."

22. Ask the students, "What might be a more scientific question that we could ask about this?"

23. Allow the students to share their ideas.

24. Ask the students, "Does anyone have any questions about the types of questions that scientists can investigate?"

25. Answer any questions that come up.

26. Tell the students, "We are now going to take a minute to talk about what went well and what didn't go so well during our investigation. We need to talk about this because you all are going to be planning and carrying out your own investigations like this a lot this year, and I want to help you all get better at it."

27. Show an image of the question "What made your investigation scientific?" on the screen. Tell the students, "Take a few minutes to talk about how you would answer this question with the other people in your group. Be ready to share with the rest of the class." Give the students two to three minutes to talk in their group.

28. Ask the students, "What do you all think? Who would like to share an idea?"

29. Allow the students to share their ideas. Be sure to expand on their ideas about what makes an investigation scientific.

30. Show an image of the question "What made your investigation not so scientific?" on the screen. Tell the students, "Take a few minutes to talk about how you would answer this question with the other people in your group. Be ready to share with the rest of the class." Give the students two to three minutes to talk in their group.

31. Ask the students, "What do you all think? Who would like to share an idea?"

32. Allow the students to share their ideas. Be sure to expand on their ideas about what makes an investigation less scientific.

33. Show an image of the question "What rules can we put into place to help us make sure our next investigation is more scientific?" on the screen. Tell the students, "Take a few minutes to talk about how you would answer this question with the other people in your group. Be ready to share with the rest of the class." Give the students two to three minutes to talk in their group.

Teacher Notes

34. Ask the students, "What do you all think? Who would like to share an idea?"

35. Allow the students to share their ideas. Once they have shared their ideas, offer a suggestion for a possible class rule.

36. Ask the students, "What do you all think? Should we make this a rule?"

37. If the students agree, write the rule on the board or on a class "Rules for Scientific Investigation" chart so you can refer to it during the next investigation.

Stage 6: Write a Draft Report (30 minutes)

Your students will use either the Investigation Handout or the Investigation Log in the student workbook when writing the draft report. When you give the directions shown in quotes in the following steps, substitute "Investigation Log in your workbook" or just "Investigation Log" (as shown in brackets) for "handout" if they are using the workbook.

1. Tell the students, "You are now going to write an investigation report to share what you have learned. Please take out a pencil and turn to the 'Draft Report' section of your handout [Investigation Log in your workbook]."

2. If possible, use a document camera to project the "Introduction" section of the draft report from the Investigation Handout (or the Investigation Log in their workbook) on a screen or board (or take a picture of it and project the picture on a screen or board).

3. Tell the students, "The first part of the report is called the 'Introduction.' In this section of the report you want to explain to the reader what you were investigating, why you were investigating it, and what question you were trying to answer. All this information can be found in the text at the beginning of your handout [Investigation Log]." Point to the image. "Here are some sentence starters to help you begin writing."

4. Ask the students, "Do you have any questions about what you need to do?"

5. Answer any questions that come up.

6. Tell the students, "Okay, let's write."

7. Give the students 10 minutes to write the "Introduction" section of the report. As they work, move from student to student to check in, ask probing questions, and offer a suggestion if a student gets stuck.

8. If possible, use a document camera to project the "Method" section of the draft report from the Investigation Handout (or the Investigation Log in their workbook) on a screen or board (or take a picture of it and project the picture on a screen or board).

9. Tell the students, "The second part of the report is called the 'Method.' In this section of the report you want to explain to the reader what you did during the investigation, what data you collected and why, and how you went

Investigation 4. Chemical Reactions: Which Pairs of the Available Liquids Produce a New Substance When They Are Mixed Together?

about analyzing your data. All this information can be found in the 'Plan Your Investigation' section of the handout [Investigation Log]. Remember that you all planned and carried out different investigations, so do not assume that the reader will know what you did." Point to the image. "Here are some sentence starters to help you begin writing."

10. Ask the students, "Do you have any questions about what you need to do?"
11. Answer any questions that come up.
12. Tell the students, "Okay, let's write."
13. Give the students 10 minutes to write the "Method" section of the report. As they work, move from student to student to check in, ask probing questions, and offer a suggestion if a student gets stuck.
14. If possible, use a document camera to project the "Argument" section of the draft report from the Investigation Handout (or the Investigation Log in their workbook) on a screen or board (or take a picture of it and project the picture on a screen or board).
15. Tell the students, "The last part of the report is called the 'Argument.' In this section of the report you want to share your claim, evidence, and justification of the evidence with the reader. All this information can be found on your whiteboard." Point to the image. "Here are some sentence starters to help you begin writing."
16. Ask the students, "Do you have any questions about what you need to do?"
17. Answer any questions that come up.
18. Tell the students, "Okay, let's write."
19. Give the students 10 minutes to write the "Argument" section of the report. As they work, move from student to student to check in, ask probing questions, and offer a suggestion if a student gets stuck.

Stage 7: Peer Review (35 minutes)

Your students will use either the Investigation Handout or their workbook when doing the peer review. Except where noted below, the directions are the same whether using the handout or the workbook.

1. Tell the students, "We are now going to review our reports to find ways to make them better. I'm going to come around and collect your draft reports. While I do that, please take out a pencil."
2. Collect the handouts or the workbooks with the draft reports from the students.
3. If possible, use a document camera to project the peer-review guide (see Appendix 4) on a screen or board (or take a picture of it and project the picture on a screen or board).

Teacher Notes

4. Tell the students, "We are going to use this peer-review guide to give each other feedback." Point to the image.

5. Tell the students, "I'm going to ask you to work with a partner to do this. I'm going to give you and your partner a draft report to read. You two will then read the report together. Once you are done reading the report, I want you to answer each of the questions on the peer-review guide." Point to the review questions on the image of the peer-review guide.

6. Tell the students, "You can check 'no,' 'almost,' or 'yes' after each question." Point to the checkboxes on the image of the peer-review guide.

7. Tell the students, "This will be your rating for this part of the report. Make sure you agree on the rating you give the author. If you mark 'no' or 'almost,' then you need to tell the author what he or she needs to do to get a 'yes.'" Point to the space for the reviewer feedback on the image of the peer-review guide.

8. Tell the students, "It is really important for you to give the authors feedback that is helpful. That means you need to tell them exactly what they need to do to make their report better."

9. Ask the students, "Do you have any questions about what you need to do?"

10. Answer any questions that come up.

11. Tell the students, "Please sit with a partner who is not in your current group." Allow the students time to sit with a partner.

12. Tell the students, "Okay, I'm now going to give you one report to read." Pass out one Investigation Handout with a draft report or one workbook to each pair. Make sure that the report you give a pair was not written by one of the students in that pair. Give each pair one peer-review guide to fill out. If the students are using workbooks, the peer-review guide is included right after the draft report so you do not need to pass out copies of the peer-review guide.

13. Tell the students, "Okay, I'm going to give you 15 minutes to read the report I gave you and to fill out the peer-review guide. Go ahead and get started."

14. Give the students 15 minutes to work. As they work, move around from pair to pair to check in and see how things are going, answer questions, and offer advice.

15. After 15 minutes pass, tell the students, "Okay, time is up. Please give me the report and the peer-review guide that you filled out."

16. Collect the Investigation Handouts and the peer-review guides, or collect the workbooks if students are using them. If the students are using the Investigation Handouts and separate peer-review guides, be sure you keep each handout with its corresponding peer-review guide.

17. Tell the students, "Okay, I am now going to give you a different report to read and a new peer-review guide to fill out." Pass out one more report to each pair.

Investigation 4. Chemical Reactions: Which Pairs of the Available Liquids Produce a New Substance When They Are Mixed Together?

Make sure that the report you give a pair was not written by one of the students in that pair. Give each pair a new peer-review guide to fill out as a group.

18. Tell the students, "Okay, I'm going to give you 15 minutes to read this new report and to fill out the peer-review guide. Go ahead and get started."

19. Give the students 15 minutes to work. As they work, move around from pair to pair to check in and see how things are going, answer questions, and offer advice.

20. After 15 minutes pass, tell the students, "Okay, time is up. Please give me the report and the peer-review guide that you filled out."

21. Collect the Investigation Handouts and the peer-review guides, or collect the workbooks if students are using them. If the students are using the Investigation Handouts and separate peer-review guides, be sure you keep each handout with its corresponding peer-review guide.

Stage 8: Revise the Report (30 minutes)

Your students will use either the Investigation Handout or their workbook when revising the report. Except where noted below, the directions are the same whether using the handout or the workbook.

1. Tell the students, "You are now going to revise your draft report based on the feedback you get from your classmates. Please take out a pencil."

2. Return the reports to the students.

    - *If the students used the Investigation Handout and a copy of the peer-review guide,* pass back the handout and the peer-review guide to each student.

    - *If the students used the workbook,* pass that back to each student.

3. Tell the students, "Please take a few minutes to read over the peer-review guide. You should use it to figure out what you need to change in your report and how you will change it."

4. Allow the students to read the peer-review guide.

5. *If the students used the workbook,* if possible use a document camera to project the "Write Your Final Report" section from the Investigation Log on a screen or board (or take a picture of it and project the picture on a screen or board).

6. Give the following directions about how to revise their reports:

    - If the students used the Investigation Handout and a copy of the peer-review guide, tell them, "Okay, let's revise our reports. Please take out a piece of paper. I would like you to rewrite your report. You can use your draft report as a starting point, but you also need to change it to make it better. Use the feedback on the peer-review guide to make it better."

Teacher Notes

- If the students used the workbook, tell them, "Okay, let's revise our reports. I would like you to rewrite your report in the section of the Investigation Log called "Write Your Final Report." You can use your draft report as a starting point, but you also need to change it to make it better. Use the feedback on the peer-review guide to make it better."

7. Ask the students, "Do you have any questions about what you need to do?"
8. Answer any questions that come up.
9. Tell the students, "Okay, let's write." Allow about 20 minutes for the students to revise their reports.
10. After about 20 minutes, give the following directions:
    - *If the students used the Investigation Handout,* tell them, "Okay, time's up. I will now come around and collect your Investigation Handout, the peer-review guide, and your final report."
    - *If the students used the workbook,* tell them, "Okay, time's up. I will now come around and collect your workbooks."
11. *If the students used the Investigation Handout,* collect all the Investigation Handouts, peer-review guides, and final reports. *If the students used the workbook,* collect all the workbooks.
12. *If the students used the Investigation Handout,* use the "Teacher Score" column in the peer-review guide to grade the final report. *If the students used the workbook,* use the "Investigation Report Grading Rubric" in the Investigation Log to grade the final report. Whether you are using the handout or the log, you can give the students feedback about their writing in the "Teacher Comments" section.

How to Use the Checkout Questions

The Checkout Questions are an optional assessment. We recommend giving them to students at the start of the next class period after the students finish stage 8 of the investigation. You can then look over the student answers to determine if you need to reteach the core idea from the investigation. Appendix 6 gives the answers to the Checkout Questions that should be given by a student who can apply the core idea correctly and can explain (1) the cause-and-effect relationship and (2) the importance of using standard units to measure or describe the physical quantities of matter.

Alignment With Standards

Table 4.3 highlights how the investigation can be used to address specific performance expectations from the *Next Generation Science Standards, Common Core State Standards for English Language Arts (CCSS ELA),* and *English Language Proficiency (ELP) Standards.*

Investigation 4. Chemical Reactions: Which Pairs of the Available Liquids Produce a New Substance When They Are Mixed Together?

TABLE 4.3

Investigation 4 alignment with standards

NGSS performance expectations	• 5-PS1-3: Make observations and measurements to identify materials based on their properties. • 5-PS1-4: Conduct an investigation to determine whether the mixing of two or more substances results in new substances.
CCSS ELA—Reading: Informational Text	Key ideas and details • CCSS.ELA-LITERACY.RI.5.1: Quote accurately from a text when explaining what the text says explicitly and when drawing inferences from the text. • CCSS.ELA-LITERACY.RI.5.2: Determine two or more main ideas of a text and explain how they are supported by key details; summarize the text. • CCSS.ELA-LITERACY.RI.5.3: Explain the relationships or interactions between two or more individuals, events, ideas, or concepts in a historical, scientific, or technical text based on specific information in the text. Craft and structure • CCSS.ELA-LITERACY.RI.5.4: Determine the meaning of general academic and domain-specific words and phrases in a text relevant to a *grade 5 topic or subject area*. • CCSS.ELA-LITERACY.RI.5.5: Compare and contrast the overall structure (e.g., chronology, comparison, cause/effect, problem/solution) of events, ideas, concepts, or information in two or more texts. • CCSS.ELA-LITERACY.RI.5.6: Analyze multiple accounts of the same event or topic, noting important similarities and differences in the point of view they represent. Integration of knowledge and ideas • CCSS.ELA-LITERACY.RI.5.7: Draw on information from multiple print or digital sources, demonstrating the ability to locate an answer to a question quickly or to solve a problem efficiently. • CCSS.ELA-LITERACY.RI.5.8: Explain how an author uses reasons and evidence to support particular points in a text, identifying which reasons and evidence support which point(s). Range of reading and level of text complexity • CCSS.ELA-LITERACY.RI.5.10: By the end of the year, read and comprehend informational texts, including history/social studies, science, and technical texts, at the high end of the grades 4–5 text complexity band independently and proficiently.

Continued

Teacher Notes

Table 4.3 (*continued*)

CCSS ELA—**Writing**	Text types and purposes
	• CCSS.ELA-LITERACY.W.5.1: Write opinion pieces on topics or texts, supporting a point of view with reasons.
	○ CCSS.ELA-LITERACY.W.5.1.A: Introduce a topic or text clearly, state an opinion, and create an organizational structure in which ideas are logically grouped to support the writer's purpose.
	○ CCSS.ELA-LITERACY.W.5.1.B: Provide logically ordered reasons that are supported by facts and details.
	○ CCSS.ELA-LITERACY.W.5.1.C: Link opinion and reasons using words, phrases, and clauses (e.g., *consequently*, *specifically*).
	○ CCSS.ELA-LITERACY.W.5.1.D: Provide a concluding statement or section related to the opinion presented.
	• CCSS.ELA-LITERACY.W.5.2: Write informative or explanatory texts to examine a topic and convey ideas and information clearly.
	○ CCSS.ELA-LITERACY.W.5.2.A: Introduce a topic clearly, provide a general observation and focus, and group related information logically; include formatting (e.g., headings), illustrations, and multimedia when useful to aiding comprehension.
	○ CCSS.ELA-LITERACY.W.5.2.B: Develop the topic with facts, definitions, concrete details, quotations, or other information and examples related to the topic.
	○ CCSS.ELA-LITERACY.W.5.2.C: Link ideas within and across categories of information using words, phrases, and clauses (e.g., *in contrast*, *especially*).
	○ CCSS.ELA-LITERACY.W.5.2.D: Use precise language and domain-specific vocabulary to inform about or explain the topic.
	○ CCSS.ELA-LITERACY.W.5.2.E: Provide a concluding statement or section related to the information or explanation presented.
	Production and distribution of writing
	• CCSS.ELA-LITERACY.W.5.4: Produce clear and coherent writing in which the development and organization are appropriate to task, purpose, and audience.
	• CCSS.ELA-LITERACY.W.5.5: With guidance and support from peers and adults, develop and strengthen writing as needed by planning, revising, editing, rewriting, or trying a new approach.
	• CCSS.ELA-LITERACY.W.5.6: With some guidance and support from adults, use technology, including the internet, to produce and publish writing as well as to interact and collaborate with others; demonstrate sufficient command of keyboarding skills to type a minimum of two pages in a single sitting.

Continued

Investigation 4. Chemical Reactions: Which Pairs of the Available Liquids Produce a New Substance When They Are Mixed Together?

Table 4.3 (*continued*)

CCSS ELA—Writing (*continued*)	Research to build and present knowledge • CCSS.ELA-LITERACY.W.5.8: Recall relevant information from experiences or gather relevant information from print and digital sources; summarize or paraphrase information in notes and finished work, and provide a list of sources. • CCSS.ELA-LITERACY.W.5.9: Draw evidence from literary or informational texts to support analysis, reflection, and research. Range of writing • CCSS.ELA-LITERACY.W.5.10: Write routinely over extended time frames (time for research, reflection, and revision) and shorter time frames (a single sitting or a day or two) for a range of discipline-specific tasks, purposes, and audiences.

CCSS ELA—Speaking and Listening	Comprehension and collaboration
	• CCSS.ELA-LITERACY.SL.5.1: Engage effectively in a range of collaborative discussions (one-on-one, in groups, and teacher-led) with diverse partners on *grade 5 topics and texts,* building on others' ideas and expressing their own clearly.
	  ○ CCSS.ELA-LITERACY.SL.5.1.A: Come to discussions prepared, having read or studied required material; explicitly draw on that preparation and other information known about the topic to explore ideas under discussion.
	  ○ CCSS.ELA-LITERACY.SL.5.1.B: Follow agreed-upon rules for discussions and carry out assigned roles.
	  ○ CCSS.ELA-LITERACY.SL.5.1.C: Pose and respond to specific questions by making comments that contribute to the discussion and elaborate on the remarks of others.
	  ○ CCSS.ELA-LITERACY.SL.5.1.D: Review the key ideas expressed and draw conclusions in light of information and knowledge gained from the discussions.
	• CCSS.ELA-LITERACY.SL.5.2: Summarize a written text read aloud or information presented in diverse media and formats, including visually, quantitatively, and orally.
	• CCSS.ELA-LITERACY.SL.5.3: Summarize the points a speaker makes and explain how each claim is supported by reasons and evidence.
	Presentation of knowledge and ideas
	• CCSS.ELA-LITERACY.SL.5.4: Report on a topic or text or present an opinion, sequencing ideas logically and using appropriate facts and relevant, descriptive details to support main ideas or themes; speak clearly at an understandable pace.
	• CCSS.ELA-LITERACY.SL.5.5: Include multimedia components (e.g., graphics, sound) and visual displays in presentations when appropriate to enhance the development of main ideas or themes.
	• CCSS.ELA-LITERACY.SL.5.6: Adapt speech to a variety of contexts and tasks, using formal English when appropriate to task and situation.

Continued

Teacher Notes

Table 4.3 (*continued*)

ELP Standards	Receptive modalities
	• ELP 1: Construct meaning from oral presentations and literary and informational text through grade-appropriate listening, reading, and viewing.
	• ELP 8: Determine the meaning of words and phrases in oral presentations and literary and informational text.
	Productive modalities
	• ELP 3: Speak and write about grade-appropriate complex literary and informational texts and topics.
	• ELP 4: Construct grade-appropriate oral and written claims and support them with reasoning and evidence.
	• ELP 7: Adapt language choices to purpose, task, and audience when speaking and writing.
	Interactive modalities
	• ELP 2: Participate in grade-appropriate oral and written exchanges of information, ideas, and analyses, responding to peer, audience, or reader comments and questions.
	• ELP 5: Conduct research and evaluate and communicate findings to answer questions or solve problems.
	• ELP 6: Analyze and critique the arguments of others orally and in writing.
	Linguistic structures of English
	• ELP 9: Create clear and coherent grade-appropriate speech and text.
	• ELP 10: Make accurate use of standard English to communicate in grade-appropriate speech and writing.

Investigation Handout

Investigation 4
Chemical Reactions: Which Pairs of the Available Liquids Produce a New Substance When They Are Mixed Together?

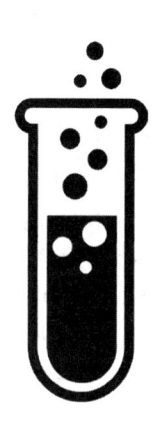

Introduction

A substance is a particular type of matter that has the same properties throughout it. There are many different types of substances, such as baking soda, calcium chloride, and water. Interesting things can happen when you mix different substances together. Watch what happens when your teacher adds baking soda, calcium chloride, and water to a plastic bag and then seals it. As you watch this demonstration, keep track of things you notice and things you wonder about in the boxes below.

Things I NOTICED …	Things I WONDER about …

A chemical reaction took place inside the bag when your teacher added the water to the baking soda and calcium chloride. A *chemical reaction* is a term that scientists use to describe what happens when two or more substances are mixed together and a new substance with different properties is formed. The rearrangement of the particles that make up the substances involved in a chemical reaction creates one or more different substances. In the case of the chemical reaction that you just observed, the particles that made up the baking soda and the particles that made up the calcium chloride separated from each

Investigation Handout

other when the water was added to the bag. The particles then recombined (mixed together again) in a different way and produced four new substances that were not in the bag before the reaction started.

The four new substances that were produced inside the bag because of the chemical reaction were water, salt, calcium carbonate, and carbon dioxide. The new water that was created mixed with the water that was already in the bag. Salt is a white powder at room temperature. Salt dissolves in water, so you could not see it being produced during the reaction. Calcium carbonate is also a white powder that dissolves in water. Carbon dioxide is an invisible gas. The bag filled up like a balloon because of all the carbon dioxide that was produced during the reaction. The bag also got hotter because this chemical reaction released energy as the particles separated from each other and then recombined in a new way.

We often need to be able to determine whether a chemical reaction has happened after we mix two or more substances together. People can tell when a new substance is formed because the new substance will have one or more physical properties that are different from the physical properties of the original substances. A *physical property* is any measurable or observable characteristic of matter. Some physical properties that people often use to identify an unknown substance are color, texture, and physical state (such as solid or liquid). Your goal in this investigation is to figure out if mixing two liquids together in a test tube results in the production of a new substance by looking for changes in the properties of the substances found inside the test tube.

Things we KNOW from what we read ...

Your Task

Use what you know about the properties of matter, how matter interacts with other matter, cause and effect, and the importance of using standard units to measure and describe objects to determine if a new substance is created when two liquids are mixed.

Investigation 4. Chemical Reactions:
Which Pairs of the Available Liquids Produce
a New Substance When They Are Mixed Together?

The *guiding question* of this investigation is, **Which pairs of the available liquids produce a new substance when they are mixed together?**

Materials

You may use any of the following materials during your investigation:

Liquids for mixing

- Calcium chloride (pool salt) in water
- Magnesium sulfate (Epsom salt) in water
- Sodium bicarbonate (baking soda) in water
- Sodium carbonate (washing soda) in water

Equipment

- Safety goggles, nonlatex apron, and vinyl gloves (required)
- Test tube rack
- 6 test tubes

Note: You can only mix two of these four liquids together in a test tube at a time.

Safety Rules

Follow all normal safety rules. In addition, be sure to follow these rules:

- Wear sanitized indirectly vented chemical-splash goggles, nonlatex aprons, and vinyl gloves during setup, investigation activity, and cleanup.
- Do not put any of the liquids in your mouth.
- Immediately clean up any spills to avoid a slip or fall hazard
- Keep water sources away from electrical receptacles to prevent shock.
- Be careful when handling glassware, because it can shatter and cut skin.
- Wash your hands with soap and water when you are done cleaning up.

Plan Your Investigation

Prepare a plan for your investigation by filling out the chart on the next page; this plan is called an *investigation proposal*. Before you start developing your plan, be sure to discuss the following question with the other members of your group:

- What **causes** a chemical reaction and what is the **effect** of that reaction?
- What information should we collect so we can **describe the properties** of a substance?

Investigation Handout

Our guiding question:

This is a picture of how we will set up the equipment:

We will collect the following data:

These are the steps we will follow to collect data:

I approve of this investigation proposal.

_____   _____
Teacher's signature   Date

Investigation 4. Chemical Reactions:
Which Pairs of the Available Liquids Produce
a New Substance When They Are Mixed Together?

Collect Your Data

Keep a record of what you measure or observe during your investigation in the space below.

Analyze Your Data

You will need to analyze the data you collected before you can develop an answer to the guiding question. To analyze the data you collected, create a table, graph, or other visual representation in the space below that shows any changes in the physical properties of the substances.

Investigation Handout

Draft Argument

Develop an argument on a whiteboard. It should include the following:

1. A *claim*: Your answer to the guiding question.
2. *Evidence*: An analysis of the data and an explanation of what the analysis means.
3. A *justification of the evidence*: Why your group thinks the evidence is important.

The Guiding Question:	
Our Claim:	
Our Evidence:	Our Justification of the Evidence:

Argumentation Session

Share your argument with your classmates. Be sure to ask them how to make your draft argument better. Keep track of their suggestions in the space below.

Ways to IMPROVE our argument …

Investigation 4. Chemical Reactions: Which Pairs of the Available Liquids Produce a New Substance When They Are Mixed Together?

Draft Report

Prepare an *investigation report* to share what you have learned. Use the information in this handout and your group's final argument to write a *draft* of your investigation report.

Introduction

We have been studying _____ in class.

Before we started this investigation, we explored _____

_____

_____

We noticed _____

_____

_____

My goal for this investigation was to figure out _____

_____

_____

The guiding question was _____

_____

_____

Method

To gather the data I needed to answer this question, I _____

_____

_____

_____

_____

Investigation Handout

I then analyzed the data I collected by _____

_____

_____

_____

Argument

My claim is _____

_____

_____

_____

The _____ below shows _____

_____

_____

_____

_____

_____

Investigation 4. Chemical Reactions:
Which Pairs of the Available Liquids Produce
a New Substance When They Are Mixed Together?

This analysis of the data I collected suggests _____

_____

_____

_____

_____

This evidence is based on several important scientific concepts. The first one is _____

_____

_____

_____

_____

Review

Your classmates need your help! Review the draft of their investigation reports and give them ideas about how to improve. Use the *peer-review guide* when doing your review.

Submit Your Final Report

Once you have received feedback from your classmates about your draft report, create your final investigation report and hand it in to your teacher.

Checkout Questions

Investigation 4. Chemical Reactions

Use the following information to answer questions 1 and 2. The picture below shows substance A and substance B before mixing (left) and what happened after these substances were mixed together in a cup (right).

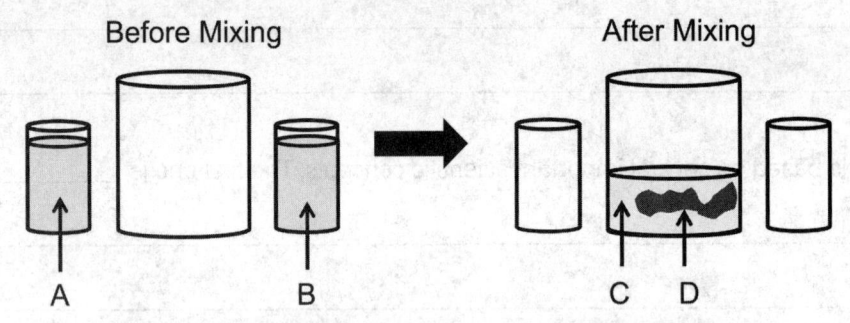

Before Mixing → **After Mixing**

A   B   C D

The table below shows the physical properties of the labeled substances in the picture.

Substance	Color	Texture	State
A	Clear	Smooth	Liquid
B	Clear	Smooth	Liquid
C	Clear	Smooth	Liquid
D	Green	Squishy	Solid

1. Was a new substance produced when substance A and substance B were mixed together?

    a. Yes

    b. No

    c. Unable to tell

2. How do you know? Use what you know about cause and effect to explain your answer.

    _____
    _____
    _____
    _____
    _____

Teacher Scoring Rubric for Checkout Questions 1 and 2

Level	Description
3	The student can apply the core idea correctly and can explain the cause-and-effect relationship.
2	The student can apply the core idea correctly but cannot explain the cause-and-effect relationship.
1	The student cannot apply the core idea correctly but can explain the cause-and-effect relationship.
0	The student cannot apply the core idea correctly and cannot explain the cause-and-effect relationship.

Investigation 4. Chemical Reactions: Which Pairs of the Available Liquids Produce a New Substance When They Are Mixed Together?

Use the following information to answer questions 3 and 4. The picture below shows substance E and substance F before mixing (left) and what happened after these substances were mixed together in a cup (right).

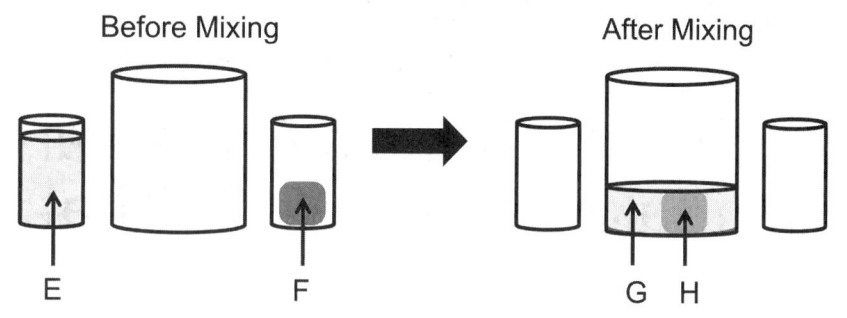

Before Mixing → After Mixing

E   F   G H

The table below shows the physical properties of the labeled substances in the picture.

Substance	Color	State	Mass	Volume
E	Clear	Liquid	11 g	10 ml
F	Gray	Solid	14 g	2 ml
G	Clear	Liquid	11 g	10 ml
H	Gray	Solid	14 g	2 ml

3. Was a new substance produced when substance E and substance F were mixed together?

    a. Yes

    b. No

    c. Unable to tell

4. How do you know? Use what you know about using standard units to measure and describe the physical quantities of matter to help explain your answer.

_____

_____

_____

_____

Teacher Scoring Rubric for Checkout Questions 3 and 4

Level	Description
3	The student can apply the core idea correctly and can explain the importance of using standard units to measure or describe the physical quantities of matter.
2	The student cannot apply the core idea correctly but can explain the importance of using standard units to measure or describe the physical quantities of matter.
1	The student can apply the core idea correctly but cannot explain the importance of using standard units to measure or describe the physical quantities of matter.
0	The student cannot apply the core idea correctly and cannot explain the importance of using standard units to measure or describe the physical quantities of matter.

Argument-Driven Inquiry in **Fifth-Grade Science:** Three-Dimensional Investigations

Teacher Notes

Investigation 5

Reactions and Weight: What Happens to the Total Weight of a Closed System When a Chemical Reaction Takes Place Within That System?

Purpose

The purpose of this investigation is to give students an opportunity to use three disciplinary cored ideas (DCIs), two crosscutting concepts (CCs), and seven scientific and engineering practices (SEPs) to figure out how the total weight of a closed system changes when a chemical reaction takes place within that system. Students will also learn about the assumptions that scientists make about order and consistency in nature.

The DCIs, CCs, and SEPs That Students Use During This Investigation to Figure Things Out

DCIs

- *PS1.A: Structure and Properties of Matter:* Matter of any type can be subdivided into particles that are too small to see, but even then, the matter still exists and can be detected by other means.
- *PS1.A: Structure and Properties of Matter:* Measurements of a variety of properties can be used to identify materials.
- *PS1.B: Chemical Reactions:* When two or more different substances ae mixed, a new substance with different properties may be formed.

CCs

- *CC 3: Scale, Proportion, and Quantity:* Standard units are used to measure and describe physical quantities such as weight, time, temperature, and volume.
- *CC 4: Systems and System Models:* A system can be described in terms of its components and their interactions. A system is a group of related parts that make up a whole and can carry out functions its individual parts cannot.

SEPs

- *SEP 1: Asking Questions and Defining Problems:* Ask questions about what would happen if a variable is changed. Ask questions that can be investigated and predict reasonable outcomes based on patterns such as cause-and-effect relationships.

Investigation 5. Reactions and Weight:
What Happens to the Total Weight of a Closed System When
a Chemical Reaction Takes Place Within That System?

- *SEP 3: Planning and Carrying Out Investigations:* Plan and conduct an investigation collaboratively to produce data to serve as the basis for evidence, using fair tests in which variables are controlled and the number of trials considered. Evaluate appropriate methods and/or tools for collecting data.

- *SEP 4: Analyzing and Interpreting Data:* Represent data in tables and/or various graphical displays (bar graphs, pictographs, and/or pie charts) to reveal patterns that indicate relationships. Analyze and interpret data to make sense of phenomena, using logical reasoning, mathematics, and/or computation. Compare and contrast data collected by different groups in order to discuss similarities and differences in their findings.

- *SEP 5: Using Mathematics and Computational Thinking:* Organize simple data sets to reveal patterns that suggest relationships. Describe, measure, estimate, and/or graph quantities (e.g., area, volume, weight, time) to address scientific and engineering questions and problems.

- *SEP 6: Constructing Explanations and Designing Solutions:* Construct an explanation of observed relationships. Use evidence to construct or support an explanation. Identify the evidence that supports particular points in an explanation.

- *SEP 7: Engaging in Argument From Evidence:* Compare and refine arguments based on an evaluation of the evidence presented. Distinguish among facts, reasoned judgment based on research findings, and speculation in an explanation. Respectfully provide and receive critiques from peers about a proposed procedure, explanation, or model by citing relevant evidence and posing specific questions.

- *SEP 8: Obtaining, Evaluating, and Communicating Information:* Read and comprehend grade-appropriate complex texts and/or other reliable media to summarize and obtain scientific and technical ideas. Combine information in written text with that contained in corresponding tables, diagrams, and/or charts to support the engagement in other scientific and/or engineering practices. Communicate scientific and/or technical information orally and/or in written formats, including various forms of media as well as tables, diagrams, and charts.

Other Concepts That Students May Use During This Investigation

Students might also use some of the following concepts:

- Matter is anything that has mass and takes up space.
- Matter is made up of particles called atoms.
- Atoms can attach to other atoms to create molecules.
- A substance is a particular type of matter with uniform properties.
- A mixture contains two or more substances.

Teacher Notes

- A chemical reaction occurs when the atoms that make up one or more substances break apart from each other and then recombine in a different way. The rearrangement of atoms during a chemical reaction creates one or more different substances.
- A closed system is an area that is separated from its surroundings by a barrier that does not let any matter transfer into or out of that system.

What Students Figure Out

The weight (amount) of matter does not change during a chemical reaction.

Background Information About This Investigation for the Teacher

Matter is anything that has mass and takes up space. A substance is a particular type of matter with uniform properties. Baking soda and water are examples of substances. At room temperature, baking soda is a white powder and water is a clear liquid. A mixture contains two or more substances that are not chemically combined. Vinegar is an example of a mixture. Vinegar contains acetic acid and water. Acetic acid is a white powder at room temperature. When acetic acid is added to water, it dissolves but does react with the water.

A chemical reaction occurs when the atoms that make up one or more substances break apart from each other and then recombine in a different way. The rearrangement of atoms during a chemical reaction creates one or more different substances. To illustrate, consider the chemical reaction that takes place when baking soda and vinegar are mixed together. The atoms that made up the baking soda and the atoms that made up the acetic acid in the vinegar will separate from each other. The water in the vinegar is not involved in the reaction. These atoms that made up the acetic acid and the baking soda will then recombine in a different way and create three new substances: water, carbon dioxide, and sodium acetate. The new water that is created during the reaction simply mixes with the water that was already in the vinegar. Carbon dioxide is an invisible gas. The gas will form bubbles in the water. Once the bubbles of gas reach the surface of the water, it will leave the water and mix with the surrounding air. Sodium acetate is a white powder at room temperature. This powder dissolves in water, just like acetic acid does, so we cannot see it being produced during the reaction.

When a chemical reaction takes place between two or more different substances, the total weight of all the substances involved in the reaction does not change during or after the reaction. The weight of substances does not change because matter is neither created nor destroyed during a chemical reaction; it simply transitions from one form to another, such as two liquids turning into a solid, a solid and a liquid turning into a gas, or two liquids turning into a gas and a solid.

The students can test three different reactions during this investigation:

Investigation 5. Reactions and Weight:
What Happens to the Total Weight of a Closed System When
a Chemical Reaction Takes Place Within That System?

1. The reaction between baking soda and vinegar; this reaction, as noted earlier, produces a gas.
2. The reaction between soap and Epsom salt; this reaction produces a precipitate (a solid).
3. The reaction between baking soda and calcium chloride; this reaction produces a gas.

To investigate how, or if, the weight of matter changes as a result of any of these chemical reactions, the students will need to make these chemical reactions happen inside a closed system. They can then measure the total weight of that system before and after the reaction. A closed system is an area that is separated from its surroundings by a barrier that does not let any matter transfer into or out of that system. An example of a closed system is a container with a sealed lid.

Figure 5.1 shows the basic steps for measuring the weight of a closed system before and after a chemical reaction takes place inside that system. The first step is to put one liquid (such as the baking soda solution) into a plastic vial and a second liquid (such as the vinegar) into a second plastic vial. Then place both vials inside the container and seal the lid. At this point, the liquids will not mix and the weight of the entire system before the reaction begins can be measured using an electronic scale. The container can be turned over so the liquids spill out of the vials and mix together inside the container; this will start the chemical reaction. The entire container can then be placed back on the electronic scale to measure the weight of the entire system after the reaction. This method is important to use for this type of investigation because some chemical reactions produce a gas (such as the reaction of baking soda and acetic acid) and the gas can escape. If the gas escapes, then the two liquids will appear to lose weight after they are mixed together instead of the weight of the two liquids staying constant.

FIGURE 5.1

How to measure the weight of a closed system before and after a chemical reaction

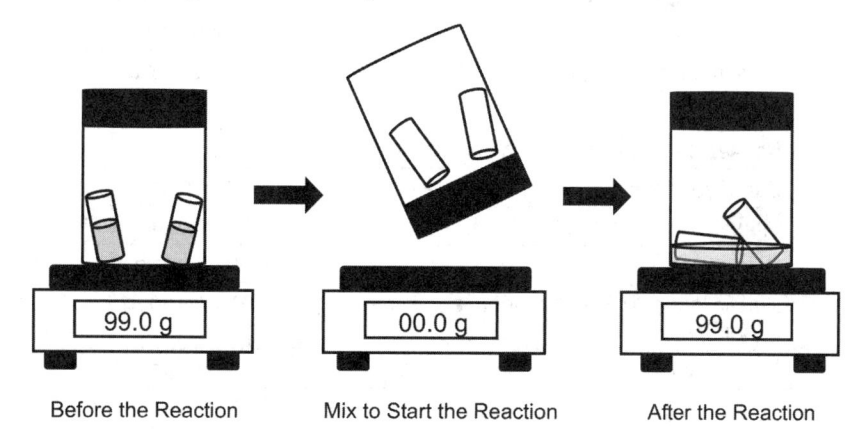

Before the Reaction — 99.0 g
Mix to Start the Reaction — 00.0 g
After the Reaction — 99.0 g

Teacher Notes

Timeline

The time needed to complete this investigation is 270 minutes (4 hours and 30 minutes). The amount of instructional time needed for each stage of the investigation is as follows:

- *Stage 1.* Introduce the task and the guiding question: 35 minutes
- *Stage 2.* Design a method and collect data: 50 minutes
- *Stage 3.* Create a draft argument: 45 minutes
- *Stage 4.* Argumentation session: 30 minutes
- *Stage 5.* Reflective discussion: 15 minutes
- *Stage 6.* Write a draft report: 30 minutes
- *Stage 7.* Peer review: 35 minutes
- *Stage 8.* Revise the report: 30 minutes

Materials and Preparation

The materials needed for this investigation are listed in Table 5.1. The items can be purchased from a big-box retail store such as Walmart or Target or through an online retailer such as Amazon. The baking soda, soap, and Epsom salt needed to make the solutions are available at most grocery stories. The calcium chloride is available at hardware stores such as Lowe's, Home Depot, and Ace Hardware. The materials for this investigation can also be purchased as a complete kit (which includes enough materials for 24 students, or six groups of four students) at *www.argumentdriveninquiry.com*.

TABLE 5.1
Materials for Investigation 5

Item	Quantity
Indirectly vented chemical-splash goggles	1 per student
Nonlatex apron	1 per student
Vinyl gloves	1 pair per student
Sodium bicarbonate (baking soda)	50 g per class
Vinegar	50 ml per class
Beaker, 1,000 ml (for mixing solutions)	1 per class
Bar of soap (Ivory or other brand)	1 per class
Magnesium sulfate (Epsom salt)	60 g per class
Calcium chloride (pool salt)	50 g per class
Plastic cups for baking soda and vinegar (for opening activity)	2 per group
PET clear plastic round wide-mouth jar, 16 oz	1 per group

Continued

Investigation 5. Reactions and Weight:
What Happens to the Total Weight of a Closed System When a Chemical Reaction Takes Place Within That System?

Table 5.1. (*continued*)

Item	Quantity
Plastic vials, 5 ml	6 per group
Electronic scale	1 per group
Sodium bicarbonate solution in a 30-ml plastic dropper bottle with cap	1 bottle per group
Vinegar in a 30-ml plastic dropper bottle with cap	1 bottle per group
Soap solution in a 30-ml plastic dropper bottle with cap	1 bottle per group
Magnesium sulfate solution in a 30-ml plastic dropper bottle with cap	1 bottle per group
Calcium chloride solution in a 30-ml plastic dropper bottle with cap	1 bottle per group
Whiteboard, 2' × 3'*	1 per group
Investigation Handout	1 per student
Peer-review guide and teacher scoring rubric	1 per student
Checkout Questions (optional)	1 per student

*As an alternative, students can use computer and presentation software such as Microsoft PowerPoint or Apple Keynote to create their arguments.

The day before you plan to start the investigation, you will need to prepare the solutions for the students to use. Prepare the solutions as follows:

- Sodium bicarbonate solution: Add 40 g of sodium bicarbonate to 500 ml of water in a beaker. Mix thoroughly to dissolve. Fill a 30-ml plastic dropper bottle with the baking soda solution. Place the cap on the bottle and label it. Make one bottle for each group.

- *Vinegar solution*: Fill a 30-ml plastic dropper bottle with vinegar. Place the cap on the bottle and label it. Make one bottle per group.

- *Soap solution*: Cut about 30 small slivers off a bar of soap (about 50 g of slivers). Add the slivers of soap to 500 ml of water in a beaker. Mix thoroughly to dissolve. Fill a 30-ml plastic dropper bottle with the soap solution. Place the cap on the bottle and label it. Make one bottle per group.

- *Magnesium sulfate solution*: Add 60 g of magnesium sulfate to 500 ml of water in a beaker. Mix thoroughly to dissolve. Fill a 30-ml plastic dropper bottle with the Epsom salt solution. Place the cap on the bottle and label it. Make one bottle per group.

- *Calcium chloride solution*: Add 50 g of calcium chloride to 500 ml of water in a beaker. Mix thoroughly to dissolve. Fill a 30-ml plastic dropper bottle with the calcium chloride solution. Place the cap on the bottle and label it. Make one bottle per group.

Teacher Notes

Excess solution may be flushed down the drain with a large quantity of water. The students may also dispose of any liquid waste from their investigations by flushing it down the drain with a large quantity of water.

Safety Precautions

Remind students to follow all normal safety rules. In addition, tell the students to take the following safety precautions:

- Wear sanitized safety goggles, nonlatex aprons, and vinyl gloves during setup, investigation activity, and cleanup.
- Do not put any liquids in their mouth.
- Immediately clean up any spills to avoid a slip or fall hazard.
- Keep water sources away from electrical receptacles to prevent shock.
- Make sure all materials are put away after completing the activity.
- Wash their hands with soap and water when they are done cleaning up.

Lesson Plan by Stage

This lesson plan is only a suggestion. It is included here to illustrate what you can say and do during each stage of ADI for this specific investigation. We encourage you to modify this lesson plan by asking different questions, using different examples, and providing different scaffolds as needed to better meet the needs of the students in your class.

Stage 1: Introduce the Task and the Guiding Question (35 minutes)

1. Ask the students to sit in six groups, with three or four students in each group.
2. Ask the students to clear off their desks except for a pencil (and their *Student Workbook for Argument-Driven Inquiry in Fifth-Grade Science* if they have one).
3. Pass out an Investigation Handout to each student (or ask students to turn to the Investigation Log for Investigation 5 in their workbook).
4. Read the first paragraph of the "Introduction" aloud to the class. Ask the students to follow along as you read.
5. Give each student chemical-splash goggles and ask them to put them on. Then remind students of the safety rules and explain the safety precautions for this investigation.
6. Pass out a cup of vinegar and a cup with a spoonful of baking soda (sodium bicarbonate) to each group.
7. Tell students to add the baking soda to the vinegar.

Investigation 5. Reactions and Weight:
What Happens to the Total Weight of a Closed System When
a Chemical Reaction Takes Place Within That System?

8. Ask the students to record their observations and questions on the "NOTICED/ WONDER" chart in the "Introduction."

9. After the students have recorded their observations and questions, ask them to share what they observed when they added the baking soda to the vinegar.

10. Ask the students to share what questions they have about what happened when they added the baking soda to the vinegar.

11. Tell the students, "Some of your questions might be answered by reading the rest of the 'Introduction.'"

12. Ask the students to read the rest of the "Introduction" on their own *or* ask them to follow along as you read it aloud.

13. Once the students have read the rest of the "Introduction," ask them to fill out the "Things we KNOW" chart on their Investigation Handout (or in their Investigation Log) as a group.

14. Ask the students to share what they learned from the reading. Add these ideas to a class "Things we KNOW" chart.

15. Tell the students, "Let's see what we will need to figure out during our investigation."

16. Read the task and the guiding question aloud.

17. Tell the students, "I have lots of materials here that you can use."

18. Introduce the students to the materials available for them to use during the investigation by holding each one up and asking how it might be used. Also, tell the students that the reactions that they can test during their investigation are limited to (1) baking soda and vinegar, (2) soap and Epsom salt, and (3) baking soda and calcium chloride. You may want to write these reactions on the board so students can refer them later.

Stage 2: Design a Method and Collect Data (50 minutes)

1. Tell the students, "I am now going to give you and the other members of your group about 15 minutes to plan your investigation. Before you begin, I want you all to take a couple of minutes to discuss the following questions with the rest of your group."

2. Show the following questions on the screen or board:
   - What are the components of the *system* that we are studying?
   - Which standard *units* should we use to describe the total weight of the system?

3. Tell the students, "Please take a few minutes to come up with an answer to these questions."

4. Give the students two to three minutes to discuss these two questions.

Teacher Notes

5. Ask two or three different groups to share their answers. Highlight or write down any important ideas on the board so students can refer to them later.

6. If possible, use a document camera to project an image of the graphic organizer for this investigation on a screen or board (or take a picture of it and project the picture on a screen or board). Tell the students, "I now want you all to plan out your investigation. To do that, you will need to fill out this investigation proposal."

7. Point to the box labeled "Our guiding question:" and tell the students, "You can put the question we are trying to answer in this box." Then ask, "Where can we find the guiding question?"

8. Wait for a student to answer.

9. Point to the box labeled "This is a picture of how we will set up the equipment:" and tell the students, "You can draw a picture in this box of how you will set up the equipment in order to carry out this investigation."

10. Point to the box labeled "We will collect the following data:" and tell the students, "You can list the measurements or observations that you will need to collect during the investigation in this box."

11. Point to the box labeled "These are the steps we will follow to collect data:" and tell the students, "You can list what you are going to do to collect the data you need and what you will do with your data once you have it. Be sure to give enough detail that I could do your investigation for you."

12. Ask the students, "Do you have any questions about what you need to do?"

13. Answer any questions that come up.

14. Tell the students, "Once you are done, raise your hand and let me know. I'll then come by and look over your proposal and give you some feedback. You may not begin collecting data until I have approved your proposal by signing it. You need to have your proposal done in the next 15 minutes."

15. Give the students 15 minutes to work in their groups on their investigation proposal. As they work, move from group to group to check in, ask probing questions, and offer a suggestion if a group gets stuck.

16. As each group finishes its investigation proposal, read it over and determine if it will be productive or not. If you feel the investigation will be productive (not necessarily what you would do or what the other groups are doing), sign your name on the proposal and let the group start collecting data. If the plan needs to be changed, offer some suggestions or ask some probing questions, and have the group make the changes before you approve it.

17. Pass out the materials or have one student from each group collect the materials they need from a central supply table or cart for the groups that have an approved proposal.

Investigation 5. Reactions and Weight:
What Happens to the Total Weight of a Closed System When
a Chemical Reaction Takes Place Within That System?

18. Remind students of the safety rules and precautions for this investigation.
19. Tell the students to collect their data and record their observations or measurements in the "Collect Your Data" box in their Investigation Handout (or the Investigation Log in their workbook).
20. Give the students 30 minutes to collect their data. Collect the materials from each group before asking them to analyze their data.

What should a student-designed investigation look like?

There are a number of different investigations that students can design to answer the question "What happens to the total weight of a closed system when a chemical reaction takes place within that system?" For example, one method might include the following steps:

1. Add 10 ml of baking soda solution to a vial.
2. Add 10 ml of vinegar to a vial.
3. Place both vials inside a container so the liquids do not mix. Seal the container.
4. Weigh the container.
5. Turn the container over several times until the liquids inside the container mix.
6. Wait for the chemical reaction to finish.
7. Weigh the container a second time.
8. Clean out the container.
9. Add 10 ml of soap solution to a vial.
10. Add 10 ml of Epsom salt solution to a vial.
11. Repeat steps 3–8.
12. Add 10 ml of baking soda solution to a vial.
13. Add 10 ml of calcium chloride solution to a vial.
14. Repeat steps 3–8.

If students use this method, they will need to collect data on (1) the weight of the container before the reaction and (2) the weight of the container after the reaction.

Teacher Notes

Stage 3: Create a Draft Argument (45 minutes)

1. Tell the students, "Now that we have all this data, we need to analyze the data so we can figure out an answer to the guiding question."

2. If possible, project an image of the "Analyze Your Data" section for this investigation on a screen or board using a document camera (or take a picture of it and project the picture on a screen or board). Point to the section and tell the students, "You can create a graph as a way to analyze your data. You can make your graph in this section."

3. Ask the students, "What information do we need to include in a graph?"

4. Tell the students, "Please take a few minutes to discuss this question with your group and be ready to share."

5. Give the students five minutes to discuss.

6. Ask two or three different groups to share their answers. Highlight or write down any important ideas on the board so students can refer to them later.

7. Tell the students, "I am now going to give you and the other members of your group about 10 minutes to create your graph." The graph they create should include the weights of the containers before and after the reaction. If the students are having trouble making a graph, you can take a few minutes to provide a mini-lesson about how to create a graph from a bunch of measurements (this strategy is called just-in-time instruction because it is offered only when students get stuck).

8. Give the students 10 minutes to analyze their data by creating a graph. As they work, move from group to group to check in, ask probing questions, and offer suggestions.

9. Tell the students, "I am now going to give you and the other members of your group about 15 minutes to create an argument to share what you have learned and convince others that they should believe you. Before you do that, we need to take a few minutes to discuss what you need to include in your argument."

10. If possible, use a document camera to project the "Argument Presentation on a Whiteboard" image from the "Draft Argument" section of the Investigation Handout (or the Investigation Log in their workbook) on a screen or board (or take a picture of it and project the picture on a screen or board).

11. Point to the box labeled "The Guiding Question:" and tell the students, "You can put the question we are trying to answer here on your whiteboard."

12. Point to the box labeled "Our Claim:" and tell the students, "You can put your claim here on your whiteboard. The claim is your answer to the guiding question."

13. Point to the box labeled "Our Evidence:" and tell the students, "You can put the evidence that you are using to support your claim here on your whiteboard.

Investigation 5. Reactions and Weight:
What Happens to the Total Weight of a Closed System When a Chemical Reaction Takes Place Within That System?

Your evidence will need to include the analysis you just did and an explanation of what your analysis means or shows. Scientists always need to support their claims with evidence."

14. Point to the box labeled "Our Justification of the Evidence:" and tell the students, "You can put your justification of your evidence here on your whiteboard. Your justification needs to explain why your evidence is important. Scientists often use core ideas to explain why the evidence they are using matters. Core ideas are important concepts that scientists use to help them make sense of what happens during an investigation."

15. Ask the students, "What are some core ideas that we read about earlier that might help us explain why the evidence we are using is important?"

16. Ask the students to share some of the core ideas from the "Introduction" section of the Investigation Handout (or the Investigation Log in the workbook). List these core ideas on the board.

17. Tell the students, "That is great. I would like to see everyone try to include these core ideas in your justification of the evidence. Your goal is to use these core ideas to help explain why your evidence matters and why the rest of us should pay attention to it."

18. Ask the students, "Do you have any questions about what you need to do?"

19. Answer any questions that come up.

20. Tell the students, "Okay, go ahead and start working on your arguments. You need to have your argument done in the next 15 minutes. It doesn't need to be perfect. We just need something down on the whiteboards so we can share our ideas."

What should the graph for this investigation look like?

There are a number of different ways that students can analyze the measurements they collect during this investigation. One of the most straightforward ways is to create a grouped bar graph with the reactions (baking soda solution and vinegar, soap solution and Epsom salt solution, and baking soda solution and calcium chloride solution) on the horizontal axis, or x-axis, and the weight on the vertical axis, or y-axis. The groups of bars can be labeled by time (before and after the reaction). An example of a grouped bar graph can be seen in Figure 5.2 (p. 212). There are other options for analyzing the collected data. Students often come up with some unique ways of analyzing their data, so be sure to give them some voice and choice during this stage.

Teacher Notes

21. Give the students 15 minutes to work in their groups on their arguments. As they work, move from group to group to check in, ask probing questions, and offer a suggestion if a group gets stuck. Figure 5.2 shows an example of an argument created by students for this investigation.

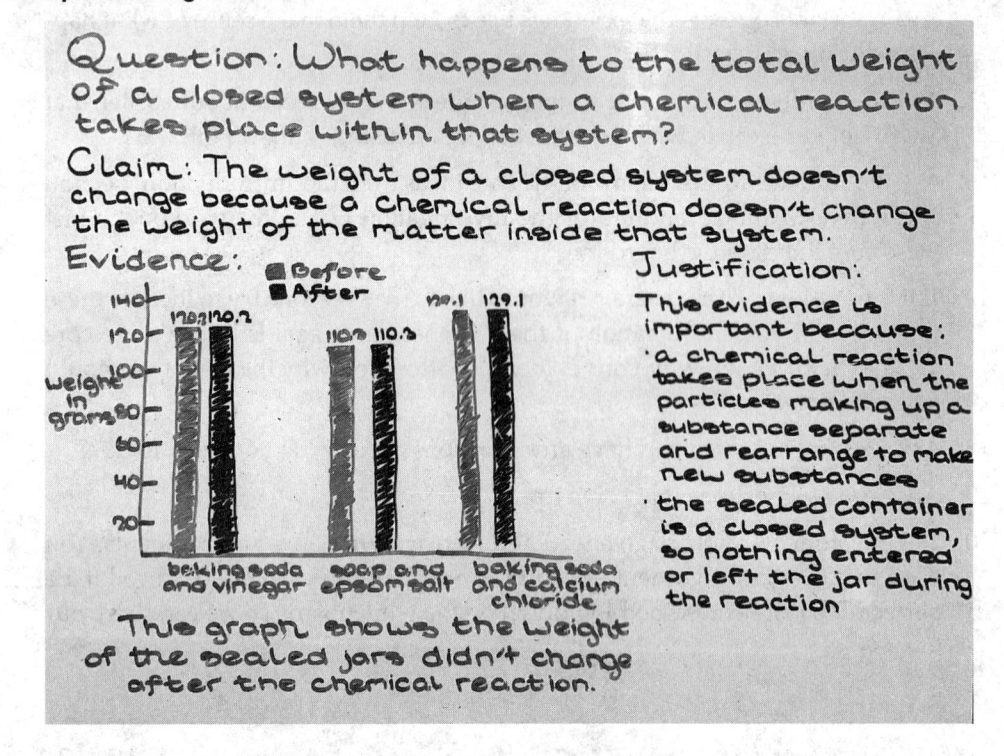

FIGURE 5.2
Example of an argument

> **Question:** What happens to the total weight of a closed system when a chemical reaction takes place within that system?
>
> **Claim:** The weight of a closed system doesn't change because a chemical reaction doesn't change the weight of the matter inside that system.
>
> **Evidence:**
>
> [Bar graph comparing Before and After weights in grams for three reactions:
> - baking soda and vinegar: 129.2, 130.2
> - soap and epsom salt: 110.3, 110.3
> - baking soda and calcium chloride: 129.1, 129.1]
>
> This graph shows the weight of the sealed jars didn't change after the chemical reaction.
>
> **Justification:**
>
> This evidence is important because:
> - a chemical reaction takes place when the particles making up a substance separate and rearrange to make new substances
> - the sealed container is a closed system, so nothing entered or left the jar during the reaction

Investigation 5. Reactions and Weight: What Happens to the Total Weight of a Closed System When a Chemical Reaction Takes Place Within That System?

Stage 4: Argumentation Session (30 minutes)

The argumentation session can be conducted in a whole-class presentation format, a gallery walk format, or a modified gallery walk format. We recommend using a whole-class presentation format for the first investigation, but try to transition to either the gallery walk or modified gallery walk format as soon as possible because that will maximize student voice and choice inside the classroom. The following list shows the steps for the three formats; unless otherwise noted, the steps are the same for all three formats.

1. Begin by introducing the use of the whiteboard.
   - *If using the whole-class presentation format,* tell the students, "We are now going to share our arguments. Please set up your whiteboards so everyone can see them."
   - *If using the gallery walk or modified gallery walk format,* tell the students, "We are now going to share our arguments. Please set up your whiteboards so they are facing the walls."

2. Allow the students to set up their whiteboards.
   - *If using the whole-class presentation format,* the whiteboards should be set up on stands or chairs so they are facing toward the center of the room.
   - *If using the gallery walk or modified gallery walk format,* the whiteboards should be set up on stands or chairs so they are facing toward the outside of the room.

3. Give the following instructions to the students:
   - *If using the whole-class presentation format,* tell the students, "Okay, before we get started I want to explain what we are going to do next. Your group will have an opportunity to share your argument with the rest of the class. After you are done, everyone else in the class will have a chance to ask questions and offer some suggestions about ways to make your group's argument better. After we have a chance to listen to each other and learn something new, I'm going to give you some time to revise your arguments and make them better."
   - *If using the gallery walk format,* tell the students, "Okay, before we get started I want to explain what we are going to do next. You are going to read the arguments that were created by other groups. When I say 'go,' your group will go to a different group's station so you can see their argument. Once you are there, I'll give your group a few minutes to read and review their argument. Your job is to offer them some suggestions about ways to make their argument better. You can use sticky notes to give them suggestions. Please be specific about what you want to change and how you think they should change it. After we have a chance to learn from each other, I'm going to give you some time to revise your arguments and make them better."
   - *If using the modified gallery walk format,* tell the students, "Okay, before we get started I want to explain what we are going to do next. I'm going to ask some of

Teacher Notes

you to present your arguments to your classmates. If you are presenting your argument, your job is to share your group's claim, evidence, and justification of the evidence. The rest of you will be travelers. If you are a traveler, your job is to listen to the presenters, ask the presenters questions if you do not understand something, and then offer them some suggestions about ways to make their argument better. After we have a chance to learn from each other, I'm going to give you some time to revise your arguments and make them better."

4. Use a document camera to project the "Ways to IMPROVE our argument ..." box from the Investigation Handout (or the Investigation Log in their workbook) on a screen or board (or take a picture of it and project the picture on a screen or board).

   - *If using the whole-class presentation format,* point to the box and tell the students, "After your group presents your argument, you can write down the suggestions you get from your classmates here. If you are listening to a presentation and you see a good idea from another group, you can write down that idea here as well. Once we are done with the presentations, I will give you a chance to use these suggestions or ideas to improve your arguments."

   - *If using the gallery walk format,* point to the box and tell the students, "If you see a good idea from another group, you can write it down here. Once we are done reviewing the different arguments, I will give you a chance to use these ideas to improve your own arguments. It is important to share ideas like this."

   - *If using the modified gallery walk format,* point to the box and tell the students, "If you are a presenter, you can write down the suggestions you get from the travelers here. If you are a traveler and you see a good idea from another group, you can write down that idea here. Once we are done with the presentations, I will give you a chance to use these suggestions or ideas to improve your arguments."

5. Ask the students, "Do you have any questions about what you need to do?"

6. Answer any questions that come up.

7. Give the following instructions:

   - *If using the whole-class presentation format,* tell the students, "Okay. Let's get started."

   - *If using the gallery walk format,* tell the students, "Okay, I'm now going to tell you which argument to go to and review."

   - *If using the modified gallery walk format,* tell the students, "Okay, I'm now going to assign you to be a presenter or a traveler." Assign one or two students from each group to be presenters and one or two students from each group to be travelers.

8. Give the students an opportunity to review the arguments.

Investigation 5. Reactions and Weight: What Happens to the Total Weight of a Closed System When a Chemical Reaction Takes Place Within That System?

- *If using the whole-class presentation format,* have each group present their argument one at a time. Give each group only two to three minutes to present their argument. Then give the class two to three minutes to ask them questions and offer suggestions. Encourage as much participation from the students as possible.

- *If using the gallery walk format,* tell the students, "Okay. Let's get started. Each group, move one argument to the left. Don't move to the next argument until I tell you to move. Once you get there, read the argument and then offer suggestions about how to make it better. I will put some sticky notes next to each argument. You can use the sticky notes to leave your suggestions." Give each group about three to four minutes to read the arguments, talk, and offer suggestions.

    a. After three to four minutes, tell the students, "Okay. Let's move on to the next argument. Please move one group to the left."

    b. Again, give each group three to four minutes to read, talk, and offer suggestions.

    c. Repeat this process until each group has had their argument read and critiqued three times.

- *If using the modified gallery walk format,* tell the students, "Okay. Let's get started. Reviewers, move one group to the left. Don't move to the next group until I tell you to move. Presenters, go ahead and share your argument with the travelers when they get there." Give each group of presenters and travelers about three to four minutes to talk.

    a. Tell the students, "Okay. Let's move on to the next argument. Travelers, move one group to the left."

    b. Again, give each group of presenters and travelers about three to four minutes to talk.

    c. Repeat this process until each group has had their argument read and critiqued three times.

9. Tell the students to return to their workstations.

10. Give the following instructions about revising the argument:

    - *If using the whole-class presentation format,* tell the students, "I'm now going to give you all about 10 minutes to revise your argument. Take a few minutes to talk in your groups and determine what you want to change to make your argument better. Once you have decided what to change, go ahead and make the changes to your whiteboard."

    - *If using the gallery walk format,* tell the students, "I'm now going to give you all about 10 minutes to revise your argument. Take a few minutes to read the suggestions that were left at your argument. Then talk in your groups and determine what you want to change to make your argument better. Once

Teacher Notes

you have decided what to change, go ahead and make the changes to your whiteboard."

- *If using the modified gallery walk format,* tell the students, "I'm now going to give you all about 10 minutes to revise your argument. Please return to your original groups." Wait for the students to move back into their original groups and then tell the students, "Okay, take a few minutes to talk in your groups and determine what you want to change to make your argument better. Once you have decided what to change, go ahead and make the changes to your whiteboard."

11. Ask the students, "Do you have any questions about what you need to do?"
12. Answer any questions that come up.
13. Tell the students, "Okay. Let's get started."
14. Give the students 10 minutes to work in their groups on their arguments. As they work, move from group to group to check in, ask probing questions, and offer a suggestion if a group gets stuck.

Stage 5: Reflective Discussion (15 minutes)

1. Tell the students, "We are now going to take a minute to talk about some of the core ideas and crosscutting concepts that we have used during our investigation."
2. Show Figure 5.3 on the screen.
3. Ask the students, "What do you all see going on here?"
4. Allow the students to share their ideas.
5. Ask the students, "How can we explain this in terms of the structure and properties of matter or chemical reactions?"

FIGURE 5.3

The reaction of baking soda and vinegar

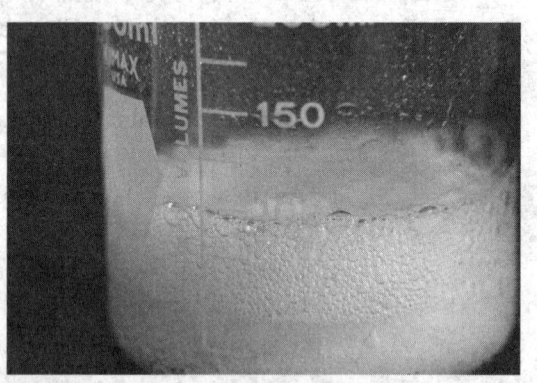

6. Allow the students to share their ideas. As they share their ideas, ask different questions to encourage them to expand on their thinking (e.g., "Can you tell me more about that?"), clarify a contribution (e.g., "Can you say that in another way?"), support an idea (e.g., "Why do you think that?"), add to an idea mentioned by a classmate (e.g., "Would anyone like to add to the idea?"), re-voice an idea offered by a classmate (e.g., "Who can explain that to me in another way?"), or critique an idea during the discussion (e.g., "Do you agree or disagree with that idea and why?") until students are able to generate an adequate explanation.
7. Show Figure 5.4 on the screen.

Investigation 5. Reactions and Weight:
What Happens to the Total Weight of a Closed System When
a Chemical Reaction Takes Place Within That System?

FIGURE 5.4

A model of a chemical reaction

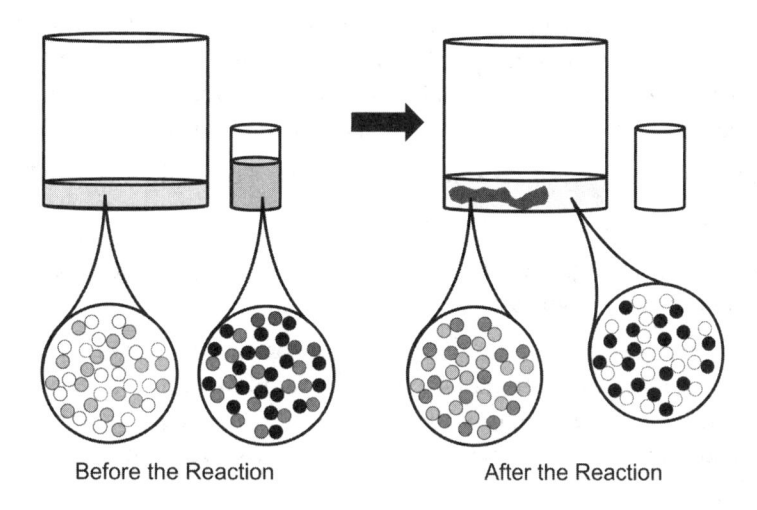

Before the Reaction → After the Reaction

Note: A full-color version of this figure is available on the book's Extras page at *www.nsta.org/adi-5th*.

8. Ask the students, "What do you all see going on here?"
9. Allow the students to share their ideas.
10. Ask the students, "How can we explain this in terms of the structure and properties of matter or chemical reactions?"
11. Allow students to share their ideas. As they share their ideas, ask different questions to encourage them to expand on their thinking (e.g., "Can you tell me more about that?"), clarify a contribution (e.g., "Can you say that in another way?"), support an idea (e.g., "Why do you think that?"), add to an idea mentioned by a classmate (e.g., "Would anyone like to add to the idea?"), re-voice an idea offered by a classmate (e.g., "Who can explain that to me in another way?"), or critique an idea during the discussion (e.g., "Do you agree or disagree with that idea and why?") until students are able to generate an adequate explanation.
12. Tell the students, "We used standard units to measure weight during our investigation." Then ask, "Can anyone tell me why this was important?"
13. Allow the students to share their ideas.
14. Tell the students, "I think using standard units to measure and describe physical quantities such as weight is important because we need to have a common understanding of what our measurements mean. This allows us to share and critique each other's ideas and evidence."

Argument-Driven Inquiry in **Fifth-Grade Science**: Three-Dimensional Investigations

Teacher Notes

15. Ask the students, "What units did you use today?"
16. Allow the students to share their ideas.
17. Show an image of the question "What do you think are the most important core ideas or crosscutting concepts that we used during this investigation to help us make sense of what we observed?" Tell the students, "Okay, let's make sure we are all on the same page. Please take a moment to discuss this question with the other people in your group." Give them a few minutes to discuss the question.
18. Ask the students, "What do you all think? Who would like to share?"
19. Allow the students to share their ideas.
20. Tell the students, "We are now going to take a minute to talk about how scientists think about the world."
21. Show an image of the question "Are the laws of nature the same everywhere?" on the screen. Tell the students, "Take a few minutes to talk about how you would answer this question with the other people in your group. Be ready to share with the rest of the class." Give the students two to three minutes to talk in their group.
22. Ask the students, "What do you all think? Who would like to share an idea?"
23. Allow the students to share their ideas.
24. Ask the students, "Why would it be important for scientists to assume that the laws of nature are the same everywhere?"
25. Allow the students to share their ideas.
26. Tell the students, "I think it is important that scientists assume that the laws of nature are the same everywhere because it allows them to make predictions based on those laws no matter where the scientists are located."
27. Ask the students, "Suppose I mixed baking soda and vinegar in Europe, in Africa, and in Australia: Would the same chemical reaction happen in all three of these locations?"
28. Allow the students to share their ideas.
29. Ask the students, "Does anyone have any questions about why scientists assume that the laws of nature are the same everywhere?"
30. Answer any questions that come up.
31. Tell the students, "We are now going to take a minute to talk about what went well and what didn't go so well during our investigation. We need to talk about this because you all are going to be planning and carrying out your own investigations like this a lot this year, and I want to help you all get better at it."
32. Show an image of the question "What made your investigation scientific?" on the screen. Tell the students, "Take a few minutes to talk about how you would

answer this question with the other people in your group. Be ready to share with the rest of the class." Give the students two to three minutes to talk in their group.

33. Ask the students, "What do you all think? Who would like to share an idea?"
34. Allow the students to share their ideas. Be sure to expand on their ideas about what makes an investigation scientific.
35. Show an image of the question "What made your investigation not so scientific?" on the screen. Tell the students, "Take a few minutes to talk about how you would answer this question with the other people in your group. Be ready to share with the rest of the class." Give the students two to three minutes to talk in their group.
36. Ask the students, "What do you all think? Who would like to share an idea?"
37. Allow the students to share their ideas. Be sure to expand on their ideas about what makes an investigation less scientific.
38. Show an image of the question "What rules can we put into place to help us make sure our next investigation is more scientific?" on the screen. Tell the students, "Take a few minutes to talk about how you would answer this question with the other people in your group. Be ready to share with the rest of the class." Give the students two to three minutes to talk in their group.
39. Ask the students, "What do you all think? Who would like to share an idea?"
40. Allow the students to share their ideas. Once they have shared their ideas, offer a suggestion for a possible class rule.
41. Ask the students, "What do you all think? Should we make this a rule?"
42. If the students agree, write the rule on the board or on a class "Rules for Scientific Investigation" chart so you can refer to it during the next investigation.

Stage 6: Write a Draft Report (30 minutes)

Your students will use either the Investigation Handout or the Investigation Log in the student workbook when writing the draft report. When you give the directions shown in quotes in the following steps, substitute "Investigation Log in your workbook" or just "Investigation Log" (as shown in brackets) for "handout" if they are using the workbook.

1. Tell the students, "You are now going to write an investigation report to share what you have learned. Please take out a pencil and turn to the 'Draft Report' section of your handout [Investigation Log in your workbook]."
2. If possible, use a document camera to project the "Introduction" section of the draft report from the Investigation Handout (or the Investigation Log in their workbook) on a screen or board (or take a picture of it and project the picture on a screen or board).

Teacher Notes

3. Tell the students, "The first part of the report is called the 'Introduction.' In this section of the report you want to explain to the reader what you were investigating, why you were investigating it, and what question you were trying to answer. All this information can be found in the text at the beginning of your handout [Investigation Log]." Point to the image. "Here are some sentence starters to help you begin writing."

4. Ask the students, "Do you have any questions about what you need to do?"

5. Answer any questions that come up.

6. Tell the students, "Okay, let's write."

7. Give the students 10 minutes to write the "Introduction" section of the report. As they work, move from student to student to check in, ask probing questions, and offer a suggestion if a student gets stuck.

8. If possible, use a document camera to project the "Method" section of the draft report from the Investigation Handout (or the Investigation Log in their workbook) on a screen or board (or take a picture of it and project the picture on a screen or board).

9. Tell the students, "The second part of the report is called the 'Method.' In this section of the report you want to explain to the reader what you did during the investigation, what data you collected and why, and how you went about analyzing your data. All this information can be found in the 'Plan Your Investigation' section of the handout [Investigation Log]. Remember that you all planned and carried out different investigations, so do not assume that the reader will know what you did." Point to the image. "Here are some sentence starters to help you begin writing."

10. Ask the students, "Do you have any questions about what you need to do?"

11. Answer any questions that come up.

12. Tell the students, "Okay, let's write."

13. Give the students 10 minutes to write the "Method" section of the report. As they work, move from student to student to check in, ask probing questions, and offer a suggestion if a student gets stuck.

14. If possible, use a document camera to project the "Argument" section of the draft report from the Investigation Handout (or the Investigation Log in their workbook) on a screen or board (or take a picture of it and project the picture on a screen or board).

15. Tell the students, "The last part of the report is called the 'Argument.' In this section of the report you want to share your claim, evidence, and justification of the evidence with the reader. All this information can be found on your whiteboard." Point to the image. "Here are some sentence starters to help you begin writing."

16. Ask the students, "Do you have any questions about what you need to do?"
17. Answer any questions that come up.
18. Tell the students, "Okay, let's write."
19. Give the students 10 minutes to write the "Argument" section of the report. As they work, move from student to student to check in, ask probing questions, and offer a suggestion if a student gets stuck.

Stage 7: Peer Review (35 minutes)

Your students will use either the Investigation Handout or their workbook when doing the peer review. Except where noted below, the directions are the same whether using the handout or the workbook.

1. Tell the students, "We are now going to review our reports to find ways to make them better. I'm going to come around and collect your draft reports. While I do that, please take out a pencil."
2. Collect the handouts or the workbooks with the draft reports from the students.
3. If possible, use a document camera to project the peer-review guide (see Appendix 4) on a screen or board (or take a picture of it and project the picture on a screen or board).
4. Tell the students, "We are going to use this peer-review guide to give each other feedback." Point to the image.
5. Tell the students, "I'm going to ask you to work with a partner to do this. I'm going to give you and your partner a draft report to read. You two will then read the report together. Once you are done reading the report, I want you to answer each of the questions on the peer-review guide." Point to the review questions on the image of the peer-review guide.
6. Tell the students, "You can check 'no,' 'almost,' or 'yes' after each question." Point to the checkboxes on the image of the peer-review guide.
7. Tell the students, "This will be your rating for this part of the report. Make sure you agree on the rating you give the author. If you mark 'no' or 'almost,' then you need to tell the author what he or she needs to do to get a 'yes.'" Point to the space for the reviewer feedback on the image of the peer-review guide.
8. Tell the students, "It is really important for you to give the authors feedback that is helpful. That means you need to tell them exactly what they need to do to make their report better."
9. Ask the students, "Do you have any questions about what you need to do?"
10. Answer any questions that come up.

Teacher Notes

11. Tell the students, "Please sit with a partner who is not in your current group." Allow the students time to sit with a partner.

12. Tell the students, "Okay, I'm now going to give you one report to read." Pass out one Investigation Handout with a draft report or one workbook to each pair. Make sure that the report you give a pair was not written by one of the students in that pair. Give each pair one peer-review guide to fill out. If the students are using workbooks, the peer-review guide is included right after the draft report so you do not need to pass out copies of the peer-review guide.

13. Tell the students, "Okay, I'm going to give you 15 minutes to read the report I gave you and to fill out the peer-review guide. Go ahead and get started."

14. Give the students 15 minutes to work. As they work, move around from pair to pair to check in and see how things are going, answer questions, and offer advice.

15. After 15 minutes pass, tell the students, "Okay, time is up. Please give me the report and the peer-review guide that you filled out."

16. Collect the Investigation Handouts and the peer-review guides, or collect the workbooks if students are using them. If the students are using the Investigation Handouts and separate peer-review guides, be sure you keep each handout with its corresponding peer-review guide.

17. Tell the students, "Okay, I am now going to give you a different report to read and a new peer-review guide to fill out." Pass out one more report to each pair. Make sure that the report you give a pair was not written by one of the students in that pair. Give each pair a new peer-review guide to fill out as a group.

18. Tell the students, "Okay, I'm going to give you 15 minutes to read this new report and to fill out the peer-review guide. Go ahead and get started."

19. Give the students 15 minutes to work. As they work, move around from pair to pair to check in and see how things are going, answer questions, and offer advice.

20. After 15 minutes pass, tell the students, "Okay, time is up. Please give me the report and the peer-review guide that you filled out."

21. Collect the Investigation Handouts and the peer-review guides, or collect the workbooks if students are using them. If the students are using the Investigation Handouts and separate peer-review guides, be sure you keep each handout with its corresponding peer-review guide.

Stage 8: Revise the Report (30 minutes)

Your students will use either the Investigation Handout or their workbook when revising the report. Except where noted below, the directions are the same whether using the handout or the workbook.

Investigation 5. Reactions and Weight:
What Happens to the Total Weight of a Closed System When
a Chemical Reaction Takes Place Within That System?

1. Tell the students, "You are now going to revise your draft report based on the feedback you get from your classmates. Please take out a pencil."

2. Return the reports to the students.
   - *If the students used the Investigation Handout and a copy of the peer-review guide,* pass back the handout and the peer-review guide to each student.
   - If the students used the workbook, pass that back to each student.

3. Tell the students, "Please take a few minutes to read over the peer-review guide. You should use it to figure out what you need to change in your report and how you will change it."

4. Allow the students to read the peer-review guide.

5. *If the students used the workbook,* if possible use a document camera to project the "Write Your Final Report" section from the Investigation Log on a screen or board (or take a picture of it and project the picture on a screen or board).

6. Give the following directions about how to revise their reports:
   - *If the students used the Investigation Handout and a copy of the peer-review guide,* tell them, "Okay, let's revise our reports. Please take out a piece of paper. I would like you to rewrite your report. You can use your draft report as a starting point, but you also need to change it to make it better. Use the feedback on the peer-review guide to make it better."
   - *If the students used the workbook,* tell them, "Okay, let's revise our reports. I would like you to rewrite your report in the section of the Investigation Log called "Write Your Final Report." You can use your draft report as a starting point, but you also need to change it to make it better. Use the feedback on the peer-review guide to make it better."

7. Ask the students, "Do you have any questions about what you need to do?"

8. Answer any questions that come up.

9. Tell the students, "Okay, let's write." Allow about 20 minutes for the students to revise their reports.

10. After about 20 minutes, give the following directions:
    - *If the students used the Investigation Handout,* tell them, "Okay, time's up. I will now come around and collect your Investigation Handout, the peer-review guide, and your final report."
    - *If the students used the workbook,* tell them, "Okay, time's up. I will now come around and collect your workbooks."

11. *If the students used the Investigation Handout,* collect all the Investigation Handouts, peer-review guides, and final reports. *If the students used the workbook,* collect all the workbooks.

Teacher Notes

12. *If the students used the Investigation Handout,* use the "Teacher Score" column in the peer-review guide to grade the final report. *If the students used the workbook,* use the "Investigation Report Grading Rubric" in the Investigation Log to grade the final report. Whether you are using the handout or the log, you can give the students feedback about their writing in the "Teacher Comments" section.

How to Use the Checkout Questions

The Checkout Questions are an optional assessment. We recommend giving them to students at the start of the next class period after the students finish stage 8 of the investigation. You can then look over the student answers to determine if you need to reteach the core idea from the investigation. Appendix 6 gives the answers to the Checkout Questions that should be given by a student who can apply the core ideas correctly and can explain the importance of (1) thinking about systems and (2) using standard units to measure or describe the physical quantities of matter.

Connections to Standards

Table 5.2 highlights how the investigation can be used to address specific performance expectations from the *NGSS, Common Core State Standards for English Language Arts (CCSS ELA)* and *Common Core State Standards for Mathematics (CCSS Mathematics),* and *English Language Proficiency (ELP) Standards.*

TABLE 5.2

Investigation 5 alignment with standards

NGSS performance expectation	5-PS1-2: Measure and graph quantities to provide evidence that regardless of the type of change that occurs when heating, cooling, or mixing substances, the total weight of matter is conserved.
CCSS ELA—Reading: Informational Text	Key ideas and details • CCSS.ELA-LITERACY.RI.5.1: Quote accurately from a text when explaining what the text says explicitly and when drawing inferences from the text. • CCSS.ELA-LITERACY.RI.5.2: Determine two or more main ideas of a text and explain how they are supported by key details; summarize the text. • CCSS.ELA-LITERACY.RI.5.3: Explain the relationships or interactions between two or more individuals, events, ideas, or concepts in a historical, scientific, or technical text based on specific information in the text. Craft and structure • CCSS.ELA-LITERACY.RI.5.4: Determine the meaning of general academic and domain-specific words and phrases in a text relevant to a *grade 5 topic or subject area*.

Continued

Investigation 5. Reactions and Weight: What Happens to the Total Weight of a Closed System When a Chemical Reaction Takes Place Within That System?

Table 5.2 (*continued*)

***CCSS ELA**—Reading: Informational Text* (*continued*)	• CCSS.ELA-LITERACY.RI.5.5: Compare and contrast the overall structure (e.g., chronology, comparison, cause/effect, problem/solution) of events, ideas, concepts, or information in two or more texts. • CCSS.ELA-LITERACY.RI.5.6: Analyze multiple accounts of the same event or topic, noting important similarities and differences in the point of view they represent. Integration of knowledge and ideas • CCSS.ELA-LITERACY.RI.5.7: Draw on information from multiple print or digital sources, demonstrating the ability to locate an answer to a question quickly or to solve a problem efficiently. • CCSS.ELA-LITERACY.RI.5.8: Explain how an author uses reasons and evidence to support particular points in a text, identifying which reasons and evidence support which point(s). Range of reading and level of text complexity • CCSS.ELA-LITERACY.RI.5.10: By the end of the year, read and comprehend informational texts, including history/social studies, science, and technical texts, at the high end of the grades 4–5 text complexity band independently and proficiently.

***CCSS ELA**—Writing*	Text types and purposes
	• CCSS.ELA-LITERACY.W.5.1: Write opinion pieces on topics or texts, supporting a point of view with reasons.
	  ○ CCSS.ELA-LITERACY.W.5.1.A: Introduce a topic or text clearly, state an opinion, and create an organizational structure in which ideas are logically grouped to support the writer's purpose.
	  ○ CCSS.ELA-LITERACY.W.5.1.B: Provide logically ordered reasons that are supported by facts and details.
	  ○ CCSS.ELA-LITERACY.W.5.1.C: Link opinion and reasons using words, phrases, and clauses (e.g., *consequently*, *specifically*).
	  ○ CCSS.ELA-LITERACY.W.5.1.D: Provide a concluding statement or section related to the opinion presented.
	• CCSS.ELA-LITERACY.W.5.2: Write informative or explanatory texts to examine a topic and convey ideas and information clearly.
	  ○ CCSS.ELA-LITERACY.W.5.2.A: Introduce a topic clearly, provide a general observation and focus, and group related information logically; include formatting (e.g., headings), illustrations, and multimedia when useful to aiding comprehension.
	  ○ CCSS.ELA-LITERACY.W.5.2.B: Develop the topic with facts, definitions, concrete details, quotations, or other information and examples related to the topic.
	  ○ CCSS.ELA-LITERACY.W.5.2.C: Link ideas within and across categories of information using words, phrases, and clauses (e.g., *in contrast*, *especially*).

Continued

Teacher Notes

Table 5.2 (*continued*)

***CCSS ELA*—Writing** (*continued*)	○ CCSS.ELA-LITERACY.W.5.2.D: Use precise language and domain-specific vocabulary to inform about or explain the topic. ○ CCSS.ELA-LITERACY.W.5.2.E: Provide a concluding statement or section related to the information or explanation presented. Production and distribution of writing • CCSS.ELA-LITERACY.W.5.4: Produce clear and coherent writing in which the development and organization are appropriate to task, purpose, and audience. • CCSS.ELA-LITERACY.W.5.5: With guidance and support from peers and adults, develop and strengthen writing as needed by planning, revising, editing, rewriting, or trying a new approach. • CCSS.ELA-LITERACY.W.5.6: With some guidance and support from adults, use technology, including the internet, to produce and publish writing as well as to interact and collaborate with others; demonstrate sufficient command of keyboarding skills to type a minimum of two pages in a single sitting. Research to build and present knowledge • CCSS.ELA-LITERACY.W.5.8: Recall relevant information from experiences or gather relevant information from print and digital sources; summarize or paraphrase information in notes and finished work, and provide a list of sources. • CCSS.ELA-LITERACY.W.5.9: Draw evidence from literary or informational texts to support analysis, reflection, and research. Range of writing • CCSS.ELA-LITERACY.W.5.10: Write routinely over extended time frames (time for research, reflection, and revision) and shorter time frames (a single sitting or a day or two) for a range of discipline-specific tasks, purposes, and audiences.
***CCSS ELA*—Speaking and Listening**	Comprehension and collaboration • CCSS.ELA-LITERACY.SL.5.1: Engage effectively in a range of collaborative discussions (one-on-one, in groups, and teacher-led) with diverse partners on *grade 5 topics and texts,* building on others' ideas and expressing their own clearly. ○ CCSS.ELA-LITERACY.SL.5.1.A: Come to discussions prepared, having read or studied required material; explicitly draw on that preparation and other information known about the topic to explore ideas under discussion. ○ CCSS.ELA-LITERACY.SL.5.1.B: Follow agreed-upon rules for discussions and carry out assigned roles. ○ CCSS.ELA-LITERACY.SL.5.1.C: Pose and respond to specific questions by making comments that contribute to the discussion and elaborate on the remarks of others. ○ CCSS.ELA-LITERACY.SL.5.1.D: Review the key ideas expressed and draw conclusions in light of information and knowledge gained from the discussions.

Continued

Investigation 5. Reactions and Weight: What Happens to the Total Weight of a Closed System When a Chemical Reaction Takes Place Within That System?

Table 5.2 (*continued*)

CCSS ELA—Speaking and Listening (*continued*)	• CCSS.ELA-LITERACY.SL.5.2: Summarize a written text read aloud or information presented in diverse media and formats, including visually, quantitatively, and orally. • CCSS.ELA-LITERACY.SL.5.3: Summarize the points a speaker makes and explain how each claim is supported by reasons and evidence. Presentation of knowledge and ideas • CCSS.ELA-LITERACY.SL.5.4: Report on a topic or text or present an opinion, sequencing ideas logically and using appropriate facts and relevant, descriptive details to support main ideas or themes; speak clearly at an understandable pace. • CCSS.ELA-LITERACY.SL.5.5: Include multimedia components (e.g., graphics, sound) and visual displays in presentations when appropriate to enhance the development of main ideas or themes. • CCSS.ELA-LITERACY.SL.5.6: Adapt speech to a variety of contexts and tasks, using formal English when appropriate to task and situation.
CCSS Mathematics—Numbers and Operations in Base Ten	Understand the place value system. • CCSS.MATH.CONTENT.5.NBT.A.1: Recognize that in a multi-digit number, a digit in one place represents 10 times as much as it represents in the place to its right and 1/10 of what it represents in the place to its left • CCSS.MATH.CONTENT.5.NBT.A.3: Read, write, and compare decimals to thousandths. • CCSS.MATH.CONTENT.5.NBT.A.4: Use place value understanding to round decimals to any place.
CCSS Mathematics—Measurement and Data	Convert like measurement units within a given measurement system. • CCSS.MATH.CONTENT.5.MD.A.1: Convert among different-sized standard measurement units within a given measurement system (e.g., convert 5 cm to 0.05 m), and use these conversions in solving multi-step, real-world problems.
ELP Standards	Receptive modalities • ELP 1: Construct meaning from oral presentations and literary and informational text through grade-appropriate listening, reading, and viewing. • ELP 8: Determine the meaning of words and phrases in oral presentations and literary and informational text. Productive modalities • ELP 3: Speak and write about grade-appropriate complex literary and informational texts and topics. • ELP 4: Construct grade-appropriate oral and written claims and support them with reasoning and evidence. • ELP 7: Adapt language choices to purpose, task, and audience when speaking and writing.

Continued

Teacher Notes

Table 5.2 (*continued*)

ELP Standards (*continued*)	Interactive modalities • ELP 2: Participate in grade-appropriate oral and written exchanges of information, ideas, and analyses, responding to peer, audience, or reader comments and questions. • ELP 5: Conduct research and evaluate and communicate findings to answer questions or solve problems. • ELP 6: Analyze and critique the arguments of others orally and in writing. Linguistic structures of English • ELP 9: Create clear and coherent grade-appropriate speech and text. • ELP 10: Make accurate use of standard English to communicate in grade-appropriate speech and writing.

Investigation Handout

Investigation 5

Reactions and Weight: What Happens to the Total Weight of a Closed System When a Chemical Reaction Takes Place Within That System?

Introduction

Matter is anything that has mass and takes up space. A substance is a type of matter with the same properties throughout it. Baking soda and water are examples of substances. Vinegar is a mixture of two different substances. The first substance is called acetic acid and the second one is water. Watch what happens when baking soda and vinegar are mixed together. As you watch this, keep track of things you notice and things you wonder about in the boxes below.

Things I NOTICED …	Things I WONDER about …

A chemical reaction took place inside the container when the baking soda and vinegar were mixed together. A *chemical reaction* is a term that scientists use to describe what happens when two or more substances are mixed together and a new substance with different properties is formed. In the case

Investigation Handout

of the chemical reaction that you just observed, the particles that made up the baking soda and the particles that made up the acetic acid in the vinegar separated from each other and then recombined (mixed together again) in a different way to create three new substances that were not in the container before the reaction started. The water in the vinegar was not involved in the reaction.

The three new substances that were created inside the container as a result of the chemical reaction were water, carbon dioxide, and sodium acetate. The new water that was created during the reaction simply mixed with the water that was already in the vinegar. Carbon dioxide is an invisible gas. The bubbles that you saw in the container during the reaction were the carbon dioxide. Once the bubbles of gas reached the surface of the liquid in the container, the gas left the container and mixed with the rest of the air in the room. Sodium acetate is a white powder at room temperature. This powder dissolves in water, just like salt or sugar does, so you could not see it being produced during the reaction.

These ideas about how the particles that make up a substance can separate from each other and then recombine in new ways during a chemical reaction can help us understand why we often see substances or mixtures turn a different color, produce bubbles, or change temperature after we mix them together. These ideas can also help us understand why some substances will disappear during a chemical reaction and other substances will appear in their place once the reaction is finished. Although useful, these ideas do not tell us anything about what happens to the weight of all the matter involved in a chemical reaction before, during, and after a chemical reaction takes place.

In this investigation, your goal is to figure out how a chemical reaction (a cause) affects the total weight of all the matter that was involved in the reaction (the effect). One way to investigate how, or if, the weight of matter changes as a result of a chemical reaction is to make several different chemical reactions happen inside a closed system. You can then measure the total weight of that system before and after the reaction. A *closed system* is an area that is separated from its surroundings by a barrier that does not let any matter transfer into or out of that system. An example of a closed system is a container with a sealed lid. Your teacher will tell you about the different reactions you can test.

Things we KNOW from what we read …

Investigation 5. Reactions and Weight:
What Happens to the Total Weight of a Closed System When a Chemical Reaction Takes Place Within That System?

Your Task

Use what you know about the structure and properties of matter, chemical reactions, tracking matter in systems, and the importance of using standard units to design and carry out an experiment to figure out how a chemical reaction inside a closed system affects the total weight of that system.

The *guiding question* of this investigation is, **What happens to the total weight of a closed system when a chemical reaction takes place within that system?**

Materials

You may use any of the following materials during your investigation:

Liquids for mixing

- Sodium bicarbonate (baking soda) in water
- Vinegar
- Soap in water
- Magnesium sulfate (Epsom salt) in water
- Calcium chloride (pool salt) in water

Equipment

- Safety goggles, nonlatex apron, and vinyl gloves (required)
- Container with lid
- 6 plastic vials (5 ml)
- Electronic scale

Safety Rules

Follow all normal safety rules. In addition, be sure to follow these rules:

- Wear sanitized indirectly vented chemical-splash goggles, nonlatex aprons, and vinyl gloves during setup, investigation activity, and cleanup.
- Do not put any liquids in your mouth.
- Immediately clean up any spills to avoid a slip or fall hazard.
- Keep water sources away from electrical receptacles to prevent shock.
- Wash your hands with soap and water when you are done cleaning up.

Plan Your Investigation

Prepare a plan for your investigation by filling out the chart on the next page; this plan is called an *investigation proposal*. Before you start developing your plan, be sure to discuss the following questions with the other members of your group:

- What are the components of the **system** that we are studying?
- Which standard **units** should we use to describe the total weight of the system?

Investigation Handout

Our guiding question:

This is a picture of how we will set up the equipment:

We will collect the following data:

These are the steps we will follow to collect data:

I approve of this investigation proposal.

_____   _____
Teacher's signature   Date

Investigation 5. Reactions and Weight: What Happens to the Total Weight of a Closed System When a Chemical Reaction Takes Place Within That System?

Collect Your Data

Keep a record of what you measure or observe during your investigation in the space below.

Analyze Your Data

You will need to analyze the data you collected before you can develop an answer to the guiding question. To analyze the data you collected, create a graph that shows what you measured before and after the reaction.

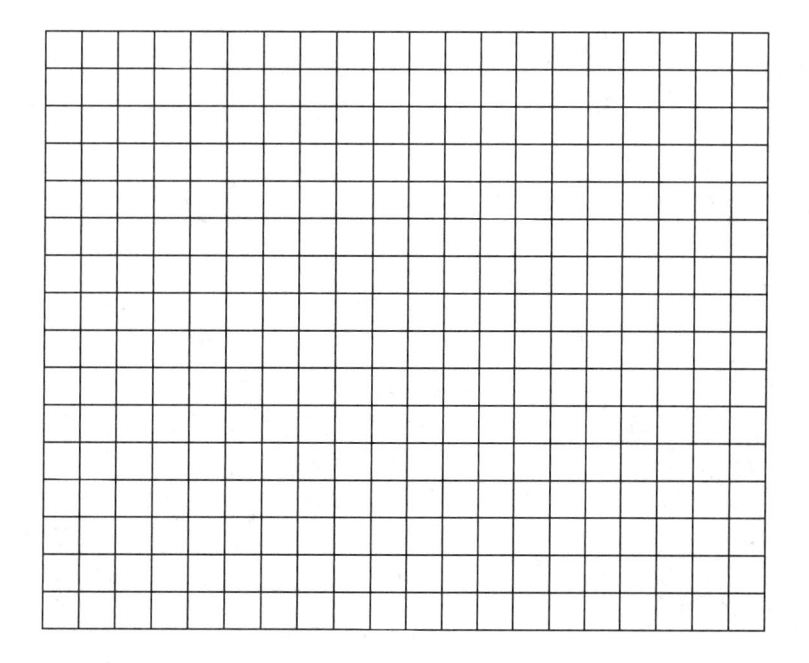

Argument-Driven Inquiry in **Fifth-Grade Science**: Three-Dimensional Investigations

Investigation Handout

Draft Argument

Develop an argument on a whiteboard. It should include the following:

1. A *claim*: Your answer to the guiding question.
2. *Evidence*: An analysis of the data and an explanation of what the analysis means.
3. A *justification of the evidence*: Why your group thinks the evidence is important.

The Guiding Question:	
Our Claim:	
Our Evidence:	Our Justification of the Evidence:

Argumentation Session

Share your argument with your classmates. Be sure to ask them how to make your draft argument better. Keep track of their suggestions in the space below.

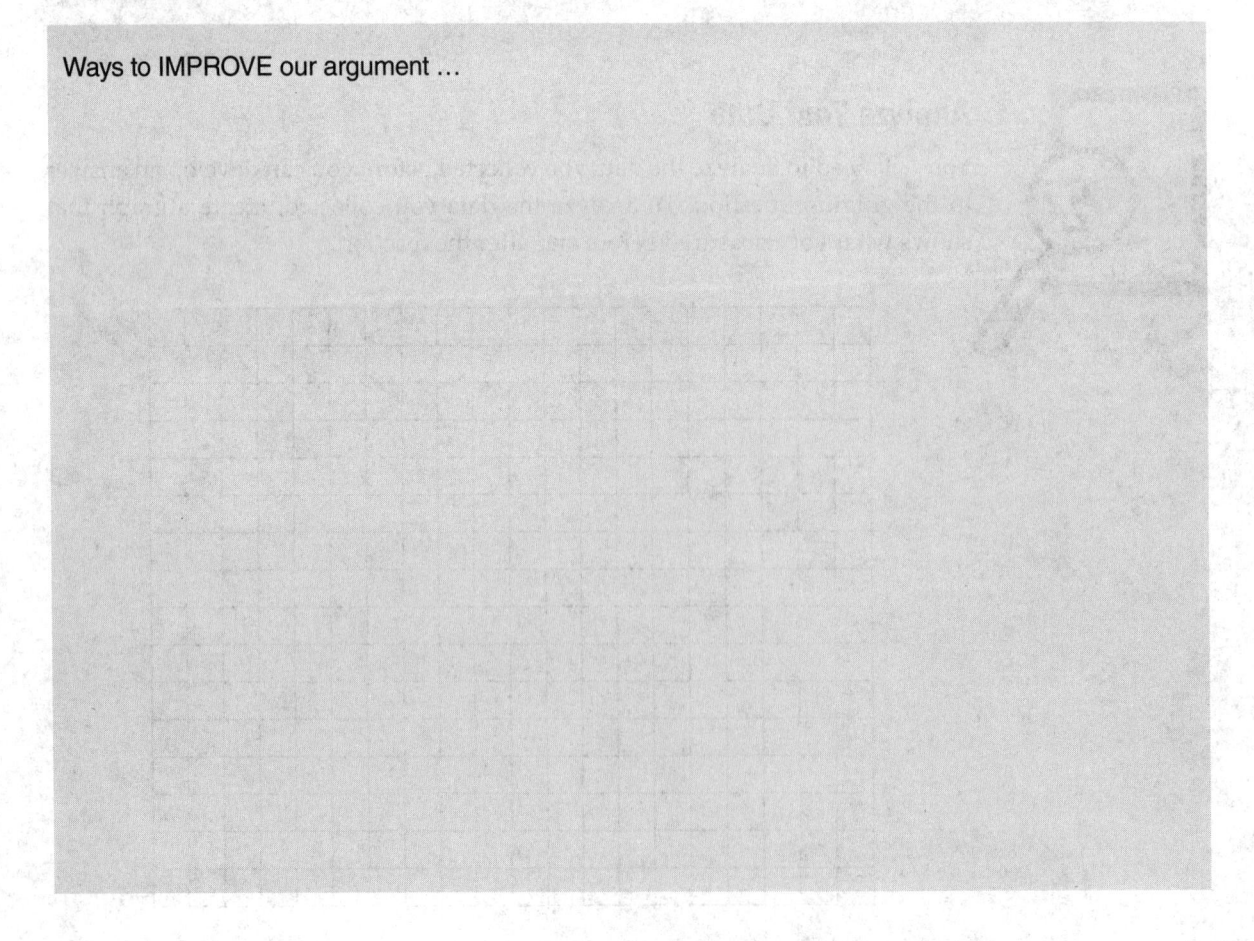

Ways to IMPROVE our argument …

Investigation 5. Reactions and Weight: What Happens to the Total Weight of a Closed System When a Chemical Reaction Takes Place Within That System?

Draft Report

Prepare an investigation report to share what you have learned. Use the information in this handout and your group's final argument to write a *draft* of your investigation report.

Introduction

We have been studying _____ in class.

Before we started this investigation, we explored _____
_____
_____

We noticed _____
_____
_____

My goal for this investigation was to figure out _____
_____
_____

The guiding question was _____
_____
_____

Method

To gather the data I needed to answer this question, I _____
_____
_____
_____
_____

Argument-Driven Inquiry in **Fifth-Grade Science**: Three-Dimensional Investigations

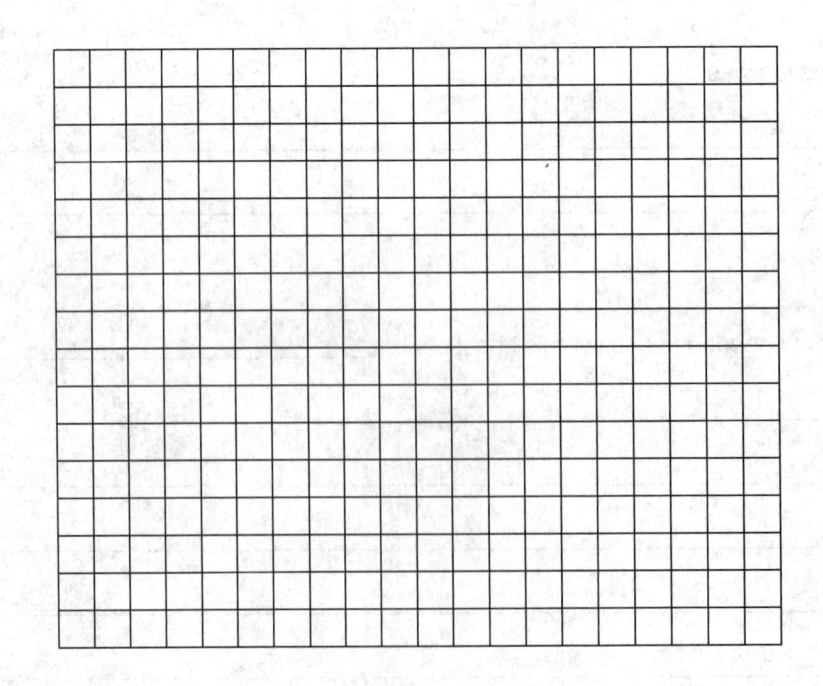

Investigation Handout

I then analyzed the data I collected by _____

_____

_____

_____

Argument

My claim is _____

_____

_____

_____

The graph below shows _____

_____

_____

_____

_____

Investigation 5. Reactions and Weight:
What Happens to the Total Weight of a Closed System When
a Chemical Reaction Takes Place Within That System?

This analysis of the data I collected suggests _____

_____

_____

_____

_____

This evidence is based on several important scientific concepts. The first one is _____

_____

_____

_____

_____

_____

Review

Your classmates need your help! Review the draft of their investigation reports and give them ideas about how to improve. Use the *peer-review guide* when doing your review.

Submit Your Final Report

Once you have received feedback from your classmates about your draft report, create your final investigation report and hand it in to your teacher.

Checkout Questions

Investigation 5. Reactions and Weight

Use the following information to answer questions 1 and 2. The picture below shows a sealed container. Inside the container are two liquids: one liquid is in a small test tube and the other liquid is at the bottom of the container. The container, the test tube, and the two liquids weigh a total of 163 grams. The sealed container is then turned over so the two liquids mix. A chemical reaction takes place inside the container. The chemical reaction creates a solid at the bottom of the container. The container remained sealed the entire time.

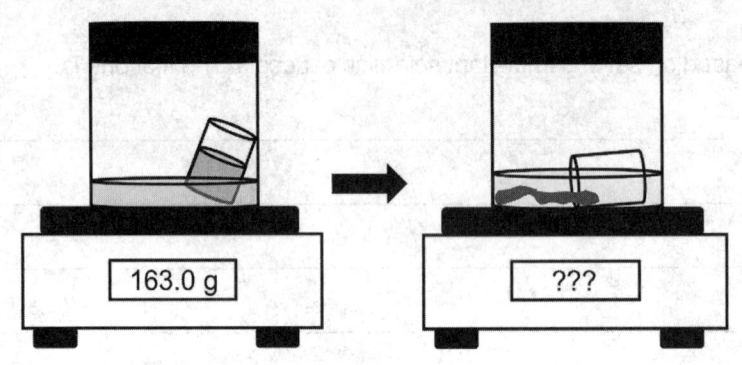

1. How much do you think the container weighed after the chemical reaction?

    a. 163 grams

    b. More than 163 grams

    c. Less than 163 grams

2. How do you know? Use what you know about systems to explain your answer.

___

Teacher Scoring Rubric for Checkout Questions 1 and 2

Level	Description
3	The student can apply the core idea correctly and can explain the importance of thinking about systems.
2	The student cannot apply the core idea correctly but can explain the importance of thinking about systems.
1	The student can apply the core idea correctly but cannot explain the importance of thinking about systems.
0	The student cannot apply the core idea correctly and cannot explain the importance of thinking about systems.

Investigation 5. Reactions and Weight: What Happens to the Total Weight of a Closed System When a Chemical Reaction Takes Place Within That System?

Use the following information to answer questions 3 and 4. The picture below shows a sealed container. Inside the container is a piece of metal and a test tube with a liquid in it. The container and everything inside of it weighs 99 grams. The container is then turned upside down so the metal and the liquid mix. A chemical reaction takes place inside the container, and the metal inside the container seem to disappear. The sealed container with everything inside of it is then weighed for a second time. The weight of the container and everything inside of it after the chemical reaction is 3.5 ounces.

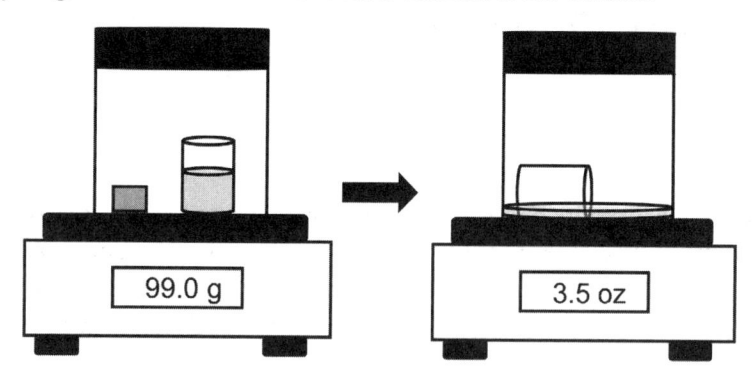

3. What do you think happened to the weight of the container and everything inside of it after the chemical reaction?

    a. It decreased.

    b. It increased.

    c. It stayed the same.

    d. Unable to tell.

4. Explain your thinking. Use what you know about using standard units during an investigation as part of your answer.

___

Teacher Scoring Rubric for Checkout Questions 3 and 4

Level	Description
3	The student can apply the core idea correctly and can explain the importance of using standard units to measure or describe the physical quantities of matter.
2	The student can apply the core idea correctly but cannot explain the importance of using standard units to measure or describe the physical quantities of matter.
1	The student cannot apply the core idea correctly but can explain the importance of using standard units to measure or describe the physical quantities of matter.
0	The student cannot apply the core idea correctly and cannot explain the importance of using standard units to measure or describe the physical quantities of matter.

Argument-Driven Inquiry in **Fifth-Grade Science:** Three-Dimensional Investigations

Teacher Notes

Investigation 6
Physical and Chemical Properties: What Are the Identities of the Unknown Powders?

Purpose

The purpose of this investigation is to give students an opportunity to use one disciplinary core idea (DCI), two crosscutting concepts (CCs), and six scientific and engineering practices (SEPs) to figure out the identities of two unknown powders. Students will also learn about how scientists use different methods to answer different types of questions.

The DCI, CCs, and SEPs That Students Use During This Investigation to Figure Things Out

DCI

- *PS1.A: Structure and Properties of Matter:* Measurements of a variety of properties can be used to identify materials.

CCs

- *CC 1: Patterns*: Patterns can be used as evidence to support an explanation.
- *CC 3: Scale, Proportion, and Quantity:* Standard units are used to measure and describe physical quantities such as weight, time, temperature, and volume.

SEPs

- *SEP 1: Asking Questions and Defining Problems:* Ask questions about what would happen if a variable is changed. Ask questions that can be investigated and predict reasonable outcomes based on patterns such as cause-and-effect relationships.
- *SEP 3: Planning and Carrying Out Investigations:* Plan and conduct an investigation collaboratively to produce data to serve as the basis for evidence, using fair tests in which variables are controlled and the number of trials considered. Evaluate appropriate methods and/or tools for collecting data.
- *SEP 4: Analyzing and Interpreting Data:* Represent data in tables and/or various graphical displays (bar graphs, pictographs, and/or pie charts) to reveal patterns that indicate relationships. Analyze and interpret data to make sense of phenomena, using logical reasoning, mathematics, and/or computation. Compare and contrast data collected by different groups in order to discuss similarities and differences in their findings.

Investigation 6. Physical and Chemical Properties: What Are the Identities of the Unknown Powders?

- *SEP 6: Constructing Explanations and Designing Solutions:* Construct an explanation of observed relationships. Use evidence to construct or support an explanation. Identify the evidence that supports particular points in an explanation.
- *SEP 7: Engaging in Argument From Evidence:* Compare and refine arguments based on an evaluation of the evidence presented. Distinguish among facts, reasoned judgment based on research findings, and speculation in an explanation. Respectfully provide and receive critiques from peers about a proposed procedure, explanation, or model by citing relevant evidence and posing specific questions.
- *SEP 8: Obtaining, Evaluating, and Communicating Information:* Read and comprehend grade-appropriate complex texts and/or other reliable media to summarize and obtain scientific and technical ideas. Combine information in written text with that contained in corresponding tables, diagrams, and/or charts to support the engagement in other scientific and/or engineering practices. Communicate scientific and/or technical information orally and/or in written formats, including various forms of media as well as tables, diagrams, and charts.

Other Concepts That Students May Use During This Investigation

Students might also use some of the following concepts:

- Physical properties are characteristics of a material that can be observed without mixing it with another type of material.
- Chemical properties are characteristics of a material that can only be observed when it is mixed with a different type of material.
- The physical and chemical properties of a material do not change.
- Materials of any type are made up of particles that are too small to see.
- Different types of materials have different physical and chemical properties because they are made of different types of particles. These particles have a unique structure.

What Students Figure Out

The unknown powders are salt and cornstarch.

Background Information About This Investigation for the Teacher

Different types of materials have different properties. A property is any measurable or observable characteristic. There are two kinds of properties that we use to describe a material. *Physical properties* are characteristics of a material that can be observed without mixing it with another type of material. Examples of physical properties are color and texture. *Chemical properties,* in contrast, are characteristics of a material that can only be observed

Teacher Notes

when it is mixed with another material. Examples of chemical properties are how a material reacts with water or how a material reacts with an acid (such as vinegar). The physical and chemical properties of a material do not change. Scientists can therefore use physical and chemical properties to tell different materials apart.

Materials are made up of particles that are too small to see. Different materials have different physical and chemical properties because they are made of different types of particles. These particles have a unique structure. For example, all particles of sugar have the same structure, but this structure is different from the particles that make up other materials. The unique structure of the particles that make up sugar causes it to be white in color, have a granular texture, and dissolve when it is added to water. Particles of iron have a different structure. The unique structure of the particles that make up iron causes it to be a shiny, silver-gray metal with a smooth texture that does not react with water. However, iron does react with acids to form a gas and with oxygen to form rust. The structure of the particles of iron also makes it magnetic. Scientists can use these physical and chemical properties to distinguish sugar from other types of powders such as salt or baking soda and to distinguish iron from other types of metals such as aluminum and tin.

Timeline

The time needed to complete this investigation is 270 minutes (4 hours and 30 minutes). The amount of instructional time needed for each stage of the investigation is as follows:

- *Stage 1.* Introduce the task and the guiding question: 35 minutes
- *Stage 2.* Design a method and collect data: 50 minutes
- *Stage 3.* Create a draft argument: 45 minutes
- *Stage 4.* Argumentation session: 30 minutes
- *Stage 5.* Reflective discussion: 15 minutes
- *Stage 6.* Write a draft report: 30 minutes
- *Stage 7.* Peer review: 35 minutes
- *Stage 8.* Revise the report: 30 minutes

Materials and Preparation

The materials needed for this investigation are listed in Table 6.1. Most of the items can be purchased from a big-box retail store such as Walmart or Target or through an online retailer such as Amazon, but you may need to purchase some items (such as the well plate) from a science education supply company such as Ward's Science (*www.wardsci.com*) or Flinn Scientific (*www.flinnsci.com*). The materials for this investigation can also be purchased as a complete kit (which includes enough materials for 24 students, or six groups of four students) at *www.argumentdriveninquiry.com*.

Investigation 6. Physical and Chemical Properties:
What Are the Identities of the Unknown Powders?

TABLE 6.1

Materials for Investigation 6

Item	Quantity
Safety goggles	1 per student
Nonlatex apron	1 per student
Vinyl gloves	1 pair per student
Alka-Seltzer tablet (crushed) (for opening activity)	1 per class
Sugar (for opening activity)	5 g per class
Clear plastic cup half-filled with water (for opening activity)	1 per class
Plastic ramekins with lids	6 per class
Baking powder	10 g per group
Baking soda	10 g per group
Cornstarch	10 g per group
Salt	10 g per group
Unknown powder A (salt), 10 g	1 per group
Unknown powder B (cornstarch), 10 g	1 per group
Iodine in 30-ml amber glass dropper bottle	1 bottle per group
Vinegar in 30-ml plastic dropper bottle	1 bottle per group
Water in 30-ml plastic dropper bottle	1 bottle per group
Hand lens	2 per group
Well plate with 6 wells	1 per group
Small plastic spoons	6 per group
Whiteboard, 2' × 3'*	1 per group
Investigation Handout	1 per student
Peer-review guide and teacher scoring rubric	1 per student
Checkout Questions (optional)	1 per student

*As an alternative, students can use computer and presentation software such as Microsoft PowerPoint or Apple Keynote to create their arguments.

We recommend putting each powder in a plastic ramekin with a lid. You can then label the ramekin with the name of the powder using a felt-tip marker. This will make it easier to pass out and collect all these different materials during stage 2. Students will only need to use a small amount of each powder and liquid for each test.

Safety Precautions

Remind students to follow all normal safety rules. In addition, tell the students to take the following safety precautions:

Teacher Notes

- Wear sanitized safety goggles, nonlatex aprons, and vinyl gloves during setup, investigation activity, and cleanup.
- Be careful when working with iodine because it can stain skin and clothes.
- Do not taste any substances used in this activity.
- Immediately clean up any spills to avoid a slip or fall hazard.
- Keep water sources away from electrical receptacles to prevent shock.
- Be careful when handling glassware, because it can shatter and cut skin.
- Make sure all materials are put away after completing the activity.
- Wash their hands with soap and water when they are done cleaning up.

Lesson Plan by Stage

This lesson plan is only a suggestion. It is included here to illustrate what you can say and do during each stage of ADI for this specific investigation. We encourage you to modify this lesson plan by asking different questions, using different examples, and providing different scaffolds as needed to better meet the needs of students in your class.

Stage 1: Introduce the Task and the Guiding Question (35 minutes)

1. Ask the students to sit in six groups, with three or four students in each group.
2. Ask the students to clear off their desks except for a pencil (and their *Student Workbook for Argument-Driven Inquiry in Fifth-Grade Science* if they have one).
3. Pass out an Investigation Handout to each student (or ask students to turn to the Investigation Log for Investigation 6 in their workbook).
4. Read the first paragraph of the "Introduction" aloud to the class. Ask the students to follow along as you read.
5. Put a clear plastic cup that is half-filled with water on a table where everyone can see it. You can also place it under a document camera and project the image on a screen.
6. Add 5 grams of sugar to the cup.
7. Ask the students to record their observations and questions on the "NOTICED/WONDER" chart in the "Introduction."
8. Add a crushed Alka-Seltzer tablet to the same cup.
9. Ask the students to record their observations and questions on the "NOTICED/WONDER" chart in the "Introduction."
10. Ask the students to share what they noticed after each powder was added to the cup.

Investigation 6. Physical and Chemical Properties:
What Are the Identities of the Unknown Powders?

11. Ask the students to share what questions they have about what happened after each powder was added to the cup.

12. Tell the students, "Some of your questions might be answered by reading the rest of the 'Introduction.'"

13. Ask the students to read the rest of the "Introduction" on their own *or* ask them to follow along as you read it aloud.

14. Once the students have read the rest of the "Introduction," ask them to fill out the "Things we KNOW" chart on their Investigation Handout (or in their Investigation Log) as a group.

15. Ask the students to share what they learned from the reading. Add these ideas to a class "Things we KNOW" chart.

16. Tell the students, "Let's see what we will need to figure out during our investigation."

17. Read the task and the guiding question aloud.

18. Tell the students, "I have lots of materials here that you can use."

19. Introduce the students to the materials available for them to use during the investigation by holding each one up and asking how it might be used.

Stage 2: Design a Method and Collect Data (50 minutes)

1. Tell the students, "I am now going to give you and the other members of your group about 15 minutes to plan your investigation. Before you begin, I want you all to take a couple of minutes to discuss the following questions with the rest of your group."

2. Show the following questions on the screen or board:
   - What information should we collect so we can *describe the properties* of a powder?
   - What types of *patterns* might we look for to help answer the guiding question?

3. Tell the students, "Please take a few minutes to come up with an answer to these questions."

4. Give the students two to three minutes to discuss these two questions.

5. Ask two or three different groups to share their answers. Highlight or write down any important ideas on the board so students can refer to them later.

6. If possible, use a document camera to project an image of the graphic organizer for this investigation on a screen or board (or take a picture of it and project the picture on a screen or board). Tell the students, "I now want you all to plan out your investigation. To do that, you will need to fill out this investigation proposal."

Argument-Driven Inquiry in **Fifth-Grade Science:** Three-Dimensional Investigations

Teacher Notes

7. Point to the box labeled "Our guiding question:" and tell the students, "You can put the question we are trying to answer in this box." Then ask, "Where can we find the guiding question?"

8. Wait for a student to answer.

9. Point to the box labeled "This is a picture of how we will set up the equipment:" and tell the students, "You can draw a picture in this box of how you will set up the equipment in order to carry out this investigation."

10. Point to the box labeled "We will collect the following data:" and tell the students, "You can list the measurements or observations that you will need to collect during the investigation in this box."

11. Point to the box labeled "These are the steps we will follow to collect data:" and tell the students, "You can list what you are going to do to collect the data you need and what you will do with your data once you have it. Be sure to give enough detail that I could do your investigation for you."

12. Ask the students, "Do you have any questions about what you need to do?"

13. Answer any questions that come up.

14. Tell the students, "Once you are done, raise your hand and let me know. I'll then come by and look over your proposal and give you some feedback. You may not begin collecting data until I have approved your proposal by signing it. You need to have your proposal done in the next 15 minutes."

15. Give the students 15 minutes to work in their groups on their investigation proposal. As they work, move from group to group to check in, ask probing questions, and offer a suggestion if a group gets stuck.

16. As each group finishes its investigation proposal, read it over and determine if it will be productive or not. If you feel the investigation will be productive (not necessarily what you would do or what the other groups are doing), sign your name on the proposal and let the group start collecting data. If the plan needs to be changed, offer some suggestions or ask some probing questions, and have the group make the changes before you approve it.

17. Pass out the materials or have one student from each group collect the materials they need from a central supply table or cart for the groups that have an approved proposal.

18. Remind students of the safety rules and precautions for this investigation.

19. Tell the students to collect their data and record their observations or measurements in the "Collect Your Data" box in their Investigation Handout (or the Investigation Log in their workbook).

20. Give the students 30 minutes to collect their data. Collect the materials from each group before asking them to analyze their data.

Investigation 6. Physical and Chemical Properties:
What Are the Identities of the Unknown Powders?

What should a student-designed investigation look like?

There are a number of different investigations that students can design to answer the question "What are the identities of the unknown powders?" For example, one method might include the following steps:

1. Examine the color and texture of the four known powders (cornstarch, baking powder, baking soda, salt).
2. Mix a few drops of iodine with each known powder.
3. Mix a few drops of vinegar with each known powder.
4. Mix a few drops of water with each known powder.
5. Examine the color and texture of the two unknown powders (A and B).
6. Mix a few drops of iodine with each unknown powder.
7. Mix a few drops of vinegar with each unknown powder.
8. Mix a few drops of water with each unknown powder.

If students use this method, they will need to collect the following data:

1. Color of each powder (a physical property)
2. Texture of each powder (a physical property)
3. Reaction of each powder with iodine (a chemical property)
4. Reaction of each powder with vinegar (a chemical property)
5. Reaction of each powder with water (a chemical property)

Stage 3: Create a Draft Argument (45 minutes)

1. Tell the students, "Now that we have all this data, we need to analyze the data so we can figure out an answer to the guiding question."

2. If possible, project an image of the "Analyze Your Data" section for this investigation on a screen or board using a document camera (or take a picture of it and project the picture on a screen or board). Point to the section and tell the students, "You can create a table, graph, or other visual representation as a way to analyze your data. You can make your table, graph, or other visual representation in this section."

3. Ask the students, "What information do we need to include in this analysis?"

4. Tell the students, "Please take a few minutes to discuss this question with your group, and be ready to share."

Teacher Notes

5. Give the students five minutes to discuss.

6. Ask two or three different groups to share their answers. Highlight or write down any important ideas on the board so students can refer to them later.

7. Tell the students, "I am now going to give you and the other members of your group about 10 minutes to analyze your data." If the students are having trouble analyzing their data, you can take a few minutes to provide a mini-lesson about possible ways to analyze the data they collected (this strategy is called just-in-time instruction because it is offered only when students get stuck).

8. Give the students 10 minutes to analyze their data. As they work, move from group to group to check in, ask probing questions, and offer suggestions.

9. Tell the students, "I am now going to give you and the other members of your group about 15 minutes to create an argument to share what you have learned and convince others that they should believe you. Before you do that, we need to take a few minutes to discuss what you need to include in your argument."

10. If possible, use a document camera to project the "Argument Presentation on a Whiteboard" image from the "Draft Argument" section of the Investigation Handout (or the Investigation Log in their workbook) on a screen or board (or take a picture of it and project the picture on a screen or board).

11. Point to the box labeled "The Guiding Question:" and tell the students, "You can put the question we are trying to answer here on your whiteboard."

12. Point to the box labeled "Our Claim:" and tell the students, "You can put your claim here on your whiteboard. The claim is your answer to the guiding question."

13. Point to the box labeled "Our Evidence:" and tell the students, "You can put the evidence that you are using to support your claim here on your whiteboard. Your evidence will need to include the analysis you just did and an explanation of what your analysis means or shows. Scientists always need to support their claims with evidence."

14. Point to the box labeled "Our Justification of the Evidence:" and tell the students, "You can put your justification of your evidence here on your whiteboard. Your justification needs to explain why your evidence is important. Scientists often use core ideas to explain why the evidence they are using matters. Core ideas are important concepts that scientists use to help them make sense of what happens during an investigation."

15. Ask the students, "What are some core ideas that we read about earlier that might help us explain why the evidence we are using is important?"

16. Ask the students to share some of the core ideas from the "Introduction" section of the Investigation Handout (or the Investigation Log in the workbook). List these core ideas on the board.

Investigation 6. Physical and Chemical Properties:
What Are the Identities of the Unknown Powders?

17. Tell the students, "That is great. I would like to see everyone try to include these core ideas in your justification of the evidence. Your goal is to use these core ideas to help explain why your evidence matters and why the rest of us should pay attention to it."

18. Ask the students, "Do you have any questions about what you need to do?"

19. Answer any questions that come up.

20. Tell the students, "Okay, go ahead and start working on your arguments. You need to have your argument done in the next 15 minutes. It doesn't need to be perfect. We just need something down on the whiteboards so we can share our ideas."

21. Give the students 15 minutes to work in their groups on their arguments. As they work, move from group to group to check in, ask probing questions, and offer a suggestion if a group gets stuck. Figure 6.1 (p. 250) shows an example of an argument created by students for this investigation.

What should the table, graph, or other visual representation for this investigation look like?

There are a number of different ways that students can analyze the qualitative data they collect during this investigation. One of the most straightforward ways is to create a seven-column table. With this approach, the heading for column 1 should be "Properties," the headings for columns 2–5 should be the names of the known powders, and the headings for columns 6 and 7 should be "Unknown powder A" and "Unknown powder B."

Another way to analyze the data is to create a column graph. With this approach, the vertical, or y-axis, should show the number of properties in common (with values between 0 and 5), and the horizontal, or x-axis, should show pairs of names such as "Unknown powder A and baking powder," "Unknown powder A and baking soda," and "Unknown powder A and cornstarch." There will be eight columns in total. The height of the columns should correspond to the number of properties that each unknown powder has in common with each of the known powders.

There are other options for analyzing the collected data (see Figure 6.1 on p. 250 for an example of an argument that includes qualitative descriptions and a picture of the group's well plate as evidence). Students often come up with some unique ways of analyzing their data, so be sure to give them some voice and choice during this stage.

Teacher Notes

Stage 4: Argumentation Session (30 minutes)

The argumentation session can be conducted in a whole-class presentation format, a gallery walk format, or a modified gallery walk format. We recommend using a whole-class presentation format for the first investigation, but try to transition to either the gallery walk or modified gallery walk format as soon as possible because that will maximize student voice and choice inside the classroom. The following list shows the steps for the three formats; unless otherwise noted, the steps are the same for all three formats.

FIGURE 6.1

Example of an argument

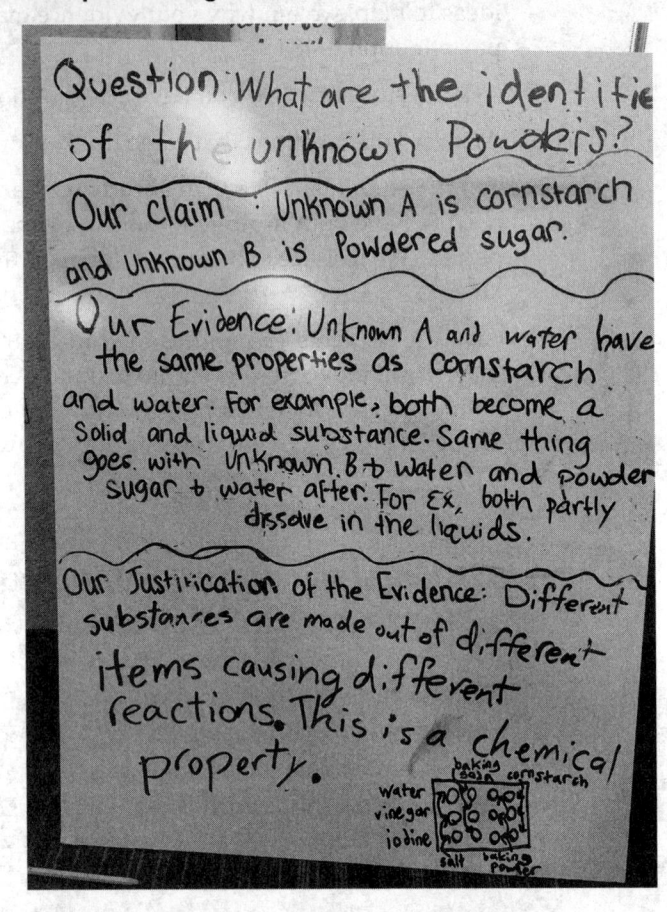

> Question: What are the identities of the unknown powders?
> Our Claim: Unknown A is cornstarch and Unknown B is powdered sugar.
> Our Evidence: Unknown A and water have the same properties as cornstarch and water. For example, both become a solid and liquid substance. Same thing goes with Unknown B to water and powder sugar to water after. For Ex, both partly dissolve in the liquids.
> Our Justification of the Evidence: Different substances are made out of different items causing different reactions. This is a chemical property.

1. Begin by introducing the use of the whiteboard.
   - *If using the whole-class presentation format*, tell the students, "We are now going to share our arguments. Please set up your whiteboards so everyone can see them."
   - *If using the gallery walk or modified gallery walk format*, tell the students, "We are now going to share our arguments. Please set up your whiteboards so they are facing the walls."

2. Allow the students to set up their whiteboards.
   - *If using the whole-class presentation format*, the whiteboards should be set up on stands or chairs so they are facing toward the center of the room.
   - *If using the gallery walk or modified gallery walk format*, the whiteboards should be set up on stands or chairs so they are facing toward the outside of the room.

3. Give the following instructions to the students:
   - *If using the whole-class presentation format*, tell the students, "Okay, before we get started I want to explain what we are going to do next. Your group will have an opportunity to share your argument with the rest of the class. After you are done, everyone else in the class will have a chance to ask questions and offer

some suggestions about ways to make your group's argument better. After we have a chance to listen to each other and learn something new, I'm going to give you some time to revise your arguments and make them better."

- *If using the gallery walk format*, tell the students, "Okay, before we get started I want to explain what we are going to do next. You are going to read the arguments that were created by other groups. When I say 'go,' your group will go to a different group's station so you can see their argument. Once you are there, I'll give your group a few minutes to read and review their argument. Your job is to offer them some suggestions about ways to make their argument better. You can use sticky notes to give them suggestions. Please be specific about what you want to change and how you think they should change it. After we have a chance to learn from each other, I'm going to give you some time to revise your arguments and make them better."

- *If using the modified gallery walk format*, tell the students, "Okay, before we get started I want to explain what we are going to do next. I'm going to ask some of you to present your arguments to your classmates. If you are presenting your argument, your job is to share your group's claim, evidence, and justification of the evidence. The rest of you will be travelers. If you are a traveler, your job is to listen to the presenters, ask the presenters questions if you do not understand something, and then offer them some suggestions about ways to make their argument better. After we have a chance to learn from each other, I'm going to give you some time to revise your arguments and make them better."

4. Use a document camera to project the "Ways to IMPROVE our argument ..." box from the Investigation Handout (or the Investigation Log in their workbook) on a screen or board (or take a picture of it and project the picture on a screen or board).

- *If using the whole-class presentation format*, point to the box and tell the students, "After your group presents your argument, you can write down the suggestions you get from your classmates here. If you are listening to a presentation and you see a good idea from another group, you can write down that idea here as well. Once we are done with the presentations, I will give you a chance to use these suggestions or ideas to improve your arguments."

- *If using the gallery walk format*, point to the box and tell the students, "If you see a good idea from another group, you can write it down here. Once we are done reviewing the different arguments, I will give you a chance to use these ideas to improve your own arguments. It is important to share ideas like this."

- *If using the modified gallery walk format*, point to the box and tell the students, "If you are a presenter, you can write down the suggestions you get from the travelers here. If you are a traveler and you see a good idea from another group, you can write down that idea here. Once we are done with the presentations,

Teacher Notes

I will give you a chance to use these suggestions or ideas to improve your arguments."

5. Ask the students, "Do you have any questions about what you need to do?"

6. Answer any questions that come up.

7. Give the following instructions:

    - *If using the whole-class presentation format,* tell the students, "Okay. Let's get started."

    - *If using the gallery walk format,* tell the students, "Okay, I'm now going to tell you which argument to go to and review."

    - *If using the modified gallery walk format,* tell the students, "Okay, I'm now going to assign you to be a presenter or a traveler." Assign one or two students from each group to be presenters and one or two students from each group to be travelers.

8. Give the students an opportunity to review the arguments.

    - *If using the whole-class presentation format,* have each group present their argument one at a time. Give each group only two to three minutes to present their argument. Then give the class two to three minutes to ask them questions and offer suggestions. Encourage as much participation from the students as possible.

    - *If using the gallery walk format,* tell the students, "Okay. Let's get started. Each group, move one argument to the left. Don't move to the next argument until I tell you to move. Once you get there, read the argument and then offer suggestions about how to make it better. I will put some sticky notes next to each argument. You can use the sticky notes to leave your suggestions." Give each group about three to four minutes to read the arguments, talk, and offer suggestions.

        a. After three to four minutes, tell the students, "Okay. Let's move on to the next argument. Please move one group to the left."

        b. Again, give each group three to four minutes to read, talk, and offer suggestions.

        c. Repeat this process until each group has had their argument read and critiqued three times.

    - *If using the modified gallery walk format,* tell the students, "Okay. Let's get started. Reviewers, move one group to the left. Don't move to the next group until I tell you to move. Presenters, go ahead and share your argument with the travelers when they get there." Give each group of presenters and travelers about three to four minutes to talk.

        a. Tell the students, "Okay. Let's move on to the next argument. Travelers, move one group to the left."

Investigation 6. Physical and Chemical Properties:
What Are the Identities of the Unknown Powders?

b. Again, give each group of presenters and travelers about three to four minutes to talk.

c. Repeat this process until each group has had their argument read and critiqued three times.

9. Tell the students to return to their workstations.

10. Give the following instructions about revising the argument:

- *If using the whole-class presentation format,* tell the students, "I'm now going to give you all about 10 minutes to revise your argument. Take a few minutes to talk in your groups and determine what you want to change to make your argument better. Once you have decided what to change, go ahead and make the changes to your whiteboard."

- *If using the gallery walk format,* tell the students, "I'm now going to give you all about 10 minutes to revise your argument. Take a few minutes to read the suggestions that were left at your argument. Then talk in your groups and determine what you want to change to make your argument better. Once you have decided what to change, go ahead and make the changes to your whiteboard."

- *If using the modified gallery walk format,* tell the students, "I'm now going to give you all about 10 minutes to revise your argument. Please return to your original groups." Wait for the students to move back into their original groups and then tell the students, "Okay, take a few minutes to talk in your groups and determine what you want to change to make your argument better. Once you have decided what to change, go ahead and make the changes to your whiteboard."

11. Ask the students, "Do you have any questions about what you need to do?"

12. Answer any questions that come up.

13. Tell the students, "Okay. Let's get started."

14. Give the students 10 minutes to work in their groups on their arguments. As they work, move from group to group to check in, ask probing questions, and offer a suggestion if a group gets stuck.

Stage 5: Reflective Discussion (15 minutes)

1. Tell the students, "We are now going to take a minute to talk about some of the core ideas and crosscutting concepts that we have used during our investigation."

2. Show Figure 6.2 (p. 254) on the screen.

3. Ask the students, "What do you all see going on here?"

Teacher Notes

4. Allow the students to share their ideas.

5. Ask the students, "How can use what we know about the structure and properties of matter to figure out the identity of this white powder?"

6. Allow the students to share their ideas. As they share their ideas, ask different questions to encourage them to expand on their thinking (e.g., "Can you tell me more about that?"), clarify a contribution (e.g., "Can you say that in another way?"), support an idea (e.g., "Why do you think that?"), add to an idea mentioned by a classmate (e.g., "Would anyone like to add to the idea?"), re-voice an idea offered by a classmate (e.g., "Who can explain that to me in another way?"), or critique an idea during the discussion (e.g., "Do you agree or disagree with that idea and why?") until students are able to generate an adequate explanation.

7. Tell the students, "We also looked for patterns during our investigation." Then ask, "Can anyone tell me why we needed to look for patterns?"

8. Allow the students to share their ideas.

9. Tell the students, "I think patterns are really important in science because differences or similarities in patterns can be used to classify and identify different materials. Patterns can also be used as evidence to support an explanation."

10. Ask the students, "What are some of the patterns you found during your investigation?"

11. Allow the students to share their ideas.

12. Tell the students, "We used standard units to describe the properties of the powders during our investigation." Then ask, "Can anyone tell me why this was important?"

13. Allow the students to share their ideas.

14. Tell the students, "I think using standard units to measure and describe physical properties such as color or texture is important because we need to have a common understanding of what our observations mean. This allows us to share and critique each other's ideas and evidence."

15. Ask the students, "How did you try to standardize your observations today?"

FIGURE 6.2

A small bowl of flour

Investigation 6. Physical and Chemical Properties:
What Are the Identities of the Unknown Powders?

16. Show an image of the question "What do you think are the most important core ideas or crosscutting concepts that we used during this investigation to help us make sense of what we observed?" Tell the students, "Okay, let's make sure we are all on the same page. Please take a moment to discuss this question with the other people in your group." Give them a few minutes to discuss the question.

17. Ask the students, "What do you all think? Who would like to share?"

18. Allow the students to share their ideas.

19. Tell the students, "We are now going to take a minute to talk about what scientists do to figure out answers to their questions."

20. Show an image of the question "Do all scientists follow the same method regardless of what they are trying to figure out?" on the screen. Tell the students, "Take a few minutes to talk about how you would answer this question with the other people in your group. Be ready to share with the rest of the class." Give the students two to three minutes to talk in their group.

21. Ask the students, "What do you all think? Who would like to share an idea?"

22. Allow the students to share their ideas.

23. Tell the students, "I think there is no universal step-by-step scientific method that all scientists follow; rather, scientists who work in different scientific disciplines such as biology or physics and fields within a discipline such as genetics or ecology use different types of methods, use different core ideas, and rely on different standards to figure out how the world works."

24. Ask the students, "What might be some different methods that scientists can use?"

25. Allow the students to share their ideas.

26. Tell the students, "There are many examples of different methods, including experiments, systematic observations, literature reviews, and analysis of existing data sets. The choice of method depends on what the scientist is trying to accomplish."

27. Ask the students, "What type of method do you think we used today?"

28. Allow students to share their ideas.

29. Tell the students, "I think we did a systematic observation because we were looking for patterns that we could use to identify the unknown powders. It wasn't an experiment because we were not looking for a cause-and-effect relationship."

30. Ask the students, "Does anyone have any questions about the different methods that scientists use or why scientists use different methods?"

31. Answer any questions that come up.

Teacher Notes

32. Tell the students, "We are now going to take a minute to talk about what went well and what didn't go so well during our investigation. We need to talk about this because you all are going to be planning and carrying out your own investigations like this a lot this year, and I want to help you all get better at it."

33. Show an image of the question "What made your investigation scientific?" on the screen. Tell the students, "Take a few minutes to talk about how you would answer this question with the other people in your group. Be ready to share with the rest of the class." Give the students two to three minutes to talk in their group.

34. Ask the students, "What do you all think? Who would like to share an idea?"

35. Allow the students to share their ideas. Be sure to expand on their ideas about what makes an investigation scientific.

36. Show an image of the question "What made your investigation not so scientific?" on the screen. Tell the students, "Take a few minutes to talk about how you would answer this question with the other people in your group. Be ready to share with the rest of the class." Give the students two to three minutes to talk in their group.

37. Ask the students, "What do you all think? Who would like to share an idea?"

38. Allow the students to share their ideas. Be sure to expand on their ideas about what makes an investigation less scientific.

39. Show an image of the question "What rules can we put into place to help us make sure our next investigation is more scientific?" on the screen. Tell the students, "Take a few minutes to talk about how you would answer this question with the other people in your group. Be ready to share with the rest of the class." Give the students two to three minutes to talk in their group.

40. Ask the students, "What do you all think? Who would like to share an idea?"

41. Allow the students to share their ideas. Once they have shared their ideas, offer a suggestion for a possible class rule.

42. Ask the students, "What do you all think? Should we make this a rule?"

43. If the students agree, write the rule on the board or on a class "Rules for Scientific Investigation" chart so you can refer to it during the next investigation.

Stage 6: Write a Draft Report (30 minutes)

Your students will use either the Investigation Handout or the Investigation Log in the student workbook when writing the draft report. When you give the directions shown in quotes in the following steps, substitute "Investigation Log in your workbook" or just "Investigation Log" (as shown in brackets) for "handout" if they are using the workbook.

Investigation 6. Physical and Chemical Properties:
What Are the Identities of the Unknown Powders?

1. Tell the students, "You are now going to write an investigation report to share what you have learned. Please take out a pencil and turn to the 'Draft Report' section of your handout [Investigation Log in your workbook]."

2. If possible, use a document camera to project the "Introduction" section of the draft report from the Investigation Handout (or the Investigation Log in their workbook) on a screen or board (or take a picture of it and project the picture on a screen or board).

3. Tell the students, "The first part of the report is called the 'Introduction.' In this section of the report you want to explain to the reader what you were investigating, why you were investigating it, and what question you were trying to answer. All this information can be found in the text at the beginning of your handout [Investigation Log]." Point to the image. "Here are some sentence starters to help you begin writing."

4. Ask the students, "Do you have any questions about what you need to do?"

5. Answer any questions that come up.

6. Tell the students, "Okay, let's write."

7. Give the students 10 minutes to write the "Introduction" section of the report. As they work, move from student to student to check in, ask probing questions, and offer a suggestion if a student gets stuck.

8. If possible, use a document camera to project the "Method" section of the draft report from the Investigation Handout (or the Investigation Log in their workbook) on a screen or board (or take a picture of it and project the picture on a screen or board).

9. Tell the students, "The second part of the report is called the 'Method.' In this section of the report you want to explain to the reader what you did during the investigation, what data you collected and why, and how you went about analyzing your data. All this information can be found in the 'Plan Your Investigation' section of the handout [Investigation Log]. Remember that you all planned and carried out different investigations, so do not assume that the reader will know what you did." Point to the image. "Here are some sentence starters to help you begin writing."

10. Ask the students, "Do you have any questions about what you need to do?"

11. Answer any questions that come up.

12. Tell the students, "Okay, let's write."

13. Give the students 10 minutes to write the "Method" section of the report. As they work, move from student to student to check in, ask probing questions, and offer a suggestion if a student gets stuck.

14. If possible, use a document camera to project the "Argument" section of the draft report from the Investigation Handout (or the Investigation Log in their

Teacher Notes

workbook) on a screen or board (or take a picture of it and project the picture on a screen or board).

15. Tell the students, "The last part of the report is called the 'Argument.' In this section of the report you want to share your claim, evidence, and justification of the evidence with the reader. All this information can be found on your whiteboard." Point to the image. "Here are some sentence starters to help you begin writing."

16. Ask the students, "Do you have any questions about what you need to do?"

17. Answer any questions that come up.

18. Tell the students, "Okay, let's write."

19. Give the students 10 minutes to write the "Argument" section of the report. As they work, move from student to student to check in, ask probing questions, and offer a suggestion if a student gets stuck.

Stage 7: Peer Review (35 minutes)

Your students will use either the Investigation Handout or their workbook when doing the peer review. Except where noted below, the directions are the same whether using the handout or the workbook.

1. Tell the students, "We are now going to review our reports to find ways to make them better. I'm going to come around and collect your draft reports. While I do that, please take out a pencil."

2. Collect the handouts or the workbooks with the draft reports from the students.

3. If possible, use a document camera to project the peer-review guide (see Appendix 4) on a screen or board (or take a picture of it and project the picture on a screen or board).

4. Tell the students, "We are going to use this peer-review guide to give each other feedback." Point to the image.

5. Tell the students, "I'm going to ask you to work with a partner to do this. I'm going to give you and your partner a draft report to read. You two will then read the report together. Once you are done reading the report, I want you to answer each of the questions on the peer-review guide." Point to the review questions on the image of the peer-review guide.

6. Tell the students, "You can check 'no,' 'almost,' or 'yes' after each question." Point to the checkboxes on the image of the peer-review guide.

7. Tell the students, "This will be your rating for this part of the report. Make sure you agree on the rating you give the author. If you mark 'no' or 'almost,' then you need to tell the author what he or she needs to do to get a 'yes.'" Point to the space for the reviewer feedback on the image of the peer-review guide.

Investigation 6. Physical and Chemical Properties:
What Are the Identities of the Unknown Powders?

8. Tell the students, "It is really important for you to give the authors feedback that is helpful. That means you need to tell them exactly what they need to do to make their report better."

9. Ask the students, "Do you have any questions about what you need to do?"

10. Answer any questions that come up.

11. Tell the students, "Please sit with a partner who is not in your current group." Allow the students time to sit with a partner.

12. Tell the students, "Okay, I'm now going to give you one report to read." Pass out one Investigation Handout with a draft report or one workbook to each pair. Make sure that the report you give a pair was not written by one of the students in that pair. Give each pair one peer-review guide to fill out. If the students are using workbooks, the peer-review guide is included right after the draft report so you do not need to pass out copies of the peer-review guide.

13. Tell the students, "Okay, I'm going to give you 15 minutes to read the report I gave you and to fill out the peer-review guide. Go ahead and get started."

14. Give the students 15 minutes to work. As they work, move around from pair to pair to check in and see how things are going, answer questions, and offer advice.

15. After 15 minutes pass, tell the students, "Okay, time is up. Please give me the report and the peer-review guide that you filled out."

16. Collect the Investigation Handouts and the peer-review guides, or collect the workbooks if students are using them. If the students are using the Investigation Handouts and separate peer-review guides, be sure you keep each handout with its corresponding peer-review guide.

17. Tell the students, "Okay, I am now going to give you a different report to read and a new peer-review guide to fill out." Pass out one more report to each pair. Make sure that the report you give a pair was not written by one of the students in that pair. Give each pair a new peer-review guide to fill out as a group.

18. Tell the students, "Okay, I'm going to give you 15 minutes to read this new report and to fill out the peer-review guide. Go ahead and get started."

19. Give the students 15 minutes to work. As they work, move around from pair to pair to check in and see how things are going, answer questions, and offer advice.

20. After 15 minutes pass, tell the students, "Okay, time is up. Please give me the report and the peer-review guide that you filled out."

21. Collect the Investigation Handouts and the peer-review guides, or collect the workbooks if students are using them. If the students are using the Investigation Handouts and separate peer-review guides, be sure you keep each handout with its corresponding peer-review guide.

Teacher Notes

Stage 8: Revise the Report (30 minutes)

Your students will use either the Investigation Handout or their workbook when revising the report. Except where noted below, the directions are the same whether using the handout or the workbook.

1. Tell the students, "You are now going to revise your draft report based on the feedback you get from your classmates. Please take out a pencil."
2. Return the reports to the students.
    - *If the students used the Investigation Handout and a copy of the peer-review guide,* pass back the handout and the peer-review guide to each student.
    - *If the students used the workbook,* pass that back to each student.
3. Tell the students, "Please take a few minutes to read over the peer-review guide. You should use it to figure out what you need to change in your report and how you will change it."
4. Allow the students to read the peer-review guide.
5. *If the students used the workbook,* if possible use a document camera to project the "Write Your Final Report" section from the Investigation Log on a screen or board (or take a picture of it and project the picture on a screen or board).
6. Give the following directions about how to revise their reports:
    - *If the students used the Investigation Handout and a copy of the peer-review guide,* tell them, "Okay, let's revise our reports. Please take out a piece of paper. I would like you to rewrite your report. You can use your draft report as a starting point, but you also need to change it to make it better. Use the feedback on the peer-review guide to make it better."
    - *If the students used the workbook,* tell them, "Okay, let's revise our reports. I would like you to rewrite your report in the section of the Investigation Log called "Write Your Final Report." You can use your draft report as a starting point, but you also need to change it to make it better. Use the feedback on the peer-review guide to make it better."
7. Ask the students, "Do you have any questions about what you need to do?"
8. Answer any questions that come up.
9. Tell the students, "Okay, let's write." Allow about 20 minutes for the students to revise their reports.
10. After about 20 minutes, give the following directions:
    - *If the students used the Investigation Handout,* tell them, "Okay, time's up. I will now come around and collect your Investigation Handout, the peer-review guide, and your final report."

Investigation 6. Physical and Chemical Properties: What Are the Identities of the Unknown Powders?

- *If the students used the workbook,* tell them, "Okay, time's up. I will now come around and collect your workbooks."

11. *If the students used the Investigation Handout,* collect all the Investigation Handouts, peer-review guides, and final reports. *If the students used the workbook,* collect all the workbooks.

12. *If the students used the Investigation Handout,* use the "Teacher Score" column in the peer-review guide to grade the final report. *If the students used the workbook,* use the "Investigation Report Grading Rubric" in the Investigation Log to grade the final report. Whether you are using the handout or the log, you can give the students feedback about their writing in the "Teacher Comments" section.

How to Use the Checkout Questions

The Checkout Questions are an optional assessment. We recommend giving them to students at the start of the next class period after the students finish stage 8 of the investigation. You can then look over the student answers to determine if you need to reteach the core idea from the investigation. Appendix 6 gives the answers to the Checkout Questions that should be given by a student who can apply the core idea correctly in all cases and can explain the importance of (a) using patterns as evidence and (b) using standard units to measure or describe the physical quantities of matter.

Alignment With Standards

Table 6.2 highlights how the investigation can be used to address specific performance expectations from the *Next Generation Science Standards, Common Core State Standards for English Language Arts (CCSS ELA),* and *English Language Proficiency (ELP) Standards.*

TABLE 6.2

Investigation 6 alignment with standards

NGSS performance expectation	Strong alignment • 5-PS1-3: Make observations and measurements to identify materials based on their properties. Moderate alignment • 5-PS1-4: Conduct an investigation to determine whether the mixing of two or more substances results in a new substance.

Continued

Table 6.2 (continued)

CCSS ELA—Reading: Informational Text	Key ideas and details • CCSS.ELA-LITERACY.RI.5.1: Quote accurately from a text when explaining what the text says explicitly and when drawing inferences from the text. • CCSS.ELA-LITERACY.RI.5.2: Determine two or more main ideas of a text and explain how they are supported by key details; summarize the text. • CCSS.ELA-LITERACY.RI.5.3: Explain the relationships or interactions between two or more individuals, events, ideas, or concepts in a historical, scientific, or technical text based on specific information in the text. Craft and structure • CCSS.ELA-LITERACY.RI.5.4: Determine the meaning of general academic and domain-specific words and phrases in a text relevant to a *grade 5 topic or subject area*. • CCSS.ELA-LITERACY.RI.5.5: Compare and contrast the overall structure (e.g., chronology, comparison, cause/effect, problem/solution) of events, ideas, concepts, or information in two or more texts. • CCSS.ELA-LITERACY.RI.5.6: Analyze multiple accounts of the same event or topic, noting important similarities and differences in the point of view they represent. Integration of knowledge and ideas • CCSS.ELA-LITERACY.RI.5.7: Draw on information from multiple print or digital sources, demonstrating the ability to locate an answer to a question quickly or to solve a problem efficiently. • CCSS.ELA-LITERACY.RI.5.8: Explain how an author uses reasons and evidence to support particular points in a text, identifying which reasons and evidence support which point(s). Range of reading and level of text complexity • CCSS.ELA-LITERACY.RI.5.10: By the end of the year, read and comprehend informational texts, including history/social studies, science, and technical texts, at the high end of the grades 4–5 text complexity band independently and proficiently.

CCSS ELA—Writing	Text types and purposes
	• CCSS.ELA-LITERACY.W.5.1: Write opinion pieces on topics or texts, supporting a point of view with reasons.
	   ○ CCSS.ELA-LITERACY.W.5.1.A: Introduce a topic or text clearly, state an opinion, and create an organizational structure in which ideas are logically grouped to support the writer's purpose.
	   ○ CCSS.ELA-LITERACY.W.5.1.B: Provide logically ordered reasons that are supported by facts and details.
	   ○ CCSS.ELA-LITERACY.W.5.1.C: Link opinion and reasons using words, phrases, and clauses (e.g., *consequently*, *specifically*).
	   ○ CCSS.ELA-LITERACY.W.5.1.D: Provide a concluding statement or section related to the opinion presented.

Continued

Investigation 6. Physical and Chemical Properties: What Are the Identities of the Unknown Powders?

Table 6.2 (*continued*)

***CCSS ELA*—Writing** (*continued*)	• CCSS.ELA-LITERACY.W.5.2: Write informative or explanatory texts to examine a topic and convey ideas and information clearly.
	  ○ CCSS.ELA-LITERACY.W.5.2.A: Introduce a topic clearly, provide a general observation and focus, and group related information logically; include formatting (e.g., headings), illustrations, and multimedia when useful to aiding comprehension.
	  ○ CCSS.ELA-LITERACY.W.5.2.B: Develop the topic with facts, definitions, concrete details, quotations, or other information and examples related to the topic.
	  ○ CCSS.ELA-LITERACY.W.5.2.C: Link ideas within and across categories of information using words, phrases, and clauses (e.g., *in contrast*, *especially*).
	  ○ CCSS.ELA-LITERACY.W.5.2.D: Use precise language and domain-specific vocabulary to inform about or explain the topic.
	  ○ CCSS.ELA-LITERACY.W.5.2.E: Provide a concluding statement or section related to the information or explanation presented.
	
	Production and distribution of writing
	• CCSS.ELA-LITERACY.W.5.4: Produce clear and coherent writing in which the development and organization are appropriate to task, purpose, and audience.
	• CCSS.ELA-LITERACY.W.5.5: With guidance and support from peers and adults, develop and strengthen writing as needed by planning, revising, editing, rewriting, or trying a new approach.
	• CCSS.ELA-LITERACY.W.5.6: With some guidance and support from adults, use technology, including the internet, to produce and publish writing as well as to interact and collaborate with others; demonstrate sufficient command of keyboarding skills to type a minimum of two pages in a single sitting.
	
	Research to build and present knowledge
	• CCSS.ELA-LITERACY.W.5.8: Recall relevant information from experiences or gather relevant information from print and digital sources; summarize or paraphrase information in notes and finished work, and provide a list of sources.
	• CCSS.ELA-LITERACY.W.5.9: Draw evidence from literary or informational texts to support analysis, reflection, and research.
	
	Range of writing
	• CCSS.ELA-LITERACY.W.5.10: Write routinely over extended time frames (time for research, reflection, and revision) and shorter time frames (a single sitting or a day or two) for a range of discipline-specific tasks, purposes, and audiences.

***CCSS ELA*—Speaking and Listening**	Comprehension and collaboration • CCSS.ELA-LITERACY.SL.5.1: Engage effectively in a range of collaborative discussions (one-on-one, in groups, and teacher-led) with diverse partners on *grade 5 topics and texts*, building on others' ideas and expressing their own clearly.

Continued

Teacher Notes

Table 6.2 (*continued*)

CCSS ELA—Speaking and Listening (*continued*)	Comprehension and collaboration (*continued*)
	
	    o CCSS.ELA-LITERACY.SL.5.1.A: Come to discussions prepared, having read or studied required material; explicitly draw on that preparation and other information known about the topic to explore ideas under discussion.
	
	    o CCSS.ELA-LITERACY.SL.5.1.B: Follow agreed-upon rules for discussions and carry out assigned roles.
	
	    o CCSS.ELA-LITERACY.SL.5.1.C: Pose and respond to specific questions by making comments that contribute to the discussion and elaborate on the remarks of others.
	
	    o CCSS.ELA-LITERACY.SL.5.1.D: Review the key ideas expressed and draw conclusions in light of information and knowledge gained from the discussions.
	
	• CCSS.ELA-LITERACY.SL.5.2: Summarize a written text read aloud or information presented in diverse media and formats, including visually, quantitatively, and orally.
	
	• CCSS.ELA-LITERACY.SL.5.3: Summarize the points a speaker makes and explain how each claim is supported by reasons and evidence.
	
	Presentation of knowledge and ideas
	
	• CCSS.ELA-LITERACY.SL.5.4: Report on a topic or text or present an opinion, sequencing ideas logically and using appropriate facts and relevant, descriptive details to support main ideas or themes; speak clearly at an understandable pace.
	
	• CCSS.ELA-LITERACY.SL.5.5: Include multimedia components (e.g., graphics, sound) and visual displays in presentations when appropriate to enhance the development of main ideas or themes.
	
	• CCSS.ELA-LITERACY.SL.5.6: Adapt speech to a variety of contexts and tasks, using formal English when appropriate to task and situation.

ELP Standards	Receptive modalities • ELP 1: Construct meaning from oral presentations and literary and informational text through grade-appropriate listening, reading, and viewing. • ELP 8: Determine the meaning of words and phrases in oral presentations and literary and informational text. Productive modalities • ELP 3: Speak and write about grade-appropriate complex literary and informational texts and topics. • ELP 4: Construct grade-appropriate oral and written claims and support them with reasoning and evidence. • ELP 7: Adapt language choices to purpose, task, and audience when speaking and writing.

Continued

Investigation 6. Physical and Chemical Properties:
What Are the Identities of the Unknown Powders?

Table 6.2 (*continued*)

ELP Standards (*continued*)	Interactive modalities
	• ELP 2: Participate in grade-appropriate oral and written exchanges of information, ideas, and analyses, responding to peer, audience, or reader comments and questions.
	• ELP 5: Conduct research and evaluate and communicate findings to answer questions or solve problems.
	• ELP 6: Analyze and critique the arguments of others orally and in writing.
	Linguistic structures of English
	• ELP 9: Create clear and coherent grade-appropriate speech and text.
	• ELP 10: Make accurate use of standard English to communicate in grade-appropriate speech and writing.

Investigation Handout

Investigation 6

Physical and Chemical Properties: What Are the Identities of the Unknown Powders?

Introduction

There are many different types of materials. Materials can include such things as powders, metals, minerals, and liquids. Two different materials can look similar to each other but will do different things when they are mixed with another type of material such as water. Watch what happens when your teacher adds two different white powders to water. As you watch this demonstration, keep track of things you notice and things you wonder about in the boxes below.

Things I NOTICED …	Things I WONDER about …

The first powder that was added to the cup of water was sugar. Sugar dissolves when it mixes with water. The second powder that was added to the cup of water is called Alka-Seltzer. This powder is a mixture of two different materials called baking soda and citric acid. The baking soda and citric acid react and create a gas when they are mixed with the water. The sugar, baking soda, and citric acid react with water in different ways because they have different properties.

A *property* is any measurable or observable characteristic of a material. There are two kinds of properties that we use to describe a material. *Physical properties* are characteristics of a material that we can see without mixing it with another material. Examples of physical properties are color and texture. *Chemical properties* are characteristics that we can see when a material is mixed with another material.

Investigation 6. Physical and Chemical Properties: What Are the Identities of the Unknown Powders?

Examples of chemical properties are how a material reacts with water or how a material reacts with an acid (such as vinegar).

Different materials have different physical and chemical properties because they are made of different types of particles. These particles have a unique structure. For example, all particles of sugar have the same structure, but this structure is different from the particles that make up other materials. The unique structure of the particles that make up sugar causes it to be white in color, have a granular texture, and dissolve when it is added to water. Particles of iron have a different structure. The unique structure of the particles that make up iron causes it to be a shiny, silver-gray metal with a smooth texture that does not react with water. However, iron does react with acids to form a gas and with oxygen to form rust. People can use the physical and chemical properties of a material to identify it because each material has a unique set of physical and chemical properties. However, before you can identify a material based on its physical and chemical properties, you must know the physical and chemical properties of that material. In this investigation, you will need to figure out how to identify unknown powders based on their unique physical and chemical properties.

Things we KNOW from what we read …

Your Task

You will be given four labeled powders. You will use what you know about the physical and chemical properties of materials and about patterns to figure out the unique properties of each of these powders. Your teacher will then give you two unknown powders that could be the same as any of the four labeled powders. Your goal is to identify the two unknown powders using what you have figured out about the properties of the labeled powders.

The *guiding question* of this investigation is, ***What are the identities of the unknown powders?***

Investigation Handout

Materials

You may use any of the following materials during your investigation:

Powders
- Cornstarch
- Baking powder
- Baking soda
- Salt
- Unknown A
- Unknown B

Liquids for mixing
- Iodine
- Vinegar
- Water

Equipment
- Safety goggles, nonlatex apron, and vinyl gloves (required)
- Hand lens
- Well plates
- Plastic spoons

Safety Rules

Follow all normal safety rules. In addition, be sure to follow these rules:

- Wear sanitized safety goggles, nonlatex aprons, and vinyl gloves during setup, investigation activity, and cleanup.
- Be careful when working with iodine because it can stain skin and clothes.
- Do not taste any substances used in this activity.
- Immediately clean up any spills to avoid a slip or fall hazard.
- Keep water sources away from electrical receptacles to prevent shock.
- Be careful when handling glassware, because it can shatter and cut skin.
- Wash your hands with soap and water when you are done cleaning up.

Plan Your Investigation

Prepare a plan for your investigation by filling out the chart on the next page; this plan is called an *investigation proposal*. Before you start developing your plan, be sure to discuss the following questions with the other members of your group:

- What information should we collect so we can **describe the properties** of a powder?
- What types of **patterns** might we look for to help answer the guiding question?

Investigation 6. Physical and Chemical Properties: What Are the Identities of the Unknown Powders?

Our guiding question:

This is a picture of how we will set up the equipment:

We will collect the following data:

These are the steps we will follow to collect data:

I approve of this investigation proposal.

_____   _____
Teacher's signature   Date

Argument-Driven Inquiry in **Fifth-Grade Science**: Three-Dimensional Investigations

Investigation Handout

Collect Your Data

Keep a record of what you measure or observe during your investigation in the space below.

Analyze Your Data

You will need to analyze the data you collected before you can develop an answer to the guiding question. To analyze the data you collected, create a table, graph, or other visual representation in the space below that compares and contrasts the properties of the different powders.

Investigation 6. Physical and Chemical Properties: What Are the Identities of the Unknown Powders?

Draft Argument

Develop an argument on a whiteboard. It should include the following:

1. A *claim*: Your answer to the guiding question.
2. *Evidence*: An analysis of the data and an explanation of what the analysis means.
3. A *justification of the evidence*: Why your group thinks the evidence is important.

The Guiding Question:	
Our Claim:	
Our Evidence:	Our Justification of the Evidence:

Argumentation Session

Share your argument with your classmates. Be sure to ask them how to make your draft argument better. Keep track of their suggestions in the space below.

Ways to IMPROVE our argument ...

Argument-Driven Inquiry in **Fifth-Grade Science:** Three-Dimensional Investigations

Investigation Handout

Draft Report

Prepare an *investigation report* to share what you have learned. Use the information in this handout and your group's final argument to write a *draft* of your investigation report.

Introduction

We have been studying _____ in class.

Before we started this investigation, we explored _____
_____
_____

We noticed _____
_____
_____

My goal for this investigation was to figure out _____
_____
_____

The guiding question was _____
_____
_____

Method

To gather the data I needed to answer this question, I _____
_____
_____
_____
_____

Investigation 6. Physical and Chemical Properties:
What Are the Identities of the Unknown Powders?

I then analyzed the data I collected by _____

_____

_____

_____

Argument

My claim is _____

_____

_____

_____

The _____ below shows _____

_____

_____

_____

_____

_____

Argument-Driven Inquiry in **Fifth-Grade Science:** Three-Dimensional Investigations

273

Investigation Handout

This analysis of the data I collected suggests _____
_____
_____
_____
_____

This evidence is based on several important scientific concepts. The first one is _____
_____
_____
_____
_____

Review

Your classmates need your help! Review the draft of their investigation reports and give them ideas about how to improve. Use the *peer-review guide* when doing your review.

Submit Your Final Report

Once you have received feedback from your classmates about your draft report, create your final investigation report and hand it in to your teacher.

Checkout Questions

Investigation 6. Physical and Chemical Properties

The table below shows the physical and chemical properties of several different powders. Use the information in the table to answer questions 1–3.

Powders	Physical Properties		Chemical Properties		
	Color	Texture	Reaction With Vinegar	Reaction With Iodine	Heated
Flour	White	Fine	No change	Turns black	No change
Sugar	White	Coarse	No change	No change	Turns black
Baking Soda	White	Fine	Bubbles	No change	No change
Unknown A	White	Coarse	No change	No change	Turns black
Unknown B	White	Fine	No change	Turns black	No change

1. What is the identity of unknown powder A?

    a. Baking soda

    b. Flour

    c. Sugar

2. What is the identity of unknown powder B?

    a. Baking soda

    b. Flour

    c. Sugar

3. How do you know? Use what you know about using patterns as evidence to explain your answer.

_____

_____

_____

_____

Teacher Scoring Rubric for Checkout Questions 1–3

Level	Description
3	The student can apply the core idea correctly in all cases and can explain the importance of using patterns as evidence.
2	The student cannot apply the core idea correctly in all cases but can explain the importance of using patterns as evidence.
1	The student can apply the core idea correctly in all cases but cannot explain the importance of using patterns as evidence.
0	The student cannot apply the core idea correctly and cannot explain the importance of using patterns as evidence.

Argument-Driven Inquiry in **Fifth-Grade Science**: Three-Dimensional Investigations

Checkout Questions

The table below shows the physical and chemical properties of several different powders. Use the information in the table to answer questions 4–6.

Powders	Physical Properties		Chemical Properties		
	Mass	Volume	Reaction With Vinegar	Reaction With Iodine	Heated
Salt	10.8 g	5 cm^3	No change	No change	No change
Cornstarch	7.3 g	5 cm^3	No change	Turns black	No change
Sugar	4.3 g	5 cm^3	No change	No change	Turns black
Unknown C	4.3 g	5 cm^3	No change	No change	Turns black
Unknown D	10.8 g	5 cm^3	No change	No change	No change

4. What is the identity of unknown powder C?

    a. Salt

    b. Starch

    c. Sugar

5. What is the identity of unknown powder D?

    a. Salt

    b. Starch

    c. Sugar

6. How do you know? Use what you know about using standard units to measure or describe the physical quantities of matter to explain your answer.

_____

_____

_____

_____

_____

Teacher Scoring Rubric for Checkout Questions 4–6

Level	Description
3	The student can apply the core idea correctly in all cases and can explain the importance of using standard units to measure or describe the physical quantities of matter.
2	The student cannot apply the core idea correctly in all cases but can explain the importance of using standard units to measure or describe the physical quantities of matter.
1	The student can apply the core idea correctly in all cases but cannot explain the importance of using standard units to measure or describe the physical quantities of matter.
0	The student cannot apply the core idea correctly and cannot explain the importance of using standard units to measure or describe the physical quantities of matter.

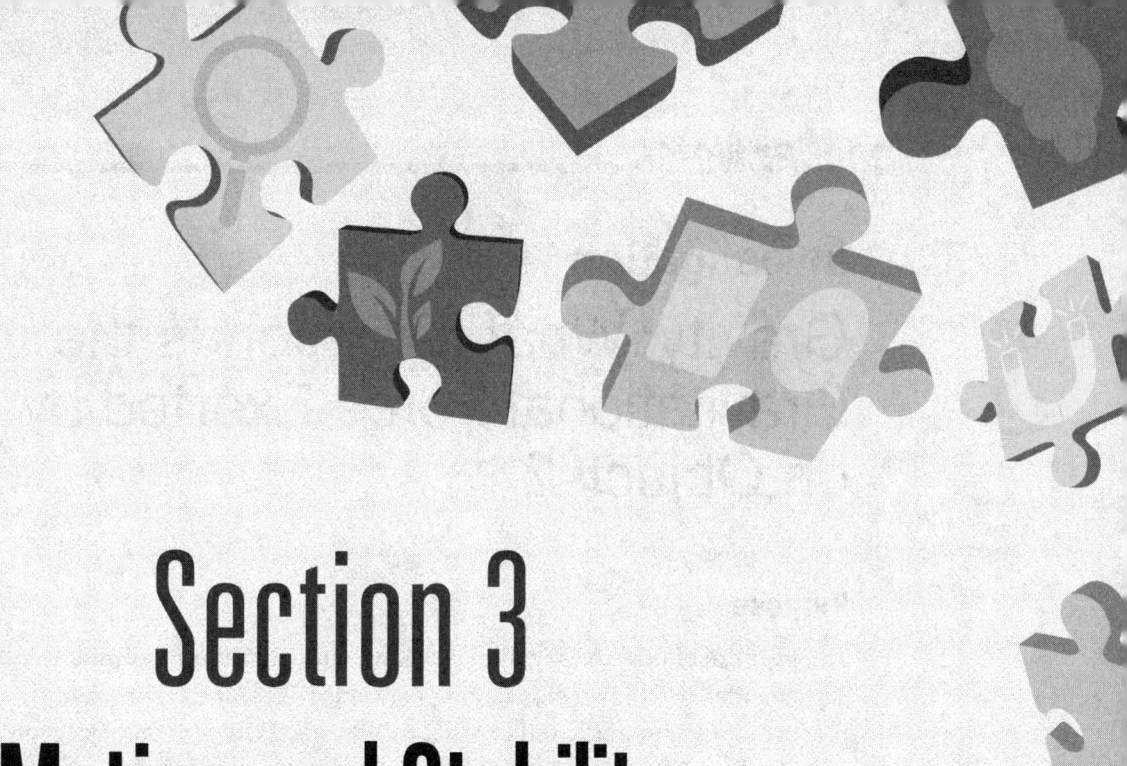

Section 3
Motion and Stability

Teacher Notes

Investigation 7
Gravity: What Direction Is the Gravitational Force Exerted by Earth on Objects?

Purpose

The purpose of this investigation is to give students an opportunity to use one disciplinary core idea (DCI), two crosscutting concepts (CCs), and eight scientific and engineering practices (SEPs) to figure out in what direction gravity acts on objects on Earth. Students will also learn about the use of models as tools for reasoning about natural phenomena.

The DCI, CCs, and SEPs That Students Use During This Investigation to Figure Things Out

DCI

- *PS2.A: Forces and Motion:* Each force acts on one particular object and has both strength and a direction. An object at rest typically has multiple forces acting on it, but they add to give zero net force on the object. Forces that do not sum to zero can cause changes in the object's speed or direction of motion.

CCs

- *CC 2: Cause and Effect:* Cause-and-effect relationships are routinely identified, tested, and used to explain change. Events that occur together with regularity might or might not be a cause-and-effect relationship.
- *CC 4: Systems and System Models:* A system can be described in terms of its components and their interactions. A system is a group of related parts that make up a whole and can carry out functions its individual parts cannot.

SEPs

- *SEP 1: Asking Questions and Defining Problems:* Ask questions about what would happen if a variable is changed. Ask questions that can be investigated and predict reasonable outcomes based on patterns such as cause-and-effect relationships.
- *SEP 2: Developing and Using Models:* Develop and/or use models to describe and/or predict phenomena.
- *SEP 3: Planning and Carrying Out Investigations*: Plan and conduct an investigation collaboratively to produce data to serve as the basis for evidence, using fair tests in which variables are controlled and the number of trials considered. Evaluate appropriate methods and/or tools for collecting data.

Investigation 7. Gravity: What Direction Is the Gravitational Force Exerted by Earth on Objects?

- *SEP 4: Analyzing and Interpreting Data:* Represent data in tables and/or various graphical displays (bar graphs, pictographs, and/or pie charts) to reveal patterns that indicate relationships. Analyze and interpret data to make sense of phenomena, using logical reasoning, mathematics, and/or computation. Compare and contrast data collected by different groups in order to discuss similarities and differences in their findings.

- *SEP 5: Using Mathematics and Computational Thinking:* Organize simple data sets to reveal patterns that suggest relationships. Describe, measure, estimate, and/or graph quantities (e.g., area, volume, weight, time) to address scientific and engineering questions and problems.

- *SEP 6: Constructing Explanations and Designing Solutions:* Construct an explanation of observed relationships. Use evidence to construct or support an explanation. Identify the evidence that supports particular points in an explanation.

- *SEP 7: Engaging in Argument From Evidence:* Compare and refine arguments based on an evaluation of the evidence presented. Distinguish among facts, reasoned judgment based on research findings, and speculation in an explanation. Respectfully provide and receive critiques from peers about a proposed procedure, explanation, or model by citing relevant evidence and posing specific questions.

- *SEP 8: Obtaining, Evaluating, and Communicating Information:* Read and comprehend grade-appropriate complex texts and/or other reliable media to summarize and obtain scientific and technical ideas. Combine information in written text with that contained in corresponding tables, diagrams, and/or charts to support the engagement in other scientific and/or engineering practices. Communicate scientific and/or technical information orally and/or in written formats, including various forms of media as well as tables, diagrams, and charts.

Other Concepts That Students May Use During This Investigation

Students might also use some of the following concepts:

- Any force is either a push or a pull.
- A contact force is a push or a pull on an object from another object that is touching it.
- A non-contact force is a push or a pull on an object that comes from something that is not touching that object.

What Students Figure Out

Students will figure out that the force of gravity always points toward the center of Earth. For a person kicking a ball, this direction is perceived as down.

Teacher Notes

Background Information About This Investigation for the Teacher

Any force is a push or a pull. Objects often have more than one force acting on them at the same time. The forces acting on an object can be either balanced or unbalanced. *Balanced forces* are two forces that are the same size but are acting on the object in opposite directions. Scientists describe balanced forces as being equal and opposite. *Unbalanced forces* are not equal and opposite. One example of an unbalanced force is two different-size forces acting on an object in opposite directions. A second example of an unbalanced force is two same-size forces acting on an object in the same direction. When you try to determine if all the forces acting on an object are balanced or unbalanced, it is important to remember that forces that act in the same direction combine by addition and forces that act in opposite directions combine by subtraction. Forces that sum to zero do not change an object's speed or direction of motion, but forces that do not sum to zero can cause a change an object's motion.

There are many ways to classify forces, and one of the more common ways is to distinguish between contact forces and non-contact forces. A *contact force* is a force that occurs when two objects are touching. An example of a contact force is when you kick a ball. The force occurs when your foot and the ball are in contact. When your foot is no longer touching the ball, there is no longer a force acting on the ball. *Non-contact forces* are forces that can act between two objects at a distance. One example of a non-contact force is the magnetic force. When you push the north poles of two magnets near each other, a force pushes back against the magnets. Another non-contact force is the *force of gravity*, which exists between two objects with mass. The force of gravity exists between the Sun and Earth—that is what keeps Earth in orbit. The force of gravity also exists between Earth and the Moon and keeps the Moon orbiting Earth.

The force of gravity also exists between Earth and objects on Earth's surface, such as balls and people. This is the force of gravity we are most familiar with. When you jump in the air, the force of gravity acts on you to pull you back toward the ground. Similarly, when you kick a ball, the force of gravity acts on the ball to pull it back toward the ground. And when you drop an object from your hand, like a cell phone, gravity causes it to fall toward the ground.

If we were to model the force of gravity, we could draw a picture of a person standing on Earth's surface and use an arrow to represent the direction in which gravity acts on the person at that location. On the surface of Earth at the top of the Northern Hemisphere, we draw the arrow pointing down as shown in Figure 7.1 because that is how we perceive the direction of gravity for a person standing in that location. People, however, live all over the world and not just in the Northern Hemisphere, so we need to be able to model how the force of gravity acts on them too. We can model the way people in different parts of the world feel the force of gravity by drawing arrows toward the center of Earth, as shown in Figures 7.2–7.4. Although the person in each figure feels the direction of gravity as down, just like the person in Figure 7.1, for those of us looking at these models we perceive gravity as acting in four different directions.

Investigation 7. Gravity: What Direction Is the Gravitational Force Exerted by Earth on Objects?

FIGURE 7.1
The direction of gravity for a person standing on the surface of Earth in the Northern Hemisphere

FIGURE 7.2
The direction of gravity for a person standing on the surface of Earth on the equator in the Western Hemisphere

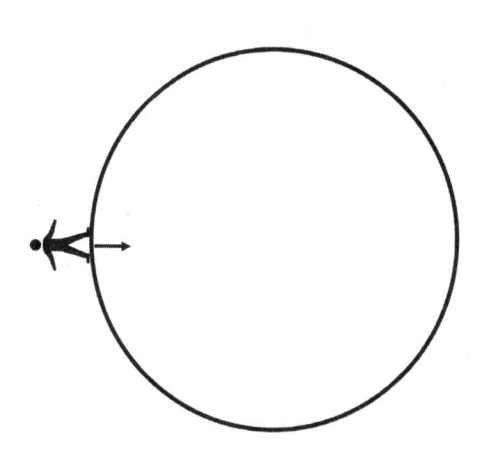

FIGURE 7.3
The direction of gravity for a person standing on the surface of Earth on the equator in the Eastern Hemisphere

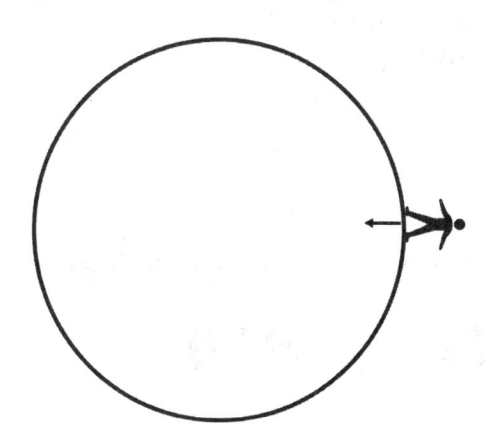

FIGURE 7.4
The direction of gravity for a person standing on the surface of Earth in the Southern Hemisphere

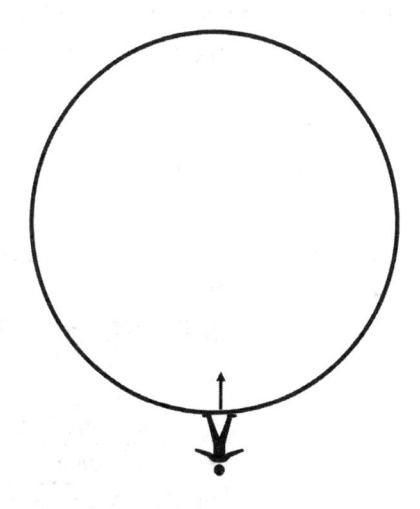

Gravity acts on all objects and people at the same time, so we can combine Figures 7.1–7.4 into one figure. Figure 7.5 (p. 282) shows all four people standing on Earth in their respective locations at the same time. This figure makes it clear that the direction in which gravity is acting on objects is toward the center of Earth.

Teacher Notes

FIGURE 7.5
The direction of gravity for people standing on the surface of Earth in various locations

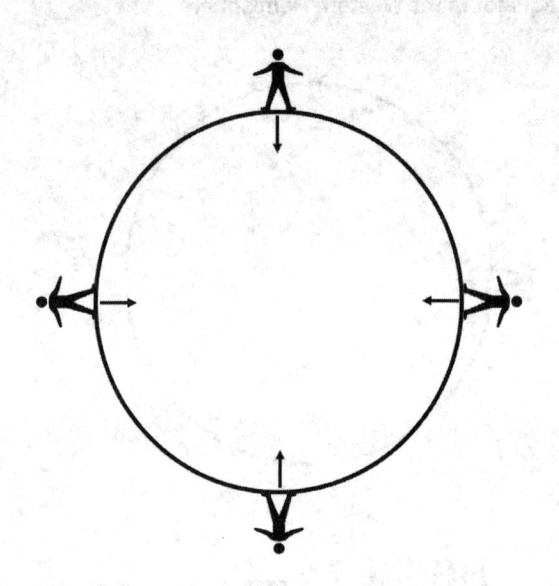

Timeline

The time needed to complete this investigation is 265 minutes (4 hours and 25 minutes). The amount of instructional time needed for each stage of the investigation is as follows:

- *Stage 1.* Introduce the task and the guiding question: 35 minutes
- *Stage 2.* Design a method and collect data: 45 minutes
- *Stage 3.* Create a draft argument: 45 minutes
- *Stage 4.* Argumentation session: 30 minutes
- *Stage 5.* Reflective discussion: 15 minutes
- *Stage 6.* Write a draft report: 30 minutes
- *Stage 7.* Peer review: 35 minutes
- *Stage 8.* Revise the report: 30 minutes

Materials and Preparation

The materials needed for this investigation are listed in Table 7.1. The items can be purchased from a big-box retail store such as Walmart or Target or through an online retailer such as Amazon. The materials for this investigation can also be purchased as a complete kit (which includes enough materials for 24 students, or six groups of four students) at *www.argumentdriveninquiry.com*.

TABLE 7.1
Materials for Investigation 7

Item	Quantity
Safety goggles	1 per student
Inflatable plastic globe	1 per group
Miniature figures of people	4 per group
Miniature soccer ball	1 per group
Dry-erase marker	1 per group
Tape	As needed
Ruler	1 per group

Continued

Investigation 7. Gravity:
What Direction Is the Gravitational Force Exerted by Earth on Objects?

Table 7.1 (*continued*)

Item	Quantity
Whiteboard, 2' × 3'*	1 per group
Investigation Handout	1 per student
Peer-review guide and teacher scoring rubric	1 per student
Checkout Questions (optional)	1 per student

*As an alternative, students can use computer and presentation software such as Microsoft PowerPoint or Apple Keynote to create their arguments.

Safety Precautions

Remind students to follow all normal safety rules. In addition, tell the students to take the following safety precautions:

- Wear sanitized safety goggles during setup, investigation activity, and cleanup.
- Make sure all materials are put away after completing the activity.
- Wash their hands with soap and water when they are done cleaning up.

Lesson Plan by Stage

This lesson plan is only a suggestion. It is included here to illustrate what you can say and do during each stage of ADI for this specific investigation. We encourage you to modify this lesson plan by asking different questions, using different examples, and providing different scaffolds as needed to better meet the needs of students in your class.

Stage 1: Introduce the Task and the Guiding Question (35 minutes)

1. Ask the students to sit in six groups, with three or four students in each group.
2. Ask the students to clear off their desks except for a pencil (and their *Student Workbook for Argument-Driven Inquiry in Fifth-Grade Science* if they have one).
3. Pass out an Investigation Handout to each student (or ask students to turn to the Investigation Log for Investigation 7 in their workbook).
4. Read the first paragraph of the "Introduction" aloud to the class. Ask the students to follow along as you read.
5. Tell the students that you will show them one video at a time, and tell them where each video was filmed before you start it. Ask the students to record their observations about how the ball moves in each video in the "Things I NOTICED …" box of the "NOTICED/WONDER" chart in the "Introduction." The videos are about three minutes each and can be found at the following URLs:

Teacher Notes

- Asia: *www.youtube.com/watch?v=W0NLDSBlWKw*
- Australia: *www.youtube.com/watch?v=nrbNyUukqmU*
- Europe: *www.youtube.com/watch?v=PxVgQrw0RVQ*
- North America: *www.youtube.com/watch?v=J_wSHYieHTg*
- Africa: *www.youtube.com/watch?v=y8Z3qjxU3Hk*
- South America: *www.youtube.com/watch?v=MLPZSoFz91o*

6. After showing all the videos, wait three to four minutes for the students to finish writing down what they observed. Also have them fill out the "Things I WONDER about …" box of the "NOTICED/WONDER" chart.
7. Ask the students to share what they observed about how the ball moved in each video.
8. Ask the students to share what questions they have about how the ball moved in each video.
9. Tell the students, "Some of your questions might be answered by reading the rest of the 'Introduction.'"
10. Ask the students to read the rest of the "Introduction" on their own *or* ask them to follow along as you read it aloud.
11. Once the students have read the rest of the "Introduction," ask them to fill out the "Things we KNOW" chart on their Investigation Handout (or in their Investigation Log) as a group.
12. Ask the students to share what they learned from the reading. Add these ideas to a class "Things we KNOW" chart.
13. Tell the students, "Let's see what we will need to figure out during our investigation."
14. Read the task and the guiding question aloud.
15. Tell the students, "I have lots of materials here that you can use."
16. Introduce the students to the materials available for them to use during the investigation by holding each one up and asking how it might be used.

Stage 2: Design a Method and Collect Data (45 minutes)

1. Tell the students, "I am now going to give you and the other members of your group about 15 minutes to plan your investigation. Before you begin, I want you all to take a couple of minutes to discuss the following questions with the rest of your group."
2. Show the following questions on the screen or board:
   - What does a force acting on a ball (a *cause*) do to the motion of that ball (the *effect*)?

Investigation 7. Gravity:
What Direction Is the Gravitational Force Exerted by Earth on Objects?

- How can you use the materials to *model the movement* of a ball when it is kicked?

3. Tell the students, "Please take a few minutes to come up with an answer to these questions."

4. Give the students two to three minutes to discuss these two questions.

5. Ask two or three different groups to share their answers. Highlight or write down any important ideas on the board so students can refer to them later.

6. If possible, use a document camera to project an image of the graphic organizer for this investigation on a screen or board (or take a picture of it and project the picture on a screen or board). Tell the students, "I now want you all to plan out your investigation. To do that, you will need to fill out this investigation proposal."

7. Point to the box labeled "Our guiding question:" and tell the students, "You can put the question we are trying to answer in this box." Then ask, "Where can we find the guiding question?"

8. Wait for a student to answer.

9. Point to the box labeled "This is a picture of how we will set up the equipment:" and tell the students, "You can draw a picture in this box of how you will set up the equipment in order to carry out this investigation."

10. Point to the box labeled "We will collect the following data:" and tell the students, "You can list the measurements or observations that you will need to collect during the investigation in this box."

11. Point to the box labeled "These are the steps we will follow to collect data:" and tell the students, "You can list what you are going to do to collect the data you need and what you will do with your data once you have it. Be sure to give enough detail that I could do your investigation for you."

12. Ask the students, "Do you have any questions about what you need to do?"

13. Answer any questions that come up.

14. Tell the students, "Once you are done, raise your hand and let me know. I'll then come by and look over your proposal and give you some feedback. You may not begin collecting data until I have approved your proposal by signing it. You need to have your proposal done in the next 15 minutes."

15. Give the students 15 minutes to work in their groups on their investigation proposal. As they work, move from group to group to check in, ask probing questions, and offer a suggestion if a group gets stuck.

16. As each group finishes its investigation proposal, read it over and determine if it will be productive or not. If you feel the investigation will be productive (not necessarily what you would do or what the other groups are doing), sign your name on the proposal and let the group start collecting data. If the plan needs to be changed, offer some suggestions or ask some probing questions, and have

Teacher Notes

FIGURE 7.6
Students collecting data

17. Pass out the materials or have one student from each group collect the materials they need from a central supply table or cart for the groups that have an approved proposal.

the group make the changes before you approve it.

18. Remind students of the safety rules and precautions for this investigation.

19. Tell the students to collect their data and record their observations or measurements in the "Collect Your Data" box in their Investigation Handout (or the Investigation Log in their workbook).

20. Give the students 30 minutes to collect their data (see Figure 7.6 for an example of students collecting data during this investigation). Collect the materials from each group before asking them to analyze their data.

What should a student-designed investigation look like?

There are a number of different investigations that students can design to answer the question "What direction is the gravitational force exerted by Earth on objects?" For example, one method might include the following steps:

1. Model a person kicking a ball on Earth by placing the miniature figure on a location on plastic globe (see Figure 7.6).
2. Mark the location with a dry-erase marker.
3. Model the path in which the ball will move if it is kicked at that location.
4. Determine in what direction gravity is acting on the ball.
5. Repeat steps 1–4 for five more locations.
6. If students use this method, they will need to collect data on (1) the location on Earth where the figure is placed and (2) the direction in which gravity acts on the ball.

Investigation 7. Gravity:
What Direction Is the Gravitational Force Exerted by Earth on Objects?

Stage 3: Create a Draft Argument (45 minutes)

1. Tell the students, "Now that we have all this data, we need to analyze the data so we can figure out an answer to the guiding question."

2. If possible, project an image of the "Analyze Your Data" section for this investigation on a screen or board using a document camera (or take a picture of it and project the picture on a screen or board). Point to the section and tell the students, "You can create a picture or model as a way to analyze your data. You can make your picture or model in this section."

3. Ask the students, "What information do we need to include in this analysis?"

4. Tell the students, "Please take a few minutes to discuss this question with your group, and be ready to share."

5. Give the students five minutes to discuss.

6. Ask two or three different groups to share their answers. Highlight or write down any important ideas on the board so students can refer to them later.

7. Tell the students, "I am now going to give you and the other members of your group about 10 minutes to create a picture or model from your data." The picture or model they create should include the locations of the person and the direction of the force of gravity. If the students are having trouble determining how to model the situation, you can take a few minutes to provide a mini-lesson about different ways to model data (this strategy is called just-in-time instruction because it is offered only when students get stuck).

8. Give the students 10 minutes to analyze their data. As they work, move from group to group to check in, ask probing questions, and offer suggestions.

9. Tell the students, "I am now going to give you and the other members of your group about 15 minutes to create an argument to share what you have learned and convince others that they should believe you. Before you do that, we need to take a few minutes to discuss what you need to include in your argument."

10. If possible, use a document camera to project the "Argument Presentation on a Whiteboard" image from the "Draft Argument" section of the Investigation Handout (or the Investigation Log in their workbook) on a screen or board (or take a picture of it and project the picture on a screen or board).

11. Point to the box labeled "The Guiding Question:" and tell the students, "You can put the question we are trying to answer here on your whiteboard."

12. Point to the box labeled "Our Claim:" and tell the students, "You can put your claim here on your whiteboard. The claim is your answer to the guiding question."

13. Point to the box labeled "Our Evidence:" and tell the students, "You can put the evidence that you are using to support your claim here on your whiteboard. Your evidence will need to include the analysis you just did and an explanation

Teacher Notes

of what your analysis means or shows. Scientists always need to support their claims with evidence."

14. Point to the box labeled "Our Justification of the Evidence:" and tell the students, "You can put your justification of your evidence here on your whiteboard. Your justification needs to explain why your evidence is important. Scientists often use core ideas to explain why the evidence they are using matters. Core ideas are important concepts that scientists use to help them make sense of what happens during an investigation."

15. Ask the students, "What are some core ideas that we read about earlier that might help us explain why the evidence we are using is important?"

16. Ask the students to share some of the core ideas from the "Introduction" section of the Investigation Handout (or the Investigation Log in the workbook). List these core ideas on the board.

17. Tell the students, "That is great. I would like to see everyone try to include these core ideas in your justification of the evidence. Your goal is to use these core ideas to help explain why your evidence matters and why the rest of us should pay attention to it."

18. Ask the students, "Do you have any questions about what you need to do?"

19. Answer any questions that come up.

20. Tell the students, "Okay, go ahead and start working on your arguments. You need to have your argument done in the next 15 minutes. It doesn't need to be perfect. We just need something down on the whiteboards so we can share our ideas."

What should the picture or model for this investigation look like?

There are a number of different ways that students can analyze the data they collect during this investigation. One of the most straightforward ways is to create a picture or a model of Earth using arrows and labels to show the direction of gravity at different locations. They can then use this picture to show that for all locations on Earth, gravity acts toward the center of Earth. An example of a picture is shown in Figure 7.7.

There are other options for analyzing the collected data. Students often come up with some unique ways of analyzing their data, so be sure to give them some voice and choice during this stage.

Investigation 7. Gravity:
What Direction Is the Gravitational Force Exerted by Earth on Objects?

FIGURE 7.7
Example of an argument

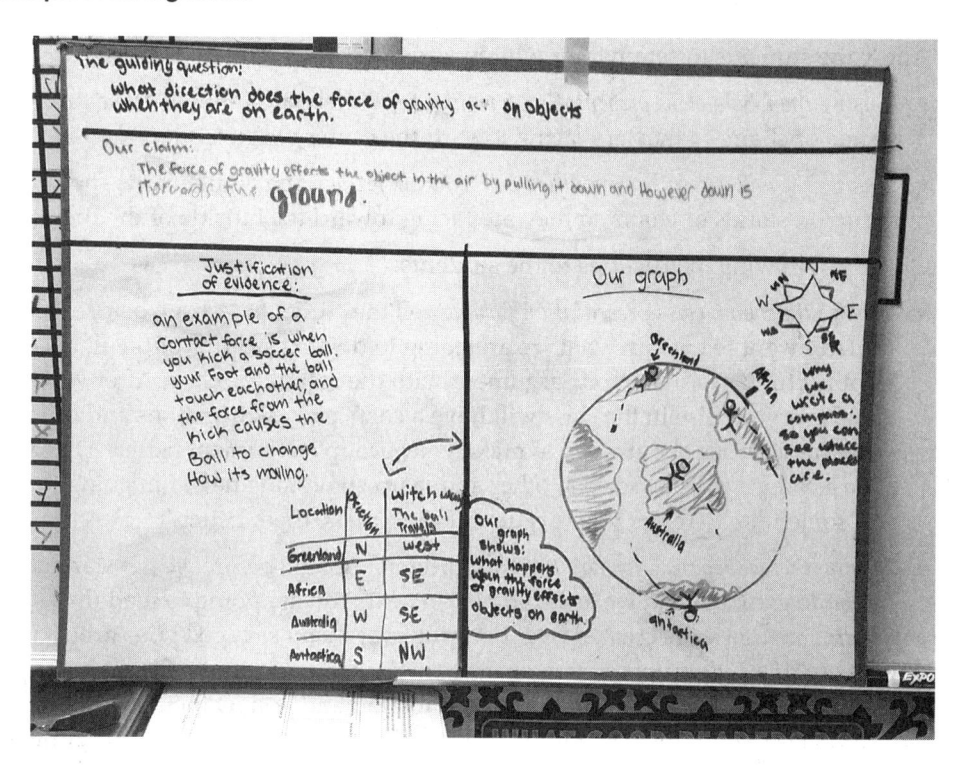

21. Give the students 15 minutes to work in their groups on their arguments. As they work, move from group to group to check in, ask probing questions, and offer a suggestion if a group gets stuck. Figure 7.7 shows an example of an argument created by students for this investigation.

Stage 4: Argumentation Session (30 minutes)

The argumentation session can be conducted in a whole-class presentation format, a gallery walk format, or a modified gallery walk format. We recommend using a whole-class presentation format for the first investigation, but try to transition to either the gallery walk or modified gallery walk format as soon as possible because that will maximize student voice and choice inside the classroom. The following list shows the steps for the three formats; unless otherwise noted, the steps are the same for all three formats.

1. Begin by introducing the use of the whiteboard.
   - *If using the whole-class presentation format,* tell the students, "We are now going to share our arguments. Please set up your whiteboards so everyone can see them."

Teacher Notes

- *If using the gallery walk or modified gallery walk format,* tell the students, "We are now going to share our arguments. Please set up your whiteboards so they are facing the walls."

2. Allow the students to set up their whiteboards.

    - *If using the whole-class presentation format,* the whiteboards should be set up on stands or chairs so they are facing toward the center of the room.

    - *If using the gallery walk or modified gallery walk format,* the whiteboards should be set up on stands or chairs so they are facing toward the outside of the room.

3. Give the following instructions to the students:

    - *If using the whole-class presentation format,* tell the students, "Okay, before we get started I want to explain what we are going to do next. Your group will have an opportunity to share your argument with the rest of the class. After you are done, everyone else in the class will have a chance to ask questions and offer some suggestions about ways to make your group's argument better. After we have a chance to listen to each other and learn something new, I'm going to give you some time to revise your arguments and make them better."

    - *If using the gallery walk format,* tell the students, "Okay, before we get started I want to explain what we are going to do next. You are going to read the arguments that were created by other groups. When I say 'go,' your group will go to a different group's station so you can see their argument. Once you are there, I'll give your group a few minutes to read and review their argument. Your job is to offer them some suggestions about ways to make their argument better. You can use sticky notes to give them suggestions. Please be specific about what you want to change and how you think they should change it. After we have a chance to learn from each other, I'm going to give you some time to revise your arguments and make them better."

    - *If using the modified gallery walk format,* tell the students, "Okay, before we get started I want to explain what we are going to do next. I'm going to ask some of you to present your arguments to your classmates. If you are presenting your argument, your job is to share your group's claim, evidence, and justification of the evidence. The rest of you will be travelers. If you are a traveler, your job is to listen to the presenters, ask the presenters questions if you do not understand something, and then offer them some suggestions about ways to make their argument better. After we have a chance to learn from each other, I'm going to give you some time to revise your arguments and make them better."

4. Use a document camera to project the "Ways to IMPROVE our argument ..." box from the Investigation Handout (or the Investigation Log in their workbook) on a screen or board (or take a picture of it and project the picture on a screen or board).

Investigation 7. Gravity:
What Direction Is the Gravitational Force Exerted by Earth on Objects?

- *If using the whole-class presentation format,* point to the box and tell the students, "After your group presents your argument, you can write down the suggestions you get from your classmates here. If you are listening to a presentation and you see a good idea from another group, you can write down that idea here as well. Once we are done with the presentations, I will give you a chance to use these suggestions or ideas to improve your arguments."

- *If using the gallery walk format,* point to the box and tell the students, "If you see a good idea from another group, you can write it down here. Once we are done reviewing the different arguments, I will give you a chance to use these ideas to improve your own arguments. It is important to share ideas like this."

- *If using the modified gallery walk format,* point to the box and tell the students, "If you are a presenter, you can write down the suggestions you get from the travelers here. If you are a traveler and you see a good idea from another group, you can write down that idea here. Once we are done with the presentations, I will give you a chance to use these suggestions or ideas to improve your arguments."

5. Ask the students, "Do you have any questions about what you need to do?"

6. Answer any questions that come up.

7. Give the following instructions:

    - *If using the whole-class presentation format,* tell the students, "Okay. Let's get started."

    - *If using the gallery walk format,* tell the students, "Okay, I'm now going to tell you which argument to go to and review."

    - *If using the modified gallery walk format,* tell the students, "Okay, I'm now going to assign you to be a presenter or a traveler." Assign one or two students from each group to be presenters and one or two students from each group to be travelers.

8. Give the students an opportunity to review the arguments.

    - *If using the whole-class presentation format,* have each group present their argument one at a time. Give each group only two to three minutes to present their argument. Then give the class two to three minutes to ask them questions and offer suggestions. Encourage as much participation from the students as possible.

    - *If using the gallery walk format,* tell the students, "Okay. Let's get started. Each group, move one argument to the left. Don't move to the next argument until I tell you to move. Once you get there, read the argument and then offer suggestions about how to make it better. I will put some sticky notes next to each argument. You can use the sticky notes to leave your suggestions." Give each group about three to four minutes to read the arguments, talk, and offer suggestions.

Teacher Notes

   a. After three to four minutes, tell the students, "Okay. Let's move on to the next argument. Please move one group to the left."

   b. Again, give each group three to four minutes to read, talk, and offer suggestions.

   c. Repeat this process until each group has had their argument read and critiqued three times.

- *If using the modified gallery walk format,* tell the students, "Okay. Let's get started. Reviewers, move one group to the left. Don't move to the next group until I tell you to move. Presenters, go ahead and share your argument with the travelers when they get there." Give each group of presenters and travelers about three to four minutes to talk.

   a. Tell the students, "Okay. Let's move on to the next argument. Travelers, move one group to the left."

   b. Again, give each group of presenters and travelers about three to four minutes to talk.

   c. Repeat this process until each group has had their argument read and critiqued three times.

9. Tell the students to return to their workstations.

10. Give the following instructions about revising the argument:

    - *If using the whole-class presentation format,* tell the students, "I'm now going to give you all about 10 minutes to revise your argument. Take a few minutes to talk in your groups and determine what you want to change to make your argument better. Once you have decided what to change, go ahead and make the changes to your whiteboard."

    - *If using the gallery walk format,* tell the students, "I'm now going to give you all about 10 minutes to revise your argument. Take a few minutes to read the suggestions that were left at your argument. Then talk in your groups and determine what you want to change to make your argument better. Once you have decided what to change, go ahead and make the changes to your whiteboard."

    - *If using the modified gallery walk format,* tell the students, "I'm now going to give you all about 10 minutes to revise your argument. Please return to your original groups." Wait for the students to move back into their original groups and then tell the students, "Okay, take a few minutes to talk in your groups and determine what you want to change to make your argument better. Once you have decided what to change, go ahead and make the changes to your whiteboard."

11. Ask the students, "Do you have any questions about what you need to do?"

12. Answer any questions that come up.

13. Tell the students, "Okay. Let's get started."

Investigation 7. Gravity:
What Direction Is the Gravitational Force Exerted by Earth on Objects?

14. Give the students 10 minutes to work in their groups on their arguments. As they work, move from group to group to check in, ask probing questions, and offer a suggestion if a group gets stuck.

Stage 5: Reflective Discussion (15 minutes)

1. Tell the students, "We are now going to take a minute to talk about some of the core ideas and crosscutting concepts that we have used during our investigation."
2. Place a ball on a table. Ask the students, "What do you all see going on here?"
3. Allow the students to share their ideas.
4. Tell the students, "Okay, watch this." Push on the ball so it starts to roll. Allow it to roll off the table and fall to the ground.
5. Ask the students, "How can we use what we know about forces to explain what happened to the ball?"
6. Allow the students to share their ideas. As they share their ideas, ask different questions to encourage them to expand on their thinking (e.g., "Can you tell me more about that?"), clarify a contribution (e.g., "Can you say that in another way?"), support an idea (e.g., "Why do you think that?"), add to an idea mentioned by a classmate (e.g., "Would anyone like to add to the idea?"), re-voice an idea offered by a classmate (e.g., "Who can explain that to me in another way?"), or critique an idea during the discussion (e.g., "Do you agree or disagree with that idea and why?") until students are able to generate an adequate explanation.
7. Tell the students, "We also looked for cause-and-effect relationships during our investigation." Then ask, "Can anyone tell me why it is useful to look for cause-and-effect relationships?"
8. Allow the students to share their ideas.
9. Tell the students, "We are now going to take a minute to talk about the use of models as tools for reasoning about natural phenomena."
10. Show an image of the question "What is a model?" on the screen. Tell the students, "Take a few minutes to talk about how you would answer this question with the other people in your group. Be ready to share with the rest of the class." Give the students two to three minutes to talk in their group.
11. Ask the students, "What do you all think? Who would like to share an idea?"
12. Allow the students to share their ideas.
13. Tell the students, "Okay, let's make sure we are all using the same definition. I think a model is something scientists use to help them reason or explain a

Teacher Notes

phenomenon. Models can be pictures. Or they can be graphs. Or they can even be physical models of things like a skeleton."

14. Show Figure 7.8 (without the caption) on a screen with the question "Is this a model?".

FIGURE 7.8
A model of a person standing on Earth

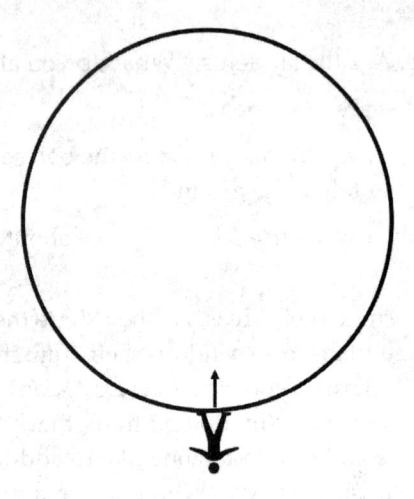

15. Ask the students, "Do you think this is a model? Why or why not?"
16. Allow the students to share their ideas.
17. Tell the students, "I think this is a model because it allows us to understand what we see happen on Earth."
18. Ask the students, "What is the system being modeled?"
19. Allow the students to share their ideas.
20. Ask the students, "What questions do you have about how scientists use models?"
21. Answer any questions that come up.
22. Allow the students to share their ideas.
23. Show an image of the question "What do you think are the most important core ideas or crosscutting concepts that we used during this investigation to help us make sense of what we observed?" Tell the students, "Okay, let's make sure we are all on the same page. Please take a moment to discuss this question with the other people in your group." Give them a few minutes to discuss the question.
24. Ask the students, "What do you all think? Who would like to share?"
25. Allow the students to share their ideas.

Investigation 7. Gravity:
What Direction Is the Gravitational Force Exerted by Earth on Objects?

26. Tell the students, "We are now going to take a minute to talk about what went well and what didn't go so well during our investigation. We need to talk about this because you all are going to be planning and carrying out your own investigations like this a lot this year, and I want to help you all get better at it."

27. Show an image of the question "What made your investigation scientific?" on the screen. Tell the students, "Take a few minutes to talk about how you would answer this question with the other people in your group. Be ready to share with the rest of the class." Give the students two to three minutes to talk in their group.

28. Ask the students, "What do you all think? Who would like to share an idea?"

29. Allow the students to share their ideas. Be sure to expand on their ideas about what makes an investigation scientific.

30. Show an image of the question "What made your investigation not so scientific?" on the screen. Tell the students, "Take a few minutes to talk about how you would answer this question with the other people in your group. Be ready to share with the rest of the class." Give the students two to three minutes to talk in their group.

31. Ask the students, "What do you all think? Who would like to share an idea?"

32. Allow the students to share their ideas. Be sure to expand on their ideas about what makes an investigation less scientific.

33. Show an image of the question "What rules can we put into place to help us make sure our next investigation is more scientific?" on the screen. Tell the students, "Take a few minutes to talk about how you would answer this question with the other people in your group. Be ready to share with the rest of the class." Give the students two to three minutes to talk in their group.

34. Ask the students, "What do you all think? Who would like to share an idea?"

35. Allow the students to share their ideas. Once they have shared their ideas, offer a suggestion for a possible class rule.

36. Ask the students, "What do you all think? Should we make this a rule?"

37. If the students agree, write the rule on the board or on a class "Rules for Scientific Investigation" chart so you can refer to it during the next investigation.

Stage 6: Write a Draft Report (30 minutes)

Your students will use either the Investigation Handout or the Investigation Log in the student workbook when writing the draft report. When you give the directions shown in quotes in the following steps, substitute "Investigation Log in your workbook" or just "Investigation Log" (as shown in brackets) for "handout" if they are using the workbook.

Teacher Notes

1. Tell the students, "You are now going to write an investigation report to share what you have learned. Please take out a pencil and turn to the 'Draft Report' section of your handout [Investigation Log in your workbook]."

2. If possible, use a document camera to project the "Introduction" section of the draft report from the Investigation Handout (or the Investigation Log in their workbook) on a screen or board (or take a picture of it and project the picture on a screen or board).

3. Tell the students, "The first part of the report is called the 'Introduction.' In this section of the report you want to explain to the reader what you were investigating, why you were investigating it, and what question you were trying to answer. All this information can be found in the text at the beginning of your handout [Investigation Log]." Point to the image. "Here are some sentence starters to help you begin writing."

4. Ask the students, "Do you have any questions about what you need to do?"

5. Answer any questions that come up.

6. Tell the students, "Okay, let's write."

7. Give the students 10 minutes to write the "Introduction" section of the report. As they work, move from student to student to check in, ask probing questions, and offer a suggestion if a student gets stuck.

8. If possible, use a document camera to project the "Method" section of the draft report from the Investigation Handout (or the Investigation Log in their workbook) on a screen or board (or take a picture of it and project the picture on a screen or board).

9. Tell the students, "The second part of the report is called the 'Method.' In this section of the report you want to explain to the reader what you did during the investigation, what data you collected and why, and how you went about analyzing your data. All this information can be found in the 'Plan Your Investigation' section of the handout [Investigation Log]. Remember that you all planned and carried out different investigations, so do not assume that the reader will know what you did." Point to the image. "Here are some sentence starters to help you begin writing."

10. Ask the students, "Do you have any questions about what you need to do?"

11. Answer any questions that come up.

12. Tell the students, "Okay, let's write."

13. Give the students 10 minutes to write the "Method" section of the report. As they work, move from student to student to check in, ask probing questions, and offer a suggestion if a student gets stuck.

14. If possible, use a document camera to project the "Argument" section of the draft report from the Investigation Handout (or the Investigation Log in their

workbook) on a screen or board (or take a picture of it and project the picture on a screen or board).

15. Tell the students, "The last part of the report is called the 'Argument.' In this section of the report you want to share your claim, evidence, and justification of the evidence with the reader. All this information can be found on your whiteboard." Point to the image. "Here are some sentence starters to help you begin writing."

16. Ask the students, "Do you have any questions about what you need to do?"

17. Answer any questions that come up.

18. Tell the students, "Okay, let's write."

19. Give the students 10 minutes to write the "Argument" section of the report. As they work, move from student to student to check in, ask probing questions, and offer a suggestion if a student gets stuck.

Stage 7: Peer Review (35 minutes)

Your students will use either the Investigation Handout or their workbook when doing the peer review. Except where noted below, the directions are the same whether using the handout or the workbook.

1. Tell the students, "We are now going to review our reports to find ways to make them better. I'm going to come around and collect your draft reports. While I do that, please take out a pencil."

2. Collect the handouts or the workbooks with the draft reports from the students.

3. If possible, use a document camera to project the peer-review guide (see Appendix 4) on a screen or board (or take a picture of it and project the picture on a screen or board).

4. Tell the students, "We are going to use this peer-review guide to give each other feedback." Point to the image.

5. Tell the students, "I'm going to ask you to work with a partner to do this. I'm going to give you and your partner a draft report to read. You two will then read the report together. Once you are done reading the report, I want you to answer each of the questions on the peer-review guide." Point to the review questions on the image of the peer-review guide.

6. Tell the students, "You can check 'no,' 'almost,' or 'yes' after each question." Point to the checkboxes on the image of the peer-review guide.

7. Tell the students, "This will be your rating for this part of the report. Make sure you agree on the rating you give the author. If you mark 'no' or 'almost,' then you need to tell the author what he or she needs to do to get a 'yes.'" Point to the space for the reviewer feedback on the image of the peer-review guide.

Teacher Notes

8. Tell the students, "It is really important for you to give the authors feedback that is helpful. That means you need to tell them exactly what they need to do to make their report better."

9. Ask the students, "Do you have any questions about what you need to do?"

10. Answer any questions that come up.

11. Tell the students, "Please sit with a partner who is not in your current group." Allow the students time to sit with a partner.

12. Tell the students, "Okay, I'm now going to give you one report to read." Pass out one Investigation Handout with a draft report or one workbook to each pair. Make sure that the report you give a pair was not written by one of the students in that pair. Give each pair one peer-review guide to fill out. If the students are using workbooks, the peer-review guide is included right after the draft report so you do not need to pass out copies of the peer-review guide.

13. Tell the students, "Okay, I'm going to give you 15 minutes to read the report I gave you and to fill out the peer-review guide. Go ahead and get started."

14. Give the students 15 minutes to work. As they work, move around from pair to pair to check in and see how things are going, answer questions, and offer advice.

15. After 15 minutes pass, tell the students, "Okay, time is up. Please give me the report and the peer-review guide that you filled out."

16. Collect the Investigation Handouts and the peer-review guides, or collect the workbooks if students are using them. If the students are using the Investigation Handouts and separate peer-review guides, be sure you keep each handout with its corresponding peer-review guide.

17. Tell the students, "Okay, I am now going to give you a different report to read and a new peer-review guide to fill out." Pass out one more report to each pair. Make sure that the report you give a pair was not written by one of the students in that pair. Give each pair a new peer-review guide to fill out as a group.

18. Tell the students, "Okay, I'm going to give you 15 minutes to read this new report and to fill out the peer-review guide. Go ahead and get started."

19. Give the students 15 minutes to work. As they work, move around from pair to pair to check in and see how things are going, answer questions, and offer advice.

20. After 15 minutes pass, tell the students, "Okay, time is up. Please give me the report and the peer-review guide that you filled out."

21. Collect the Investigation Handouts and the peer-review guides, or collect the workbooks if students are using them. If the students are using the Investigation Handouts and separate peer-review guides, be sure you keep each handout with its corresponding peer-review guide.

Investigation 7. Gravity:
What Direction Is the Gravitational Force Exerted by Earth on Objects?

Stage 8: Revise the Report (30 minutes)

Your students will use either the Investigation Handout or their workbook when revising the report. Except where noted below, the directions are the same whether using the handout or the workbook.

1. Tell the students, "You are now going to revise your draft report based on the feedback you get from your classmates. Please take out a pencil."

2. Return the reports to the students.
   - *If the students used the Investigation Handout and a copy of the peer-review guide,* pass back the handout and the peer-review guide to each student.
   - *If the students used the workbook,* pass that back to each student.

3. Tell the students, "Please take a few minutes to read over the peer-review guide. You should use it to figure out what you need to change in your report and how you will change it."

4. Allow the students to read the peer-review guide.

5. *If the students used the workbook,* if possible use a document camera to project the "Write Your Final Report" section from the Investigation Log on a screen or board (or take a picture of it and project the picture on a screen or board).

6. Give the following directions about how to revise their reports:
   - *If the students used the Investigation Handout and a copy of the peer-review guide,* tell them, "Okay, let's revise our reports. Please take out a piece of paper. I would like you to rewrite your report. You can use your draft report as a starting point, but you also need to change it to make it better. Use the feedback on the peer-review guide to make it better."
   - *If the students used the workbook,* tell them, "Okay, let's revise our reports. I would like you to rewrite your report in the section of the Investigation Log called "Write Your Final Report." You can use your draft report as a starting point, but you also need to change it to make it better. Use the feedback on the peer-review guide to make it better."

7. Ask the students, "Do you have any questions about what you need to do?"

8. Answer any questions that come up.

9. Tell the students, "Okay, let's write." Allow about 20 minutes for the students to revise their reports.

10. After about 20 minutes, give the following directions:
    - *If the students used the Investigation Handout,* tell them, "Okay, time's up. I will now come around and collect your Investigation Handout, the peer-review guide, and your final report."

Teacher Notes

- *If the students used the workbook,* tell them, "Okay, time's up. I will now come around and collect your workbooks."

11. *If the students used the Investigation Handout,* collect all the Investigation Handouts, peer-review guides, and final reports. *If the students used the workbook,* collect all the workbooks.

12. *If the students used the Investigation Handout,* use the "Teacher Score" column in the peer-review guide to grade the final report. *If the students used the workbook,* use the "Investigation Report Grading Rubric" in the Investigation Log to grade the final report. Whether you are using the handout or the log, you can give the students feedback about their writing in the "Teacher Comments" section.

How to Use the Checkout Questions

The Checkout Questions are an optional assessment. We recommend giving them to students at the start of the next class period after the students finish stage 8 of the investigation. You can then look over the student answers to determine if you need to reteach the core idea from the investigation. Appendix 6 gives the answers to the Checkout Questions that should be given by a student who can apply the core idea correctly in all cases and can explain (1) the cause-and-effect relationship and (2) the importance of systems and system models.

Alignment With Standards

Table 7.2 highlights how the investigation can be used to address specific performance expectations from the *Next Generation Science Standards, Common Core State Standards for English Language Arts (CCSS ELA),* and *English Language Proficiency (ELP) Standards.*

TABLE 7.2

Investigation 7 alignment with standards

NGSS performance expectation	5-PS2-1: Support an argument that the gravitational force exerted by Earth on objects is directed down.
CCSS ELA—Reading: Informational Text	Key ideas and details • CCSS.ELA-LITERACY.RI.5.1: Quote accurately from a text when explaining what the text says explicitly and when drawing inferences from the text. • CCSS.ELA-LITERACY.RI.5.2: Determine two or more main ideas of a text and explain how they are supported by key details; summarize the text. • CCSS.ELA-LITERACY.RI.5.3: Explain the relationships or interactions between two or more individuals, events, ideas, or concepts in a historical, scientific, or technical text based on specific information in the text.

Continued

Investigation 7. Gravity: What Direction Is the Gravitational Force Exerted by Earth on Objects?

Table 7.2 (*continued*)

***CCSS ELA*—Reading: Informational Text** (*continued*)	Craft and structure	
	• CCSS.ELA-LITERACY.RI.5.4: Determine the meaning of general academic and domain-specific words and phrases in a text relevant to a *grade 5 topic or subject area*.	
	• CCSS.ELA-LITERACY.RI.5.5: Compare and contrast the overall structure (e.g., chronology, comparison, cause/effect, problem/solution) of events, ideas, concepts, or information in two or more texts.	
	• CCSS.ELA-LITERACY.RI.5.6: Analyze multiple accounts of the same event or topic, noting important similarities and differences in the point of view they represent.	
	Integration of knowledge and ideas	
	• CCSS.ELA-LITERACY.RI.5.7: Draw on information from multiple print or digital sources, demonstrating the ability to locate an answer to a question quickly or to solve a problem efficiently.	
	• CCSS.ELA-LITERACY.RI.5.8: Explain how an author uses reasons and evidence to support particular points in a text, identifying which reasons and evidence support which point(s).	
	Range of reading and level of text complexity	
	• CCSS.ELA-LITERACY.RI.5.10: By the end of the year, read and comprehend informational texts, including history/social studies, science, and technical texts, at the high end of the grades 4–5 text complexity band independently and proficiently.	
***CCSS ELA*—Writing**	Text types and purposes	
	• CCSS.ELA-LITERACY.W.5.1: Write opinion pieces on topics or texts, supporting a point of view with reasons.	
	○ CCSS.ELA-LITERACY.W.5.1.A: Introduce a topic or text clearly, state an opinion, and create an organizational structure in which ideas are logically grouped to support the writer's purpose.	
	○ CCSS.ELA-LITERACY.W.5.1.B: Provide logically ordered reasons that are supported by facts and details.	
	○ CCSS.ELA-LITERACY.W.5.1.C: Link opinion and reasons using words, phrases, and clauses (e.g., *consequently*, *specifically*).	
	○ CCSS.ELA-LITERACY.W.5.1.D: Provide a concluding statement or section related to the opinion presented.	
	• CCSS.ELA-LITERACY.W.5.2: Write informative or explanatory texts to examine a topic and convey ideas and information clearly.	
	• CCSS.ELA-LITERACY.W.5.2.A: Introduce a topic clearly, provide a general observation and focus, and group related information logically; include formatting (e.g., headings), illustrations, and multimedia when useful to aiding comprehension.	
	• CCSS.ELA-LITERACY.W.5.2.B: Develop the topic with facts, definitions, concrete details, quotations, or other information and examples related to the topic.	
	• CCSS.ELA-LITERACY.W.5.2.C: Link ideas within and across categories of information using words, phrases, and clauses (e.g., *in contrast*, *especially*).	

Continued

Teacher Notes

Table 7.2 *(continued)*

CCSS ELA—Writing *(continued)*	• CCSS.ELA-LITERACY.W.5.2.D: Use precise language and domain-specific vocabulary to inform about or explain the topic. • CCSS.ELA-LITERACY.W.5.2.E: Provide a concluding statement or section related to the information or explanation presented. Production and distribution of writing • CCSS.ELA-LITERACY.W.5.4: Produce clear and coherent writing in which the development and organization are appropriate to task, purpose, and audience. • CCSS.ELA-LITERACY.W.5.5: With guidance and support from peers and adults, develop and strengthen writing as needed by planning, revising, editing, rewriting, or trying a new approach. • CCSS.ELA-LITERACY.W.5.6: With some guidance and support from adults, use technology, including the internet, to produce and publish writing as well as to interact and collaborate with others; demonstrate sufficient command of keyboarding skills to type a minimum of two pages in a single sitting. Research to build and present knowledge • CCSS.ELA-LITERACY.W.5.8: Recall relevant information from experiences or gather relevant information from print and digital sources; summarize or paraphrase information in notes and finished work, and provide a list of sources. • CCSS.ELA-LITERACY.W.5.9: Draw evidence from literary or informational texts to support analysis, reflection, and research. Range of writing • CCSS.ELA-LITERACY.W.5.10: Write routinely over extended time frames (time for research, reflection, and revision) and shorter time frames (a single sitting or a day or two) for a range of discipline-specific tasks, purposes, and audiences.
CCSS ELA—Speaking and Listening	Comprehension and collaboration • CCSS.ELA-LITERACY.SL.5.1: Engage effectively in a range of collaborative discussions (one-on-one, in groups, and teacher-led) with diverse partners on *grade 5 topics and texts,* building on others' ideas and expressing their own clearly. • CCSS.ELA-LITERACY.SL.5.1.A: Come to discussions prepared, having read or studied required material; explicitly draw on that preparation and other information known about the topic to explore ideas under discussion. • CCSS.ELA-LITERACY.SL.5.1.B: Follow agreed-upon rules for discussions and carry out assigned roles. • CCSS.ELA-LITERACY.SL.5.1.C: Pose and respond to specific questions by making comments that contribute to the discussion and elaborate on the remarks of others. • CCSS.ELA-LITERACY.SL.5.1.D: Review the key ideas expressed and draw conclusions in light of information and knowledge gained from the discussions.

Continued

Table 7.2 (*continued*)

CCSS ELA—Speaking and Listening (*continued*)	• CCSS.ELA-LITERACY.SL.5.2: Summarize a written text read aloud or information presented in diverse media and formats, including visually, quantitatively, and orally. • CCSS.ELA-LITERACY.SL.5.3: Summarize the points a speaker makes and explain how each claim is supported by reasons and evidence. Presentation of knowledge and ideas • CCSS.ELA-LITERACY.SL.5.4: Report on a topic or text or present an opinion, sequencing ideas logically and using appropriate facts and relevant, descriptive details to support main ideas or themes; speak clearly at an understandable pace. • CCSS.ELA-LITERACY.SL.5.5: Include multimedia components (e.g., graphics, sound) and visual displays in presentations when appropriate to enhance the development of main ideas or themes. • CCSS.ELA-LITERACY.SL.5.6: Adapt speech to a variety of contexts and tasks, using formal English when appropriate to task and situation.
ELP Standards	Receptive modalities • ELP 1: Construct meaning from oral presentations and literary and informational text through grade-appropriate listening, reading, and viewing. • ELP 8: Determine the meaning of words and phrases in oral presentations and literary and informational text. Productive modalities • ELP 3: Speak and write about grade-appropriate complex literary and informational texts and topics. • ELP 4: Construct grade-appropriate oral and written claims and support them with reasoning and evidence. • ELP 7: Adapt language choices to purpose, task, and audience when speaking and writing. Interactive modalities • ELP 2: Participate in grade-appropriate oral and written exchanges of information, ideas, and analyses, responding to peer, audience, or reader comments and questions. • ELP 5: Conduct research and evaluate and communicate findings to answer questions or solve problems. • ELP 6: Analyze and critique the arguments of others orally and in writing. Linguistic structures of English • ELP 9: Create clear and coherent grade-appropriate speech and text. • ELP 10: Make accurate use of standard English to communicate in grade-appropriate speech and writing.

Investigation Handout

Investigation 7
Gravity: What Direction Is the Gravitational Force Exerted by Earth on Objects?

Introduction

Many people enjoy playing and watching sports. Globally, the most popular sport for people to watch is soccer. Other sports that people enjoying watching include football in the United States and rugby in Australia and Europe. What these three sports have in common is that, at various times during a game, players are kicking a ball. Your teacher will show you several videos of people kicking balls when they play soccer, football, or rugby. As you watch what happens, keep track of things you notice and things you wonder about in the boxes below.

Things I NOTICED …	Things I WONDER about …

Objects, such as the soccer ball, football, and rugby balls you just saw in the videos, often have more than one force acting on them at the same time. The forces acting on an object can be either balanced or unbalanced. Balanced forces are two forces that are the same size but are acting on the object in opposite directions. Scientists describe balanced forces as being equal and opposite. *Unbalanced forces* are not equal and opposite. One example of an unbalanced force is two different-size forces acting on an object in opposite directions. A second example of an unbalanced force is two same-size forces

Investigation 7. Gravity:
What Direction Is the Gravitational Force Exerted by Earth on Objects?

acting on an object in the same direction. When you try to determine if all the forces acting on an object are balanced or unbalanced, it is important to remember that forces that act in the same direction combine by addition and forces that act in opposite directions combine by subtraction. Forces that sum to zero do not change an object's speed or direction of motion, but forces that do not sum to zero can cause a change in an object's motion.

One type of force that can act on an object is called a *contact force*. A contact force is a push or a pull on an object from another object that is touching it. An example of a contact force is when you kick a soccer ball, football, or rugby ball. Your foot and the ball touch each other, and the force from your foot hitting the ball causes the ball to change how it is moving (change from not moving to moving, or start moving in a different direction). Once the ball leaves your foot, your foot no longer exerts a force on the ball.

A second type of force that can act on an object is called a *non-contact force*. A non-contact force is a push or a pull on an object that comes from something that is not touching that object. *Gravity* is an example of a non-contact force. Gravity is a pulling force between objects. When you kick a ball in the air, the pulling force of gravity from Earth makes it return to the ground. It is a non-contact force because the ball does not need to be touching the surface of Earth for the force of gravity to act on it. When the ball is kicked high up into the air, the force of gravity is still acting on the ball.

To be able to explain and predict the motion of the soccer ball, football, and rugby balls in each of the different videos, you need to first identify all the forces (contact and non-contact) that are acting on each ball at different points in time and the direction of each of these forces. When a player kicks a ball during a match or game, for example, there is contact force that acts on the ball in the up and forward directions. An important question for you to figure out, however, is in what direction the force of gravity acts on the ball in each of the videos. This is important because the videos were filmed in different locations on Earth, but gravity was acting on the ball in each one of the videos.

Things we KNOW from what we read …

Investigation Handout

Your Task

Use what you know about forces and interactions, cause and effect, and how scientists use models to design and carry out an investigation to determine the direction of the force of gravity. You will need to figure out in what direction gravity acts on the ball in each location shown in the videos.

The *guiding question* of this investigation is, **What direction is the gravitational force exerted by Earth on objects?**

Materials

You may use any of the following materials during your investigation:

- Safety goggles (required)
- Inflatable plastic globe
- Miniature figures
- Miniature soccer ball
- Dry-erase marker
- Tape
- Ruler

Safety Rules

Follow all normal safety rules. In addition, be sure to follow these rules:

- Wear sanitized safety goggles during setup, investigation activity, and cleanup.
- Wash your hands with soap and water when you are done cleaning up.

Plan Your Investigation

Prepare a plan for your investigation by filling out the chart on the next page; this plan is called an *investigation proposal*. Before you start developing your plan, be sure to discuss the following questions with the other members of your group:

- What does a force acting on a ball (a **cause**) do to the motion of that ball (the **effect**)?
- How can you use the materials to **model the movement** of a ball when it is kicked?

Investigation 7. Gravity:
What Direction Is the Gravitational Force Exerted by Earth on Objects?

Our guiding question:

This is a picture of how we will set up the equipment:

We will collect the following data:

These are the steps we will follow to collect data:

I approve of this investigation proposal.

_____   _____
Teacher's signature   Date

Argument-Driven Inquiry in **Fifth-Grade Science**: Three-Dimensional Investigations

Investigation Handout

Collect Your Data

Keep a record of what you measure or observe during your investigation in the space below.

Analyze Your Data

You will need to analyze the data you collected before you can develop an answer to the guiding question. To analyze the data you collected, create a picture or model.

Investigation 7. Gravity:
What Direction Is the Gravitational Force Exerted by Earth on Objects?

Draft Argument

Develop an argument on a whiteboard. It should include the following:

1. A *claim*: Your answer to the guiding question.
2. *Evidence*: An analysis of the data and an explanation of what the analysis means.
3. A *justification of the evidence*: Why your group thinks the evidence is important.

The Guiding Question:	
Our Claim:	
Our Evidence:	Our Justification of the Evidence:

Argumentation Session

Share your argument with your classmates. Be sure to ask them how to make your draft argument better. Keep track of their suggestions in the space below.

Ways to IMPROVE our argument …

Investigation Handout

Draft Report

Prepare an *investigation report* to share what you have learned. Use the information in this handout and your group's final argument to write a *draft* of your investigation report.

Introduction

We have been studying _____ in class.

Before we started this investigation, we explored _____
_____
_____

We noticed _____
_____
_____

My goal for this investigation was to figure out _____
_____
_____

The guiding question was _____
_____
_____

Method

To gather the data I needed to answer this question, I _____
_____
_____
_____
_____

Investigation 7. Gravity:
What Direction Is the Gravitational Force Exerted by Earth on Objects?

I then analyzed the data I collected by _____

_____

_____

_____

Argument

My claim is _____

_____

_____

_____

The _____ below shows _____

_____

_____

_____

_____

_____

Argument-Driven Inquiry in **Fifth-Grade Science:** Three-Dimensional Investigations

Investigation Handout

This analysis of the data I collected suggests _____

_____

_____

_____

_____

This evidence is based on several important scientific concepts. The first one is _____

_____

_____

_____

_____

_____

Review

Your classmates need your help! Review the draft of their investigation reports and give them ideas about how to improve. Use the peer-review guide when doing your review.

Submit Your Final Report

Once you have received feedback from your classmates about your draft report, create your final investigation report and hand it in to your teacher.

Checkout Questions

Investigation 7. Gravity

1. The picture at right shows a person in the Northern Hemisphere standing on the surface of Earth. Which arrow represents the direction of the gravitational force exerted on a ball kicked at this location?

    a. Arrow A
    b. Arrow B
    c. Arrow C
    d. Arrow D

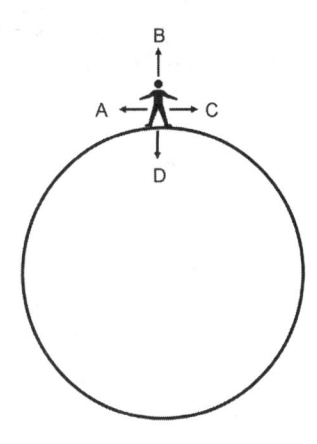

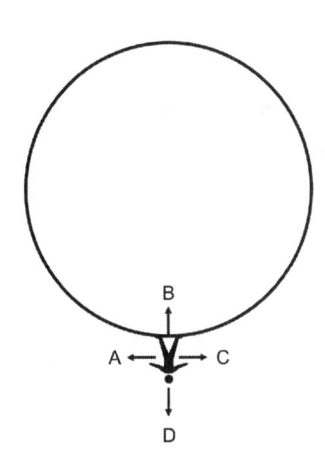

2. The picture at left shows a person in the Southern Hemisphere standing on the surface of Earth. Which arrow represents the direction of the gravitational force exerted on a ball kicked at this location?

    a. Arrow A
    b. Arrow B
    c. Arrow C
    d. Arrow D

3. How do you know? Use what you know about cause and effect to explain your answer.

_____
_____
_____
_____
_____
_____
_____

Teacher Scoring Rubric for Checkout Questions 1–3

Level	Description
3	The student can apply the core idea correctly in all cases and can explain the cause-and-effect relationship.
2	The student cannot apply the core idea correctly in all cases but can explain the cause-and-effect relationship.
1	The student can apply the core idea correctly in all cases but cannot explain the cause-and-effect relationship.
0	The student cannot apply the core idea correctly and cannot explain the cause-and-effect relationship.

Argument-Driven Inquiry in **Fifth-Grade Science**: Three-Dimensional Investigations

Checkout Questions

The picture below shows a ball sitting on a table (Before), then being pushed by someone (During), and rolling off the table and falling to the floor (After). Use this picture to answer questions 4 and 5.

Before | During | After

4. Which model (A, B, C, or D) best illustrates all the forces that are acting on the ball before, during, and after someone pushes it off a table? The arrows in the models represent forces. Circle the answer that best represents your thinking.

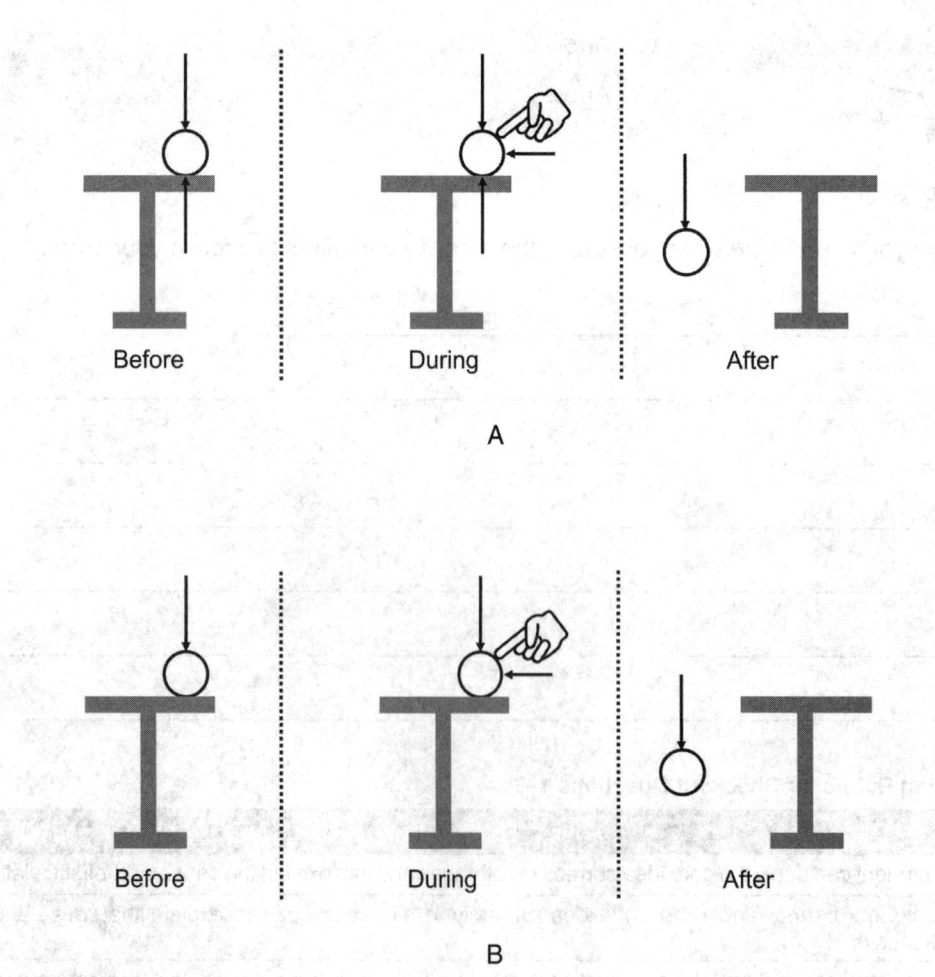

Before | During | After

A

Before | During | After

B

314 National Science Teaching Association

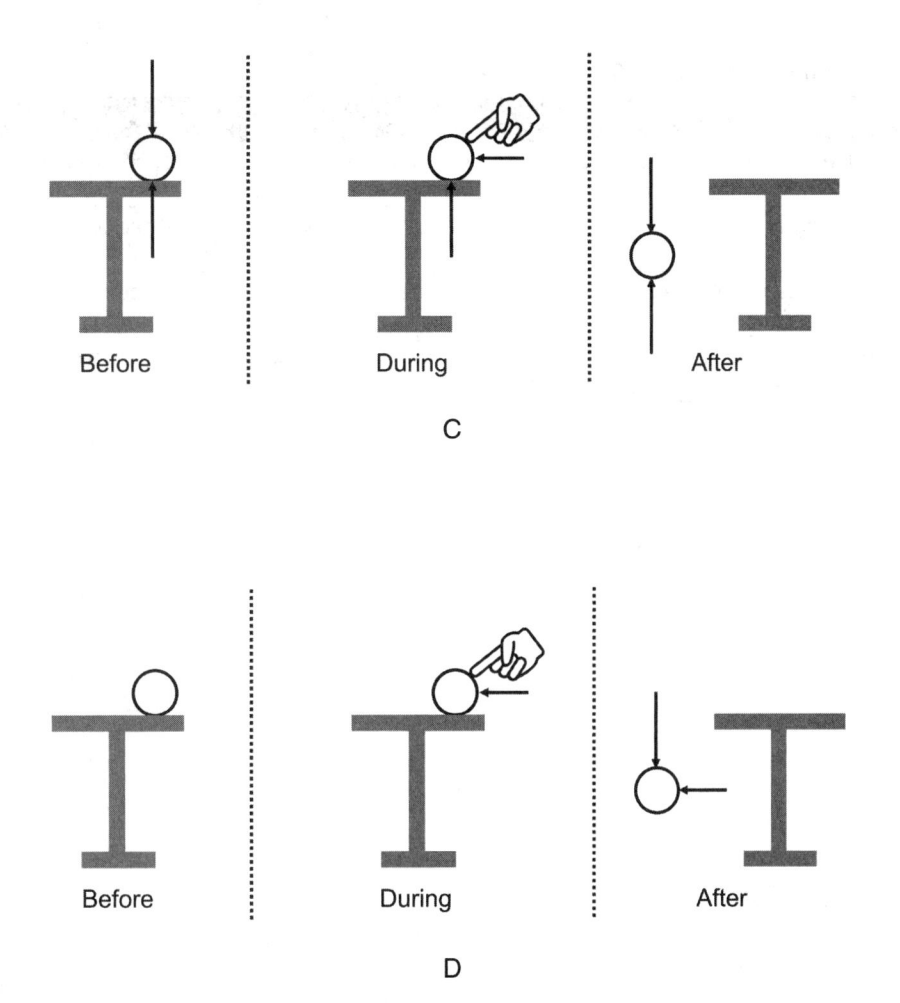

Investigation 7. Gravity:
What Direction Is the Gravitational Force Exerted by Earth on Objects?

Before | During | After

C

Before | During | After

D

5. How do you know? Use what you know about systems and system models to explain your answer.

___

Argument-Driven Inquiry in **Fifth-Grade Science**: Three-Dimensional Investigations

Checkout Questions

Teacher Scoring Rubric for Checkout Questions 4 and 5

Level	Description
3	The student can apply the core idea correctly and can explain the importance of systems and system models.
2	The student cannot apply the core idea correctly but can explain the importance of systems and system models.
1	The student can apply the core idea correctly but cannot explain the importance of systems and system models.
0	The student cannot apply the core idea correctly and cannot explain the importance of systems and system models.

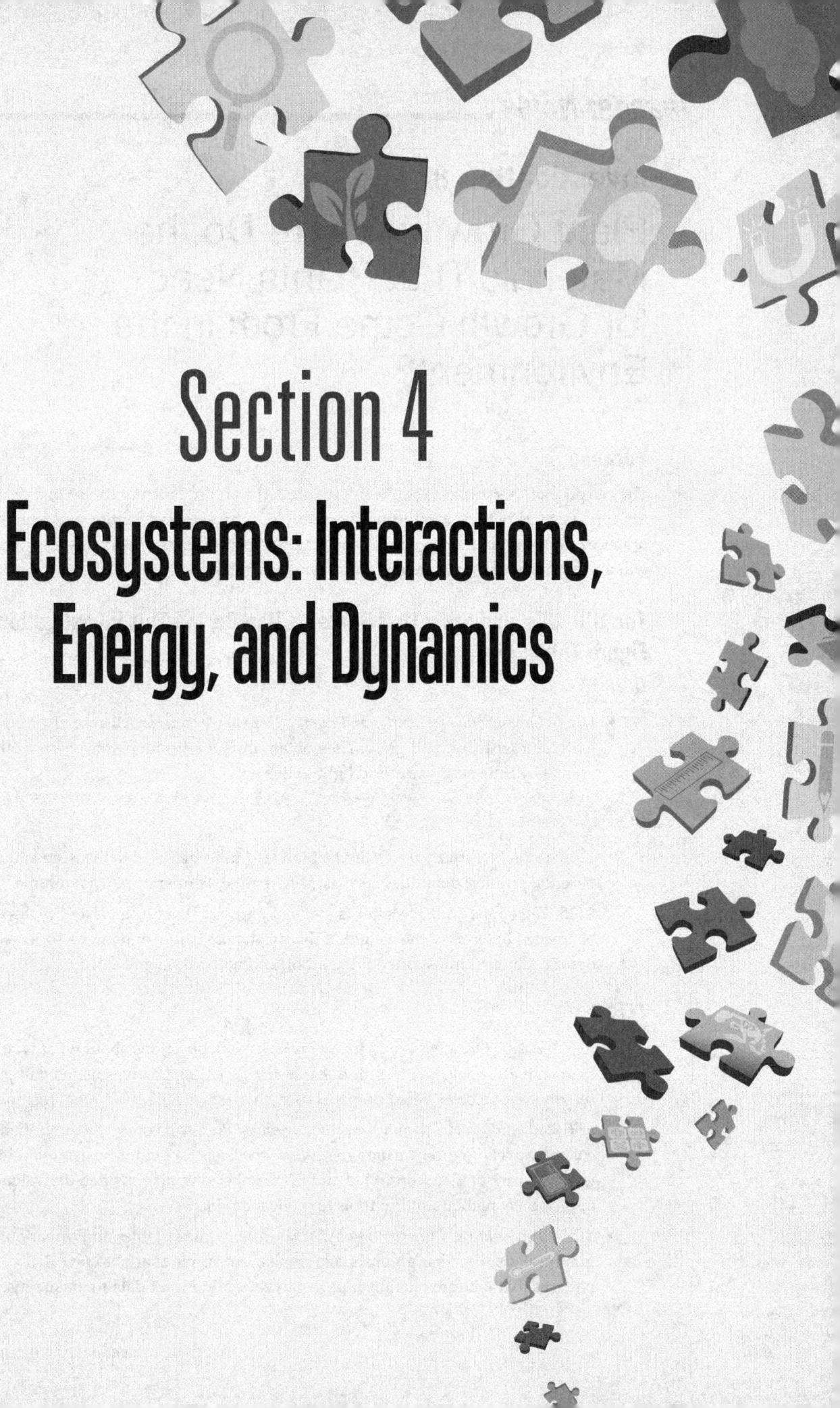

Section 4
Ecosystems: Interactions, Energy, and Dynamics

Teacher Notes

Investigation 8
Plant Growth: Where Do the Materials That Plants Need for Growth Come From in the Environment?

Purpose

The purpose of this investigation is to give students an opportunity to use one disciplinary core idea (DCI), two crosscutting concepts (CCs), and seven scientific and engineering practices (SEPs) to figure out where plants get the materials they need to grow. Students will also learn about the nature and role of experiments in science.

The DCI, CCs, and SEPs That Students Use During This Investigation to Figure Things Out

DCI

- *LS1.C: Organization for Matter and Energy Flow in Organisms:* All animals need food in order to live and grow. They obtain their food from plants or from other animals. Plants need water and light to live and grow.

CCs

- *CC 3: Scale, Proportion, and Quantity:* Standard units are used to measure and describe physical quantities such as weight, time, temperature, and volume.
- *CC 5: Energy and Matter:* Matter is made of particles. Matter flows and cycles can be tracked in terms of the weight of the substances before and after a process occurs. Matter is transported into, out of, and within systems.

SEPs

- *SEP 1: Asking Questions and Defining Problems:* Ask questions about what would happen if a variable is changed. Ask questions that can be investigated and predict reasonable outcomes based on patterns such as cause-and-effect relationships.
- *SEP 3: Planning and Carrying Out Investigations:* Plan and conduct an investigation collaboratively to produce data to serve as the basis for evidence, using fair tests in which variables are controlled and the number of trials considered. Evaluate appropriate methods and/or tools for collecting data.
- *SEP 4: Analyzing and Interpreting Data:* Represent data in tables and/or various graphical displays (bar graphs, pictographs, and/or pie charts) to reveal patterns that indicate relationships. Analyze and interpret data to make sense of

phenomena, using logical reasoning, mathematics, and/or computation. Compare and contrast data collected by different groups in order to discuss similarities and differences in their findings.

- *SEP 5: Using Mathematics and Computational Thinking:* Organize simple data sets to reveal patterns that suggest relationships. Describe, measure, estimate, and/or graph quantities (e.g., area, volume, weight, time) to address scientific and engineering questions and problems.
- *SEP 6: Constructing Explanations and Designing Solutions:* Construct an explanation of observed relationships. Use evidence to construct or support an explanation. Identify the evidence that supports particular points in an explanation.
- *SEP 7: Engaging in Argument From Evidence:* Compare and refine arguments based on an evaluation of the evidence presented. Distinguish among facts, reasoned judgment based on research findings, and speculation in an explanation. Respectfully provide and receive critiques from peers about a proposed procedure, explanation, or model by citing relevant evidence and posing specific questions.
- *SEP 8: Obtaining, Evaluating, and Communicating Information:* Read and comprehend grade-appropriate complex texts and/or other reliable media to summarize and obtain scientific and technical ideas. Combine information in written text with that contained in corresponding tables, diagrams, and/or charts to support the engagement in other scientific and/or engineering practices. Communicate scientific and/or technical information orally and/or in written formats, including various forms of media as well as tables, diagrams, and charts.

Other Concepts That Students May Use During This Investigation

Students might also use some of the following concepts:

- The Sun provides energy for plants.
- Plants go through a life cycle that may include a seed stage.
- Seeds contain material that provides matter and energy to the germinating plant.
- Air, water, and soil are all types of matter.

What Students Figure Out

Plants acquire their material for growth chiefly from air and water. Soil is not a requirement for most plant growth, nor is it sufficient for growth.

Teacher Notes

Background Information About This Investigation for the Teacher

Plants require both matter and energy to survive and grow. Plants use the energy from light to make sugar from carbon dioxide and water through a process called photosynthesis. Plants can use sugar immediately for the energy they need to survive, or the sugar can be stored for later use. Within a plant, sugar moves through a series of chemical reactions in which the sugar is broken down and rearranged to form new molecules. This process releases energy and allows the plants to create new cells, which are needed for the plant to increase in size (growth).

The majority of the matter, by weight, that plants need to increase in size comes from water and carbon dioxide in the air. Although some plants require soil to grow because the soil provides nutrients such as nitrogen, potassium, and phosphorus that are essential for many of life's processes, these nutrients do not contribute much to the overall mass of a plant. Initial germination of a seed and growth of a plant are aided by the stored material inside the seed itself. This is also why many animals eat seeds to live—they serve as a source of food and are often high in caloric energy.

Timeline

The time needed to complete this investigation is about 400 minutes (6 hours and 40 minutes) spread out across several weeks. The amount of instructional time needed for each stage of the investigation is as follows:

- *Stage 1.* Introduce the task and the guiding question: 35 minutes
- *Stage 2.* Design a method: 15 minutes; collect data: 15 minutes for setup and then 5–15 minutes a day for at least 10 days
- *Stage 3.* Create a draft argument: 45 minutes
- *Stage 4.* Argumentation session: 30 minutes
- *Stage 5.* Reflective discussion: 15 minutes
- *Stage 6.* Write a draft report: 30 minutes
- *Stage 7.* Peer review: 35 minutes
- *Stage 8.* Revise the report: 30 minutes

Materials and Preparation

The materials needed for this investigation are listed in Table 8.1. The items can be purchased from a big-box retail store such as Walmart or Target or through an online retailer such as Amazon. The materials for this investigation can also be purchased as a complete kit (which includes enough materials for 24 students, or six groups of four students) at *www.argumentdriveninquiry.com*.

Investigation 8. Plant Growth:
Where Do the Materials That Plants Need for Growth Come From in the Environment?

TABLE 8.1
Materials for Investigation 8

Item	Quantity
Safety goggles	1 per student
Seeds (radish, mustard, or alfalfa)	36 per group (minimum)
Clear plastic containers (for use as growing chambers)	6 per group
Dry potting soil, 100 ml	1 per group
Airsoft pellets	60 per group
Water beads (such as Orbeez, MarvelBeads, or UMIKU)	20 per group
Plastic graduated cylinder (10 ml)	1 per group
Electronic scale	1 per group
Ruler	1 per group
Tray or plastic container large enough to hold six growing chambers	1 per group
Large plastic pitcher for soaking water beads	1 per class
Blue painter's tape (for labeling growing chambers)	1 roll per class
Permanent marker	1 per class
Paper towels	1 roll per class
Whiteboard, 2' × 3'*	1 per group
Investigation Handout	1 per student
Peer-review guide and teacher scoring rubric	1 per student
Checkout Questions (optional)	1 per student

*As an alternative, students can use computer and presentation software such as Microsoft PowerPoint or Apple Keynote to create their arguments.

You will need to dry out the potting soil before you begin the investigation. You can dry out the soil by creating a one-inch-deep layer of soil in a large plastic tray. You can then leave it to dry for four to five weeks. Be sure to turn the soil over every few days so it dries throughout. If you live in an area with low humidity, you may be able to leave the trays outside in a covered area to allow the soil to dry. If you live in a humid area, you will need to store the trays inside as the soil dries.

About one week before you begin the investigation, you will need to gather all the other materials in Table 8.1. Be sure to pick out a suitable location for the plants to grow. An area with indirect sunlight that is not too hot or too cold will work best. Students will need to access their plants daily, so this area should be within the classroom and easily accessible. The containers will have water and soil in them, so consider an area that is

Teacher Notes

easily cleaned at the end of the investigation. We also recommend giving each group a tray or small plastic tube so they can keep all their chambers together and easily carry them to and from their desks.

One day before the students are to begin their investigation, you will need to soak the seeds in a cup of water overnight. You will also need to soak the water beads in a large pitcher of water. The water beads will absorb all or almost all the water. If your water beads come with directions for soaking, be sure to follow those directions. Once the students have the water beads they need, you can dry out the remaining beads on a clean plastic tray so you can reuse them.

Safety Precautions

Remind students to follow all normal safety rules. In addition, tell the students to take the following safety precautions:

- Wear sanitized safety goggles during setup, investigation activity, and cleanup.
- Do not put anything used in this activity in their mouth.
- Immediately clean up any spills to avoid a slip or fall hazard.
- Make sure all materials are put away after completing the activity.
- Wash their hands with soap and water when they are done cleaning up.

Lesson Plan by Stage

This lesson plan is only a suggestion. It is included here to illustrate what you can say and do during each stage of ADI for this specific investigation. We encourage you to modify this lesson plan by asking different questions, using different examples, and providing different scaffolds as needed to better meet the needs of students in your class.

Stage 1: Introduce the Task and the Guiding Question (35 minutes)

1. Ask the students to sit in six groups, with three or four students in each group.
2. Ask the students to clear off their desks except for a pencil (and their *Student Workbook for Argument-Driven Inquiry in Fifth-Grade Science* if they have one).
3. Pass out an Investigation Handout to each student (or ask students to turn to the Investigation Log for Investigation 8 in their workbook).
4. Read the first paragraph of the "Introduction" aloud to the class. Ask the students to follow along as you read.
5. Show the students a video clip of a plant germinating and sprouting, such as the one at *www.youtube.com/watch?v=w77zPAtVTuI&t=17s*.
6. Ask the students to record what they noticed in the video and their questions about it in the "NOTICED/WONDER" chart in the "Introduction."

Investigation 8. Plant Growth:
Where Do the Materials That Plants Need for Growth Come From in the Environment?

7. After the students have recorded their observations and questions about the video, ask them to share what they noticed.

8. Ask the students to share what questions they have about the video.

9. Tell the students, "Some of your questions might be answered by reading the rest of the 'Introduction.'"

10. Ask the students to read the rest of the "Introduction" on their own *or* ask them to follow along as you read it aloud.

11. Once the students have read the rest of the "Introduction," ask them to fill out the "Things we KNOW" chart on their Investigation Handout (or in their Investigation Log) as a group.

12. Ask the students to share what they learned from the reading. Add these ideas to a class "Things we KNOW" chart.

13. Tell the students, "Let's see what we will need to figure out during our investigation."

14. Read the task and the guiding question aloud.

15. Tell the students, "I have lots of materials here that you can use."

16. Introduce the students to the materials available for them to use during the investigation by either (a) holding each one up and then asking what it might be used for or (b) giving them a kit with an example of each of the materials in it and giving them three to four minutes to play with them. If you give the students an opportunity to play with the materials, collect them from each group before moving on to stage 2.

Stage 2: Design a Method and Collect Data (180 minutes)

1. Tell the students, "I am now going to give you and the other members of your group about 15 minutes to plan your investigation. Before you begin, I want you to take a couple of minutes to discuss the following questions with the rest of your group."

2. Show the following questions on the screen or board:
   - How can we *track the movement of matter* in a plant?
   - What *units* or *scale* should we use to track the movement of matter?

3. Tell the students, "Please take a few minutes to come up with an answer to these questions."

4. Give the students two or three minutes to discuss these two questions.

5. Ask two or three different groups to share their answers. Highlight or write down any important ideas on the board so students can refer to them later.

Teacher Notes

6. If possible, use a document camera to project an image of the graphic organizer for this investigation on a screen or board (or take a picture of it and project the picture on a screen or board). Tell the students, "I now want you all to plan out your investigation. To do that, you will need to fill out this investigation proposal."

7. Point to the box labeled "Our guiding question:" and tell the students, "You can put the question we are trying to answer in this box." Then ask, "Where can we find the guiding question?"

8. Wait for a student to answer.

9. Point to the box labeled "This is a picture of how we will set up the equipment:" and tell the students, "You can draw a picture in this box of how you will set up the equipment in order to carry out this investigation."

10. Point to the box labeled "We will collect the following data:" and tell the students, "You can list the measurements or observations that you will need to collect during the investigation in this box."

11. Point to the box labeled "These are the steps we will follow to collect data:" and tell the students, "You can list what you are going to do to collect the data you need and what you will do with your data once you have it. Be sure to give enough detail that I could do your investigation for you."

12. Ask the students, "Do you have any questions about what you need to do?"

13. Answer any questions that come up.

14. Tell the students, "Once you are done, raise your hand and let me know. I'll then come by and look over your proposal and give you some feedback. You may not begin collecting data until I have approved your proposal by signing it. You need to have your proposal done in the next 15 minutes."

15. Give the students 15 minutes to work in their groups on their investigation proposal. As they work, move from group to group to check in, ask probing questions, and offer a suggestion if a group gets stuck.

16. As each group finishes its investigation proposal, read it over and determine if it will be productive or not. If you feel the investigation will be productive (not necessarily what you would do or what the other groups are doing), sign your name on the proposal and let the group start collecting data. If the plan needs to be changed, offer some suggestions or ask some probing questions, and have the group make the changes before you approve it.

17. Pass out the materials or have one student from each group collect the materials they need from a central supply table or cart for the groups that have an approved proposal.

18. Remind students of the safety rules and precautions for this investigation.

Investigation 8. Plant Growth:
Where Do the Materials That Plants Need for Growth Come From in the Environment?

What should a student-designed investigation look like?

There are a number of different investigations that students can design to answer the question "Where do the materials that plants need for growth come from in the environment?" For example, one method might include the following steps:

1. Weigh 6 seeds.
2. Place 30 plastic pellets and the 6 seeds in a growing chamber.
3. Weigh 6 seeds.
4. Place 20 water beads and the 6 seeds in a second growing chamber.
5. Weigh 6 seeds.
6. Place dry soil and the 6 seeds in a third growing chamber.
7. Weigh 6 seeds.
8. Place 30 plastic pellets, the 6 seeds, and 10 ml of water in a fourth growing chamber.
9. Weigh 6 seeds.
10. Place dry soil, the 6 seeds, and 10 ml of water in a fifth growing chamber.
11. Allow plants to grow for two weeks.
12. Remove the plants from each growing chamber. Remove any soil or other materials from the plants and weigh.

If students use this method, they will need to collect data on (1) the weight of the seeds before being added to the growing chamber and (2) the weight of the plant in each growing chamber after two weeks. Students can also collect data on (1) the number of plants that sprout, (2) plant size (each day), (3) the number of leaves, and (4) the size of leaves. Students may also choose to draw pictures of their plants each day.

19. Tell the students to set up their investigation and record their initial observations or measurements in the "Collect Your Data" box in their Investigation Handout (or the Investigation Log in their workbook). Give the students about 15 minutes to set up their investigation. As students complete their setup, assist them in using the blue painter's tape and permanent marker to label their group's materials and place them in the area set aside for housing the plants.

20. For at least 10 days, allow students 5–15 minutes each day for the students to measure their plants and make observations.

Teacher Notes

Stage 3: Create a Draft Argument (45 minutes)

1. Tell the students, "Now that we have all this data, we need to analyze the data so we can figure out an answer to the guiding question."

2. If possible, project an image of the "Analyze Your Data" section for this investigation on a screen or board using a document camera (or take a picture of it and project the picture on a screen or board). Point to the section and tell the students, "You can create a graph as a way to analyze your data. You can make your graph in this section."

3. Ask the students, "What information do we need to include in a graph?"

4. Tell the students, "Please take a few minutes to discuss this question with your group and be ready to share."

5. Give the students five minutes to discuss.

6. Ask two or three different groups to share their answers. Highlight or write down any important ideas on the board so students can refer to them later.

7. Tell the students, "I am now going to give you and the other members of your group about 10 minutes to create your graph." The graph they create should include each of their chambers and the size of the plants. If the students are having trouble making a graph, you can take a few minutes to provide a mini-lesson about how to create a graph from a bunch of measurements (this strategy is called just-in-time instruction because it is offered only when students get stuck).

8. Give the students 10 minutes to analyze their data by creating a graph. As they work, move from group to group to check in, ask probing questions, and offer suggestions.

9. Tell the students, "I am now going to give you and the other members of your group about 15 minutes to create an argument to share what you have learned and convince others that they should believe you. Before you do that, we need to take a few minutes to discuss what you need to include in your argument."

10. If possible, use a document camera to project the "Argument Presentation on a Whiteboard" image from the "Draft Argument" section of the Investigation Handout (or the Investigation Log in their workbook) on a screen or board (or take a picture of it and project the picture on a screen or board).

11. Point to the box labeled "The Guiding Question:" and tell the students, "You can put the question we are trying to answer here on your whiteboard."

12. Point to the box labeled "Our Claim:" and tell the students, "You can put your claim here on your whiteboard. The claim is your answer to the guiding question."

13. Point to the box labeled "Our Evidence:" and tell the students, "You can put the evidence that you are using to support your claim here on your whiteboard. Your evidence will need to include the analysis you just did and an explanation

Investigation 8. Plant Growth:
Where Do the Materials That Plants Need for Growth Come From in the Environment?

of what your analysis means or shows. Scientists always need to support their claims with evidence."

14. Point to the box labeled "Our Justification of the Evidence:" and tell the students, "You can put your justification of your evidence here on your whiteboard. Your justification needs to explain why your evidence is important. Scientists often use core ideas to explain why the evidence they are using matters. Core ideas are important concepts that scientists use to help them make sense of what happens during an investigation."

15. Ask the students, "What are some core ideas that we read about earlier that might help us explain why the evidence we are using is important?"

16. Ask the students to share some of the core ideas from the "Introduction" section of the Investigation Handout (or the Investigation Log in the workbook). List these core ideas on the board.

17. Tell the students, "That is great. I would like to see everyone try to include these core ideas in your justification of the evidence. Your goal is to use these core ideas to help explain why your evidence matters and why the rest of us should pay attention to it."

18. Ask the students, "Do you have any questions about what you need to do?"

19. Answer any questions that come up.

20. Tell the students, "Okay, go ahead and start working on your arguments. You need to have your argument done in the next 15 minutes. It doesn't need to be perfect. We just need something down on the whiteboards so we can share our ideas."

21. Give the students 15 minutes to work in their groups on their arguments. As they work, move from group to group to check in, ask probing questions, and offer a suggestion if a group gets stuck. Figure 8.1 (p. 328) shows an example of an argument created by students for this investigation.

What should the graph for this investigation look like?

There are a number of different ways that students can analyze the measurements they collect during this investigation. One of the most straightforward ways is to create a bar graph with the contents of the growing chambers on the horizontal, or x-axis, and the increase in weight of the plants in each chamber after two weeks on the vertical, or y-axis. Figure 8.1 (p. 328) includes an example of this type of graph. There are other options for analyzing the collected data. Students often come up with some unique ways of analyzing their data, so be sure to give them some voice and choice during this stage.

Teacher Notes

FIGURE 8.1

Example of an argument

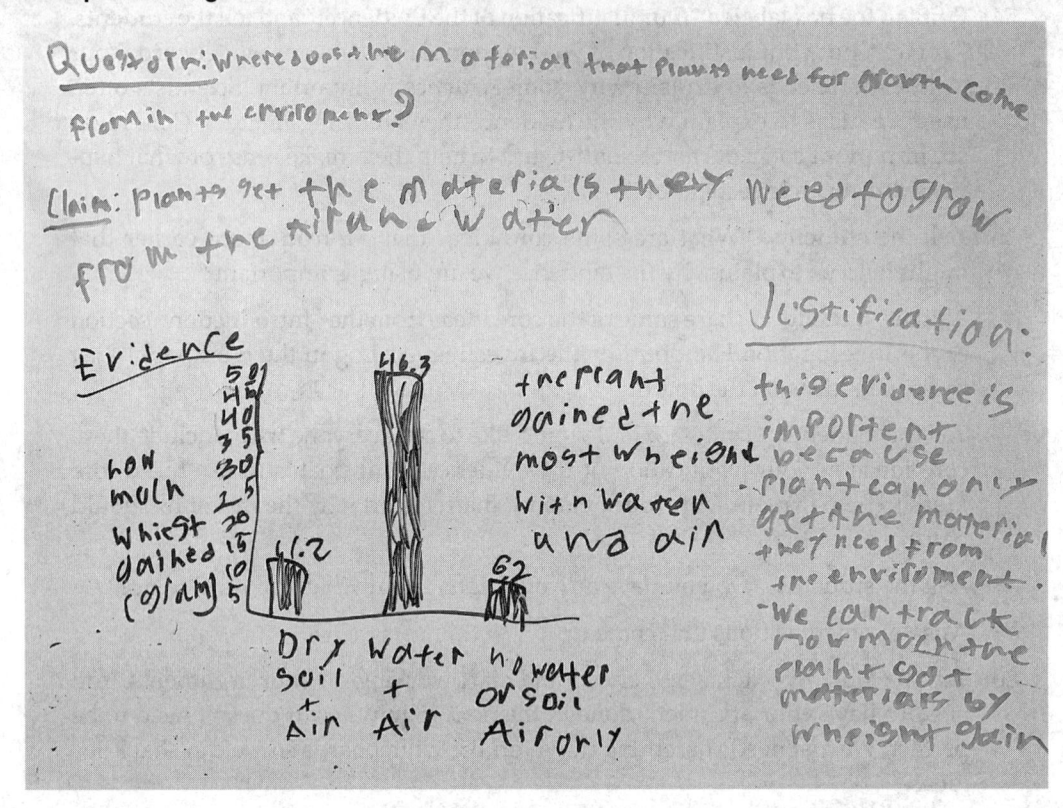

[Handwritten whiteboard content:]

Question: Where does the material that plants need for growth come from in the environment?

Claim: Plants get the materials they need to grow from the air and water

Evidence: [bar graph showing "how much weight gained (grams)" with bars for Dry soil + Air (41.2), Water + Air (46.3), No water or soil Air only (6.2)] The plant gained the most weight with water and air

Justification: This evidence is important because
- Plant can only get the materials they need from the environment.
- We can track how much of the plant got materials by weight gain

Stage 4: Argumentation Session (30 minutes)

The argumentation session can be conducted in a whole-class presentation format, a gallery walk format, or a modified gallery walk format. We recommend using a whole-class presentation format for the first investigation, but try to transition to either the gallery walk or modified gallery walk format as soon as possible because that will maximize student voice and choice inside the classroom. The following list shows the steps for the three formats; unless otherwise noted, the steps are the same for all three formats.

1. Begin by introducing the use of the whiteboard.
   - *If using the whole-class presentation format,* tell the students, "We are now going to share our arguments. Please set up your whiteboards so everyone can see them."
   - *If using the gallery walk or modified gallery walk format,* tell the students, "We are now going to share our arguments. Please set up your whiteboards so they are facing the walls."

2. Allow the students to set up their whiteboards.
   - *If using the whole-class presentation format,* the whiteboards should be set up on stands or chairs so they are facing toward the center of the room.
   - *If using the gallery walk or modified gallery walk format,* the whiteboards should be set up on stands or chairs so they are facing toward the outside of the room.

3. Give the following instructions to the students:
   - *If using the whole-class presentation format,* tell the students, "Okay, before we get started I want to explain what we are going to do next. Your group will have an opportunity to share your argument with the rest of the class. After you are done, everyone else in the class will have a chance to ask questions and offer some suggestions about ways to make your group's argument better. After we have a chance to listen to each other and learn something new, I'm going to give you some time to revise your arguments and make them better."
   - *If using the gallery walk format,* tell the students, "Okay, before we get started I want to explain what we are going to do next. You are going to read the arguments that were created by other groups. When I say 'go,' your group will go to a different group's station so you can see their argument. Once you are there, I'll give your group a few minutes to read and review their argument. Your job is to offer them some suggestions about ways to make their argument better. You can use sticky notes to give them suggestions. Please be specific about what you want to change and how you think they should change it. After we have a chance to learn from each other, I'm going to give you some time to revise your arguments and make them better."
   - *If using the modified gallery walk format,* tell the students, "Okay, before we get started I want to explain what we are going to do next. I'm going to ask some of you to present your arguments to your classmates. If you are presenting your argument, your job is to share your group's claim, evidence, and justification of the evidence. The rest of you will be travelers. If you are a traveler, your job is to listen to the presenters, ask the presenters questions if you do not understand something, and then offer them some suggestions about ways to make their argument better. After we have a chance to learn from each other, I'm going to give you some time to revise your arguments and make them better."

4. Use a document camera to project the "Ways to IMPROVE our argument ..." box from the Investigation Handout (or the Investigation Log in their workbook) on a screen or board (or take a picture of it and project the picture on a screen or board).
   - *If using the whole-class presentation format,* point to the box and tell the students, "After your group presents your argument, you can write down the suggestions you get from your classmates here. If you are listening to a presentation and you see a good idea from another group, you can write down that idea here as

Teacher Notes

well. Once we are done with the presentations, I will give you a chance to use these suggestions or ideas to improve your arguments."

- *If using the gallery walk format,* point to the box and tell the students, "If you see a good idea from another group, you can write it down here. Once we are done reviewing the different arguments, I will give you a chance to use these ideas to improve your own arguments. It is important to share ideas like this."

- *If using the modified gallery walk format,* point to the box and tell the students, "If you are a presenter, you can write down the suggestions you get from the travelers here. If you are a traveler and you see a good idea from another group, you can write down that idea here. Once we are done with the presentations, I will give you a chance to use these suggestions or ideas to improve your arguments."

5. Ask the students, "Do you have any questions about what you need to do?"

6. Answer any questions that come up.

7. Give the following instructions:

   - *If using the whole-class presentation format,* tell the students, "Okay. Let's get started."

   - *If using the gallery walk format,* tell the students, "Okay, I'm now going to tell you which argument to go to and review."

   - *If using the modified gallery walk format,* tell the students, "Okay, I'm now going to assign you to be a presenter or a traveler." Assign one or two students from each group to be presenters and one or two students from each group to be travelers.

8. Give the students an opportunity to review the arguments.

   - *If using the whole-class presentation format,* have each group present their argument one at a time. Give each group only two to three minutes to present their argument. Then give the class two to three minutes to ask them questions and offer suggestions. Encourage as much participation from the students as possible.

   - *If using the gallery walk format,* tell the students, "Okay. Let's get started. Each group, move one argument to the left. Don't move to the next argument until I tell you to move. Once you get there, read the argument and then offer suggestions about how to make it better. I will put some sticky notes next to each argument. You can use the sticky notes to leave your suggestions." Give each group about three to four minutes to read the arguments, talk, and offer suggestions.

     a. After three to four minutes, tell the students, "Okay. Let's move on to the next argument. Please move one group to the left."

     b. Again, give each group three to four minutes to read, talk, and offer suggestions.

c. Repeat this process until each group has had their argument read and critiqued three times.

- *If using the modified gallery walk format,* tell the students, "Okay. Let's get started. Reviewers, move one group to the left. Don't move to the next group until I tell you to move. Presenters, go ahead and share your argument with the travelers when they get there." Give each group of presenters and travelers about three to four minutes to talk.

  a. Tell the students, "Okay. Let's move on to the next argument. Travelers, move one group to the left."

  b. Again, give each group of presenters and travelers about three to four minutes to talk.

  c. Repeat this process until each group has had their argument read and critiqued three times.

9. Tell the students to return to their workstations.

10. Give the following instructions about revising the argument:
    - *If using the whole-class presentation format,* tell the students, "I'm now going to give you all about 10 minutes to revise your argument. Take a few minutes to talk in your groups and determine what you want to change to make your argument better. Once you have decided what to change, go ahead and make the changes to your whiteboard."
    - *If using the gallery walk format,* tell the students, "I'm now going to give you all about 10 minutes to revise your argument. Take a few minutes to read the suggestions that were left at your argument. Then talk in your groups and determine what you want to change to make your argument better. Once you have decided what to change, go ahead and make the changes to your whiteboard."
    - *If using the modified gallery walk format,* tell the students, "I'm now going to give you all about 10 minutes to revise your argument. Please return to your original groups." Wait for the students to move back into their original groups and then tell the students, "Okay, take a few minutes to talk in your groups and determine what you want to change to make your argument better. Once you have decided what to change, go ahead and make the changes to your whiteboard."

11. Ask the students, "Do you have any questions about what you need to do?"

12. Answer any questions that come up.

13. Tell the students, "Okay. Let's get started."

14. Give the students 10 minutes to work in their groups on their arguments. As they work, move from group to group to check in, ask probing questions, and offer a suggestion if a group gets stuck.

Teacher Notes

Stage 5: Reflective Discussion (15 minutes)

1. Tell the students, "We are now going to take a minute to talk about some of the core ideas and crosscutting concepts that we have used during our investigation."
2. Show Figure 8.2 on the screen.
3. Ask the students, "What do you all see going on here?"
4. Allow the students to share their ideas.

FIGURE 8.2
Growing plants

5. Show Figure 8.3 on the screen.
6. Ask the students, "What do you all see going on here?"
7. Allow the students to share their ideas.
8. Ask the students, "How can we explain this in terms of the movement of matter into, out of, or within systems?"
9. Allow the students to share their ideas. As they share their ideas, ask different questions to encourage them to expand on their thinking (e.g., "Can you tell me more about that?"), clarify a contribution (e.g., "Can you say that in another way?"), support an idea (e.g., "Why do you think that?"), add to an idea mentioned by a classmate (e.g., "Would anyone like to add to the idea?"), re-voice an idea offered by a classmate (e.g., "Who can explain that to me in another way?"), or critique an idea during the discussion (e.g., "Do you agree or disagree with that idea and why?") until students are able to generate an adequate explanation.

Investigation 8. Plant Growth:
Where Do the Materials That Plants Need for Growth Come From in the Environment?

FIGURE 8.3

A plant on a scale at two different times

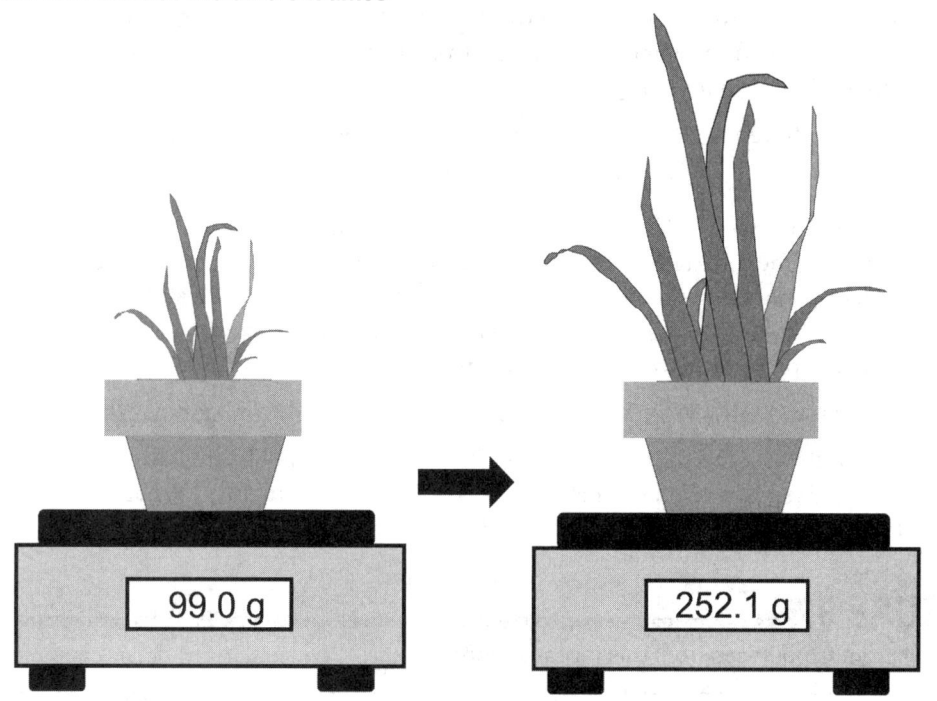

[Illustration: A potted plant on a scale reading 99.0 g, with an arrow pointing to a larger potted plant on a scale reading 252.1 g]

10. Tell the students, "We used standard units to measure our plants during our investigation." Then ask, "Can anyone tell me why this was important?"

11. Allow the students to share their ideas.

12. Tell the students, "I think using standard units to measure and describe physical quantities such as height or weight is important because we need to have a common understanding of what our measurements mean. This allows us to share and critique each other's ideas and evidence."

13. Ask the students, "What units did you use today?"

14. Allow the students to share their ideas.

15. Show an image of the question "What do you think are the most important core ideas or crosscutting concepts that we used during this investigation to help us make sense of what we observed?" Tell the students, "Okay, let's make sure we are all on the same page. Please take a moment to discuss this question with the other people in your group." Give them a few minutes to discuss the question.

16. Ask the students, "What do you all think? Who would like to share?"

17. Allow the students to share their ideas.

18. Tell the students, "We are now going to talk about how scientists use experiments."

Teacher Notes

19. Show an image of the question "How are experiments different from other types of investigations?" on the screen. Tell the students, "Take a few minutes to talk about how you would answer this question with the other people in your group. Be ready to share with the rest of the class." Give the students two to three minutes to talk in their group.

20. Ask the students, "What do you all think? Who would like to share an idea?"

21. Allow the students to share their ideas.

22. Tell the students, "Okay, let's make sure we are all on the same page. Scientists use experiments to test their ideas about how or why things happen. Experiments include making one or more hypotheses, designing a way to test these hypotheses, and then making predictions based on the tests. A *hypothesis* is a possible explanation for how or why things happen."

23. Ask the students, "What was an example of a hypothesis from our investigation?"

24. Allow students to share their ideas. Show the image in Figure 8.4. Tell the students, "Here are two possible hypotheses." Ask the students, "What did we do to test these ideas?

FIGURE 8.4
Two different hypotheses from this investigation

Hypothesis 1: Plants get the matter they need to grow chiefly from the air and water.	**Hypothesis 2:** Plants get the matter they need to grow chiefly from the soil and water.

25. Allow the students to share their ideas. Show the image in Figure 8.5. Tell the students, "Here is the test." Ask the students, "What do we predict will happen to the plants based on each hypothesis?

FIGURE 8.5
The test from the investigation

Hypothesis 1: Plants get the matter they need to grow chiefly from the air and water.	**Hypothesis 2:** Plants get the matter they need to grow chiefly from the soil and water.

The Test: Plant seeds in containers with (1) soil and water, (2) water but no soil, (3) soil but no water, and (4) no soil and no water. Wait two weeks and see which containers have healthy plants in them.

26. Allow the students to share their ideas. Show the image in Figure 8.6. Tell the students, "Here are two possible predictions."

Investigation 8. Plant Growth:
Where Do the Materials That Plants Need for Growth Come From in the Environment?

FIGURE 8.6

Two predictions based on these hypotheses and the test

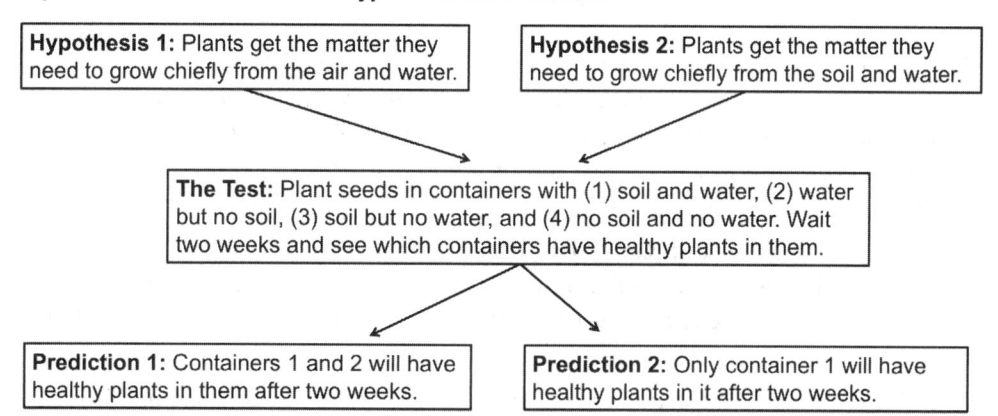

- **Hypothesis 1:** Plants get the matter they need to grow chiefly from the air and water.
- **Hypothesis 2:** Plants get the matter they need to grow chiefly from the soil and water.

The Test: Plant seeds in containers with (1) soil and water, (2) water but no soil, (3) soil but no water, and (4) no soil and no water. Wait two weeks and see which containers have healthy plants in them.

- **Prediction 1:** Containers 1 and 2 will have healthy plants in them after two weeks.
- **Prediction 2:** Only container 1 will have healthy plants in it after two weeks.

27. Tell the students, "Once we have a prediction, we can carry out the test. If the results match a prediction, then the hypothesis is supported. If the results do not match the prediction, then the hypothesis is not supported."

28. Show the image in Figure 8.7. Tell the students, "In this investigation the results matched the predictions for hypothesis 1, so that hypothesis is supported. This is how experiments work in science."

FIGURE 8.7

The results from this investigation

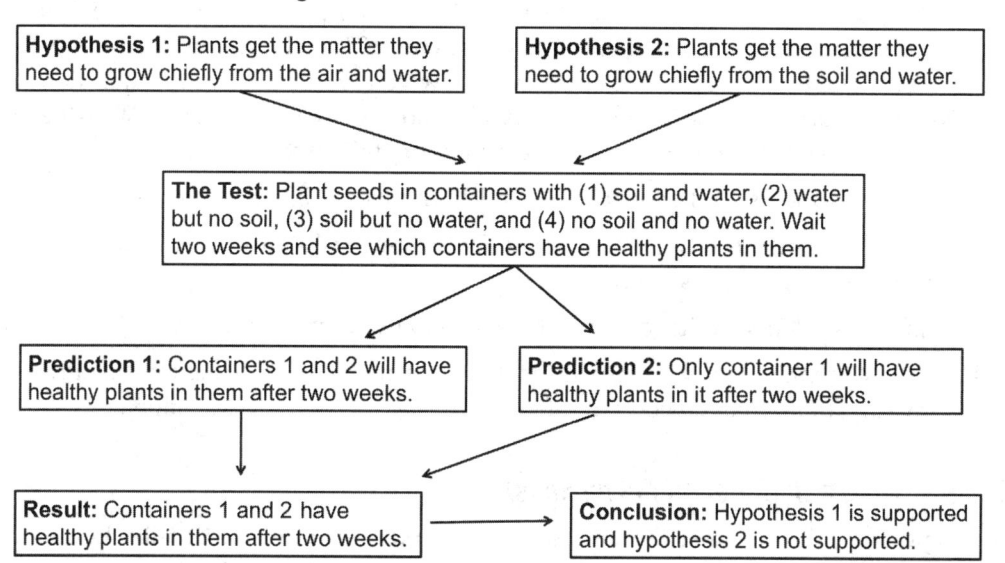

- **Hypothesis 1:** Plants get the matter they need to grow chiefly from the air and water.
- **Hypothesis 2:** Plants get the matter they need to grow chiefly from the soil and water.

The Test: Plant seeds in containers with (1) soil and water, (2) water but no soil, (3) soil but no water, and (4) no soil and no water. Wait two weeks and see which containers have healthy plants in them.

- **Prediction 1:** Containers 1 and 2 will have healthy plants in them after two weeks.
- **Prediction 2:** Only container 1 will have healthy plants in it after two weeks.

- **Result:** Containers 1 and 2 have healthy plants in them after two weeks.
- **Conclusion:** Hypothesis 1 is supported and hypothesis 2 is not supported.

Teacher Notes

29. Ask the students, "Does anyone have any questions about how experiments work?"

30. Answer any questions that come up.

31. Tell the students, "We are now going to take a minute to talk about what went well and what didn't go so well during our investigation. We need to talk about this because you all are going to be planning and carrying out your own investigations like this a lot this year, and I want to help you all get better at it."

32. Show an image of the question "What made your investigation scientific?" on the screen. Tell the students, "Take a few minutes to talk about how you would answer this question with the other people in your group. Be ready to share with the rest of the class." Give the students two to three minutes to talk in their group.

33. Ask the students, "What do you all think? Who would like to share an idea?"

34. Allow the students to share their ideas. Be sure to expand on their ideas about what makes an investigation scientific.

35. Show an image of the question "What made your investigation not so scientific?" on the screen. Tell the students, "Take a few minutes to talk about how you would answer this question with the other people in your group. Be ready to share with the rest of the class." Give the students two to three minutes to talk in their group.

36. Ask the students, "What do you all think? Who would like to share an idea?"

37. Allow the students to share their ideas. Be sure to expand on their ideas about what makes an investigation less scientific.

38. Show an image of the question "What rules can we put into place to help us make sure our next investigation is more scientific?" on the screen. Tell the students, "Take a few minutes to talk about how you would answer this question with the other people in your group. Be ready to share with the rest of the class." Give the students two to three minutes to talk in their group.

39. Ask the students, "What do you all think? Who would like to share an idea?"

40. Allow the students to share their ideas. Once they have shared their ideas, offer a suggestion for a possible class rule.

41. Ask the students, "What do you all think? Should we make this a rule?"

42. If the students agree, write the rule on the board or on a class "Rules for Scientific Investigation" chart so you can refer to it during the next investigation.

Stage 6: Write a Draft Report (30 minutes)

Your students will use either the Investigation Handout or the Investigation Log in the student workbook when writing the draft report. When you give the directions shown

Investigation 8. Plant Growth:
Where Do the Materials That Plants Need for Growth Come From in the Environment?

in quotes in the following steps, substitute "Investigation Log in your workbook" or just "Investigation Log" (as shown in brackets) for "handout" if they are using the workbook.

1. Tell the students, "You are now going to write an investigation report to share what you have learned. Please take out a pencil and turn to the 'Draft Report' section of your handout [Investigation Log in your workbook]."

2. If possible, use a document camera to project the "Introduction" section of the draft report from the Investigation Handout (or the Investigation Log in their workbook) on a screen or board (or take a picture of it and project the picture on a screen or board).

3. Tell the students, "The first part of the report is called the 'Introduction.' In this section of the report you want to explain to the reader what you were investigating, why you were investigating it, and what question you were trying to answer. All this information can be found in the text at the beginning of your handout [Investigation Log]." Point to the image. "Here are some sentence starters to help you begin writing."

4. Ask the students, "Do you have any questions about what you need to do?"

5. Answer any questions that come up.

6. Tell the students, "Okay, let's write."

7. Give the students 10 minutes to write the "Introduction" section of the report. As they work, move from student to student to check in, ask probing questions, and offer a suggestion if a student gets stuck.

8. If possible, use a document camera to project the "Method" section of the draft report from the Investigation Handout (or the Investigation Log in their workbook) on a screen or board (or take a picture of it and project the picture on a screen or board).

9. Tell the students, "The second part of the report is called the 'Method.' In this section of the report you want to explain to the reader what you did during the investigation, what data you collected and why, and how you went about analyzing your data. All this information can be found in the 'Plan Your Investigation' section of the handout [Investigation Log]. Remember that you all planned and carried out different investigations, so do not assume that the reader will know what you did." Point to the image. "Here are some sentence starters to help you begin writing."

10. Ask the students, "Do you have any questions about what you need to do?"

11. Answer any questions that come up.

12. Tell the students, "Okay, let's write."

Teacher Notes

13. Give the students 10 minutes to write the "Method" section of the report. As they work, move from student to student to check in, ask probing questions, and offer a suggestion if a student gets stuck.

14. If possible, use a document camera to project the "Argument" section of the draft report from the Investigation Handout (or the Investigation Log in their workbook) on a screen or board (or take a picture of it and project the picture on a screen or board).

15. Tell the students, "The last part of the report is called the 'Argument.' In this section of the report you want to share your claim, evidence, and justification of the evidence with the reader. All this information can be found on your whiteboard." Point to the image. "Here are some sentence starters to help you begin writing."

16. Ask the students, "Do you have any questions about what you need to do?"

17. Answer any questions that come up.

18. Tell the students, "Okay, let's write."

19. Give the students 10 minutes to write the "Argument" section of the report. As they work, move from student to student to check in, ask probing questions, and offer a suggestion if a student gets stuck.

Stage 7: Peer Review (35 minutes)

Your students will use either the Investigation Handout or their workbook when doing the peer review. Except where noted below, the directions are the same whether using the handout or the workbook.

1. Tell the students, "We are now going to review our reports to find ways to make them better. I'm going to come around and collect your draft reports. While I do that, please take out a pencil."

2. Collect the handouts or the workbooks with the draft reports from the students.

3. If possible, use a document camera to project the peer-review guide (see Appendix 4) on a screen or board (or take a picture of it and project the picture on a screen or board).

4. Tell the students, "We are going to use this peer-review guide to give each other feedback." Point to the image.

5. Tell the students, "I'm going to ask you to work with a partner to do this. I'm going to give you and your partner a draft report to read. You two will then read the report together. Once you are done reading the report, I want you to answer each of the questions on the peer-review guide." Point to the review questions on the image of the peer-review guide.

Investigation 8. Plant Growth:
Where Do the Materials That Plants Need for Growth Come From in the Environment?

6. Tell the students, "You can check 'no,' 'almost,' or 'yes' after each question." Point to the checkboxes on the image of the peer-review guide.

7. Tell the students, "This will be your rating for this part of the report. Make sure you agree on the rating you give the author. If you mark 'no' or 'almost,' then you need to tell the author what he or she needs to do to get a 'yes.'" Point to the space for the reviewer feedback on the image of the peer-review guide.

8. Tell the students, "It is really important for you to give the authors feedback that is helpful. That means you need to tell them exactly what they need to do to make their report better."

9. Ask the students, "Do you have any questions about what you need to do?"

10. Answer any questions that come up.

11. Tell the students, "Please sit with a partner who is not in your current group." Allow the students time to sit with a partner.

12. Tell the students, "Okay, I'm now going to give you one report to read." Pass out one Investigation Handout with a draft report or one workbook to each pair. Make sure that the report you give a pair was not written by one of the students in that pair. Give each pair one peer-review guide to fill out. If the students are using workbooks, the peer-review guide is included right after the draft report so you do not need to pass out copies of the peer-review guide.

13. Tell the students, "Okay, I'm going to give you 15 minutes to read the report I gave you and to fill out the peer-review guide. Go ahead and get started."

14. Give the students 15 minutes to work. As they work, move around from pair to pair to check in and see how things are going, answer questions, and offer advice.

15. After 15 minutes pass, tell the students, "Okay, time is up. Please give me the report and the peer-review guide that you filled out."

16. Collect the Investigation Handouts and the peer-review guides, or collect the workbooks if students are using them. If the students are using the Investigation Handouts and separate peer-review guides, be sure you keep each handout with its corresponding peer-review guide.

17. Tell the students, "Okay, I am now going to give you a different report to read and a new peer-review guide to fill out." Pass out one more report to each pair. Make sure that the report you give a pair was not written by one of the students in that pair. Give each pair a new peer-review guide to fill out as a group.

18. Tell the students, "Okay, I'm going to give you 15 minutes to read this new report and to fill out the peer-review guide. Go ahead and get started."

19. Give the students 15 minutes to work. As they work, move around from pair to pair to check in and see how things are going, answer questions, and offer advice.

Teacher Notes

20. After 15 minutes pass, tell the students, "Okay, time is up. Please give me the report and the peer-review guide that you filled out."

21. Collect the Investigation Handouts and the peer-review guides, or collect the workbooks if students are using them. If the students are using the Investigation Handouts and separate peer-review guides, be sure you keep each handout with its corresponding peer-review guide.

Stage 8: Revise the Report (30 minutes)

Your students will use either the Investigation Handout or their workbook when revising the report. Except where noted below, the directions are the same whether using the handout or the workbook.

1. Tell the students, "You are now going to revise your draft report based on the feedback you get from your classmates. Please take out a pencil."

2. Return the reports to the students.
   - *If the students used the Investigation Handout and a copy of the peer-review guide,* pass back the handout and the peer-review guide to each student.
   - *If the students used the workbook,* pass that back to each student.

3. Tell the students, "Please take a few minutes to read over the peer-review guide. You should use it to figure out what you need to change in your report and how you will change it."

4. Allow the students to read the peer-review guide.

5. *If the students used the workbook,* if possible use a document camera to project the "Write Your Final Report" section from the Investigation Log on a screen or board (or take a picture of it and project the picture on a screen or board).

6. Give the following directions about how to revise their reports:
   - *If the students used the Investigation Handout and a copy of the peer-review guide,* tell them, "Okay, let's revise our reports. Please take out a piece of paper. I would like you to rewrite your report. You can use your draft report as a starting point, but you also need to change it to make it better. Use the feedback on the peer-review guide to make it better."
   - *If the students used the workbook,* tell them, "Okay, let's revise our reports. I would like you to rewrite your report in the section of the Investigation Log called "Write Your Final Report." You can use your draft report as a starting point, but you also need to change it to make it better. Use the feedback on the peer-review guide to make it better."

7. Ask the students, "Do you have any questions about what you need to do?"

8. Answer any questions that come up.

9. Tell the students, "Okay, let's write." Allow about 20 minutes for the students to revise their reports.

10. After about 20 minutes, give the following directions:
    - *If the students used the Investigation Handout,* tell them, "Okay, time's up. I will now come around and collect your Investigation Handout, the peer-review guide, and your final report."
    - *If the students used the workbook,* tell them, "Okay, time's up. I will now come around and collect your workbooks."

11. *If the students used the Investigation Handout,* collect all the Investigation Handouts, peer-review guides, and final reports. *If the students used the workbook,* collect all the workbooks.

12. *If the students used the Investigation Handout,* use the "Teacher Score" column in the peer-review guide to grade the final report. *If the students used the workbook,* use the "Investigation Report Grading Rubric" in the Investigation Log to grade the final report. Whether you are using the handout or the log, you can give the students feedback about their writing in the "Teacher Comments" section.

How to Use the Checkout Questions

The Checkout Questions are an optional assessment. We recommend giving them to students at the start of the next class period after the students finish stage 8 of the investigation. You can then look over the student answers to determine if you need to reteach the core idea from the investigation. Appendix 6 gives the answers to the Checkout Questions that should be given by a student who can apply the core idea correctly and can explain how matter moves into, within, and out of systems.

Alignment With Standards

Table 8.2 highlights how the investigation can be used to address specific performance expectations from the *Next Generation Science Standards, Common Core State Standards for English Language Arts (CCSS ELA)* and *Common Core State Standards for Mathematics (CCSS Mathematics),* and *English Language Proficiency (ELP) Standards.*

TABLE 8.2

Investigation 8 alignment with standards

NGSS performance expectation	5-LS1-1: Support an argument that plants get the materials they need for growth chiefly from air and water.

Continued

Teacher Notes

Table 8.2 (*continued*)

***CCSS ELA*—Reading: Informational Text**	Key ideas and details • CCSS.ELA-LITERACY.RI.5.1: Quote accurately from a text when explaining what the text says explicitly and when drawing inferences from the text. • CCSS.ELA-LITERACY.RI.5.2: Determine two or more main ideas of a text and explain how they are supported by key details; summarize the text. • CCSS.ELA-LITERACY.RI.5.3: Explain the relationships or interactions between two or more individuals, events, ideas, or concepts in a historical, scientific, or technical text based on specific information in the text. Craft and structure • CCSS.ELA-LITERACY.RI.5.4: Determine the meaning of general academic and domain-specific words and phrases in a text relevant to a *grade 5 topic or subject area*. • CCSS.ELA-LITERACY.RI.5.5: Compare and contrast the overall structure (e.g., chronology, comparison, cause/effect, problem/solution) of events, ideas, concepts, or information in two or more texts. • CCSS.ELA-LITERACY.RI.5.6: Analyze multiple accounts of the same event or topic, noting important similarities and differences in the point of view they represent. Integration of knowledge and ideas • CCSS.ELA-LITERACY.RI.5.7: Draw on information from multiple print or digital sources, demonstrating the ability to locate an answer to a question quickly or to solve a problem efficiently. • CCSS.ELA-LITERACY.RI.5.8: Explain how an author uses reasons and evidence to support particular points in a text, identifying which reasons and evidence support which point(s). Range of reading and level of text complexity • CCSS.ELA-LITERACY.RI.5.10: By the end of the year, read and comprehend informational texts, including history/social studies, science, and technical texts, at the high end of the grades 4–5 text complexity band independently and proficiently.
***CCSS ELA*—Writing**	• CCSS.ELA-LITERACY.W.5.1: Write opinion pieces on topics or texts, supporting a point of view with reasons. • CCSS.ELA-LITERACY.W.5.1.A: Introduce a topic or text clearly, state an opinion, and create an organizational structure in which ideas are logically grouped to support the writer's purpose. • CCSS.ELA-LITERACY.W.5.1.B: Provide logically ordered reasons that are supported by facts and details.

Continued

Table 8.2 (*continued*)

CCSS ELA—**Writing** (*continued*)	Text types and purposes
	• CCSS.ELA-LITERACY.W.5.1.C: Link opinion and reasons using words, phrases, and clauses (e.g., *consequently*, *specifically*).
	• CCSS.ELA-LITERACY.W.5.1.D: Provide a concluding statement or section related to the opinion presented.
	• CCSS.ELA-LITERACY.W.5.2: Write informative or explanatory texts to examine a topic and convey ideas and information clearly.
	○ CCSS.ELA-LITERACY.W.5.2.A: Introduce a topic clearly, provide a general observation and focus, and group related information logically; include formatting (e.g., headings), illustrations, and multimedia when useful to aiding comprehension.
	○ CCSS.ELA-LITERACY.W.5.2.B: Develop the topic with facts, definitions, concrete details, quotations, or other information and examples related to the topic.
	○ CCSS.ELA-LITERACY.W.5.2.C: Link ideas within and across categories of information using words, phrases, and clauses (e.g., *in contrast*, *especially*).
	○ CCSS.ELA-LITERACY.W.5.2.D: Use precise language and domain-specific vocabulary to inform about or explain the topic.
	○ CCSS.ELA-LITERACY.W.5.2.E: Provide a concluding statement or section related to the information or explanation presented.
	Production and distribution of writing
	• CCSS.ELA-LITERACY.W.5.4: Produce clear and coherent writing in which the development and organization are appropriate to task, purpose, and audience.
	• CCSS.ELA-LITERACY.W.5.5: With guidance and support from peers and adults, develop and strengthen writing as needed by planning, revising, editing, rewriting, or trying a new approach.
	• CCSS.ELA-LITERACY.W.5.6: With some guidance and support from adults, use technology, including the internet, to produce and publish writing as well as to interact and collaborate with others; demonstrate sufficient command of keyboarding skills to type a minimum of two pages in a single sitting.
	Research to build and present knowledge
	• CCSS.ELA-LITERACY.W.5.8: Recall relevant information from experiences or gather relevant information from print and digital sources; summarize or paraphrase information in notes and finished work, and provide a list of sources.
	• CCSS.ELA-LITERACY.W.5.9: Draw evidence from literary or informational texts to support analysis, reflection, and research.
	Range of writing
	• CCSS.ELA-LITERACY.W.5.10: Write routinely over extended time frames (time for research, reflection, and revision) and shorter time frames (a single sitting or a day or two) for a range of discipline-specific tasks, purposes, and audiences.

Continued

Teacher Notes

Table 8.2 (*continued*)

CCSS ELA—Speaking and Listening	Comprehension and collaboration
	• CCSS.ELA-LITERACY.SL.5.1: Engage effectively in a range of collaborative discussions (one-on-one, in groups, and teacher-led) with diverse partners on *grade 5 topics and texts,* building on others' ideas and expressing their own clearly.
	    ○ CCSS.ELA-LITERACY.SL.5.1.A: Come to discussions prepared, having read or studied required material; explicitly draw on that preparation and other information known about the topic to explore ideas under discussion.
	    ○ CCSS.ELA-LITERACY.SL.5.1.B: Follow agreed-upon rules for discussions and carry out assigned roles.
	    ○ CCSS.ELA-LITERACY.SL.5.1.C: Pose and respond to specific questions by making comments that contribute to the discussion and elaborate on the remarks of others.
	    ○ CCSS.ELA-LITERACY.SL.5.1.D: Review the key ideas expressed and draw conclusions in light of information and knowledge gained from the discussions.
	• CCSS.ELA-LITERACY.SL.5.2: Summarize a written text read aloud or information presented in diverse media and formats, including visually, quantitatively, and orally.
	• CCSS.ELA-LITERACY.SL.5.3: Summarize the points a speaker makes and explain how each claim is supported by reasons and evidence.
	Presentation of knowledge and ideas
	• CCSS.ELA-LITERACY.SL.5.4: Report on a topic or text or present an opinion, sequencing ideas logically and using appropriate facts and relevant, descriptive details to support main ideas or themes; speak clearly at an understandable pace.
	• CCSS.ELA-LITERACY.SL.5.5: Include multimedia components (e.g., graphics, sound) and visual displays in presentations when appropriate to enhance the development of main ideas or themes.
	• CCSS.ELA-LITERACY.SL.5.6: Adapt speech to a variety of contexts and tasks, using formal English when appropriate to task and situation.

CCSS Mathematics—Numbers and Operations in Base Ten	Understand the place value system. • CCSS.MATH.CONTENT.5.NBT.A.3: Read, write, and compare decimals to thousandths. • CCSS.MATH.CONTENT.5.NBT.A.4: Use place value understanding to round decimals to any place. Perform operations with multi-digit whole numbers and with decimals to hundredths. • CCSS.MATH.CONTENT.5.NBT.B.7: Add, subtract, multiply, and divide decimals to hundredths.

Continued

Investigation 8. Plant Growth: Where Do the Materials That Plants Need for Growth Come From in the Environment?

Table 8.2 (*continued*)

CCSS Mathematics—Measurement and Data	Convert like measurement units within a given measurement system. • CCSS.MATH.CONTENT.5.MD.A.1: Convert among different-sized standard measurement units within a given measurement system (e.g., convert 5 cm to 0.05 m), and use these conversions in solving multi-step, real-world problems.
ELP Standards	Receptive modalities • ELP 1: Construct meaning from oral presentations and literary and informational text through grade-appropriate listening, reading, and viewing. • ELP 8: Determine the meaning of words and phrases in oral presentations and literary and informational text. Productive modalities • ELP 3: Speak and write about grade-appropriate complex literary and informational texts and topics. • ELP 4: Construct grade-appropriate oral and written claims and support them with reasoning and evidence. • ELP 7: Adapt language choices to purpose, task, and audience when speaking and writing. Interactive modalities • ELP 2: Participate in grade-appropriate oral and written exchanges of information, ideas, and analyses, responding to peer, audience, or reader comments and questions. • ELP 5: Conduct research and evaluate and communicate findings to answer questions or solve problems. • ELP 6: Analyze and critique the arguments of others orally and in writing. Linguistic structures of English • ELP 9: Create clear and coherent grade-appropriate speech and text. • ELP 10: Make accurate use of standard English to communicate in grade-appropriate speech and writing.

Investigation Handout

Investigation 8

Plant Growth: Where Do the Materials That Plants Need for Growth Come From in the Environment?

Introduction

Plants increase in size over time as they grow. Your teacher will show you a time-lapse video of a plant that starts as a seed and then grows into a seedling over several days. As you watch the video, keep track of things you notice and things you wonder about in the boxes below.

Things I NOTICED …	Things I WONDER about …

The plant in the video, like all plants, created new structures as it grew. These new structures included roots, stems, and leaves. The plant created these structures using materials that came from the environment. As the plant grew, it took in materials from the environment, broke these materials down into smaller particles, and then rearranged those particles to create roots, stems, and leaves.

Investigation 8. Plant Growth:
Where Do the Materials That Plants Need for Growth Come From in the Environment?

We can actually track the movement of materials into, or out of, a plant by tracking changes in the weight of a plant over time. The weight of a plant will increase when it takes in materials from the environment and then uses these materials to add new structures such as leaves, flowers, fruits, or seeds. The weight of a plant will decrease when some of the structures that make up a plant fall off the plant or are removed from the plant. For example, when a part of a plant is eaten by an animal, when it sheds leaves in the fall, or when we pick fruits off a plant in the summer, the weight of a plant will decrease because some of the materials move from the plant back into the environment.

Animals, like plants, take in matter from the environment in order to grow. Animals get the matter from the environment they need to grow by eating plants or other animals. Plants, however, do not eat, so plants must take in the matter they need to increase in size from the soil, water, and/or air that is found in the environment around them. Your goal in this investigation is to figure out where plants get the matter they need to grow. Here are three different possibilities:

1. Plants get the materials they need to grow chiefly from the soil and air.
2. Plants get the materials they need to grow chiefly from the air and water.
3. Plants get the materials they need to grow chiefly from the soil and water.

To determine which of these three explanations is the best one, you will need to grow seeds in different environmental conditions. For example, one condition might have air and soil but no water. Another condition might have air and water but no soil. A third condition might have air but no water or soil. You can track the movement of materials into the plant by seeing how much the plant grows or changes in weight over time. If the plant grows or weighs more over time, then you will know that materials moved from the environment into the plant, because all matter has mass. If the plants do not grow, then you will know that the plant did not get the materials it needed from the environment.

Things we KNOW from what we read ...

Investigation Handout

Your Task

Use what you know about how plants; how matter moves into, out of, and within systems; and the importance of using standard units to design and carry out an investigation to figure out where plants get the matter they need to grow.

The *guiding question* of this investigation is, **Where do the materials that plants need for growth come from in the environment?**

Materials

You may use any of the following materials during your investigation:

- Safety goggles (required)
- 36 seeds (presoaked in water)
- 6 clear plastic containers for growing the plants
- Dry potting soil
- 60 small plastic pellets
- 20 water beads
- Plastic graduated cylinder (10 ml)
- Ruler
- Tray or plastic container to hold growth chambers

Safety Rules

Follow all normal safety rules. In addition, be sure to follow these rules:

- Wear sanitized safety goggles during setup, investigation activity, and cleanup.
- Do not put anything used in this activity in your mouth.
- Immediately clean up any spills to avoid a slip or fall hazard.
- Wash your hands with soap and water after you are done cleaning up.

Plan Your Investigation

Prepare a plan for your investigation by filling out the chart on the next page; this plan is called an *investigation proposal*. Before you start developing your plan, be sure to discuss the following questions with the other members of your group:

- How can we **track the movement of matter** in a plant?
- What **units** or **scale** should we use to track the movement of matter?

Investigation 8. Plant Growth:
Where Do the Materials That Plants Need for Growth Come From in the Environment?

Our guiding question:

This is a picture of how we will set up the equipment:

We will collect the following data:

These are the steps we will follow to collect data:

I approve of this investigation proposal.

_____   _____
Teacher's signature   Date

Argument-Driven Inquiry in **Fifth-Grade Science**: Three-Dimensional Investigations

Investigation Handout

Collect Your Data
Keep a record of what you measure or observe during your investigation in the space below.

Analyze Your Data
You will need to analyze the data you collected before you can develop an answer to the guiding question. To analyze the data you collected, create a graph that shows what you measured before and after the reaction.

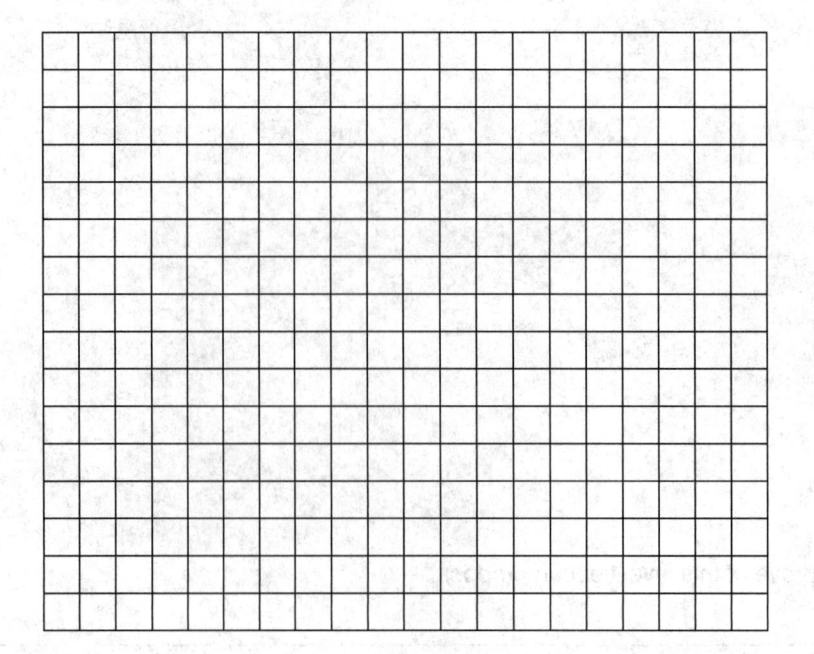

Investigation 8. Plant Growth:
Where Do the Materials That Plants Need for Growth Come From in the Environment?

Draft Argument

Develop an argument on a whiteboard. It should include the following:

1. A *claim*: Your answer to the guiding question.
2. *Evidence*: An analysis of the data and an explanation of what the analysis means.
3. A *justification of the evidence*: Why your group thinks the evidence is important.

The Guiding Question:	
Our Claim:	
Our Evidence:	Our Justification of the Evidence:

Argumentation Session

Share your argument with your classmates. Be sure to ask them how to make your draft argument better. Keep track of their suggestions in the space below.

Ways to IMPROVE our argument …

Argument-Driven Inquiry in **Fifth-Grade Science:** Three-Dimensional Investigations

Investigation Handout

Draft Report

Prepare an investigation report to share what you have learned. Use the information in this handout and your group's final argument to write a *draft* of your investigation report.

Introduction

We have been studying _____ in class.

Before we started this investigation, we explored _____
_____
_____

We noticed _____
_____
_____

My goal for this investigation was to figure out _____
_____
_____

The guiding question was _____
_____
_____

Method

To gather the data I needed to answer this question, I _____
_____
_____
_____
_____

Investigation 8. Plant Growth:
Where Do the Materials That Plants Need for Growth Come From in the Environment?

I then analyzed the data I collected by _____

_____

_____

_____

Argument

My claim is _____

_____

_____

_____

The graph below shows _____

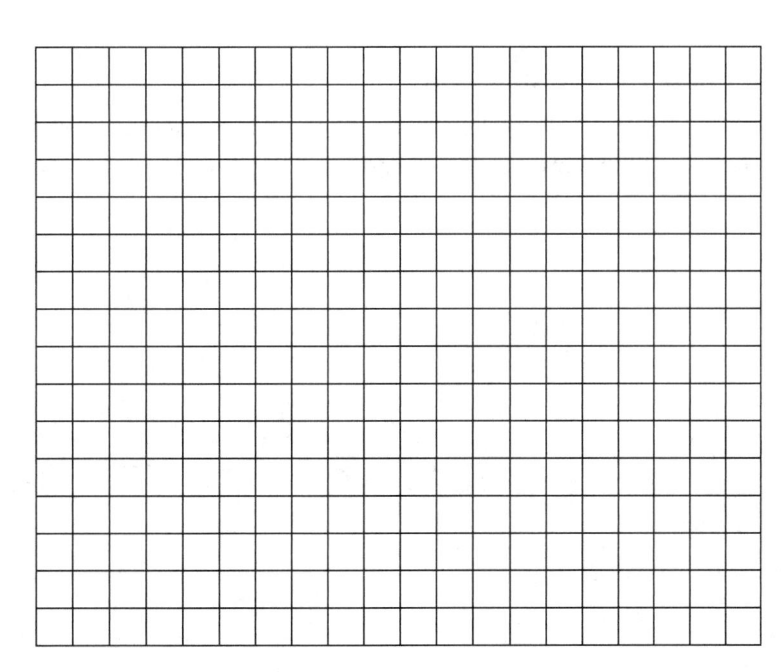

_____

_____

_____

_____

_____

Investigation Handout

This analysis of the data I collected suggests _____

_____
_____
_____
_____

This evidence is based on several important scientific concepts. The first one is _____

_____
_____
_____
_____

Review

Your classmates need your help! Review the draft of their investigation reports and give them ideas about how to improve. Use the peer-review guide when doing your review.

Submit Your Final Report

Once you have received feedback from your classmates about your draft report, create your final investigation report and hand it in to your teacher.

Checkout Questions

Investigation 8. Plant Growth

Use the following information to answer questions 1 and 2. The picture below shows an aquatic plant called duckweed. This plant floats on the surface of freshwater.

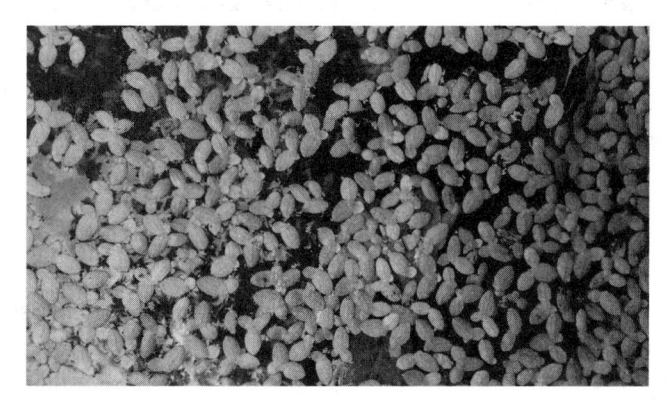

1. How is duckweed similar to the plants we studied in our plant growth investigation?

    a. Duckweed gets the matter it needs to grow chiefly from the soil and air.

    b. Duckweed gets the matter it needs to grow chiefly from the air and water.

    c. Duckweed gets the matter it needs to grow chiefly from the soil and water.

2. How do you know? Use what you know about the transfer of matter into, within, and out of systems to explain your answer.

_____

_____

_____

_____

_____

Teacher Scoring Rubric for the Checkout Questions

Level	Description
3	The student can apply the core idea correctly and can explain how matter moves into, within, and out of systems.
2	The student cannot apply the core idea correctly but can explain how matter moves into, within, and out of systems.
1	The student can apply the core idea correctly but cannot explain how matter moves into, within, and out of systems.
0	The student cannot apply the core idea correctly and cannot explain how matter moves into, within, and out of systems.

Argument-Driven Inquiry in **Fifth-Grade Science**: Three-Dimensional Investigations

Teacher Notes

Investigation 9

Energy in Ecosystems: How Do We Best Model the Transfer of Energy Into and Within the Living Things That Are Found in the Arctic Ocean?

Purpose

The purpose of this investigation is to give students an opportunity to use two disciplinary core ideas (DCIs), two crosscutting concepts (CCs), and six scientific and engineering practices (SEPs) to figure out how we can model the transfer of energy into and within living things. Students will also learn about models as tools for reasoning about phenomena.

The DCIs, CCs, and SEPs That Students Use During This Investigation to Figure Things Out

DCIs

- *PS3.D: Energy in Chemical Processes and Everyday Life:* The energy released [from] food was once energy from the Sun that was captured by plants in the chemical process that forms plant matter (from air and water).
- *LS1.C: Organization for Matter and Energy Flow in Organisms:* Food provides animals with the materials they need for body repair and growth and the energy they need to maintain body warmth and for motion.

CCs

- *CC 4: Systems and System Models:* A system can be described in terms of its components and their interactions. A system is a group of related parts that make up a whole and can carry out functions its individual parts cannot.
- *CC 5: Energy and Matter:* Energy can be transferred in various ways and between objects and organisms.

SEPs

- *SEP 1: Asking Questions and Defining Problems:* Ask questions about what would happen if a variable is changed. Ask questions that can be investigated and predict reasonable outcomes based on patterns such as cause-and-effect relationships.
- *SEP 2: Developing and Using Models:* Develop and/or use models to describe and/or predict phenomena.

Investigation 9. Energy in Ecosystems: How Do We Best Model the Transfer of Energy Into and Within the Living Things That Are Found in the Arctic Ocean?

- *SEP 3: Planning and Carrying Out Investigations:* Plan and conduct an investigation collaboratively to produce data to serve as the basis for evidence, using fair tests in which variables are controlled and the number of trials considered. Evaluate appropriate methods and/or tools for collecting data.
- *SEP 6: Constructing Explanations and Designing Solutions:* Construct an explanation of observed relationships. Use evidence to construct or support an explanation. Identify the evidence that supports particular points in an explanation.
- *SEP 7: Engaging in Argument From Evidence:* Compare and refine arguments based on an evaluation of the evidence presented. Distinguish among facts, reasoned judgment based on research findings, and speculation in an explanation. Respectfully provide and receive critiques from peers about a proposed procedure, explanation, or model by citing relevant evidence and posing specific questions.
- *SEP 8: Obtaining, Evaluating, and Communicating Information:* Read and comprehend grade-appropriate complex texts and/or other reliable media to summarize and obtain scientific and technical ideas. Combine information in written text with that contained in corresponding tables, diagrams, and/or charts to support the engagement in other scientific and/or engineering practices. Communicate scientific and/or technical information orally and/or in written formats, including various forms of media as well as tables, diagrams, and charts.

Other Concepts That Students May Use During This Investigation

Students might also use some of the following concepts:

- The Sun provides energy for plants and other producers.
- Organisms need both energy and matter to survive.
- Feeding relationships are complex, and most organisms use multiple sources of food to survive and grow.
- Different models contain different amounts of information about a phenomenon and represent that information in ways that provide more or less clarity.

Note: Do not teach students about food chains and food webs before beginning this investigation. Students will be developing their own models to show feeding relationships between the organisms that live in the Arctic Ocean and how energy transfers between these organisms, so teaching them about food chains and food webs before they begin will limit opportunities for students to make sense of this phenomenon. You can use what the students figure out about how to model energy transfer in ecosystems in this investigation to formally introduce food chains and food webs in another lesson.

Teacher Notes

What Students Figure Out

Students determine that there are multiple ways to model how energy is transferred from organism to organism within an ecosystem. However, any model should show a one-way transfer of energy from the Sun, to the producers, to the animals that eat the producers, and then to the animals that eat other animals, and it should include all producers and consumers that live in the Arctic ecosystem.

Background Information About This Investigation for the Teacher

The energy needed to sustain most life on Earth comes from the Sun. Plants, algae (including phytoplankton), and some types of microorganisms use energy from light to make sugar from carbon dioxide and water through a process called photosynthesis. These organisms can then break down the sugar to produce the energy they need to survive, or they can combine the sugar to create other materials such as starch and store them for later use. Some of this sugar is also used to create the materials that are needed to create new cells. Animals cannot make sugar from carbon dioxide and water. Animals therefore need to eat in order to get the sugar they need to survive and grow. Some animals eat plants, and some animals eat other animals.

FIGURE 9.1

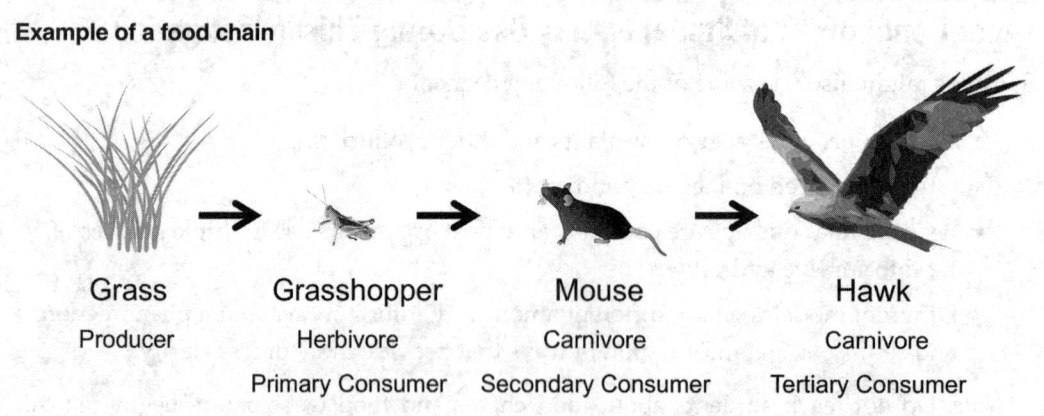

Example of a food chain

Grass → Grasshopper → Mouse → Hawk

Grass	Grasshopper	Mouse	Hawk
Producer	Herbivore	Carnivore	Carnivore
	Primary Consumer	Secondary Consumer	Tertiary Consumer

To illustrate how energy moves from the Sun to different organisms, consider the food chain illustrated in Figure 9.1. This food chain begins with grasshoppers eating grass. The materials that make up grass, as a result, transfer from grass to grasshoppers. Grasshoppers use some of these materials to generate the energy they need to move around and find more food, and they use some of these materials to grow. Figure 9.1 also shows that mice eat grasshoppers. The materials that make up grasshoppers, as a result, transfer to mice. Mice use the materials that make up grasshoppers to generate the energy they need to move

Investigation 9. Energy in Ecosystems:
How Do We Best Model the Transfer of Energy Into and
Within the Living Things That Are Found in the Arctic Ocean?

around and stay warm; these materials are also used for growth and to repair injuries. Finally, Figure 9.1 shows that hawks eat mice. The materials that make up mice transfer to the hawks and are used to generate the energy the hawks need to grow and survive. All the energy in this food chain, however, originally came from the Sun.

FIGURE 9.2
Example of a food web

Hawk — **Carnivores**

Flycatchers, Mouse — **Carnivores**

Cutworm, Grasshopper, Beetle — **Herbivores**

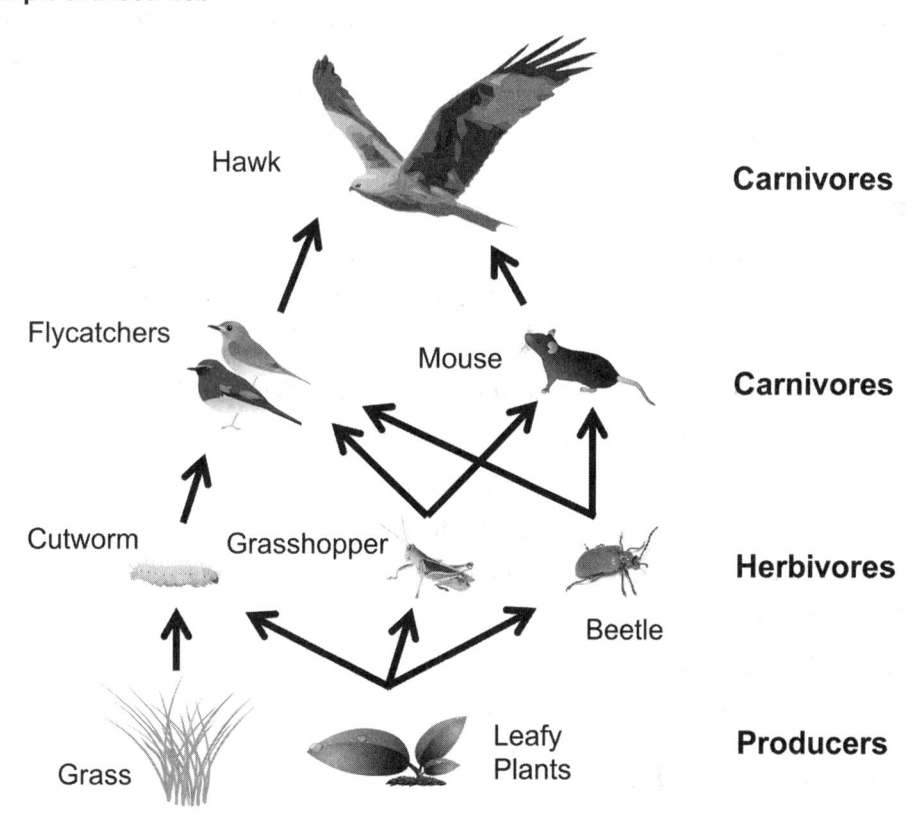

Grass, Leafy Plants — **Producers**

Figure 9.2 shows a food web. As in Figure 9.1, the arrows in the food web show how materials and energy move from one type of living thing to another type of living thing. Food webs also show all the possible things an animal can eat. Food webs can include many different food chains; some of these chains in the web can include many different organisms, whereas other chains in the web can include only a few organisms. The food web in Figure 9.2 shows several different food chains. One example of a food chain in this food web is leafy plants, grasshopper, flycatcher, and hawk. Another example of a food chain is leafy plants, beetles, mouse, and hawk.

Teacher Notes

This investigation asks students to use models to show how the energy in animals' food was once energy from the Sun. Students may develop models that closely resemble traditional food webs, or they may develop other, innovative ways of representing the relationships within an ecosystem. Regardless of the model chosen, students should be prompted to define the system, the inputs and outputs in terms of matter and energy, and the processes and interactions represented in their model. The students should also be asked to label and define the symbols they use. Traditionally, scientists do not include the Sun itself in food webs because it does not provide matter to the producers. If students include the Sun in their model, they should be prompted to indicate that it only imparts energy—for example, by using a different-color arrow than the one used to show both matter and energy moving from one organism.

Timeline

The time needed to complete this investigation is 270 minutes (4 hours and 30 minutes) spread out across several days. The amount of instructional time needed for each stage of the investigation is as follows:

- *Stage 1.* Introduce the task and the guiding question: 35 minutes
- *Stage 2.* Design a method and collect data: 50 minutes
- *Stage 3.* Create a draft argument: 45 minutes
- *Stage 4.* Argumentation session: 30 minutes
- *Stage 5.* Reflective discussion: 15 minutes
- *Stage 6.* Write a draft report: 30 minutes
- *Stage 7.* Peer review: 35 minutes
- *Stage 8.* Revise the report: 30 minutes

Materials and Preparation

The materials needed for this investigation are listed in Table 9.1. The organism cards can be accessed online at *www.nsta.org/adi-5th*. The materials for this investigation can also be purchased as a kit (which includes enough materials for 24 students, or six groups of four students) at *www.argumentdriveninquiry.com*.

Investigation 9. Energy in Ecosystems: How Do We Best Model the Transfer of Energy Into and Within the Living Things That Are Found in the Arctic Ocean?

TABLE 9.1
Materials for Investigation 9

Item	Quantity
Cards with information on arctic organisms	1 set per group
Sticky notes	1 pad per group
Marker pens	1 set per group
Dry-erase markers	2 or 3 per group
Colored pencils	1 set per group
White chart paper	5–10 sheets per group
Whiteboard, 2' × 3'*	1 per group
Investigation Handout	1 per student
Peer-review guide and teacher scoring rubric	1 per student
Checkout Questions (optional)	1 per student

*As an alternative, students can use computer and presentation software such as Microsoft PowerPoint or Apple Keynote to create their arguments.

Safety Precautions

Remind students to follow all normal safety rules.

Lesson Plan by Stage

This lesson plan is only a suggestion. It is included here to illustrate what you can say and do during each stage of ADI for this specific investigation. We encourage you to modify this lesson plan by asking different questions, using different examples, and providing different scaffolds as needed to better meet the needs of students in your class.

Stage 1: Introduce the Task and the Guiding Question (35 minutes)

1. Ask the students to sit in six groups, with three or four students in each group.
2. Ask the students to clear off their desks except for a pencil (and their *Student Workbook for Argument-Driven Inquiry in Fifth-Grade Science* if they have one).
3. Pass out an Investigation Handout to each student (or ask students to turn to the Investigation Log for Investigation 9 in their workbook).
4. Read the first paragraph of the "Introduction" aloud to the class. Ask the students to follow along as you read.
5. Show the students a video clip of orca whales hunting different types of animals; good examples of videos can be found at *www.youtube.com/watch?v=mhxeqBISsDk* and *www.youtube.com/watch?v=pEP0sMO-nUQ*.

Teacher Notes

6. Ask the students to record what they noticed in the video and their questions about it in the first "NOTICED/WONDER" chart in the "Introduction."

7. After the students have recorded their observations and questions about the video, ask them to share what they noticed. Then ask the students to share what questions they have about the video.

8. Tell the students, "Some of your questions might be answered by reading the next four paragraphs of the 'Introduction.'"

9. Ask the students to read the next four paragraphs of the "Introduction" on their own *or* ask them to follow along as you read these paragraphs aloud.

10. Pass out a set of cards with information about arctic organisms to each group.

11. Ask the students to record what they noticed about the organisms and any questions they have about them in the second "NOTICED/WONDER" chart in the "Introduction."

12. After the students have recorded their observations and questions about the organisms, ask them to share what they noticed. Then ask the students to share what questions they have about the organisms.

13. Tell the students, "Some of your questions might be answered by reading the rest of the 'Introduction.'"

14. Ask the students to read the last paragraph of the "Introduction" on their own *or* ask them to follow along as you read it aloud.

15. Once the students have read the last paragraph of the "Introduction," ask them to fill out the "Things we KNOW" chart on their Investigation Handout (or in their Investigation Log) as a group.

16. Ask the students to share what they learned from the reading. Add these ideas to a class "Things we KNOW" chart.

17. Tell the students, "Let's see what we will need to figure out during our investigation."

18. Read the task and the guiding question aloud.

19. Tell the students, "I have lots of materials here that you can use."

20. Introduce the students to the materials available for them to use during the investigation by holding each one up and asking how it might be used.

Stage 2: Design a Method and Collect Data (50 minutes)

1. Tell the students, "I am now going to give you and the other members of your group about 10 minutes to plan your investigation. Before you begin, I want you to take a couple of minutes to discuss the following questions with the rest of your group."

Investigation 9. Energy in Ecosystems:
How Do We Best Model the Transfer of Energy Into and
Within the Living Things That Are Found in the Arctic Ocean?

2. Show the following questions on the screen or board:
   - What are the *components of this system* and how do they *interact*?
   - How can we *track the transfer of energy* within a system?
3. Tell the students, "Please take a few minutes to come up with an answer to these questions."
4. Give the students two or three minutes to discuss these two questions.
5. Ask two or three different groups to share their answers. Highlight or write down any important ideas on the board so students can refer to them later.
6. If possible, use a document camera to project an image of the graphic organizer for this investigation on a screen or board (or take a picture of it and project the picture on a screen or board). Tell the students, "I now want you all to plan out your investigation. To do that, you will need to fill out this investigation proposal."
7. Point to the box labeled "Our guiding question:" and tell the students, "You can put the question we are trying to answer in this box." Then ask, "Where can we find the guiding question?"
8. Wait for a student to answer.
9. Point to the box labeled "We will collect the following data:" and tell the students, "You can list any information about the arctic organisms that is important to collect during the investigation in this box." Have them write out what steps they will follow to collect the data.
10. Ask the students, "Do you have any questions about what you need to do?"
11. Wait for questions. Answer any questions that come up.
12. Tell the students, "Once you are done, raise your hand and let me know. I'll then come by and look over your proposal and give you some feedback. You may not begin your investigation until I have approved your proposal by signing it. You need to have your proposal done in the next 10 minutes."
13. Give the students 10 minutes to work in their groups on their investigation proposal. As they work, move from group to group to check in, ask probing questions, and offer a suggestion if a group gets stuck.

What should a student-designed investigation look like?

The students' investigation proposal should include the following information:
- The guiding question is "How do we best model the transfer of energy into and within the living things that are found in the Arctic Ocean?"
- The data that the students should collect are (1) which organisms they are including in their model, (2) the feeding relationships that exist among the organisms, and (3) how they can model these relationships.

Teacher Notes

14. As each group finishes its investigation proposal, read it over and determine if it will be productive or not. If you feel the investigation will be productive (not necessarily what you would do or what the other groups are doing), sign your name on the proposal and let the group start collecting data. If the plan needs to be changed, offer some suggestions or ask some probing questions, and have the group make the changes before you approve it.

15. Pass out the materials or have one student from each group collect the materials they need from a central supply table or cart for the groups that have an approved proposal.

16. Tell the students to start their investigation. Remind students to keep track of any important information they find as they read over the cards with information about arctic organisms in the "Collect Your Data" box in their Investigation Handout (or the Investigation Log in their workbook).

17. Allow students 30 minutes to create different models. As they work together to develop and revise a model, check in with each group and ask questions such as

    - How are you defining the system?
    - How is energy entering this system? Where does that energy come from?
    - How are you representing _____ in your model?
    - How else might you model these same relationships?

18. Remind students to make a copy of their model in the "Create a Model" section in their Investigation Handout (or the Investigation Log in their workbook).

What should the model for this investigation look like?

There are a number of different ways that students can represent the transfer of energy among the organisms. The models should include labels or a key to help others understand how energy is transferring between the organisms.

Stage 3: Create a Draft Argument (45 minutes)

1. Tell the students, "I am now going to give you and the other members of your group about 15 minutes to create an argument to share what you have learned and convince others that they should believe you. Before you do that, we need to take a few minutes to discuss what you need to include in your argument."

2. If possible, use a document camera to project the "Argument Presentation on a Whiteboard" image from the "Draft Argument" section of the Investigation Handout (or the Investigation Log in their workbook) on a screen or board (or take a picture of it and project the picture on a screen or board).

Investigation 9. Energy in Ecosystems:
How Do We Best Model the Transfer of Energy Into and
Within the Living Things That Are Found in the Arctic Ocean?

3. Point to the box labeled "The Guiding Question:" and tell the students, "You can put the question we are trying to answer here on your whiteboard."

4. Point to the box labeled "Our Claim:" and tell the students, "You can put your model here on your whiteboard. The claim is your model because it is your answer to the guiding question."

5. Point to the box labeled "Our Evidence:" and tell the students, "You can put the evidence that you are using to support your claim here on your whiteboard. You can use evidence to convince us that all the components of your model are acceptable and that there are no mistakes in it. Scientists always need to support their claims with evidence."

6. Point to the box labeled "Our Justification of the Evidence:" and tell the students, "You can put your justification of your evidence here on your whiteboard. Your justification needs to explain why your evidence is important. Scientists often use core ideas to explain why the evidence they are using matters. Core ideas are important concepts that scientists use to help them make sense of what happens during an investigation."

7. Ask the students, "What are some core ideas that we read about earlier that might help us explain why the evidence we are using is important?"

8. Ask the students to share some of the core ideas from the "Introduction" section of the Investigation Handout (or the Investigation Log in the workbook). List these core ideas on the board.

9. Tell the students, "That is great. I would like to see everyone try to include these core ideas in your justification of the evidence. Your goal is to use these core ideas to help explain why your evidence matters and why the rest of us should pay attention to it."

10. Ask the students, "Do you have any questions about what you need to do?"

11. Answer any questions that come up.

12. Tell the students, "Okay, go ahead and start working on your arguments. You need to have your argument done in the next 15 minutes. It doesn't need to be perfect. We just need something down on the whiteboards so we can share our ideas."

13. Give the students 15 minutes to work in their groups on their arguments. As they work, move from group to group to check in, ask probing questions, and offer a suggestion if a group gets stuck. Figure 9.3 (p. 366) shows an example of an argument created by a group of students as part of this investigation. In this example, the students show how a food web can be used to illustrate changes in energy transfer if different types of organisms in the arctic ecosystem went extinct as evidence for their claim that a food web is the best way to model the transfer of energy between organisms. There are other ways to present an argument. For example, in the argument shown in Figure 9.4 (p. 367), the model

Teacher Notes

serves as the claim. The evidence includes a table that shows what each organism eats as a way to support the accuracy of the feeding relationships included in the model. The justification of the evidence then explains that plants use energy from the Sun and that animals eat plants or other animals to obtain the energy they need to survive.

FIGURE 9.3
Example of an argument

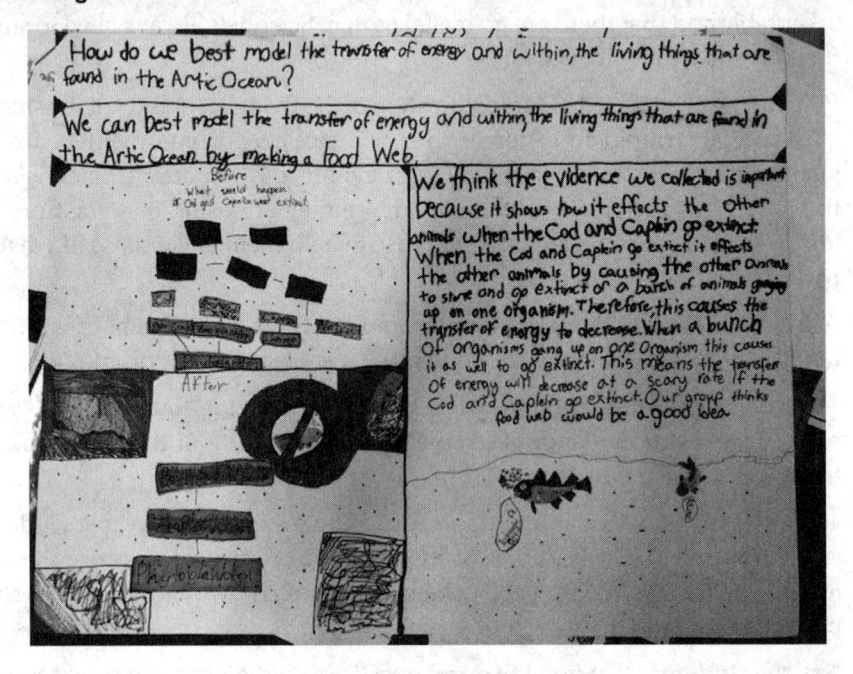

Investigation 9. Energy in Ecosystems: How Do We Best Model the Transfer of Energy Into and Within the Living Things That Are Found in the Arctic Ocean?

FIGURE 9.4

A second example of an argument

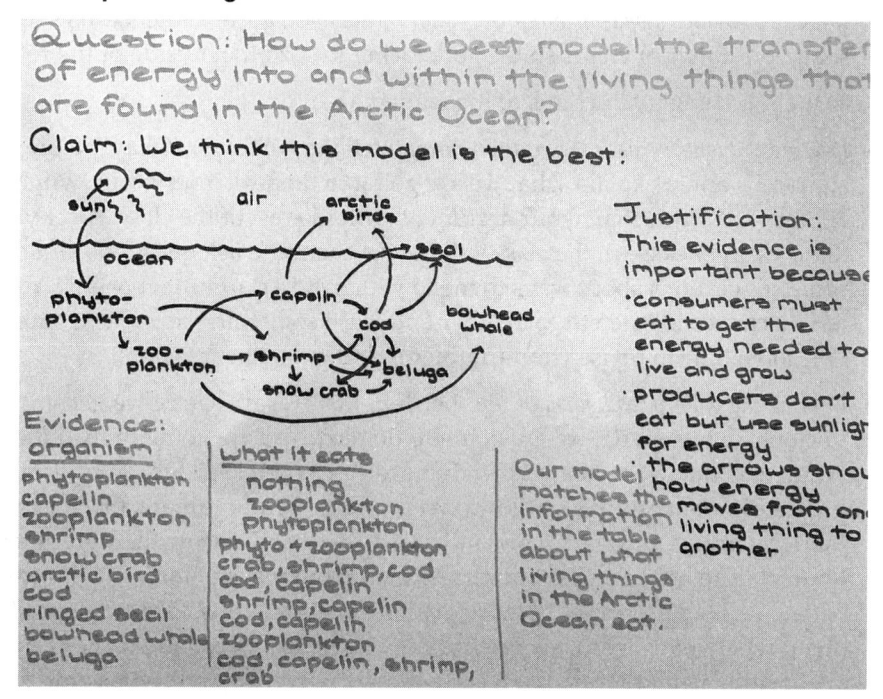

> Question: How do we best model the transfer of energy into and within the living things that are found in the Arctic Ocean?
>
> Claim: We think this model is the best:
>
> [Diagram showing energy transfer: sun → air, ocean; phytoplankton → zooplankton, capelin; zooplankton → shrimp; shrimp → snow crab, cod; capelin → cod, arctic birds, seal; cod → bowhead whale, beluga, seal; snow crab → beluga]
>
> Evidence:
>
> | organism | what it eats |
> |---|---|
> | phytoplankton | nothing |
> | capelin | zooplankton |
> | zooplankton | phytoplankton |
> | shrimp | phyto + zooplankton |
> | snow crab | crab, shrimp, cod |
> | arctic bird | cod, capelin |
> | cod | shrimp, capelin |
> | ringed seal | cod, capelin |
> | bowhead whale | zooplankton |
> | beluga | cod, capelin, shrimp, crab |
>
> Justification: This evidence is important because
> - consumers must eat to get the energy needed to live and grow
> - producers don't eat but use sunlight for energy
> - the arrows show how energy moves from one living thing to another
>
> Our model matches the information in the table about what living things in the Arctic Ocean eat.

Stage 4: Argumentation Session (30 minutes)

The argumentation session can be conducted in a whole-class presentation format, a gallery walk format, or a modified gallery walk format. We recommend using a whole-class presentation format for the first investigation, but try to transition to either the gallery walk or modified gallery walk format as soon as possible because that will maximize student voice and choice inside the classroom. The following list shows the steps for the three formats; unless otherwise noted, the steps are the same for all three formats.

1. Begin by introducing the use of the whiteboard.

    - *If using the whole-class presentation format,* tell the students, "We are now going to share our arguments. Please set up your whiteboards so everyone can see them."

    - *If using the gallery walk or modified gallery walk format,* tell the students, "We are now going to share our arguments. Please set up your whiteboards so they are facing the walls."

2. Allow the students to set up their whiteboards.

Argument-Driven Inquiry in **Fifth-Grade Science**: Three-Dimensional Investigations

Teacher Notes

- *If using the whole-class presentation format,* the whiteboards should be set up on stands or chairs so they are facing toward the center of the room.
- *If using the gallery walk or modified gallery walk format,* the whiteboards should be set up on stands or chairs so they are facing toward the outside of the room.

3. Give the following instructions to the students:

   - *If using the whole-class presentation format,* tell the students, "Okay, before we get started I want to explain what we are going to do next. Your group will have an opportunity to share your argument with the rest of the class. After you are done, everyone else in the class will have a chance to ask questions and offer some suggestions about ways to make your group's argument better. After we have a chance to listen to each other and learn something new, I'm going to give you some time to revise your arguments and make them better."

   - *If using the gallery walk format,* tell the students, "Okay, before we get started I want to explain what we are going to do next. You are going to read the arguments that were created by other groups. When I say 'go,' your group will go to a different group's station so you can see their argument. Once you are there, I'll give your group a few minutes to read and review their argument. Your job is to offer them some suggestions about ways to make their argument better. You can use sticky notes to give them suggestions. Please be specific about what you want to change and how you think they should change it. After we have a chance to learn from each other, I'm going to give you some time to revise your arguments and make them better."

   - *If using the modified gallery walk format,* tell the students, "Okay, before we get started I want to explain what we are going to do next. I'm going to ask some of you to present your arguments to your classmates. If you are presenting your argument, your job is to share your group's claim, evidence, and justification of the evidence. The rest of you will be travelers. If you are a traveler, your job is to listen to the presenters, ask the presenters questions if you do not understand something, and then offer them some suggestions about ways to make their argument better. After we have a chance to learn from each other, I'm going to give you some time to revise your arguments and make them better."

4. Use a document camera to project the "Ways to IMPROVE our argument …" box from the Investigation Handout (or the Investigation Log in their workbook) on a screen or board (or take a picture of it and project the picture on a screen or board).

   - *If using the whole-class presentation format,* point to the box and tell the students, "After your group presents your argument, you can write down the suggestions you get from your classmates here. If you are listening to a presentation and you see a good idea from another group, you can write down that idea here as well. Once we are done with the presentations, I will give you a chance to use these suggestions or ideas to improve your arguments."

Investigation 9. Energy in Ecosystems:
How Do We Best Model the Transfer of Energy Into and
Within the Living Things That Are Found in the Arctic Ocean?

- *If using the gallery walk format,* point to the box and tell the students, "If you see a good idea from another group, you can write it down here. Once we are done reviewing the different arguments, I will give you a chance to use these ideas to improve your own arguments. It is important to share ideas like this."
- *If using the modified gallery walk format,* point to the box and tell the students, "If you are a presenter, you can write down the suggestions you get from the travelers here. If you are a traveler and you see a good idea from another group, you can write down that idea here. Once we are done with the presentations, I will give you a chance to use these suggestions or ideas to improve your arguments."

5. Ask the students, "Do you have any questions about what you need to do?"
6. Answer any questions that come up.
7. Give the following instructions:
   - *If using the whole-class presentation format,* tell the students, "Okay. Let's get started."
   - *If using the gallery walk format,* tell the students, "Okay, I'm now going to tell you which argument to go to and review."
   - *If using the modified gallery walk format,* tell the students, "Okay, I'm now going to assign you to be a presenter or a traveler." Assign one or two students from each group to be presenters and one or two students from each group to be travelers.
8. Give the students an opportunity to review the arguments.
   - *If using the whole-class presentation format,* have each group present their argument one at a time. Give each group only two to three minutes to present their argument. Then give the class two to three minutes to ask them questions and offer suggestions. Encourage as much participation from the students as possible.
   - *If using the gallery walk format,* tell the students, "Okay. Let's get started. Each group, move one argument to the left. Don't move to the next argument until I tell you to move. Once you get there, read the argument and then offer suggestions about how to make it better. I will put some sticky notes next to each argument. You can use the sticky notes to leave your suggestions." Give each group about three to four minutes to read the arguments, talk, and offer suggestions.
     a. After three to four minutes, tell the students, "Okay. Let's move on to the next argument. Please move one group to the left."
     b. Again, give each group three to four minutes to read, talk, and offer suggestions.
     c. Repeat this process until each group has had their argument read and critiqued three times.

Teacher Notes

- *If using the modified gallery walk format,* tell the students, "Okay. Let's get started. Reviewers, move one group to the left. Don't move to the next group until I tell you to move. Presenters, go ahead and share your argument with the travelers when they get there." Give each group of presenters and travelers about three to four minutes to talk.

    a. Tell the students, "Okay. Let's move on to the next argument. Travelers, move one group to the left."

    b. Again, give each group of presenters and travelers about three to four minutes to talk.

    c. Repeat this process until each group has had their argument read and critiqued three times.

9. Tell the students to return to their workstations.

10. Give the following instructions about revising the argument:

    - *If using the whole-class presentation format,* tell the students, "I'm now going to give you all about 10 minutes to revise your argument. Take a few minutes to talk in your groups and determine what you want to change to make your argument better. Once you have decided what to change, go ahead and make the changes to your whiteboard."

    - *If using the gallery walk format,* tell the students, "I'm now going to give you all about 10 minutes to revise your argument. Take a few minutes to read the suggestions that were left at your argument. Then talk in your groups and determine what you want to change to make your argument better. Once you have decided what to change, go ahead and make the changes to your whiteboard."

    - *If using the modified gallery walk format,* tell the students, "I'm now going to give you all about 10 minutes to revise your argument. Please return to your original groups." Wait for the students to move back into their original groups and then tell the students, "Okay, take a few minutes to talk in your groups and determine what you want to change to make your argument better. Once you have decided what to change, go ahead and make the changes to your whiteboard."

11. Ask the students, "Do you have any questions about what you need to do?"

12. Answer any questions that come up.

13. Tell the students, "Okay. Let's get started."

14. Give the students 10 minutes to work in their groups on their arguments. As they work, move from group to group to check in, ask probing questions, and offer a suggestion if a group gets stuck.

Investigation 9. Energy in Ecosystems:
How Do We Best Model the Transfer of Energy Into and
Within the Living Things That Are Found in the Arctic Ocean?

FIGURE 9.5
Kelp and fish

FIGURE 9.6
Orca feeding on a seal

Stage 5: Reflective Discussion (15 minutes)

1. Tell the students, "We are now going to take a minute to talk about some of the core ideas and crosscutting concepts that we have used during our investigation."

2. Show Figure 9.5 on the screen.

3. Ask the students, "What do you all see going on here?"

4. Allow the students to share their ideas.

5. Show Figure 9.6 on the screen.

6. Ask the students, "What do you all see going on here?"

7. Allow the students to share their ideas.

8. Show Figures 9.5 and 9.6 on the screen at the same time.

9. Ask the students, "How can we explain what is happening in these two pictures in terms of the movement of energy into, out of, or within this system?"

10. Allow students to share their ideas. As they share their ideas, ask different questions to encourage them to expand on their thinking (e.g., "Can you tell me more about that?"), clarify a contribution (e.g., "Can you say that in another way?"), support an idea (e.g., "Why do you think that?"), add to an idea mentioned by a classmate (e.g., "Would anyone like to add to the idea?"), re-voice an idea offered by a classmate (e.g., "Who can explain that to me in another

Teacher Notes

way?"), or critique an idea during the discussion (e.g., "Do you agree or disagree with that idea and why?") until students are able to generate an adequate explanation.

11. Ask the students, "Does anyone have any questions about the way energy transfers between living things in an ecosystem?"

12. Answer any questions that come up.

13. Show an image of the question "What do you think are the most important core ideas or crosscutting concepts that we used during this investigation to help us make sense of what we observed?" Tell the students, "Okay, let's make sure we are all on the same page. Please take a moment to discuss this question with the other people in your group." Give them a few minutes to discuss the question.

14. Ask the students, "What do you all think? Who would like to share?"

15. Allow the students to share their ideas.

16. Tell the students, "We needed to develop a model during our investigation." Then ask, "Can anyone tell me why this was important?"

17. Allow the students to share their ideas.

18. Tell the students, "Developing models that represent processes and interactions allows us to better explain what we can and cannot see happening."

19. Ask the students, "How did your model make it possible for us to understand how energy transfers within an ecosystem?"

20. Allow the students to share their ideas.

21. Ask the students, "What made some of the models better than others?"

22. Allow the students to share their ideas.

23. Tell the students, "We are now going to take a minute to talk about what went well and what didn't go so well during our investigation. We need to talk about this because you all are going to be planning and carrying out your own investigations like this a lot this year, and I want to help you all get better at it."

24. Show an image of the question "What made your investigation scientific?" on the screen. Tell the students, "Take a few minutes to talk about how you would answer this question with the other people in your group. Be ready to share with the rest of the class." Give the students two to three minutes to talk in their group.

25. Ask the students, "What do you all think? Who would like to share an idea?"

26. Allow the students to share their ideas. Be sure to expand on their ideas about what makes an investigation scientific.

27. Show an image of the question "What made your investigation not so scientific?" on the screen. Tell the students, "Take a few minutes to talk about how

Investigation 9. Energy in Ecosystems:
How Do We Best Model the Transfer of Energy Into and
Within the Living Things That Are Found in the Arctic Ocean?

you would answer this question with the other people in your group. Be ready to share with the rest of the class." Give the students two to three minutes to talk in their group.

28. Ask the students, "What do you all think? Who would like to share an idea?"

29. Allow the students to share their ideas. Be sure to expand on their ideas about what makes an investigation less scientific.

30. Show an image of the question "What rules can we put into place to help us make sure our next investigation is more scientific?" on the screen. Tell the students, "Take a few minutes to talk about how you would answer this question with the other people in your group. Be ready to share with the rest of the class." Give the students two to three minutes to talk in their group.

31. Ask the students, "What do you all think? Who would like to share an idea?"

32. Allow the students to share their ideas. Once they have shared their ideas, offer a suggestion for a possible class rule.

33. Ask the students, "What do you all think? Should we make this a rule?"

34. If the students agree, write the rule on the board or on a class "Rules for Scientific Investigation" chart so you can refer to it during the next investigation.

Stage 6: Write a Draft Report (30 minutes)

Your students will use either the Investigation Handout or the Investigation Log in the student workbook when writing the draft report. When you give the directions shown in quotes in the following steps, substitute "Investigation Log in your workbook" or just "Investigation Log" (as shown in brackets) for "handout" if they are using the workbook.

1. Tell the students, "You are now going to write an investigation report to share what you have learned. Please take out a pencil and turn to the 'Draft Report' section of your handout [Investigation Log in your workbook]."

2. If possible, use a document camera to project the "Introduction" section of the draft report from the Investigation Handout (or the Investigation Log in their workbook) on a screen or board (or take a picture of it and project the picture on a screen or board).

3. Tell the students, "The first part of the report is called the 'Introduction.' In this section of the report you want to explain to the reader what you were investigating, why you were investigating it, and what question you were trying to answer. All this information can be found in the text at the beginning of your handout [Investigation Log]." Point to the image. "Here are some sentence starters to help you begin writing."

4. Ask the students, "Do you have any questions about what you need to do?"

5. Answer any questions that come up.

Teacher Notes

6. Tell the students, "Okay, let's write."

7. Give the students 10 minutes to write the "Introduction" section of the report. As they work, move from student to student to check in, ask probing questions, and offer a suggestion if a student gets stuck.

8. If possible, use a document camera to project the "Method" section of the draft report from the Investigation Handout (or the Investigation Log in their workbook) on a screen or board (or take a picture of it and project the picture on a screen or board).

9. Tell the students, "The second part of the report is called the 'Method.' In this section of the report you want to explain to the reader what you did during the investigation. All this information can be found in the 'Plan Your Investigation' section of the handout [Investigation Log]. Remember that you all planned and carried out different investigations, so do not assume that the reader will know what you did." Point to the image. "Here are some sentence starters to help you begin writing."

10. Ask the students, "Do you have any questions about what you need to do?"

11. Answer any questions that come up.

12. Tell the students, "Okay, let's write."

13. Give the students 10 minutes to write the "Method" section of the report. As they work, move from student to student to check in, ask probing questions, and offer a suggestion if a student gets stuck.

14. If possible, use a document camera to project the "Argument" section of the draft report from the Investigation Handout (or the Investigation Log in their workbook) on a screen or board (or take a picture of it and project the picture on a screen or board).

15. Tell the students, "The last part of the report is called the 'Argument.' In this section of the report you want to share your claim, evidence, and justification of the evidence with the reader. All this information can be found on your whiteboard." Point to the image. "Here are some sentence starters to help you begin writing."

16. Ask the students, "Do you have any questions about what you need to do?"

17. Answer any questions that come up.

18. Tell the students, "Okay, let's write."

19. Give the students 10 minutes to write the "Argument" section of the report. As they work, move from student to student to check in, ask probing questions, and offer a suggestion if a student gets stuck.

Investigation 9. Energy in Ecosystems: How Do We Best Model the Transfer of Energy Into and Within the Living Things That Are Found in the Arctic Ocean?

Stage 7: Peer Review (35 minutes)

Your students will use either the Investigation Handout or their workbook when doing the peer review. Except where noted below, the directions are the same whether using the handout or the workbook.

1. Tell the students, "We are now going to review our reports to find ways to make them better. I'm going to come around and collect your draft reports. While I do that, please take out a pencil."

2. Collect the handouts or the workbooks with the draft reports from the students.

3. If possible, use a document camera to project the peer-review guide (see Appendix 4) on a screen or board (or take a picture of it and project the picture on a screen or board).

4. Tell the students, "We are going to use this peer-review guide to give each other feedback." Point to the image.

5. Tell the students, "I'm going to ask you to work with a partner to do this. I'm going to give you and your partner a draft report to read. You two will then read the report together. Once you are done reading the report, I want you to answer each of the questions on the peer-review guide." Point to the review questions on the image of the peer-review guide.

6. Tell the students, "You can check 'no,' 'almost,' or 'yes' after each question." Point to the checkboxes on the image of the peer-review guide.

7. Tell the students, "This will be your rating for this part of the report. Make sure you agree on the rating you give the author. If you mark 'no' or 'almost,' then you need to tell the author what he or she needs to do to get a 'yes.'" Point to the space for the reviewer feedback on the image of the peer-review guide.

8. Tell the students, "It is really important for you to give the authors feedback that is helpful. That means you need to tell them exactly what they need to do to make their report better."

9. Ask the students, "Do you have any questions about what you need to do?"

10. Answer any questions that come up.

11. Tell the students, "Please sit with a partner who is not in your current group." Allow the students time to sit with a partner.

12. Tell the students, "Okay, I'm now going to give you one report to read." Pass out one Investigation Handout with a draft report or one workbook to each pair. Make sure that the report you give a pair was not written by one of the students in that pair. Give each pair one peer-review guide to fill out. If the students are using workbooks, the peer-review guide is included right after the draft report so you do not need to pass out copies of the peer-review guide.

Teacher Notes

13. Tell the students, "Okay, I'm going to give you 15 minutes to read the report I gave you and to fill out the peer-review guide. Go ahead and get started."
14. Give the students 15 minutes to work. As they work, move around from pair to pair to check in and see how things are going, answer questions, and offer advice.
15. After 15 minutes pass, tell the students, "Okay, time is up. Please give me the report and the peer-review guide that you filled out."
16. Collect the Investigation Handouts and the peer-review guides, or collect the workbooks if students are using them. If the students are using the Investigation Handouts and separate peer-review guides, be sure you keep each handout with its corresponding peer-review guide.
17. Tell the students, "Okay, I am now going to give you a different report to read and a new peer-review guide to fill out." Pass out one more report to each pair. Make sure that the report you give a pair was not written by one of the students in that pair. Give each pair a new peer-review guide to fill out as a group.
18. Tell the students, "Okay, I'm going to give you 15 minutes to read this new report and to fill out the peer-review guide. Go ahead and get started."
19. Give the students 15 minutes to work. As they work, move around from pair to pair to check in and see how things are going, answer questions, and offer advice.
20. After 15 minutes pass, tell the students, "Okay, time is up. Please give me the report and the peer-review guide that you filled out."
21. Collect the Investigation Handouts and the peer-review guides, or collect the workbooks if students are using them. If the students are using the Investigation Handouts and separate peer-review guides, be sure you keep each handout with its corresponding peer-review guide.

Stage 8: Revise the Report (30 minutes)

Your students will use either the Investigation Handout or their workbook when revising the report. Except where noted below, the directions are the same whether using the handout or the workbook.

1. Tell the students, "You are now going to revise your draft report based on the feedback you get from your classmates. Please take out a pencil."
2. Return the reports to the students.
   - *If the students used the Investigation Handout and a copy of the peer-review guide,* pass back the handout and the peer-review guide to each student.
   - *If the students used the workbook,* pass that back to each student.

Investigation 9. Energy in Ecosystems:
How Do We Best Model the Transfer of Energy Into and
Within the Living Things That Are Found in the Arctic Ocean?

3. Tell the students, "Please take a few minutes to read over the peer-review guide. You should use it to figure out what you need to change in your report and how you will change it."

4. Allow the students to read the peer-review guide.

5. *If the students used the workbook,* if possible use a document camera to project the "Write Your Final Report" section from the Investigation Log on a screen or board (or take a picture of it and project the picture on a screen or board).

6. Give the following directions about how to revise their reports:

   - *If the students used the Investigation Handout and a copy of the peer-review guide,* tell them, "Okay, let's revise our reports. Please take out a piece of paper. I would like you to rewrite your report. You can use your draft report as a starting point, but you also need to change it to make it better. Use the feedback on the peer-review guide to make it better."

   - *If the students used the workbook,* tell them, "Okay, let's revise our reports. I would like you to rewrite your report in the section of the Investigation Log called "Write Your Final Report." You can use your draft report as a starting point, but you also need to change it to make it better. Use the feedback on the peer-review guide to make it better."

7. Ask the students, "Do you have any questions about what you need to do?"

8. Answer any questions that come up.

9. Tell the students, "Okay, let's write." Allow about 20 minutes for the students to revise their reports.

10. After about 20 minutes, give the following directions:

    - *If the students used the Investigation Handout,* tell them, "Okay, time's up. I will now come around and collect your Investigation Handout, the peer-review guide, and your final report."

    - *If the students used the workbook,* tell them, "Okay, time's up. I will now come around and collect your workbooks."

11. *If the students used the Investigation Handout, collect all the Investigation Handouts, peer-review guides, and final reports. If the students used the workbook,* collect all the workbooks.

12. *If the students used the Investigation Handout,* use the "Teacher Score" column in the peer-review guide to grade the final report. *If the students used the workbook,* use the "Investigation Report Grading Rubric" in the Investigation Log to grade the final report. Whether you are using the handout or the log, you can give the students feedback about their writing in the "Teacher Comments" section.

Teacher Notes

How to Use the Checkout Questions

The Checkout Questions are an optional assessment. We recommend giving them to students at the start of the next class period after the students finish stage 8 of the investigation. You can then look over the student answers to determine if you need to reteach the core idea from the investigation. Appendix 6 gives the answers to the Checkout Questions that should be given by a student who can apply the core idea correctly and can explain the transfer of energy in the system.

Alignment With Standards

Table 9.2 highlights how the investigation can be used to address specific performance expectations from the *Next Generation Science Standards*, *Common Core State Standards for English Language Arts (CCSS ELA)*, and *English Language Proficiency (ELP) Standards*.

TABLE 9.2
Investigation 9 alignment with standards

NGSS performance expectations	Strong alignment • 5-PS3-1: Use models to describe that energy in animals' food (used for body repair, growth, and motion and to maintain body warmth) was once energy from the Sun. Moderate alignment • 5-LS2-1: Develop a model to describe the movement of matter among plants, animals, decomposers, and the environment.
CCSS ELA—Reading: Informational Text	Key ideas and details • CCSS.ELA-LITERACY.RI.5.1: Quote accurately from a text when explaining what the text says explicitly and when drawing inferences from the text. • CCSS.ELA-LITERACY.RI.5.2: Determine two or more main ideas of a text and explain how they are supported by key details; summarize the text. • CCSS.ELA-LITERACY.RI.5.3: Explain the relationships or interactions between two or more individuals, events, ideas, or concepts in a historical, scientific, or technical text based on specific information in the text. Craft and structure • CCSS.ELA-LITERACY.RI.5.4: Determine the meaning of general academic and domain-specific words and phrases in a text relevant to a *grade 5 topic or subject area*. • CCSS.ELA-LITERACY.RI.5.5: Compare and contrast the overall structure (e.g., chronology, comparison, cause/effect, problem/solution) of events, ideas, concepts, or information in two or more texts. • CCSS.ELA-LITERACY.RI.5.6: Analyze multiple accounts of the same event or topic, noting important similarities and differences in the point of view they represent.

Continued

Investigation 9. Energy in Ecosystems: How Do We Best Model the Transfer of Energy Into and Within the Living Things That Are Found in the Arctic Ocean?

Table 9.2 (*continued*)

***CCSS ELA*—Reading: Informational Text** (*continued*)	Integration of knowledge and ideas • CCSS.ELA-LITERACY.RI.5.7: Draw on information from multiple print or digital sources, demonstrating the ability to locate an answer to a question quickly or to solve a problem efficiently. • CCSS.ELA-LITERACY.RI.5.8: Explain how an author uses reasons and evidence to support particular points in a text, identifying which reasons and evidence support which point(s). Range of reading and level of text complexity • CCSS.ELA-LITERACY.RI.5.10: By the end of the year, read and comprehend informational texts, including history/social studies, science, and technical texts, at the high end of the grades 4–5 text complexity band independently and proficiently.

***CCSS ELA*—Writing**	Text types and purposes
	• CCSS.ELA-LITERACY.W.5.1: Write opinion pieces on topics or texts, supporting a point of view with reasons.
	  ○ CCSS.ELA-LITERACY.W.5.1.A: Introduce a topic or text clearly, state an opinion, and create an organizational structure in which ideas are logically grouped to support the writer's purpose.
	  ○ CCSS.ELA-LITERACY.W.5.1.B: Provide logically ordered reasons that are supported by facts and details.
	  ○ CCSS.ELA-LITERACY.W.5.1.C: Link opinion and reasons using words, phrases, and clauses (e.g., *consequently*, *specifically*).
	  ○ CCSS.ELA-LITERACY.W.5.1.D: Provide a concluding statement or section related to the opinion presented.
	• CCSS.ELA-LITERACY.W.5.2: Write informative or explanatory texts to examine a topic and convey ideas and information clearly.
	  ○ CCSS.ELA-LITERACY.W.5.2.A: Introduce a topic clearly, provide a general observation and focus, and group related information logically; include formatting (e.g., headings), illustrations, and multimedia when useful to aiding comprehension.
	  ○ CCSS.ELA-LITERACY.W.5.2.B: Develop the topic with facts, definitions, concrete details, quotations, or other information and examples related to the topic.
	  ○ CCSS.ELA-LITERACY.W.5.2.C: Link ideas within and across categories of information using words, phrases, and clauses (e.g., *in contrast*, *especially*).
	  ○ CCSS.ELA-LITERACY.W.5.2.D: Use precise language and domain-specific vocabulary to inform about or explain the topic.
	  ○ CCSS.ELA-LITERACY.W.5.2.E: Provide a concluding statement or section related to the information or explanation presented.
	Production and distribution of writing
	• CCSS.ELA-LITERACY.W.5.4: Produce clear and coherent writing in which the development and organization are appropriate to task, purpose, and audience.

Continued

Teacher Notes

Table 9.2 (*continued*)

CCSS ELA—Writing (*continued*)	• CCSS.ELA-LITERACY.W.5.5: With guidance and support from peers and adults, develop and strengthen writing as needed by planning, revising, editing, rewriting, or trying a new approach. • CCSS.ELA-LITERACY.W.5.6: With some guidance and support from adults, use technology, including the internet, to produce and publish writing as well as to interact and collaborate with others; demonstrate sufficient command of keyboarding skills to type a minimum of two pages in a single sitting. Research to build and present knowledge • CCSS.ELA-LITERACY.W.5.8: Recall relevant information from experiences or gather relevant information from print and digital sources; summarize or paraphrase information in notes and finished work, and provide a list of sources. • CCSS.ELA-LITERACY.W.5.9: Draw evidence from literary or informational texts to support analysis, reflection, and research. Range of writing • CCSS.ELA-LITERACY.W.5.10: Write routinely over extended time frames (time for research, reflection, and revision) and shorter time frames (a single sitting or a day or two) for a range of discipline-specific tasks, purposes, and audiences.

CCSS ELA—Speaking and Listening	Comprehension and collaboration
	• CCSS.ELA-LITERACY.SL.5.1: Engage effectively in a range of collaborative discussions (one-on-one, in groups, and teacher-led) with diverse partners on *grade 5 topics and texts,* building on others' ideas and expressing their own clearly.
	   o CCSS.ELA-LITERACY.SL.5.1.A: Come to discussions prepared, having read or studied required material; explicitly draw on that preparation and other information known about the topic to explore ideas under discussion.
	   o CCSS.ELA-LITERACY.SL.5.1.B: Follow agreed-upon rules for discussions and carry out assigned roles.
	   o CCSS.ELA-LITERACY.SL.5.1.C: Pose and respond to specific questions by making comments that contribute to the discussion and elaborate on the remarks of others.
	   o CCSS.ELA-LITERACY.SL.5.1.D: Review the key ideas expressed and draw conclusions in light of information and knowledge gained from the discussions.
	• CCSS.ELA-LITERACY.SL.5.2: Summarize a written text read aloud or information presented in diverse media and formats, including visually, quantitatively, and orally.
	• CCSS.ELA-LITERACY.SL.5.3: Summarize the points a speaker makes and explain how each claim is supported by reasons and evidence.

Continued

Investigation 9. Energy in Ecosystems: How Do We Best Model the Transfer of Energy Into and Within the Living Things That Are Found in the Arctic Ocean?

Table 9.2 (*continued*)

CCSS ELA— Speaking and Listening (*continued*)	Presentation of knowledge and ideas • CCSS.ELA-LITERACY.SL.5.4: Report on a topic or text or present an opinion, sequencing ideas logically and using appropriate facts and relevant, descriptive details to support main ideas or themes; speak clearly at an understandable pace. • CCSS.ELA-LITERACY.SL.5.5: Include multimedia components (e.g., graphics, sound) and visual displays in presentations when appropriate to enhance the development of main ideas or themes. • CCSS.ELA-LITERACY.SL.5.6: Adapt speech to a variety of contexts and tasks, using formal English when appropriate to task and situation.
ELP Standards	Receptive modalities • ELP 1: Construct meaning from oral presentations and literary and informational text through grade-appropriate listening, reading, and viewing. • ELP 8: Determine the meaning of words and phrases in oral presentations and literary and informational text. Productive modalities • ELP 3: Speak and write about grade-appropriate complex literary and informational texts and topics. • ELP 4: Construct grade-appropriate oral and written claims and support them with reasoning and evidence. • ELP 7: Adapt language choices to purpose, task, and audience when speaking and writing. Interactive modalities • ELP 2: Participate in grade-appropriate oral and written exchanges of information, ideas, and analyses, responding to peer, audience, or reader comments and questions. • ELP 5: Conduct research and evaluate and communicate findings to answer questions or solve problems. • ELP 6: Analyze and critique the arguments of others orally and in writing. Linguistic structures of English • ELP 9: Create clear and coherent grade-appropriate speech and text. • ELP 10: Make accurate use of standard English to communicate in grade-appropriate speech and writing.

Investigation Handout

Investigation 9

Energy in Ecosystems: How Do We Best Model the Transfer of Energy Into and Within the Living Things That Are Found in the Arctic Ocean?

Introduction

Orcas are the largest of the dolphins. Most orcas are found in the Arctic and Antarctic Oceans because they prefer to be in cold water. They eat a lot of different things such as fish, seals, sea lions, squid, and seabirds. Your teacher will show you a video of orcas hunting. As you watch the video, keep track of things you notice and things you wonder about in the boxes below.

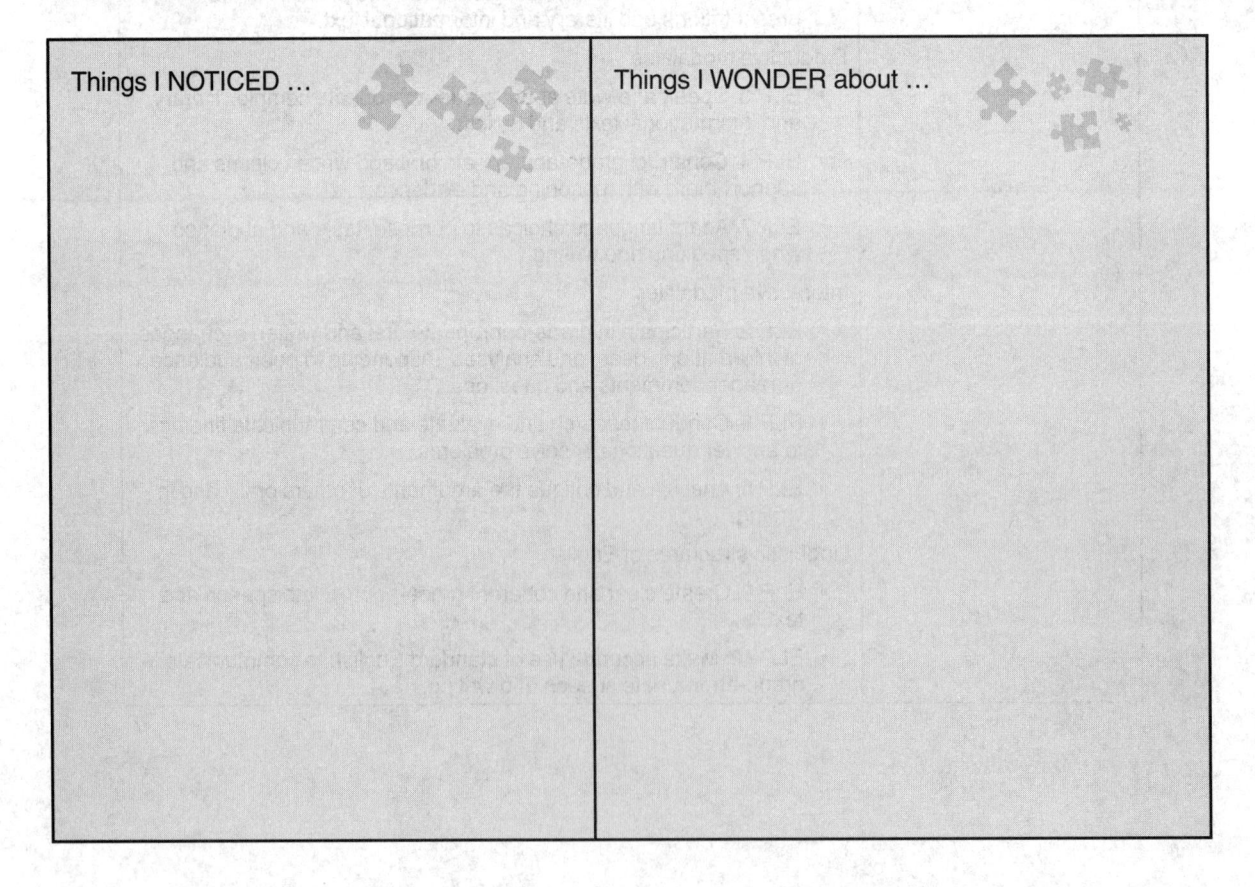

Things I NOTICED …	Things I WONDER about …

Organisms get the energy they need to live by breaking down molecules that store energy. An example of a molecule that stores energy is sugar. There are many ways that an organism can get sugar

Investigation 9. Energy in Ecosystems: How Do We Best Model the Transfer of Energy Into and Within the Living Things That Are Found in the Arctic Ocean?

into its body. One way for an organism to get sugar into its body is to eat other organisms (or parts of organisms) that have sugar in their bodies. An organism that eats other organisms is called a *consumer*. Consumers eat the bodies of animals and plants because the tissues that make up animals and plants contain many sugar molecules. Animals that only eat other animals are called *carnivores*; orcas are an example of a carnivore. Animals that only eat plants are called *herbivores*. Animals that eat both animals and plants are called *omnivores*.

When a consumer eats an animal or a plant, the sugar molecules found inside the body of animal or plant transfer into the consumer. The consumer then breaks down the sugar molecules. The energy that is released by breaking down sugar molecules is used to do things such as move, stay warm, grow, and repair damage to their bodies.

Producers are a different type of organism than consumers. Producers do not eat other organisms and are able to make sugar molecules. Plants, such as trees, grass, and carrots, are examples of organisms that are producers. Producers capture energy from the Sun and then use this energy to make sugar from air and water instead of eating things that contain sugar. Producers can then break down the sugar they make to do things like grow and repair damage to their bodies.

Your teacher will give you some cards that include some of the many different organisms that are found in the Arctic Ocean. Some of these organisms are consumers, such as the orca and the ringed seal, and some are producers, such as phytoplankton. Phytoplankton are tiny organisms that float near the surface of the ocean where they can capture energy from sunlight. Take a minute to examine the cards. Make sure that you read the information on the back of the cards and think about where each organism gets the energy it needs to live. As you examine the cards, keep track of things you notice and things you wonder about in the boxes below.

Things I NOTICED …	Things I WONDER about …

Investigation Handout

There are many ways to use a model to explain how energy in food, which is used by animals for body repair, growth, and motion and to maintain body warmth, was once energy from the Sun. Your goal in this investigation is to figure out the best way to model how energy transfers into and within the organisms that are found in the Arctic Ocean. To accomplish this goal, you will need to think of the Arctic Ocean as a system and then define it in terms of inputs (things that enter the system) and outputs (things that leave the system) or processes (things that happen within the system) and interactions (parts of the system that act on other parts of the system). You can then make a model of the system as a tool for developing a better understanding of how energy transfers into and within the living things of the Arctic Ocean. Keep in mind that no model is perfect because it only represents certain parts of the actual phenomenon that you are trying to understand.

Things we KNOW from what we read …

Investigation 9. Energy in Ecosystems:
How Do We Best Model the Transfer of Energy Into and
Within the Living Things That Are Found in the Arctic Ocean?

Your Task

Use what you know about energy in everyday life and systems to create a model that shows how the energy found in the bodies of organisms was once energy from the Sun and how this energy can transfer between different organisms that live in the Arctic Ocean.

The *guiding question* of this investigation is, **How do we best model the transfer of energy into and within the living things that are found in the Arctic Ocean?**

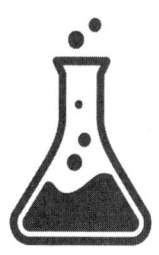

Materials

You may use any of the following materials during your investigation:

- Set of cards with information on arctic organisms
- Sticky notes
- Marker pens
- Colored pencils
- White chart paper

Safety Rules

Follow all normal safety rules.

Plan Your Investigation

Prepare a plan for your investigation by filling out the chart on the next page; this plan is called an *investigation proposal*. Before you start developing your plan, be sure to discuss the following questions with the other members of your group:

- What are the **components of this system** and how do they **interact**?
- How can we **track the transfer of energy** within a system?

Argument-Driven Inquiry in **Fifth-Grade Science**: Three-Dimensional Investigations

Investigation Handout

Our guiding question:

We will collect the following data:

These are the steps we will follow to collect data:

I approve of this investigation proposal.

_____   _____
Teacher's signature   Date

Investigation 9. Energy in Ecosystems:
How Do We Best Model the Transfer of Energy Into and
Within the Living Things That Are Found in the Arctic Ocean?

Collect Your Data

Keep a record of any important information in the space below.

Create a Model

Keep a record of your model in the space below.

Investigation Handout

Draft Argument

Develop an argument on a whiteboard. It should include the following:

1. A *claim*: Your answer to the guiding question.
2. *Evidence*: An analysis of the data and an explanation of what the analysis means.
3. A *justification of the evidence*: Why your group thinks the evidence is important.

The Guiding Question:	
Our Claim:	
Our Evidence:	Our Justification of the Evidence:

Argumentation Session

Share your argument with your classmates. Be sure to ask them how to make your draft argument better. Keep track of their suggestions in the space below.

Ways to IMPROVE our argument …

Investigation 9. Energy in Ecosystems: How Do We Best Model the Transfer of Energy Into and Within the Living Things That Are Found in the Arctic Ocean?

Draft Report

Prepare an *investigation report* to share what you have learned. Use the information in this handout and your group's final argument to write a *draft* of your investigation report.

Introduction

We have been studying _____ in class.

Before we started this investigation, we explored _____
_____
_____

We noticed _____
_____
_____

My goal for this investigation was to figure out _____
_____
_____

The guiding question was _____
_____
_____

Method

To gather the data I needed to answer this question, I _____
_____
_____
_____
_____

Argument-Driven Inquiry in **Fifth-Grade Science:** Three-Dimensional Investigations

Investigation Handout

I then analyzed the data I collected by _____

_____

_____

_____

Argument

The model below shows _____

_____

_____

_____

I know this model is accurate because _____

_____

_____

_____

_____

Investigation 9. Energy in Ecosystems:
How Do We Best Model the Transfer of Energy Into and
Within the Living Things That Are Found in the Arctic Ocean?

This evidence is based on several important scientific concepts. The first one is _____

_____
_____
_____
_____
_____
_____
_____
_____
_____
_____
_____
_____
_____
_____
_____
_____
_____

Review

Your classmates need your help! Review the draft of their investigation reports and give them ideas about how to improve. Use the *peer-review guide* when doing your review.

Submit Your Final Report

Once you have received feedback from your classmates about your draft report, create your final investigation report and hand it in to your teacher.

Checkout Questions

Investigation 9. Energy in Ecosystems

Use the following information to answer questions 1 and 2. The table below lists seven different living things that are often found in a forest ecosystem.

Living thing	What it eats
Rabbit	Grass and wildflowers
Deer	Grass and seeds from trees
Grasshopper	Grass and wildflowers
Chipmunk	Grasshoppers and seeds from trees
Robin (a type of bird)	Grasshoppers and seeds from trees
Owl (a type of bird)	Robins, chipmunks, and rabbits
Fox	Chipmunks and rabbits

1. Create a model to describe how the energy in a fox's food (used for motion, body repair, growth, and maintaining body warmth) was once energy from the Sun.

Investigation 9. Energy in Ecosystems: How Do We Best Model the Transfer of Energy Into and Within the Living Things That Are Found in the Arctic Ocean?

2. Explain your thinking. How does energy transfer into, within, or out of this system?

___

Teacher Scoring Rubric for the Checkout Questions

Level	Description
3	The student can apply the core idea correctly and can explain the transfer of energy in the system.
2	The student cannot apply the core idea correctly but can explain the transfer of energy in the system.
1	The student can apply the core idea correctly but cannot explain the transfer of energy in the system.
0	The student cannot apply the core idea correctly and cannot explain the transfer of energy in the system.

Teacher Notes

Investigation 10

Movement of Carbon in Ecosystems: How Does the Amount of Dissolved Carbon Dioxide Gas Found in Water Change Over Time When Aquatic Plants and Animals Are Present?

Purpose

The purpose of this investigation is to give students an opportunity to use one disciplinary core idea (DCI), one crosscutting concept (CC), and eight scientific and engineering practices (SEPs) to figure out how plants and animals contribute to or remove carbon dioxide (CO_2) from an aquatic environment. Students will also learn about the difference between data and evidence.

The DCI, CC, and SEPs That Students Use During This Investigation to Figure Things Out

DCI

- *LS2.B: Cycles of Matter and Energy Transfer in Ecosystems:* Matter cycles between the air and soil and among plants, animals, and microbes as these organisms live and die. Organisms obtain gases, and water, from the environment, and release waste matter (gas, liquid, or solid) back into the environment.

CC

- *CC 4: Systems and System Models:* A system can be described in terms of its components and their interactions.

SEPs

- *SEP 1: Asking Questions and Defining Problems:* Ask questions about what would happen if a variable is changed. Ask questions that can be investigated and predict reasonable outcomes based on patterns such as cause-and-effect relationships.
- *SEP 2: Developing and Using Models:* Develop and/or use models to describe and/or predict phenomena.
- *SEP 3: Planning and Carrying Out Investigations:* Plan and conduct an investigation collaboratively to produce data to serve as the basis for evidence, using fair tests in which variables are controlled and the number of trials considered. Evaluate appropriate methods and/or tools for collecting data.

Investigation 10. Movement of Carbon in Ecosystems: How Does the Amount of Dissolved Carbon Dioxide Gas Found in Water Change Over Time When Aquatic Plants and Animals Are Present?

- *SEP 4: Analyzing and Interpreting Data:* Represent data in tables and/or various graphical displays (bar graphs, pictographs, and/or pie charts) to reveal patterns that indicate relationships. Analyze and interpret data to make sense of phenomena, using logical reasoning, mathematics, and/or computation. Compare and contrast data collected by different groups in order to discuss similarities and differences in their findings.

- *SEP 5: Using Mathematics and Computational Thinking:* Organize simple data sets to reveal patterns that suggest relationships. Describe, measure, estimate, and/or graph quantities (e.g., area, volume, weight, time) to address scientific and engineering questions and problems.

- *SEP 6: Constructing Explanations and Designing Solutions:* Construct an explanation of observed relationships. Use evidence to construct or support an explanation. Identify the evidence that supports particular points in an explanation.

- *SEP 7: Engaging in Argument From Evidence:* Compare and refine arguments based on an evaluation of the evidence presented. Distinguish among facts, reasoned judgment based on research findings, and speculation in an explanation. Respectfully provide and receive critiques from peers about a proposed procedure, explanation, or model by citing relevant evidence and posing specific questions.

- *SEP 8: Obtaining, Evaluating, and Communicating Information:* Read and comprehend grade-appropriate complex texts and/or other reliable media to summarize and obtain scientific and technical ideas. Combine information in written text with that contained in corresponding tables, diagrams, and/or charts to support the engagement in other scientific and/or engineering practices. Communicate scientific and/or technical information orally and/or in written formats, including various forms of media as well as tables, diagrams, and charts.

Other Concepts That Students May Use During This Investigation

Students might also use some of the following concepts:

- An ecosystem is a community of interacting organisms and their physical environment.
- A closed system is an area that is separated from its surroundings by a barrier that does not let any matter transfer into or out of that system.
- Chemical indicators are used by scientists to determine if a substance is present or not.

Teacher Notes

What Students Figure Out

Students determine that (1) plants that live in the water move CO_2 from the water into their bodies when they are in the light and (2) animals that live in the water, like snails, move CO_2 out of the bodies as waste.

Background Information About This Investigation for the Teacher

An ecosystem is a community of interacting organisms and their physical environment. In an ecosystem, matter moves between the air, water, and soil and among plants, animals, and microbes as they live and die. Organisms obtain the matter they need from the environment and release waste matter back into the environment. The matter that moves into and out of the bodies of organisms can be a solid, liquid, or a gas. Animals, for example, take in solid parts of plants or other animals when they eat. This solid food is then broken down inside the body of an animal and used to supply the matter needed to grow, to repair damaged parts, and to reproduce. Animals also release solid waste matter, which is called feces, into the environment. Animals also take in water from the environment and release liquid waste back into the environment where they live.

Animals use oxygen gas from the air or dissolved in water to convert sugar into the energy that these animals need to move and carry out other functions. The series of chemical reactions that convert sugar into a usable form of energy is called cellular respiration. Cellular respiration also produces CO_2. Animals release CO_2 into the air or water as a waste product.

CO_2, from the air or dissolved in water, is taken in by plants and used to produce sugar that may then be turned into starch or other molecules such as cellulose. The process of turning CO_2 and water into sugar through a series of chemical reactions is called photosynthesis. Photosynthesis also produces oxygen. Plants use oxygen to break down the sugar they produce to create the energy they need to grow and carry out other functions, just like animals do. Plants therefore also produce CO_2. However, most plants take in more CO_2 from the environment than they give off.

One way to track how CO_2 moves into or out of a living thing that lives in the water is to place that living thing and some water inside a closed system. A closed system is an area that is separated from its surroundings by a barrier that does not let any matter transfer into or out of that system. An example of a closed system is a container with a sealed lid. You can then use a nontoxic chemical indicator, such as bromthymol blue (BTB), to track how the amount of CO_2 in the water found within that closed system changes over time.

When CO_2 mixes with water, it reacts with the water to form carbonic acid. The carbonic acid in the water decreases the pH of the sample (i.e., makes it more acidic). Therefore, the more CO_2 that is added to a sample of water, the more acidic that sample of water will become.

Investigation 10. Movement of Carbon in Ecosystems: How Does the Amount of Dissolved Carbon Dioxide Gas Found in Water Change Over Time When Aquatic Plants and Animals Are Present?

BTB changes color as the pH of the water increases or decreases:

- BTB is blue when the pH of the water is above 7.5 (i.e., basic). This color indicates that there is no CO_2 in the water.
- BTB is greenish-blue when the pH of the water is between 7 and 7.5 (i.e., slightly basic). This color also indicates there is no CO_2 in the water.
- BTB is green when the pH of the water is 7 (i.e., neutral). This color indicates there is very little or no CO_2 in the water.
- BTB is yellowish-green when the pH of the water is between 6.5 and 7 (i.e., slightly acidic). This color indicates there is some CO_2 in the water.

BTB is therefore a good way to track the movement of CO_2 into or out of an organism that lives in the water because BTB will turn the water different colors depending on how much CO_2 there is in the water.

The change in the color of the water with BTB in it will allow you to see if the CO_2 is moving into or out of a living thing. If the amount of CO_2 in the water goes up, then the living thing moved CO_2 out of its body and into the water. You will know that the amount of CO_2 went up in the water if the water with BTB in it changes from blue to green or yellow or if the water with BTB in it changes from green to yellow. If the amount of CO_2 in the water goes down, then the living thing moved CO_2 from the water and into its body. You will know that the amount of CO_2 in the water went down if the water with BTB in it changes from yellow to green or blue or it changes from green to blue. Finally, if the amount of CO_2 in the water does not change, then the living thing did not move CO_2 into or out of its body.

One way for students to set up the available materials for this investigation is shown in Figure 10.1 (p. 398). This approach includes a control condition (tube 1) and allows students to see if (a) a snail takes in or gives off CO_2 (tube 2) and (b) if an aquatic plant takes in or gives off CO_2 (tube 3). When the tubes are exposed to light the entire time, the water in tube 1 will stay blue or green, the water in tube 2 will turn from blue or green to yellow, and the water in tube 3 will stay blue or turn from green to greenish-blue or blue. If the tubes are not exposed to light the entire time, tube 3 will give different results. In the morning (assuming that the plant was kept in the dark all night), the water in tube 3 will likely be green, greenish-yellow, or yellow because the plant was converting stored sugar to energy and releasing CO_2 in the process during the night. Then in the afternoon, the water in tube 3 will likely be greenish-blue or blue because the plant was using light to convert CO_2 into sugar through the process of photosynthesis. This setup also allows students to see how CO_2 cycles between plants and animals (tube 4). The water in tube 4 will likely be green or greenish-blue because the plant in the tube uses some of the CO_2 given off by the snail, so there will be less CO_2 in tube 4 than there will be in the tube with just a snail (tube 2).

Teacher Notes

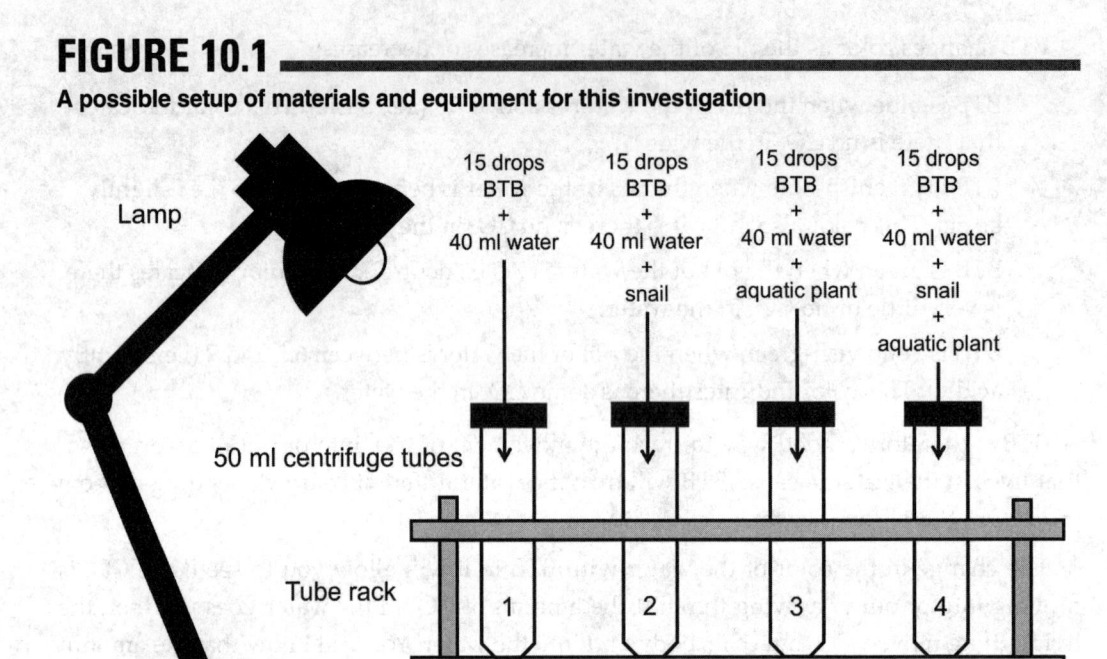

FIGURE 10.1
A possible setup of materials and equipment for this investigation

Lamp

15 drops BTB + 40 ml water

15 drops BTB + 40 ml water + snail

15 drops BTB + 40 ml water + aquatic plant

15 drops BTB + 40 ml water + snail + aquatic plant

50 ml centrifuge tubes

Tube rack 1 2 3 4

Timeline

The time needed to complete this investigation is 290 minutes (4 hours and 50 minutes). The amount of instructional time needed for each stage of the investigation is as follows:

- *Stage 1.* Introduce the task and the guiding question: 35 minutes
- *Stage 2.* Design a method and collect data: 70 minutes
- *Stage 3.* Create a draft argument: 45 minutes
- *Stage 4.* Argumentation session: 30 minutes
- *Stage 5.* Reflective discussion: 15 minutes
- *Stage 6.* Write a draft report: 30 minutes
- *Stage 7.* Peer review: 35 minutes
- *Stage 8.* Revise the report: 30 minutes

Materials and Preparation

The materials needed for this investigation are listed in Table 10.1. Some of these items can be purchased through retailers such as Walmart or Target. Snails and the aquatic plants can be purchased from a local pet supply store or from a science education supply company such as Carolina or Ward's Science. The solution of 0.04% aqueous BTB can be purchased

Investigation 10. Movement of Carbon in Ecosystems: How Does the Amount of Dissolved Carbon Dioxide Gas Found in Water Change Over Time When Aquatic Plants and Animals Are Present?

from Carolina or Ward's Science (or another science education supply company). The materials, other than snails and aquatic plants, for this investigation can also be purchased as a complete kit (which includes enough materials for 24 students, or six groups of four students) at *www.argumentdriveninquiry.com*.

TABLE 10.1
Materials for Investigation 10

Item	Quantity
Safety goggles	1 per student
Nonlatex apron	1 per student
Vinyl gloves	1 pair per student
Small (1 gallon or less) plastic aquarium	1 per class
Clear plastic cup	1 per class
Roll of paper towels	1 per class
Roll of blue painter's tape	1 per class
Straw	1 per class
Spring water	2 gallons per class
Clamp lamp with reflector	1 per group
Polypropylene centrifuge tubes with screw lids, 50 ml each	4 per group
Tube rack for 50-ml centrifuge tubes	1 per group
Aquatic plants such as *Egeria najas*, *Elodea densa*, or *Anubias nana*	2 per group
Pond snails	2 per group
0.04% Aqueous bromthymol blue (BTB) in 30-ml plastic dropper bottle	3 bottles per class
Permanent marker, fine point	1 per group
Whiteboard, 2' × 3'*	1 per group
Investigation Handout	1 per student
Peer-review guide and teacher scoring rubric	1 per student
Checkout Questions (optional)	1 per student

*As an alternative, students can use computer and presentation software such as Microsoft PowerPoint or Apple Keynote to create their arguments.

Four to five weeks (or even more) before the investigation start date, you will need to find a source for purchasing the snails and plants. Aquatic plants such as *Egeria najas* or *Elodea densa* work best for this investigation because they are hardy and tend to start the process of photosynthesis quickly when exposed to light. Unfortunately, they are only

Teacher Notes

available for purchase from a science education supply company. *Anubias nana* is available at most local pet stores but does not work as well as the *Egeria najas* or *Elodea densa*.

Once the snails and plants arrive at your school, transfer them to a small plastic aquarium to keep them alive. The snails and plants can be kept alive for several weeks if you follow the supplier's instructions. You will use the aquarium filled with the snails and plants during the opening activity in stage 1 of this investigation.

We recommend using spring water for the investigation. You may use tap water instead of spring water unless you live in an area with water that is very acidic or basic. To test the pH of your tap water, add a few drops of the BTB solution to it. If the water turns bright yellow or bright blue, then you will need to use purchased spring water instead of tap water. Two gallons of spring water should be enough for the introduction of the investigations and for six groups of students to carry out their investigation.

One week before the students set up their investigation, decide where they will keep their centrifuge tubes for the days it will take for the aquatic plants and snails to take in or give off enough CO_2 to change the color of the water with BTB blue in it. An area with indirect sunlight that is not too hot or too cold will work best. You can also set up a lamp to provide light. We recommend having the students keep their tubes in a tube rack. The tube rack can then be stored on an easily accessible counter within the classroom.

Safety Precautions

Remind students to follow all normal safety rules. In addition, tell the students to take the following safety precautions:

- Wear sanitized safety goggles, nonlatex aprons, and vinyl gloves during setup, investigation activity, and cleanup.
- Be careful when handling the water with BTB. Although BTB is considered nontoxic, ingesting large doses may cause stomach upset, and splashes on the skin or in the eyes may cause slight irritation. Do not drink water with BTB in it.
- Immediately clean up any spills to avoid a slip or fall hazard.
- Keep water sources away from electrical receptacles to prevent shock.
- Be careful when handling the organisms, because they are delicate.
- Make sure all materials are put away after completing the activity.
- Wash their hands with soap and water when they are done cleaning up.

Water with BTB may be poured down a drain and then flushed with excess water to dispose of it. Do not discard live plants or snails in a pond, stream, river, or lake, because the introduction of these organisms can damage a local ecosystem.

Investigation 10. Movement of Carbon in Ecosystems: How Does the Amount of Dissolved Carbon Dioxide Gas Found in Water Change Over Time When Aquatic Plants and Animals Are Present?

Lesson Plan by Stage

This lesson plan is only a suggestion. It is included here to illustrate what you can say and do during each stage of ADI for this specific investigation. We encourage you to modify this lesson plan by asking different questions, using different examples, and providing different scaffolds as needed to better meet the needs of students in your class.

Stage 1: Introduce the Task and the Guiding Question (35 minutes)

1. Ask the students to sit in six groups, with three or four students in each group.
2. Ask the students to clear off their desks except for a pencil (and their *Student Workbook for Argument-Driven Inquiry in Fifth-Grade Science* if they have one).
3. Pass out an Investigation Handout to each student (or ask students to turn to the Investigation Log for Investigation 10 in their workbook).
4. Read the first paragraph of the "Introduction" aloud to the class. Ask the students to follow along as you read.
5. Place a small plastic aquarium that contains water and all the snails and aquatic plants on a table. Ask the students to record what they notice about the snails and plants and the questions they have in the first "NOTICED/WONDER" chart in the "Introduction."
6. Ask the students to share what they observed.
7. Ask the students to share what questions they have.
8. Tell the students, "Some of your questions might be answered by reading the next two paragraphs of the 'Introduction.' Please follow along with me as I read it."
9. Read the next two paragraphs of the "Introduction" aloud to the class.
10. Hold up a clear plastic cup with a small amount of water in it (no more than one-quarter full), then add about 15 drops of the BTB solution to the water.
11. Tell the students, "Okay, watch what happens when I blow air into this water. Please record anything you notice or wonder about as I blow air into the water with this straw."
12. Blow air gently into the cup until the water turns from blue to yellow (about 30–45 seconds). Students should then record their observations and questions in the second "NOTICED/WONDER" table in the "Introduction."
13. Ask the students to share what they observed.
14. Ask the students to share what questions they have.
15. Ask the students to read the rest of the "Introduction" on their own *or* ask them to follow along as you read it aloud.

Teacher Notes

16. Once the students have read the rest of the "Introduction," ask them to fill out the "Things we KNOW" chart on their Investigation Handout (or in their Investigation Log) as a group.

17. Ask the students to share what they learned from the reading. Add these ideas to a class "Things we KNOW" chart.

18. Tell the students, "Let's see what we will need to figure out during our investigation."

19. Read the task and the guiding question aloud.

20. Tell the students, "I have lots of materials here that you can use."

21. Introduce the students to the materials available for them to use during the investigation by holding each one up and then asking what it might be used for. Remind students to handle the snails and plants with care, and demonstrate how to remove these organisms from the aquarium, hold them, and gently place them in a centrifuge tube.

Stage 2: Design a Method and Collect Data (70 minutes)

1. Tell the students, "I am now going to give you and the other members of your group about 15 minutes to plan your investigation. Before you begin, I want you to take a couple of minutes to discuss the following questions with the rest of your group."

2. Show the following questions on the screen or board:
   - What are the *components of this system* and how do they *interact*?
   - How can we *track the transfer of carbon* within a system?

3. Tell the students, "Please take a few minutes to come up with an answer to these questions."

4. Give the students two or three minutes to discuss these two questions.

5. Ask two or three different groups to share their answers. Highlight or write down any important ideas on the board so students can refer to them later.

6. If possible, use a document camera to project an image of the graphic organizer for this investigation on a screen or board (or take a picture of it and project the picture on a screen or board). Tell the students, "I now want you all to plan out your investigation. To do that, you will need to fill out this investigation proposal."

7. Point to the box labeled "Our guiding question:" and tell the students, "You can put the question we are trying to answer in this box." Then ask, "Where can we find the guiding question?"

8. Wait for a student to answer.

Investigation 10. Movement of Carbon in Ecosystems: How Does the Amount of Dissolved Carbon Dioxide Gas Found in Water Change Over Time When Aquatic Plants and Animals Are Present?

9. Point to the box labeled "This is a picture of how we will set up the equipment:" and tell the students, "You can draw a picture in this box of how you will set up the equipment in order to carry out this investigation."

10. Point to the box labeled "We will collect the following data:" and tell the students, "You can list the observations that you will need to collect during the investigation in this box."

11. Point to the box labeled "These are the steps we will follow to collect data:" and tell the students, "You can list what you are going to do to collect the data you need and what you will do with your data once you have it. Be sure to give enough detail that I could do your investigation for you."

12. Ask the students, "Do you have any questions about what you need to do?"

13. Answer any questions that come up.

14. Tell the students, "Once you are done, raise your hand and let me know. I'll then come by and look over your proposal and give you some feedback. You may not begin collecting data until I have approved your proposal by signing it. You need to have your proposal done in the next 15 minutes."

15. Give the students 15 minutes to work in their groups on their investigation proposal. As they work, move from group to group to check in, ask probing questions, and offer a suggestion if a group gets stuck.

16. As each group finishes its investigation proposal, read it over and determine if it will be productive or not. If you feel the investigation will be productive (not necessarily what you would do or what the other groups are doing), sign your name on the proposal and let the group start collecting data. If the plan needs to be changed, offer some suggestions or ask some probing questions, and have the group make the changes before you approve it.

17. Pass out the materials or have one student from each group collect the materials they need from a central supply table or cart for the groups that have an approved proposal.

18. Remind students of the safety rules and precautions for this investigation.

Teacher Notes

What should a student-designed investigation look like?

There are a number of different investigations that students can design to answer the question "How does the amount of dissolved carbon dioxide gas found in water change over time when aquatic plants and animals are present?" For example, one method might include the following steps (see Figure 10.1 on p. 398 for an illustration of the materials setup for this method):

1. Set up tube 1 with water and 15 drops of BTB.
2. Record the color of the water.
3. Set up tube 2 with water, 15 drops of BTB, and a snail.
4. Record the color of the water.
5. Set up tube 3 with water, 15 drops of BTB, and an aquatic plant.
6. Record the color of the water.
7. Set up tube 4 with water, 15 drops of BTB, a snail, and an aquatic plant.
8. Record the color of the water.
9. Wait 24 hours and record the color of the water for all four tubes.
10. Wait another 24 hours and record the color of the water for all four tubes.

If students use this method, they will need to collect data on the contents of each tube and the color of the water. Ideally students will keep the tubes closed throughout the investigation. If some groups choose not to keep the tubes closed, allow them to keep the tubes open because this will spark a good conversation about open versus closed systems during later stages. If any groups choose not to include a control (tube 1), that is okay but should be discussed during stage 5.

19. Give the students 25 minutes to set up the tubes and record their initial observations in the "Collect Your Data" box in their Investigation Handout (or the Investigation Log in their workbook).

20. Once the students have recorded their initial observations, help them use the blue painter's tape and permanent marker to label their tubes. Then place their tubes in the area set aside for housing the investigations.

21. The next day, give students 10 minutes to make their observations.

22. Give the students another 10 minutes on the following day to make their final observations. After they are done making their final observations, give them another 5 minutes to clean up and put the equipment away. The water with BTB may be poured down a drain and then flushed with excess water to dispose of it. The students should place any live plants or snails back in the aquarium.

Investigation 10. Movement of Carbon in Ecosystems: How Does the Amount of Dissolved Carbon Dioxide Gas Found in Water Change Over Time When Aquatic Plants and Animals Are Present?

Stage 3: Create a Draft Argument (45 minutes)

1. Tell the students, "Now that we have all this data, we need to analyze the data so we can figure out an answer to the guiding question."

2. If possible, project an image of the "Analyze Your Data" section for this investigation on a screen or board using a document camera (or take a picture of it and project the picture on a screen or board). Point to the section and tell the students, "You can create a table, graph, or picture as a way to analyze your data. You can make your table, graph, or picture in this section."

3. Ask the students, "What information do we need to include in a table, graph, or picture?"

4. Tell the students, "Please take a few minutes to discuss this question with your group and be ready to share."

5. Give the students five minutes to discuss.

6. Ask two or three different groups to share their answers. Highlight or write down any important ideas on the board so students can refer to them later.

7. Tell the groups of students, "I am now going to give you and the other members of your group about 10 minutes to create your table, graph, or picture." The table, graph, or picture they create should include information about the contents of each tube and their observations at different points in time. If the students are having trouble making a table or graph, you can take a few minutes to provide a mini-lesson about how to create a table or graph (this strategy is called just-in-time instruction because it is offered only when students get stuck).

8. Give the students 10 minutes to analyze their data. As they work, move from group to group to check in, ask probing questions, and offer suggestions.

9. Tell the students, "I am now going to give you and the other members of your group about 15 minutes to create an argument to share what you have learned and convince others that they should believe you. Before you do that, we need to take a few minutes to discuss what you need to include in your argument."

10. If possible, use a document camera to project the "Argument Presentation on a Whiteboard" image from the "Draft the Argumentation" section of the Investigation Handout (or the Investigation Log in their workbook) on a screen or board (or take a picture of it and project the picture on a screen or board).

11. Point to the box labeled "The Guiding Question:" and tell the students, "You can put the question we are trying to answer here on your whiteboard."

12. Point to the box labeled "Our Claim:" and tell the students, "You can put your claim here on your whiteboard. The claim is your answer to the guiding question."

13. Point to the box labeled "Our Evidence:" and tell the students, "You can put the evidence that you are using to support your claim here on your whiteboard. Your evidence will need to include the analysis you just did and an explanation

Teacher Notes

of what your analysis means or shows. Scientists always need to support their claims with evidence."

14. Point to the box labeled "Our Justification of the Evidence:" and tell the students, "You can put your justification of your evidence here on your whiteboard. Your justification needs to explain why your evidence is important. Scientists often use core ideas to explain why the evidence they are using matters. Core ideas are important concepts that scientists use to help them make sense of what happens during an investigation."

15. Ask the students, "What are some core ideas that we read about earlier that might help us explain why the evidence we are using is important?"

16. Ask the students to share some of the core ideas from the "Introduction" section of the Investigation Handout (or the Investigation Log in the workbook). List these core ideas on the board.

17. Tell the students, "That is great. I would like to see everyone try to include these core ideas in your justification of the evidence. Your goal is to use these core ideas to help explain why your evidence matters and why the rest of us should pay attention to it."

18. Ask the students, "Do you have any questions about what you need to do?"

19. Answer any questions that come up.

20. Tell the students, "Okay, go ahead and start working on your arguments. You need to have your argument done in the next 15 minutes. It doesn't need to be perfect. We just need something down on the whiteboards so we can share our ideas."

21. Give the students 15 minutes to work in their groups on their arguments. As they work, move from group to group to check in, ask probing questions, and offer a suggestion if a group gets stuck. Figure 10.2 shows an example of an argument created by students for this investigation.

What should the table, graph, or picture for this investigation look like?

There are a number of different ways that students can analyze the observations they collect during this investigation, but most students will create a table with four columns. The heading for the first column should be the contents of the test tube and the headings for the next three columns should be the color of the water by day. An example of this type of table is shown in Figure 10.2. There are other options for analyzing the collected data. Students often come up with some unique ways of analyzing their data, so be sure to give them some voice and choice during this stage.

Investigation 10. Movement of Carbon in Ecosystems:
How Does the Amount of Dissolved Carbon Dioxide Gas Found in Water
Change Over Time When Aquatic Plants and Animals Are Present?

FIGURE 10.2

Example of an argument

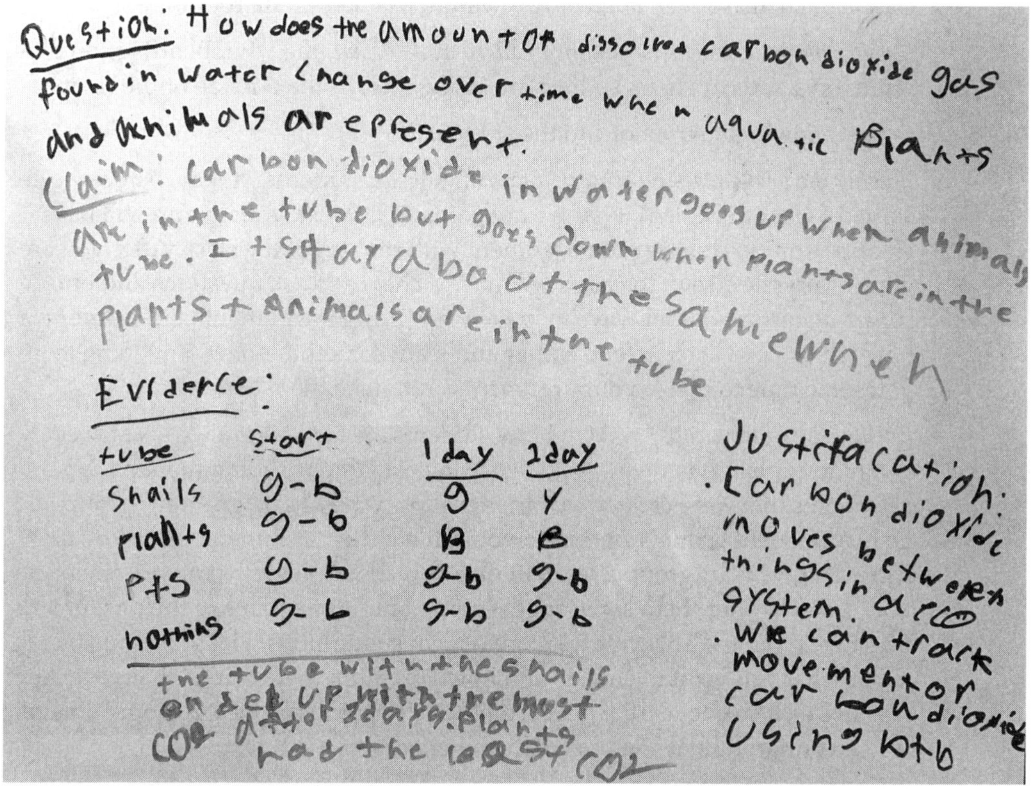

> **Question:** How does the amount of dissolved carbon dioxide gas found in water change over time when aquatic plants and animals are present.
>
> **Claim:** Carbon dioxide in water goes up when animals are in the tube but goes down when plants are in the tube. It stay about the same when plants + animals are in the tube
>
> **Evidence:**
>
> | tube | start | 1 day | 2 day |
> |------|-------|-------|-------|
> | snails | 9-b | 9 | Y |
> | plants | 9-b | B | B |
> | P+S | 9-b | 9-b | 9-b |
> | nothing | 9-b | 9-b | 9-b |
>
> the tube with the snails had the most CO2 after 2 days. Plants had the least CO2
>
> **Justification:**
> • Carbon dioxide moves between things in a system.
> • We can track movement of carbon dioxide using BTB

Stage 4: Argumentation Session (30 minutes)

The argumentation session can be conducted in a whole-class presentation format, a gallery walk format, or a modified gallery walk format. We recommend using a whole-class presentation format for the first investigation, but try to transition to either the gallery walk or modified gallery walk format as soon as possible because that will maximize student voice and choice inside the classroom. The following list shows the steps for the three formats; unless otherwise noted, the steps are the same for all three formats.

1. Begin by introducing the use of the whiteboard.
   - *If using the whole-class presentation format,* tell the students, "We are now going to share our arguments. Please set up your whiteboards so everyone can see them."
   - *If using the gallery walk or modified gallery walk format,* tell the students, "We are now going to share our arguments. Please set up your whiteboards so they are facing the walls."

Teacher Notes

2. Allow the students to set up their whiteboards.

   - *If using the whole-class presentation format*, the whiteboards should be set up on stands or chairs so they are facing toward the center of the room.

   - *If using the gallery walk or modified gallery walk format*, the whiteboards should be set up on stands or chairs so they are facing toward the outside of the room.

3. Give the following instructions to the students:

   - *If using the whole-class presentation format,* tell the students, "Okay, before we get started I want to explain what we are going to do next. Your group will have an opportunity to share your argument with the rest of the class. After you are done, everyone else in the class will have a chance to ask questions and offer some suggestions about ways to make your group's argument better. After we have a chance to listen to each other and learn something new, I'm going to give you some time to revise your arguments and make them better."

   - *If using the gallery walk format,* tell the students, "Okay, before we get started I want to explain what we are going to do next. You are going to read the arguments that were created by other groups. When I say 'go,' your group will go to a different group's station so you can see their argument. Once you are there, I'll give your group a few minutes to read and review their argument. Your job is to offer them some suggestions about ways to make their argument better. You can use sticky notes to give them suggestions. Please be specific about what you want to change and how you think they should change it. After we have a chance to learn from each other, I'm going to give you some time to revise your arguments and make them better."

   - *If using the modified gallery walk format,* tell the students, "Okay, before we get started I want to explain what we are going to do next. I'm going to ask some of you to present your arguments to your classmates. If you are presenting your argument, your job is to share your group's claim, evidence, and justification of the evidence. The rest of you will be travelers. If you are a traveler, your job is to listen to the presenters, ask the presenters questions if you do not understand something, and then offer them some suggestions about ways to make their argument better. After we have a chance to learn from each other, I'm going to give you some time to revise your arguments and make them better."

4. Use a document camera to project the "Ways to IMPROVE our argument …" box from the Investigation Handout (or the Investigation Log in their workbook) on a screen or board (or take a picture of it and project the picture on a screen or board).

   - *If using the whole-class presentation format,* point to the box and tell the students, "After your group presents your argument, you can write down the suggestions you get from your classmates here. If you are listening to a presentation and you see a good idea from another group, you can write down that idea here as

Investigation 10. Movement of Carbon in Ecosystems: How Does the Amount of Dissolved Carbon Dioxide Gas Found in Water Change Over Time When Aquatic Plants and Animals Are Present?

- well. Once we are done with the presentations, I will give you a chance to use these suggestions or ideas to improve your arguments."
- *If using the gallery walk format,* point to the box and tell the students, "If you see a good idea from another group, you can write it down here. Once we are done reviewing the different arguments, I will give you a chance to use these ideas to improve your own arguments. It is important to share ideas like this."
- *If using the modified gallery walk format,* point to the box and tell the students, "If you are a presenter, you can write down the suggestions you get from the travelers here. If you are a traveler and you see a good idea from another group, you can write down that idea here. Once we are done with the presentations, I will give you a chance to use these suggestions or ideas to improve your arguments."

5. Ask the students, "Do you have any questions about what you need to do?"
6. Answer any questions that come up.
7. Give the following instructions:
   - *If using the whole-class presentation format,* tell the students, "Okay. Let's get started."
   - *If using the gallery walk format,* tell the students, "Okay, I'm now going to tell you which argument to go to and review."
   - *If using the modified gallery walk format,* tell the students, "Okay, I'm now going to assign you to be a presenter or a traveler." Assign one or two students from each group to be presenters and one or two students from each group to be travelers.
8. Give the students an opportunity to review the arguments.
   - *If using the whole-class presentation format,* have each group present their argument one at a time. Give each group only two to three minutes to present their argument. Then give the class two to three minutes to ask them questions and offer suggestions. Encourage as much participation from the students as possible.
   - *If using the gallery walk format,* tell the students, "Okay. Let's get started. Each group, move one argument to the left. Don't move to the next argument until I tell you to move. Once you get there, read the argument and then offer suggestions about how to make it better. I will put some sticky notes next to each argument. You can use the sticky notes to leave your suggestions." Give each group about three to four minutes to read the arguments, talk, and offer suggestions.
     a. After three to four minutes, tell the students, "Okay. Let's move on to the next argument. Please move one group to the left."
     b. Again, give each group three to four minutes to read, talk, and offer suggestions.

Teacher Notes

   c. Repeat this process until each group has had their argument read and critiqued three times.

- *If using the modified gallery walk format,* tell the students, "Okay. Let's get started. Reviewers, move one group to the left. Don't move to the next group until I tell you to move. Presenters, go ahead and share your argument with the travelers when they get there." Give each group of presenters and travelers about three to four minutes to talk.

   a. Tell the students, "Okay. Let's move on to the next argument. Travelers, move one group to the left."

   b. Again, give each group of presenters and travelers about three to four minutes to talk.

   c. Repeat this process until each group has had their argument read and critiqued three times.

9. Tell the students to return to their workstations.

10. Give the following instructions about revising the argument:

    - *If using the whole-class presentation format,* tell the students, "I'm now going to give you all about 10 minutes to revise your argument. Take a few minutes to talk in your groups and determine what you want to change to make your argument better. Once you have decided what to change, go ahead and make the changes to your whiteboard."

    - *If using the gallery walk format,* tell the students, "I'm now going to give you all about 10 minutes to revise your argument. Take a few minutes to read the suggestions that were left at your argument. Then talk in your groups and determine what you want to change to make your argument better. Once you have decided what to change, go ahead and make the changes to your whiteboard."

    - *If using the modified gallery walk format,* tell the students, "I'm now going to give you all about 10 minutes to revise your argument. Please return to your original groups." Wait for the students to move back into their original groups and then tell the students, "Okay, take a few minutes to talk in your groups and determine what you want to change to make your argument better. Once you have decided what to change, go ahead and make the changes to your whiteboard."

11. Ask the students, "Do you have any questions about what you need to do?"

12. Answer any questions that come up.

13. Tell the students, "Okay. Let's get started."

14. Give the students 10 minutes to work in their groups on their arguments. As they work, move from group to group to check in, ask probing questions, and offer a suggestion if a group gets stuck.

Investigation 10. Movement of Carbon in Ecosystems: How Does the Amount of Dissolved Carbon Dioxide Gas Found in Water Change Over Time When Aquatic Plants and Animals Are Present?

Stage 5: Reflective Discussion (15 minutes)

1. Tell the students, "We are now going to take a minute to talk about some of the core ideas and crosscutting concepts that we have used during our investigation."
2. Show Figure 10.3 on the screen.
3. Ask the students, "What do you all see going on here?"
4. Allow the students to share their ideas.

FIGURE 10.3
A cow moving matter into and out of its body

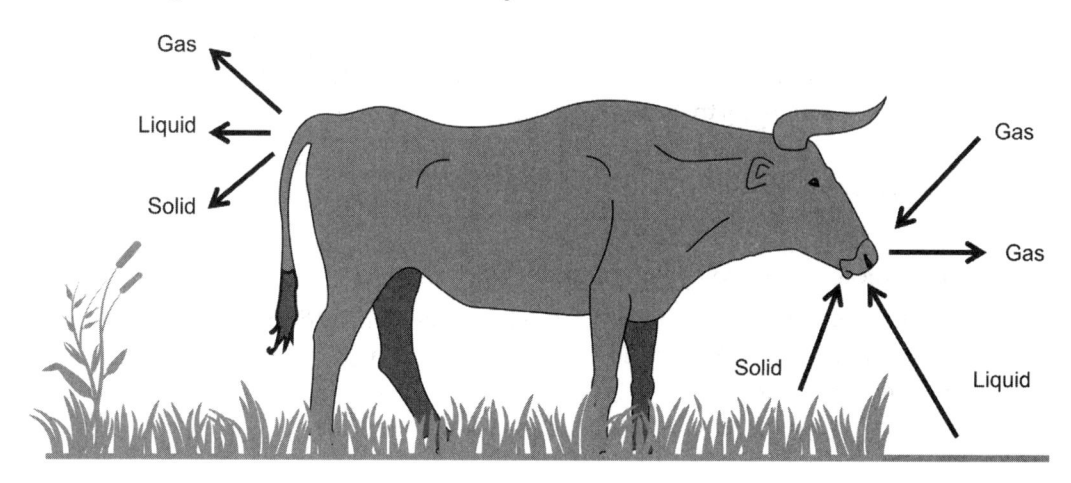

5. Ask the students, "How can we use what we know about how matter cycles between living things and the environment and the needs of living things to help explain why these different types of matter are moving into and out of this cow?"
6. Allow students to share their ideas. As they share their ideas, ask different questions to encourage them to expand on their thinking (e.g., "Can you tell me more about that?"), clarify a contribution (e.g., "Can you say that in another way?"), support an idea (e.g., "Why do you think that?"), add to an idea mentioned by a classmate (e.g., "Would anyone like to add to the idea?"), re-voice an idea offered by a classmate (e.g., "Who can explain that to me in another way?"), or critique an idea during the discussion (e.g., "Do you agree or disagree with that idea and why?") until students are able to generate an adequate explanation.
7. Ask the students, "How does this compare to what happened in your investigation?"

Teacher Notes

8. Allow students to share their ideas. As they share their ideas, ask different questions to encourage them to expand on their thinking, clarify a contribution, support an idea, add to an idea mentioned by a classmate, re-voice an idea offered by a classmate, or critique an idea until the students are able to generate an adequate explanation.

9. Ask the students, "What questions do you have about the cycling of matter between living things and the environment?"

10. Answer any questions that come up.

11. Tell the students, "We had to think about each of the tubes as a system during our investigation." Then ask, "Can anyone tell me why this was important?"

12. Allow students to share their ideas.

13. Tell the students, "I think that thinking about each tube as a system helped us describe it in terms of its components and their interactions. A system is just a group of related parts that make up a whole and can carry out functions that its individual parts cannot."

14. Ask the students, "What were some of the components and interactions of the tube systems that you examined today?"

15. Allow the students to share their ideas.

16. Tell the students, "We also had to track the movement of matter during our investigation." Then ask, "Can anyone tell me why we needed to track the movement of matter?"

17. Allow the students to share their ideas.

18. Tell the students, "I think tracking the movement of matter is important because it allows us to figure out how matter movers into, out of, and within living things."

19. Ask the students, "What are some challenges of tracking the movement of matter in an ecosystem, and how did we overcome these challenges?"

20. Allow the students to share their ideas.

21. Show an image of the question "What do you think are the most important core ideas or crosscutting concepts that we used during this investigation to help us make sense of what we observed?" Tell the students, "Okay, let's make sure we are all on the same page. Please take a moment to discuss this question with the other people in your group." Give them a few minutes to discuss the question.

22. Ask the students, "What do you all think? Who would like to share?"

23. Allow the students to share their ideas.

24. Tell the students, "We are now going to take a minute to talk about what scientists do to figure out answers to their questions."

Investigation 10. Movement of Carbon in Ecosystems: How Does the Amount of Dissolved Carbon Dioxide Gas Found in Water Change Over Time When Aquatic Plants and Animals Are Present?

25. Show an image of the question "What is the difference between data and evidence?" on the screen.
26. Tell the students, "Take a few minutes to talk about how you would answer this question with the other people in your group. Be ready to share with the rest of the class." Give the students two to three minutes to talk in their group.
27. Ask the students, "What do you all think? Who would like to share an idea?"
28. Allow the students to share their ideas.
29. Tell the students, "Okay, let's make sure we are all using the same definition. I think data are observations, measurements, and findings that we collect during an investigation, and evidence is an analysis of the data we collect and an interpretation of that analysis."
30. Show Figure 10.4 on a screen with the statement "The water in the tube with snails turned yellow after 24 hours."

FIGURE 10.4
Examples of tubes from this investigation

31. Ask the students, "Do you think this statement is data or evidence, and why do you think that?"
32. Allow the students to share their ideas.
33. Tell the students, "I think that statement is data because it is an observation."
34. Show Figure 10.4 on a screen with the statement "The tube with snails turned yellow after 24 hours, but the tube with just the plant and the tube with nothing in it stayed blue. This suggests that the snails were giving off carbon dioxide but the plant was not."
35. Ask the students, "Do you think this statement is data or evidence, and why do you think that?"

Teacher Notes

36. Allow the students to share their ideas.

37. Tell the students, "I think that statement is evidence because it includes a comparison, which is an analysis of data, and explains what the comparison means."

38. Ask the students, "Does anyone have any questions about the difference between data and evidence?"

39. Answer any questions that come up.

40. Tell the students, "We are now going to take a minute to talk about what went well and what didn't go so well during our investigation. We need to talk about this because you all are going to be planning and carrying out your own investigations like this a lot this year, and I want to help you all get better at it."

41. Show an image of the question "What made your investigation scientific?" on the screen. Tell the students, "Take a few minutes to talk about how you would answer this question with the other people in your group. Be ready to share with the rest of the class." Give the students two to three minutes to talk in their group.

42. Ask the students, "What do you all think? Who would like to share an idea?"

43. Allow the students to share their ideas. Be sure to expand on their ideas about what makes an investigation scientific.

44. Show an image of the question "What made your investigation not so scientific?" on the screen. Tell the students, "Take a few minutes to talk about how you would answer this question with the other people in your group. Be ready to share with the rest of the class." Give the students two to three minutes to talk in their group.

45. Ask the students, "What do you all think? Who would like to share an idea?"

46. Allow the students to share their ideas. Be sure to expand on their ideas about what makes an investigation less scientific.

47. Show an image of the question "What rules can we put into place to help us make sure our next investigation is more scientific?" on the screen. Tell the students, "Take a few minutes to talk about how you would answer this question with the other people in your group. Be ready to share with the rest of the class." Give the students two to three minutes to talk in their group.

48. Ask the students, "What do you all think? Who would like to share an idea?"

49. Allow the students to share their ideas. Once they have shared their ideas, offer a suggestion for a possible class rule.

50. Ask the students, "What do you all think? Should we make this a rule?"

51. If the students agree, write the rule on the board or on a class "Rules for Scientific Investigation" chart so you can refer to it during the next investigation.

Investigation 10. Movement of Carbon in Ecosystems: How Does the Amount of Dissolved Carbon Dioxide Gas Found in Water Change Over Time When Aquatic Plants and Animals Are Present?

Stage 6: Write a Draft Report (30 minutes)

Your students will use either the Investigation Handout or the Investigation Log in the student workbook when writing the draft report. When you give the directions shown in quotes in the following steps, substitute "Investigation Log in your workbook" or just "Investigation Log" (as shown in brackets) for "handout" if they are using the workbook.

1. Tell the students, "You are now going to write an investigation report to share what you have learned. Please take out a pencil and turn to the 'Draft Report' section of your handout [Investigation Log in your workbook]."

2. If possible, use a document camera to project the "Introduction" section of the draft report from the Investigation Handout (or the Investigation Log in their workbook) on a screen or board (or take a picture of it and project the picture on a screen or board).

3. Tell the students, "The first part of the report is called the 'Introduction.' In this section of the report you want to explain to the reader what you were investigating, why you were investigating it, and what question you were trying to answer. All this information can be found in the text at the beginning of your handout [Investigation Log]." Point to the image. "Here are some sentence starters to help you begin writing."

4. Ask the students, "Do you have any questions about what you need to do?"

5. Answer any questions that come up.

6. Tell the students, "Okay, let's write."

7. Give the students 10 minutes to write the "Introduction" section of the report. As they work, move from student to student to check in, ask probing questions, and offer a suggestion if a student gets stuck.

8. If possible, use a document camera to project the "Method" section of the draft report from the Investigation Handout (or the Investigation Log in their workbook) on a screen or board (or take a picture of it and project the picture on a screen or board).

9. Tell the students, "The second part of the report is called the 'Method.' In this section of the report you want to explain to the reader what you did during the investigation, what data you collected and why, and how you went about analyzing your data. All this information can be found in the 'Plan Your Investigation' section of the handout [Investigation Log]. Remember that you all planned and carried out different investigations, so do not assume that the reader will know what you did." Point to the image. "Here are some sentence starters to help you begin writing."

10. Ask the students, "Do you have any questions about what you need to do?"

11. Answer any questions that come up.

Teacher Notes

12. Tell the students, "Okay, let's write."

13. Give the students 10 minutes to write the "Method" section of the report. As they work, move from student to student to check in, ask probing questions, and offer a suggestion if a student gets stuck.

14. If possible, use a document camera to project the "Argument" section of the draft report from the Investigation Handout (or the Investigation Log in their workbook) on a screen or board (or take a picture of it and project the picture on a screen or board).

15. Tell the students, "The last part of the report is called the 'Argument.' In this section of the report you want to share your claim, evidence, and justification of the evidence with the reader. All this information can be found on your whiteboard." Point to the image. "Here are some sentence starters to help you begin writing."

16. Ask the students, "Do you have any questions about what you need to do?"

17. Answer any questions that come up.

18. Tell the students, "Okay, let's write."

19. Give the students 10 minutes to write the "Argument" section of the report. As they work, move from student to student to check in, ask probing questions, and offer a suggestion if a student gets stuck.

Stage 7: Peer Review (35 minutes)

Your students will use either the Investigation Handout or their workbook when doing the peer review. Except where noted below, the directions are the same whether using the handout or the workbook.

1. Tell the students, "We are now going to review our reports to find ways to make them better. I'm going to come around and collect your draft reports. While I do that, please take out a pencil."

2. Collect the handouts or the workbooks with the draft reports from the students.

3. If possible, use a document camera to project the peer-review guide (see Appendix 4) on a screen or board (or take a picture of it and project the picture on a screen or board).

4. Tell the students, "We are going to use this peer-review guide to give each other feedback." Point to the image.

5. Tell the students, "I'm going to ask you to work with a partner to do this. I'm going to give you and your partner a draft report to read. You two will then read the report together. Once you are done reading the report, I want you to answer each of the questions on the peer-review guide." Point to the review questions on the image of the peer-review guide.

Investigation 10. Movement of Carbon in Ecosystems: How Does the Amount of Dissolved Carbon Dioxide Gas Found in Water Change Over Time When Aquatic Plants and Animals Are Present?

6. Tell the students, "You can check 'no,' 'almost,' or 'yes' after each question." Point to the checkboxes on the image of the peer-review guide.

7. Tell the students, "This will be your rating for this part of the report. Make sure you agree on the rating you give the author. If you mark 'no' or 'almost,' then you need to tell the author what he or she needs to do to get a 'yes.'" Point to the space for the reviewer feedback on the image of the peer-review guide.

8. Tell the students, "It is really important for you to give the authors feedback that is helpful. That means you need to tell them exactly what they need to do to make their report better."

9. Ask the students, "Do you have any questions about what you need to do?"

10. Answer any questions that come up.

11. Tell the students, "Please sit with a partner who is not in your current group." Allow the students time to sit with a partner.

12. Tell the students, "Okay, I'm now going to give you one report to read." Pass out one Investigation Handout with a draft report or one workbook to each pair. Make sure that the report you give a pair was not written by one of the students in that pair. Give each pair one peer-review guide to fill out. If the students are using workbooks, the peer-review guide is included right after the draft report so you do not need to pass out copies of the peer-review guide.

13. Tell the students, "Okay, I'm going to give you 15 minutes to read the report I gave you and to fill out the peer-review guide. Go ahead and get started."

14. Give the students 15 minutes to work. As they work, move around from pair to pair to check in and see how things are going, answer questions, and offer advice.

15. After 15 minutes pass, tell the students, "Okay, time is up. Please give me the report and the peer-review guide that you filled out."

16. Collect the Investigation Handouts and the peer-review guides, or collect the workbooks if students are using them. If the students are using the Investigation Handouts and separate peer-review guides, be sure you keep each handout with its corresponding peer-review guide.

17. Tell the students, "Okay, I am now going to give you a different report to read and a new peer-review guide to fill out." Pass out one more report to each pair. Make sure that the report you give a pair was not written by one of the students in that pair. Give each pair a new peer-review guide to fill out as a group.

18. Tell the students, "Okay, I'm going to give you 15 minutes to read this new report and to fill out the peer-review guide. Go ahead and get started."

19. Give the students 15 minutes to work. As they work, move around from pair to pair to check in and see how things are going, answer questions, and offer advice.

Teacher Notes

20. After 15 minutes pass, tell the students, "Okay, time is up. Please give me the report and the peer-review guide that you filled out."

21. Collect the Investigation Handouts and the peer-review guides, or collect the workbooks if students are using them. If the students are using the Investigation Handouts and separate peer-review guides, be sure you keep each handout with its corresponding peer-review guide.

Stage 8: Revise the Report (30 minutes)

Your students will use either the Investigation Handout or their workbook when revising the report. Except where noted below, the directions are the same whether using the handout or the workbook.

1. Tell the students, "You are now going to revise your draft report based on the feedback you get from your classmates. Please take out a pencil."

2. Return the reports to the students.
   - *If the students used the Investigation Handout and a copy of the peer-review guide,* pass back the handout and the peer-review guide to each student.
   - *If the students used the workbook,* pass that back to each student.

3. Tell the students, "Please take a few minutes to read over the peer-review guide. You should use it to figure out what you need to change in your report and how you will change it."

4. Allow the students to read the peer-review guide.

5. *If the students used the workbook,* if possible use a document camera to project the "Write Your Final Report" section from the Investigation Log on a screen or board (or take a picture of it and project the picture on a screen or board).

6. Give the following directions about how to revise their reports:
   - *If the students used the Investigation Handout and a copy of the peer-review guide,* tell them, "Okay, let's revise our reports. Please take out a piece of paper. I would like you to rewrite your report. You can use your draft report as a starting point, but you also need to change it to make it better. Use the feedback on the peer-review guide to make it better."
   - *If the students used the workbook,* tell them, "Okay, let's revise our reports. I would like you to rewrite your report in the section of the Investigation Log called "Write Your Final Report." You can use your draft report as a starting point, but you also need to change it to make it better. Use the feedback on the peer-review guide to make it better."

7. Ask the students, "Do you have any questions about what you need to do?"

8. Answer any questions that come up.

Investigation 10. Movement of Carbon in Ecosystems: How Does the Amount of Dissolved Carbon Dioxide Gas Found in Water Change Over Time When Aquatic Plants and Animals Are Present?

9. Tell the students, "Okay, let's write." Allow about 20 minutes for the students to revise their reports.

10. After about 20 minutes, give the following directions:

    - *If the students used the Investigation Handout,* tell them, "Okay, time's up. I will now come around and collect your Investigation Handout, the peer-review guide, and your final report."

    - *If the students used the workbook,* tell them, "Okay, time's up. I will now come around and collect your workbooks."

11. *If the students used the Investigation Handout,* collect all the Investigation Handouts, peer-review guides, and final reports. *If the students used the workbook,* collect all the workbooks.

12. *If the students used the Investigation Handout,* use the "Teacher Score" column in the peer-review guide to grade the final report. *If the students used the workbook,* use the "Investigation Report Grading Rubric" in the Investigation Log to grade the final report. Whether you are using the handout or the log, you can give the students feedback about their writing in the "Teacher Comments" section.

How to Use the Checkout Questions

The Checkout Questions are an optional assessment. We recommend giving them to students at the start of the next class period after the students finish stage 8 of the investigation. You can then look over the student answers to determine if you need to reteach the core idea from the investigation. Appendix 6 gives the answers to the Checkout Questions that should be given by a student who can apply the core idea correctly in all cases and can explain the movement of matter into, within, and out of systems.

Alignment With Standards

Table 10.2 highlights how the investigation can be used to address specific performance expectations from the *Next Generation Science Standards, Common Core State Standards for English Language Arts* (CCSS ELA), and *English Language Proficiency (ELP) Standards.*

TABLE 10.2

Investigation 10 alignment with standards

NGSS performance expectation	5-LS2-1: Develop a model to describe the movement of matter among plants, animals, decomposers, and the environment.
***CCSS ELA*—Reading: Informational Text**	Key ideas and details • CCSS.ELA-LITERACY.RI.5.1: Quote accurately from a text when explaining what the text says explicitly and when drawing inferences from the text.

Continued

Table 10.2 (*continued*)

***CCSS ELA*—Reading: Informational Text** (*continued*)	• CCSS.ELA-LITERACY.RI.5.2: Determine two or more main ideas of a text and explain how they are supported by key details; summarize the text. • CCSS.ELA-LITERACY.RI.5.3: Explain the relationships or interactions between two or more individuals, events, ideas, or concepts in a historical, scientific, or technical text based on specific information in the text. Craft and structure • CCSS.ELA-LITERACY.RI.5.4: Determine the meaning of general academic and domain-specific words and phrases in a text relevant to a *grade 5 topic or subject area*. • CCSS.ELA-LITERACY.RI.5.5: Compare and contrast the overall structure (e.g., chronology, comparison, cause/effect, problem/solution) of events, ideas, concepts, or information in two or more texts. • CCSS.ELA-LITERACY.RI.5.6: Analyze multiple accounts of the same event or topic, noting important similarities and differences in the point of view they represent. Integration of knowledge and ideas • CCSS.ELA-LITERACY.RI.5.7: Draw on information from multiple print or digital sources, demonstrating the ability to locate an answer to a question quickly or to solve a problem efficiently. • CCSS.ELA-LITERACY.RI.5.8: Explain how an author uses reasons and evidence to support particular points in a text, identifying which reasons and evidence support which point(s). Range of reading and level of text complexity • CCSS.ELA-LITERACY.RI.5.10: By the end of the year, read and comprehend informational texts, including history/social studies, science, and technical texts, at the high end of the grades 4–5 text complexity band independently and proficiently.

***CCSS ELA*—Writing**	Text types and purposes
	• CCSS.ELA-LITERACY.W.5.1: Write opinion pieces on topics or texts, supporting a point of view with reasons.
	    ○ CCSS.ELA-LITERACY.W.5.1.A: Introduce a topic or text clearly, state an opinion, and create an organizational structure in which ideas are logically grouped to support the writer's purpose.
	    ○ CCSS.ELA-LITERACY.W.5.1.B: Provide logically ordered reasons that are supported by facts and details.
	    ○ CCSS.ELA-LITERACY.W.5.1.C: Link opinion and reasons using words, phrases, and clauses (e.g., *consequently*, *specifically*).
	    ○ CCSS.ELA-LITERACY.W.5.1.D: Provide a concluding statement or section related to the opinion presented.

Continued

Investigation 10. Movement of Carbon in Ecosystems: How Does the Amount of Dissolved Carbon Dioxide Gas Found in Water Change Over Time When Aquatic Plants and Animals Are Present?

Table 10.2 (*continued*)

CCSS ELA—Writing (*continued*)	• CCSS.ELA-LITERACY.W.5.2: Write informative or explanatory texts to examine a topic and convey ideas and information clearly.
	○ CCSS.ELA-LITERACY.W.5.2.A: Introduce a topic clearly, provide a general observation and focus, and group related information logically; include formatting (e.g., headings), illustrations, and multimedia when useful to aiding comprehension.
	○ CCSS.ELA-LITERACY.W.5.2.B: Develop the topic with facts, definitions, concrete details, quotations, or other information and examples related to the topic.
	○ CCSS.ELA-LITERACY.W.5.2.C: Link ideas within and across categories of information using words, phrases, and clauses (e.g., *in contrast*, *especially*).
	○ CCSS.ELA-LITERACY.W.5.2.D: Use precise language and domain-specific vocabulary to inform about or explain the topic.
	○ CCSS.ELA-LITERACY.W.5.2.E: Provide a concluding statement or section related to the information or explanation presented.
	Production and distribution of writing
	• CCSS.ELA-LITERACY.W.5.4: Produce clear and coherent writing in which the development and organization are appropriate to task, purpose, and audience.
	• CCSS.ELA-LITERACY.W.5.5: With guidance and support from peers and adults, develop and strengthen writing as needed by planning, revising, editing, rewriting, or trying a new approach.
	• CCSS.ELA-LITERACY.W.5.6: With some guidance and support from adults, use technology, including the internet, to produce and publish writing as well as to interact and collaborate with others; demonstrate sufficient command of keyboarding skills to type a minimum of two pages in a single sitting.
	Research to build and present knowledge
	• CCSS.ELA-LITERACY.W.5.8: Recall relevant information from experiences or gather relevant information from print and digital sources; summarize or paraphrase information in notes and finished work, and provide a list of sources.
	• CCSS.ELA-LITERACY.W.5.9: Draw evidence from literary or informational texts to support analysis, reflection, and research.
	Range of writing
	• CCSS.ELA-LITERACY.W.5.10: Write routinely over extended time frames (time for research, reflection, and revision) and shorter time frames (a single sitting or a day or two) for a range of discipline-specific tasks, purposes, and audiences.

Continued

Teacher Notes

Table 10.2 (*continued*)

CCSS ELA—Speaking and Listening	Comprehension and collaboration
	• CCSS.ELA-LITERACY.SL.5.1: Engage effectively in a range of collaborative discussions (one-on-one, in groups, and teacher-led) with diverse partners on *grade 5 topics and texts,* building on others' ideas and expressing their own clearly.
	    ○ CCSS.ELA-LITERACY.SL.5.1.A: Come to discussions prepared, having read or studied required material; explicitly draw on that preparation and other information known about the topic to explore ideas under discussion.
	    ○ CCSS.ELA-LITERACY.SL.5.1.B: Follow agreed-upon rules for discussions and carry out assigned roles.
	    ○ CCSS.ELA-LITERACY.SL.5.1.C: Pose and respond to specific questions by making comments that contribute to the discussion and elaborate on the remarks of others.
	    ○ CCSS.ELA-LITERACY.SL.5.1.D: Review the key ideas expressed and draw conclusions in light of information and knowledge gained from the discussions.
	• CCSS.ELA-LITERACY.SL.5.2: Summarize a written text read aloud or information presented in diverse media and formats, including visually, quantitatively, and orally.
	• CCSS.ELA-LITERACY.SL.5.3: Summarize the points a speaker makes and explain how each claim is supported by reasons and evidence.
	Presentation of knowledge and ideas
	• CCSS.ELA-LITERACY.SL.5.4: Report on a topic or text or present an opinion, sequencing ideas logically and using appropriate facts and relevant, descriptive details to support main ideas or themes; speak clearly at an understandable pace.
	• CCSS.ELA-LITERACY.SL.5.5: Include multimedia components (e.g., graphics, sound) and visual displays in presentations when appropriate to enhance the development of main ideas or themes.
	• CCSS.ELA-LITERACY.SL.5.6: Adapt speech to a variety of contexts and tasks, using formal English when appropriate to task and situation.

ELP Standards	Receptive modalities • ELP 1: Construct meaning from oral presentations and literary and informational text through grade-appropriate listening, reading, and viewing. • ELP 8: Determine the meaning of words and phrases in oral presentations and literary and informational text. Productive modalities • ELP 3: Speak and write about grade-appropriate complex literary and informational texts and topics. • ELP 4: Construct grade-appropriate oral and written claims and support them with reasoning and evidence. • ELP 7: Adapt language choices to purpose, task, and audience when speaking and writing.

Continued

Investigation 10. Movement of Carbon in Ecosystems: How Does the Amount of Dissolved Carbon Dioxide Gas Found in Water Change Over Time When Aquatic Plants and Animals Are Present?

Table 10.2 (*continued*)

ELP Standards (*continued*)	Interactive modalities • ELP 2: Participate in grade-appropriate oral and written exchanges of information, ideas, and analyses, responding to peer, audience, or reader comments and questions. • ELP 5: Conduct research and evaluate and communicate findings to answer questions or solve problems. • ELP 6: Analyze and critique the arguments of others orally and in writing. Linguistic structures of English • ELP 9: Create clear and coherent grade-appropriate speech and text. • ELP 10: Make accurate use of standard English to communicate in grade-appropriate speech and writing.

Investigation Handout

Investigation 10

Movement of Carbon in Ecosystems: How Does the Amount of Dissolved Carbon Dioxide Gas Found in Water Change Over Time When Aquatic Plants and Animals Are Present?

Introduction

There are many different kinds of living things on Earth. Some of these things live on land, and some live in water. Your teacher will place an aquarium filled with some plants and snails that live in water on a table for you to look at. Take a few minutes to look at the plants and snails in the aquarium. As you look at them, keep track of things you notice and things you wonder about in the boxes below.

Things I NOTICED ...	Things I WONDER about ...

Investigation 10. Movement of Carbon in Ecosystems: How Does the Amount of Dissolved Carbon Dioxide Gas Found in Water Change Over Time When Aquatic Plants and Animals Are Present?

The plants and snails in the aquarium are found in freshwater ponds. A pond is an example of an ecosystem, which is a community of interacting organisms and their physical environment. In an ecosystem, matter moves between the air, water, and soil and among plants, animals, and microbes as they live and die. Living things obtain the matter they need from the environment and release waste matter back into the environment. The matter that moves into and out of the bodies of living things can be a solid, liquid, or a gas. Snails, for example, take in solid parts of plants when they eat. They also release solid waste matter, which is called feces, into the environment. Snails also take in water from the pond and release liquid waste back into the environment where they live.

Animals that live on land also move gases from the air into and out of their bodies each time they take a breath. You can see how carbon dioxide gas moves out of a person's body when they breathe out by using some water with a chemical called bromthymol blue (or BTB) in it. Watch what happens when your teacher blows air into some water with BTB in it. As your teacher blows air into the water, write down things you notice and things you wonder about in the boxes below.

Things I NOTICED …	Things I WONDER about …

The water in the plastic cup changed color because your teacher added carbon dioxide to it. The carbon dioxide came from your teacher's body. The carbon dioxide that was in your teacher's body mixed with the air that was in your teacher's lungs. Your teacher then blew this air through the straw and into the water. As the air moved through the water, some of the carbon dioxide in the air mixed

Investigation Handout

with the water. The more air your teacher blew into the water, the more carbon dioxide mixed with the water. The BTB changes color based on how much carbon dioxide is in the water. BTB is blue when there is little or no carbon dioxide in the water, turns green or greenish-yellow when there is some carbon dioxide in the water, and turns yellow when there is a lot of carbon dioxide in the water. BTB is therefore a good way to track the movement of carbon dioxide into or out of an organism that lives in the water, because BTB turns different colors depending on how much carbon dioxide there is in the water.

Your goal in this investigation is to determine if snails and plants that live in the water move carbon dioxide into and out their bodies like animals that live on land do. One way to track how carbon dioxide moves into or out of a living thing that lives in the water is to place that living thing and some water inside a *closed system*. A closed system is an area that is separated from its surroundings by a barrier that does not let any matter transfer into or out of that system. An example of a closed system is a container with a sealed lid. You can then use an indicator, such as BTB, to track how the amount of carbon dioxide in the water inside that closed system changes over time.

The change in the color of the water with BTB in it will allow you to see if the carbon dioxide is moving into or out of a living thing. If the amount of carbon dioxide in the water goes up, then the living thing moved carbon dioxide out of its body and into the water. You will know that the amount of carbon dioxide went up in the water if the water with BTB in it changes from blue to green or yellow or from green to yellow. If the amount of carbon dioxide in the water goes down, then the living thing moved carbon dioxide from the water into its body. You will know that the amount of carbon dioxide in the water went down if the water with BTB in it changes from yellow to green, yellow to blue, or green to blue. Finally, if the amount of carbon dioxide in the water does not change, then the living thing did not move carbon dioxide into or out of its body.

Things we KNOW from what we read …

Investigation 10. Movement of Carbon in Ecosystems: How Does the Amount of Dissolved Carbon Dioxide Gas Found in Water Change Over Time When Aquatic Plants and Animals Are Present?

Your Task

Use what you know about the way matter cycles between living things and the environment and the importance of tracking the movement of matter within systems to design and carry out an investigation to figure out if carbon dioxide is used by plants and animals that live in ponds and, if so, how it is used.

The *guiding question* of this investigation is, **How does the amount of dissolved carbon dioxide gas found in water change over time when aquatic plants and animals are present?**

Materials

You may use any of the following materials during your investigation:

- Safety goggles, nonlatex apron, and vinyl gloves (required)
- 2 pond snails
- 2 aquatic plants
- 4 polypropylene centrifuge tubes (50 ml)
- Tube rack
- BTB in a plastic dropper bottle
- Blue painter's tape
- Permanent marker

Safety Rules

Follow all normal safety rules. In addition, be sure to follow these rules:

- Wear sanitized safety goggles, nonlatex aprons, and vinyl gloves during setup, investigation activity, and cleanup.
- Be careful when handling the water with BTB. Although BTB is considered nontoxic, ingesting large doses may cause stomach upset, and splashes on the skin or in the eyes may cause slight irritation. Do not drink water with BTB in it.
- Immediately clean up any spills to avoid a slip or fall hazard, and notify the teacher.
- Keep water sources away from electrical receptacles to prevent shock.
- Be careful when handling the organisms, because they are delicate.
- Wash your hands with soap and water when you are done cleaning up.

Plan Your Investigation

Prepare a plan for your investigation by filling out the chart on the next page; this plan is called an *investigation proposal*. Before you start developing your plan, be sure to discuss the following questions with the other members of your group:

- What are the **components of this system** and how do they **interact**?
- How can we **track the transfer of carbon dioxide** within a system?

Investigation Handout

Our guiding question:

This is a picture of how we will set up the equipment:

We will collect the following data:

These are the steps we will follow to collect data:

I approve of this investigation proposal.

Teacher's signature

Date

Investigation 10. Movement of Carbon in Ecosystems: How Does the Amount of Dissolved Carbon Dioxide Gas Found in Water Change Over Time When Aquatic Plants and Animals Are Present?

Collect Your Data

Keep a record of what you measure or observe during your investigation in the space below.

Analyze Your Data

You will need to analyze the data you collected before you can develop an answer to the guiding question. To analyze the data you collected, create a table, graph, or picture.

Argument-Driven Inquiry in **Fifth-Grade Science:** Three-Dimensional Investigations

Investigation Handout

Draft Argument

Develop an argument on a whiteboard. It should include the following:

1. A *claim*: Your answer to the guiding question.
2. *Evidence*: An analysis of the data and an explanation of what the analysis means.
3. A *justification of the evidence*: Why your group thinks the evidence is important.

The Guiding Question:	
Our Claim:	
Our Evidence:	Our Justification of the Evidence:

Argumentation Session

Share your argument with your classmates. Be sure to ask them how to make your draft argument better. Keep track of their suggestions in the space below.

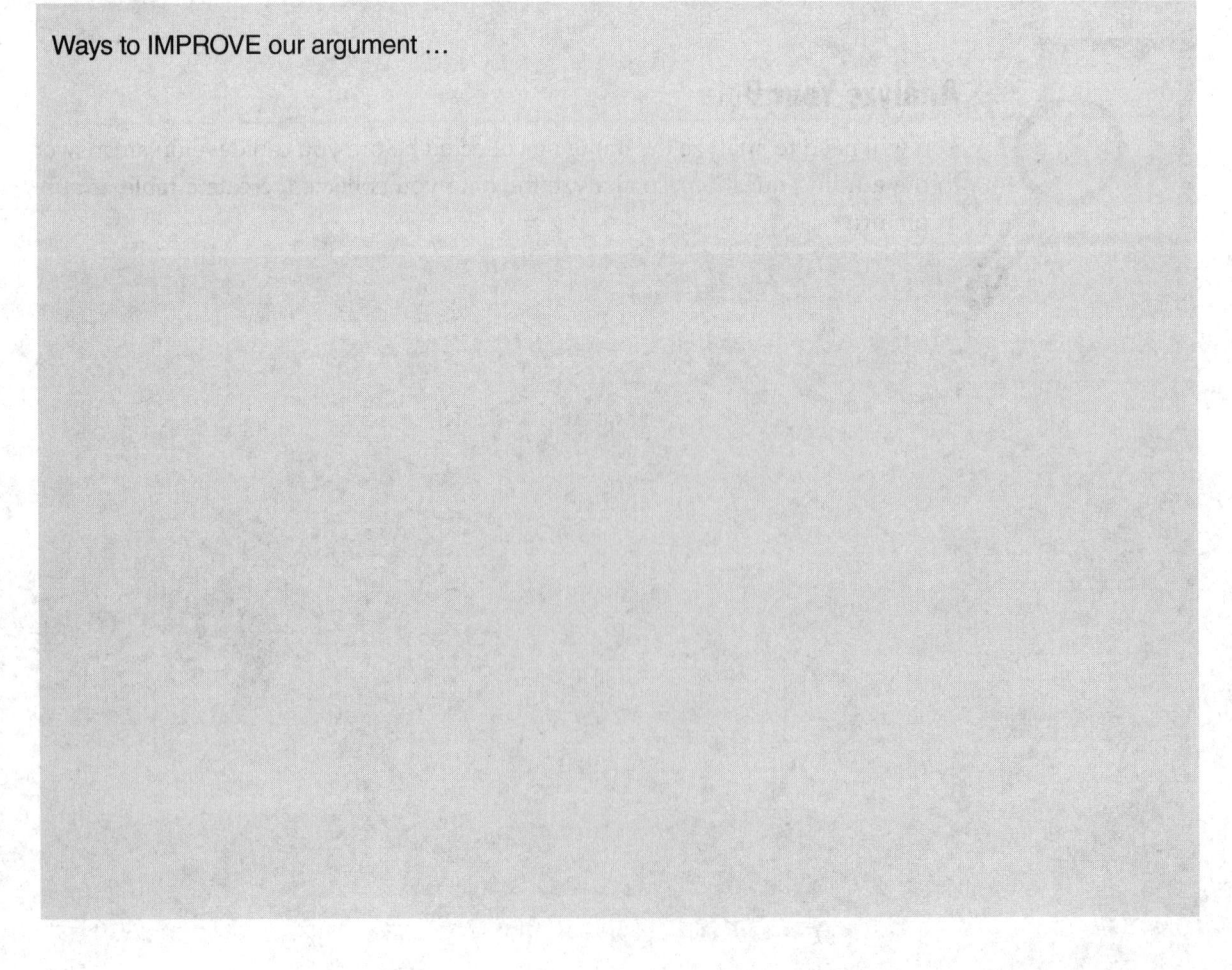

Ways to IMPROVE our argument …

Investigation 10. Movement of Carbon in Ecosystems:
How Does the Amount of Dissolved Carbon Dioxide Gas Found in Water
Change Over Time When Aquatic Plants and Animals Are Present?

Draft Report

Prepare an *investigation report* to share what you have learned. Use the information in this handout and your group's final argument to write a draft of your investigation report.

Introduction

We have been studying _____ in class.

Before we started this investigation, we explored _____

_____

_____

We noticed _____

_____

_____

My goal for this investigation was to figure out _____

_____

_____

The guiding question was _____

_____

_____

Method

To gather the data I needed to answer this question, I _____

_____

_____

_____

_____

Investigation Handout

I then analyzed the data I collected by _____

_____

_____

_____

Argument

My claim is _____

_____

_____

The _____ below shows _____

_____

_____

_____

_____

Investigation 10. Movement of Carbon in Ecosystems: How Does the Amount of Dissolved Carbon Dioxide Gas Found in Water Change Over Time When Aquatic Plants and Animals Are Present?

This analysis of the data I collected suggests _____

_____

_____

_____

_____

This evidence is based on several important scientific concepts. The first one is _____

_____

_____

_____

_____

Review

Your classmates need your help! Review the draft of their investigation reports and give them ideas about how to improve. Use the *peer-review guide* when doing your review.

Submit Your Final Report

Once you have received feedback from your friends about your draft report, create your final investigation report and hand it in to your teacher.

Checkout Questions

Investigation 10. Movement of Carbon in Ecosystems

Use the following information to answer questions 1–3. The picture below shows three containers. Container A has water and two pond snails in it. Container B has water and an aquatic plant in it. Container C has water, a pond snail, and an aquatic plant in it. All three containers are left in the light for 24 hours.

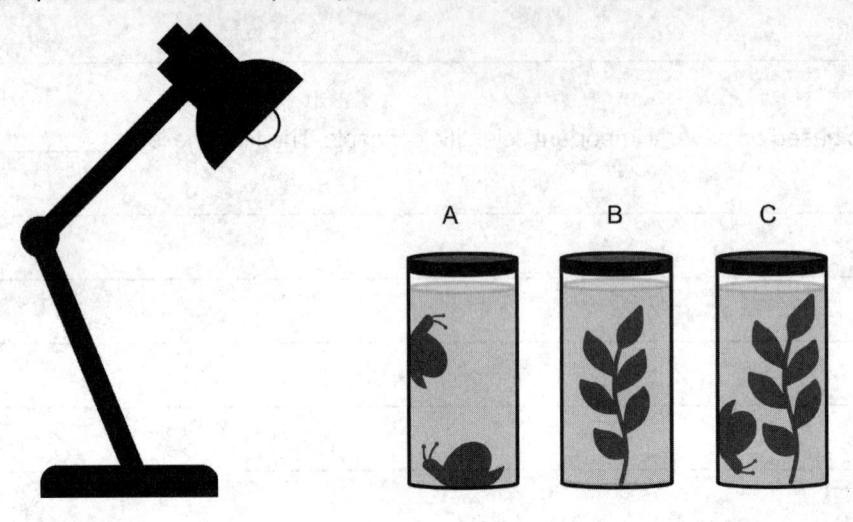

1. In which container will the carbon dioxide level in the water go up the most over 24 hours?

    a. Container A

    b. Container B

    c. Container C

    d. Unsure

2. In which container will the carbon dioxide level in the water go up the least over 24 hours?

    a. Container A

    b. Container B

    c. Container C

    d. Unsure

3. How do you know? Use what you know about the way matter, such as carbon dioxide, moves into, within, and out of living things to explain your answer.

_____
_____
_____
_____
_____

434   National Science Teaching Association

Investigation 10. Movement of Carbon in Ecosystems: How Does the Amount of Dissolved Carbon Dioxide Gas Found in Water Change Over Time When Aquatic Plants and Animals Are Present?

Teacher Scoring Rubric for the Checkout Questions

Level	Description
3	The student can apply the core idea correctly in all cases and can explain the movement of matter into, within, and out of systems.
2	The student cannot apply the core idea correctly in all cases but can explain the movement of matter into, within, and out of systems.
1	The student can apply the core idea correctly in all cases but cannot explain the movement of matter into, within, and out of systems.
0	The student cannot apply the core idea correctly and cannot explain the movement of matter into, within, and out of systems.

Section 5
Earth's Place in the Universe

Teacher Notes

Investigation 11

Patterns in Shadows: How Does the Location of the Sun in the Sky Affect the Direction and Length of Shadows?

Purpose

The purpose of this investigation is to give students an opportunity to use one disciplinary core idea (DCI), two crosscutting concepts (CCs), and eight scientific and engineering practices (SEPs) to figure out how the location of the Sun in the sky affects the direction and length of shadows. Students will also learn about how scientists use different methods to answer different types of questions.

The DCI, CCs, and SEPs That Students Use During This Investigation to Figure Things Out

DCI

- *ESS1.B: Earth and the Solar System:* The orbits of Earth around the Sun and of the Moon around Earth, together with the rotation of Earth about an axis between its North and South Poles, cause observable patterns. These include day and night; daily changes in the length and direction of shadows; and different positions of the Sun, Moon, and stars at different times of the day, month, and year.

CCs

- *CC 1: Patterns:* Similarities and differences in patterns can be used to sort, classify, communicate, and analyze simple rates of change for natural phenomena and designed products. Patterns of change can be used to make predictions. Patterns can be used as evidence to support an explanation.
- *CC 4: Systems and System Models:* A system can be described in terms of its components and their interactions. A system is a group of related parts that make up a whole and can carry out functions its individual parts cannot.

SEPs

- *SEP 1: Asking Questions and Defining Problems:* Ask questions about what would happen if a variable is changed. Ask questions that can be investigated and predict reasonable outcomes based on patterns such as cause-and-effect relationships.
- *SEP 2: Developing and Using Models:* Develop and/or use models to describe and/or predict phenomena.

Investigation 11. Patterns in Shadows:
How Does the Location of the Sun in the Sky Affect the Direction and Length of Shadows?

- *SEP 3: Planning and Carrying Out Investigations:* Plan and conduct an investigation collaboratively to produce data to serve as the basis for evidence, using fair tests in which variables are controlled and the number of trials considered. Evaluate appropriate methods and/or tools for collecting data.

- *SEP 4: Analyzing and Interpreting Data:* Represent data in tables and/or various graphical displays (bar graphs, pictographs, and/or pie charts) to reveal patterns that indicate relationships. Analyze and interpret data to make sense of phenomena, using logical reasoning, mathematics, and/or computation. Compare and contrast data collected by different groups in order to discuss similarities and differences in their findings.

- *SEP 5: Using Mathematics and Computational Thinking:* Organize simple data sets to reveal patterns that suggest relationships. Describe, measure, estimate, and/or graph quantities (e.g., area, volume, weight, time) to address scientific and engineering questions and problems.

- *SEP 6: Constructing Explanations and Designing Solutions:* Construct an explanation of observed relationships. Use evidence to construct or support an explanation. Identify the evidence that supports particular points in an explanation.

- *SEP 7: Engaging in Argument From Evidence:* Compare and refine arguments based on an evaluation of the evidence presented. Distinguish among facts, reasoned judgment based on research findings, and speculation in an explanation. Respectfully provide and receive critiques from peers about a proposed procedure, explanation, or model by citing relevant evidence and posing specific questions.

- *SEP 8: Obtaining, Evaluating, and Communicating Information:* Read and comprehend grade-appropriate complex texts and/or other reliable media to summarize and obtain scientific and technical ideas. Combine information in written text with that contained in corresponding tables, diagrams, and/or charts to support the engagement in other scientific and/or engineering practices. Communicate scientific and/or technical information orally and/or in written formats, including various forms of media as well as tables, diagrams, and charts.

Other Concepts That Students May Use During This Investigation

Students might also use some of the following concepts:

- A shadow is a dark area that is created when an object blocks the light from a light source.
- An object that blocks light is called an opaque object.
- *Rotation* is the term used to describe an object spinning on its axis.
- Earth rotates or spins counterclockwise on an axis once every 24 hours.

Teacher Notes

- The axis is a line that runs between the North and South Poles through the center of Earth.
- The Sun appears to come up from under the horizon in the east at sunrise, moves to its highest point in the sky at noon, and then disappears behind the horizon in the west at sunset.

What Students Figure Out

The students figure out that when the Sun is higher in the sky the shadows will be the smallest, and that the shadows stretch out longer when the Sun is lower. They also figure out that the shadows point in the same direction as the Sun's rays.

Background Information About This Investigation for the Teacher

The length and angle of a shadow depend on the angle at which sunlight hits Earth. The angle at which sunlight hits Earth depends on the time of day and the time of year. The angle of sunlight changes throughout the day because of Earth's rotation on its axis. Every 24 hours the Earth makes one counterclockwise spin. Because of this we see the Sun "rise" in the east and "set" in the west. When the path the Sun takes, its "day arc," reaches its apex or highest point, it is "solar noon." This happens around noon each day but might be a little before or after noon depending on how far east or west a location is from the center of a time zone.

The angle of sunlight changes throughout the year because of the tilt of Earth as it revolves around the Sun. In the winter, the Northern Hemisphere of Earth is tilted away from the Sun, so its day arc takes a lower path. In the summer, the Northern Hemisphere of Earth is tilted toward the Sun, so its day arc takes a higher path. During the spring and fall equinoxes, the day arc will fall somewhere in the middle. The path of the day arc is also dependent on a viewer's longitude, with locations closer to the equator having a higher apex than locations closer to the poles.

Timeline

The time needed to complete this investigation is 270 minutes (4 hours and 30 minutes). The amount of instructional time needed for each stage of the investigation is as follows:

- *Stage 1.* Introduce the task and the guiding question: 35 minutes
- *Stage 2.* Design a method and collect data: 50 minutes
- *Stage 3.* Create a draft argument: 45 minutes
- *Stage 4.* Argumentation session: 30 minutes
- *Stage 5.* Reflective discussion: 15 minutes
- *Stage 6.* Write a draft report: 30 minutes

Investigation 11. Patterns in Shadows: How Does the Location of the Sun in the Sky Affect the Direction and Length of Shadows?

- *Stage 7.* Peer review: 35 minutes
- *Stage 8.* Revise the report: 30 minutes

Materials and Preparation

The materials needed for this investigation are listed in Table 11.1. The items can be purchased from a big-box retail store such as Walmart or Target or through an online retailer such as Amazon. The materials for this investigation can also be purchased as a complete kit (which includes enough materials for 24 students, or six groups of four students) at *www.argumentdriveninquiry.com*.

TABLE 11.1
Materials for Investigation 11

Item	Quantity
Safety goggles	1 per student
Wooden dowel, 6 inches	1 per group
One-hole rubber stopper, size 6	1 per group
Sheet of 1-cm grid paper	5 per group
Piece of string, 1 m	1 per group
Meterstick	1 per group
Protractor	1 per group
Flashlight	1 per group
Compass	1 per group
Whiteboard, 2' × 3'*	1 per group
Investigation Handout	1 per student
Peer-review guide and teacher scoring rubric	1 per student
Checkout Questions (optional)	1 per student

*As an alternative, students can use computer and presentation software such as Microsoft PowerPoint or Apple Keynote to create their arguments.

Safety Precautions

Remind students to follow all normal safety rules. In addition, tell the students to take the following safety precautions:

- Wear sanitized safety goggles during setup, investigation activity, and cleanup.
- Do not put any of the materials used in the investigation in their mouth.
- Make sure all materials are put away after completing the activity.
- Wash their hands with soap and water when they are done cleaning up.

Teacher Notes

Lesson Plan by Stage

This lesson plan is only a suggestion. It is included here to illustrate what you can say and do during each stage of ADI for this specific investigation. We encourage you to modify this lesson plan by asking different questions, using different examples, and providing different scaffolds as needed to better meet the needs of students in your class.

Stage 1: Introduce the Task and the Guiding Question (35 minutes)

1. Ask the students to sit in six groups, with three or four students in each group.
2. Ask the students to clear off their desks except for a pencil (and their *Student Workbook for Argument-Driven Inquiry in Fifth-Grade Science* if they have one).
3. Pass out an Investigation Handout to each student (or ask students to turn to the Investigation Log for Investigation 11 in their workbook).
4. Read the first paragraph of the "Introduction" aloud to the class. Ask the students to follow along as you read.
5. Show a time-lapse video of shadows changing over the course over a day, and ask the students to record their observations and questions on the "NOTICED/WONDER" chart in the "Introduction." You can use one of the following videos:
   - *www.youtube.com/watch?v=yJo674TYGMs*
   - *www.youtube.com/watch?v=f9TFoKmEsUM*
   - *www.youtube.com/watch?v=cdn6Muc2lcQ*
   - *www.youtube.com/watch?v=ZGEEx7Kv5Go*
   - *www.youtube.com/watch?v=LgZbhogv9Q8*
   - *www.youtube.com/watch?v=loVGi_bftdU*
   - *www.youtube.com/watch?v=Lvhjbrr5GI8*
6. Ask the students to share what they observed in the video.
7. Ask the students to share what questions they have about the shadows.
8. Tell the students, "Some of your questions might be answered by reading the rest of the 'Introduction.'"
9. Ask the students to read the rest of the "Introduction" on their own *or* ask them to follow along as you read it aloud.
10. Once the students have read the rest of the "Introduction," ask them to fill out the "Things we KNOW" chart on their Investigation Handout (or in their Investigation Log) as a group.
11. Ask the students to share what they learned from the reading. Add these ideas to a class "Things we KNOW" chart.

Investigation 11. Patterns in Shadows:
How Does the Location of the Sun in the Sky Affect the Direction and Length of Shadows?

12. Tell the students, "Let's see what we will need to figure out during our investigation."
13. Read the task and the guiding question aloud.
14. Tell the students, "I have lots of materials here that you can use."
15. Introduce the students to the materials available for them to use during the investigation by either (a) holding each one up and then asking what it might be used for or (b) giving them a kit with an example of each of the materials in it and giving them three to four minutes to play with them. If you give the students an opportunity to play with the materials, collect them from each group before moving on to stage 2.
16. Show students the diagram in Figure 11.1 and tell them, "This is one way that you can measure the different angles of the flashlight."

FIGURE 11.1
How to measure the angle of a flashlight

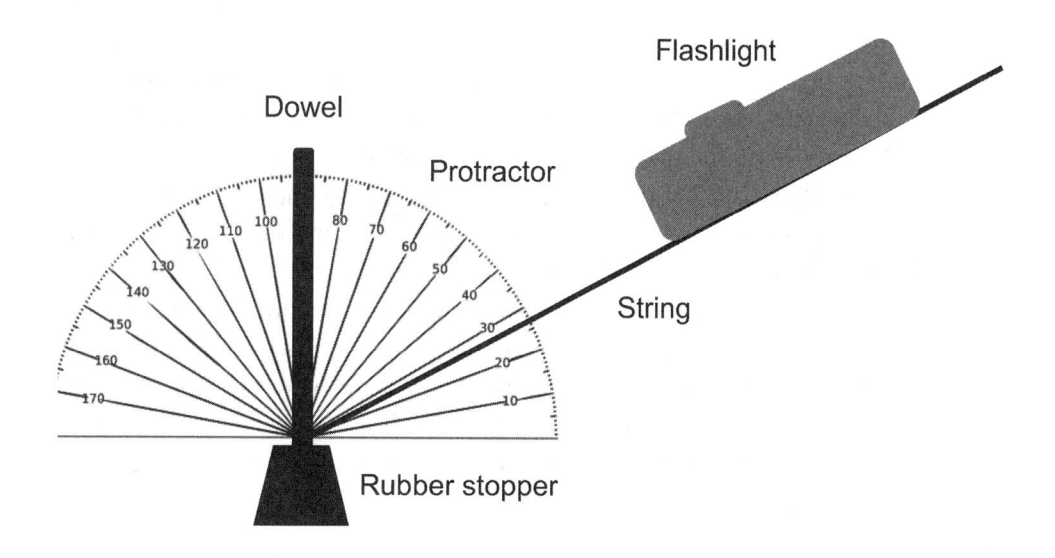

Stage 2: Design a Method and Collect Data (50 minutes)

1. Tell the students, "I am now going to give you and the other members of your group about 15 minutes to plan your investigation. Before you begin, I want you all to take a couple of minutes to discuss the following questions with the rest of your group."
2. Show the following questions on the screen or board:

Teacher Notes

- How can we use *patterns* to help us determine a *cause-and-effect relationship*?
- What are the components of the *system* we are studying, how do the components *interact*, and how could we create a *model* of it?

3. Tell the students, "Please take a few minutes to come up with an answer to these questions."

4. Give the students two or three minutes to discuss these two questions.

5. Ask two or three different groups to share their answers. Highlight or write down any important ideas on the board so students can refer to them later.

6. If possible, use a document camera to project an image of the graphic organizer for this investigation on a screen or board (or take a picture of it and project the picture on a screen or board). Tell the students, "I now want you all to plan out your investigation. To do that, you will need to fill out this investigation proposal."

7. Point to the box labeled "Our guiding question:" and tell the students, "You can put the question we are trying to answer in this box." Then ask, "Where can we find the guiding question?"

8. Wait for a student to answer.

9. Point to the box labeled "This is a picture of how we will set up the equipment:" and tell the students, "You can draw a picture in this box of how you will set up the equipment in order to carry out this investigation."

10. Point to the box labeled "We will collect the following data:" and tell the students, "You can list the measurements or observations that you will need to collect during the investigation in this box."

11. Point to the box labeled "These are the steps we will follow to collect data:" and tell the students, "You can list what you are going to do to collect the data you need and what you will do with your data once you have it. Be sure to give enough detail that I could do your investigation for you."

12. Ask the students, "Do you have any questions about what you need to do?"

13. Answer any questions that come up.

14. Tell the students, "Once you are done, raise your hand and let me know. I'll then come by and look over your proposal and give you some feedback. You may not begin collecting data until I have approved your proposal by signing it. You need to have your proposal done in the next 15 minutes."

15. Give the students 15 minutes to work in their groups on their investigation proposal. As they work, move from group to group to check in, ask probing questions, and offer a suggestion if a group gets stuck.

16. As each group finishes its investigation proposal, read it over and determine if it will be productive or not. If you feel the investigation will be productive (not

Investigation 11. Patterns in Shadows:
How Does the Location of the Sun in the Sky Affect the Direction and Length of Shadows?

necessarily what you would do or what the other groups are doing), sign your name on the proposal and let the group start collecting data. If the plan needs to be changed, offer some suggestions or ask some probing questions, and have the group make the changes before you approve it.

17. Pass out the materials or have one student from each group collect the materials they need from a central supply table or cart for the groups that have an approved proposal.
18. Remind students of the safety rules and precautions for this investigation.
19. Tell the students to collect their data and record their observations or measurements in the "Collect Your Data" box in their Investigation Handout (or the Investigation Log in their workbook).
20. Give the students 30 minutes to collect their data. Collect the materials from each group before asking them to analyze their data.

What should a student-designed investigation look like?

There are a number of different investigations that students can design to answer the question "How does the location of the Sun in the sky affect the direction and length of shadows?" For example, one method might include the following steps (see Figure 11.1 [p. 443] for an illustration of how to measure the angle of the light for this method):

1. Place the wooden dowel in a rubber stopper on the floor.
2. Pick a location in space to hold the flashlight.
3. Point the flashlight at the wooden dowel.
4. Measure the angle of the light shining on the wooden dowel.
5. Measure the length of the shadow using a ruler.
6. Use the compass to determine the direction of the flashlight and the direction of the shadow.
7. Repeat steps 2–5 five more times, but change the location of the flashlight each time.

If students use this method, they will need to collect data on (1) the angle of the light; (2) how long the shadows are, using a ruler; and (3) in which direction the shadows are pointing relative to the flashlight.

Teacher Notes

Stage 3: Create a Draft Argument (45 minutes)

1. Tell the students, "Now that we have all this data, we need to analyze the data so we can figure out an answer to the guiding question."

2. If possible, project an image of the "Analyze Your Data" section for this investigation on a screen or board using a document camera (or take a picture of it and project the picture on a screen or board). Point to the section and tell the students, "You can create a table or graph as a way to analyze your data. You can make your table or graph in this section."

3. Ask the students, "What information do we need to include in a table or graph?"

4. Tell the students, "Please take a few minutes to discuss this question with your group and be ready to share."

5. Give the students five minutes to discuss.

6. Ask two or three different groups to share their answers. Highlight or write down any important ideas on the board so students can refer to them later.

7. Tell the students, "I am now going to give you and the other members of your group about 10 minutes to create your table or graph." If the students are having trouble making a table or graph, you can take a few minutes to provide a mini-lesson about how to create a table or graph from a bunch of measurements (this strategy is called just-in-time instruction because it is offered only when students get stuck).

8. Give the students 10 minutes to analyze their data by creating a table or graph. As they work, move from group to group to check in, ask probing questions, and offer suggestions.

9. Tell the students, "I am now going to give you and the other members of your group about 15 minutes to create an argument to share what you have learned and convince others that they should believe you. Before you do that, we need to take a few minutes to discuss what you need to include in your argument."

10. If possible, use a document camera to project the "Argument Presentation on a Whiteboard" image from the "Draft Argument" section of the Investigation Handout (or the Investigation Log in their workbook) on a screen or board (or take a picture of it and project the picture on a screen or board).

11. Point to the box labeled "The Guiding Question:" and tell the students, "You can put the question we are trying to answer here on your whiteboard."

12. Point to the box labeled "Our Claim:" and tell the students, "You can put your claim here on your whiteboard. The claim is your answer to the guiding question."

13. Point to the box labeled "Our Evidence:" and tell the students, "You can put the evidence that you are using to support your claim here on your whiteboard. Your evidence will need to include the analysis you just did and an explanation

Investigation 11. Patterns in Shadows:
How Does the Location of the Sun in the Sky Affect the Direction and Length of Shadows?

of what your analysis means or shows. Scientists always need to support their claims with evidence."

14. Point to the box labeled "Our Justification of the Evidence:" and tell the students, "You can put your justification of your evidence here on your whiteboard. Your justification needs to explain why your evidence is important. Scientists often use core ideas to explain why the evidence they are using matters. Core ideas are important concepts that scientists use to help them make sense of what happens during an investigation."

15. Ask the students, "What are some core ideas that we read about earlier that might help us explain why the evidence we are using is important?"

16. Ask the students to share some of the core ideas from the "Introduction" section of the Investigation Handout (or the Investigation Log in the workbook). List these core ideas on the board.

17. Tell the students, "That is great. I would like to see everyone try to include these core ideas in your justification of the evidence. Your goal is to use these core ideas to help explain why your evidence matters and why the rest of us should pay attention to it."

18. Ask the students, "Do you have any questions about what you need to do?"

19. Answer any questions that come up.

20. Tell the students, "Okay, go ahead and start working on your arguments. You need to have your argument done in the next 15 minutes. It doesn't need to be perfect. We just need something down on the whiteboards so we can share our ideas."

21. Give the students 15 minutes to work in their groups on their arguments. As they work, move from group to group to check in, ask probing questions, and offer a suggestion if a group gets stuck. Figure 11.2 shows an example of an argument created by students for this investigation.

FIGURE 11.2

Example of an argument

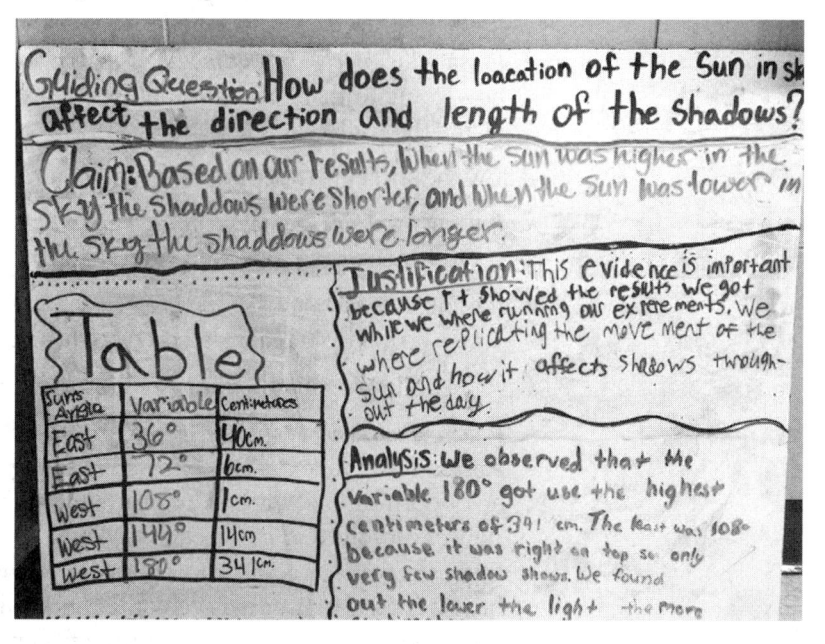

Teacher Notes

> ### What should the table or graph for this investigation look like?
>
> There are a number of different ways that students can analyze the measurements they collect during this investigation. The students might create a table or graph using the vertical angle of the flashlight as the independent variable and the length of the shadow as the dependent variable. They might also make a table or graph that shows a relationship between the direction the flashlight and the direction of the shadow.

Stage 4: Argumentation Session (30 minutes)

The argumentation session can be conducted in a whole-class presentation format, a gallery walk format, or a modified gallery walk format. We recommend using a whole-class presentation format for the first investigation, but try to transition to either the gallery walk or modified gallery walk format as soon as possible because that will maximize student voice and choice inside the classroom. The following list shows the steps for the three formats; unless otherwise noted, the steps are the same for all three formats.

1. Begin by introducing the use of the whiteboard.
   - *If using the whole-class presentation format,* tell the students, "We are now going to share our arguments. Please set up your whiteboards so everyone can see them."
   - *If using the gallery walk or modified gallery walk format,* tell the students, "We are now going to share our arguments. Please set up your whiteboards so they are facing the walls."
2. Allow the students to set up their whiteboards.
   - *If using the whole-class presentation format,* the whiteboards should be set up on stands or chairs so they are facing toward the center of the room.
   - *If using the gallery walk or modified gallery walk format,* the whiteboards should be set up on stands or chairs so they are facing toward the outside of the room.
3. Give the following instructions to the students:
   - *If using the whole-class presentation format,* tell the students, "Okay, before we get started I want to explain what we are going to do next. Your group will have an opportunity to share your argument with the rest of the class. After you are done, everyone else in the class will have a chance to ask questions and offer some suggestions about ways to make your group's argument better. After we have a chance to listen to each other and learn something new, I'm going to give you some time to revise your arguments and make them better."

Investigation 11. Patterns in Shadows:
How Does the Location of the Sun in the Sky Affect the Direction and Length of Shadows?

- *If using the gallery walk format,* tell the students, "Okay, before we get started I want to explain what we are going to do next. You are going to read the arguments that were created by other groups. When I say 'go,' your group will go to a different group's station so you can see their argument. Once you are there, I'll give your group a few minutes to read and review their argument. Your job is to offer them some suggestions about ways to make their argument better. You can use sticky notes to give them suggestions. Please be specific about what you want to change and how you think they should change it. After we have a chance to learn from each other, I'm going to give you some time to revise your arguments and make them better."

- *If using the modified gallery walk format,* tell the students, "Okay, before we get started I want to explain what we are going to do next. I'm going to ask some of you to present your arguments to your classmates. If you are presenting your argument, your job is to share your group's claim, evidence, and justification of the evidence. The rest of you will be travelers. If you are a traveler, your job is to listen to the presenters, ask the presenters questions if you do not understand something, and then offer them some suggestions about ways to make their argument better. After we have a chance to learn from each other, I'm going to give you some time to revise your arguments and make them better."

4. Use a document camera to project the "Ways to IMPROVE our argument ..." box from the Investigation Handout (or the Investigation Log in their workbook) on a screen or board (or take a picture of it and project the picture on a screen or board).

    - *If using the whole-class presentation format,* point to the box and tell the students, "After your group presents your argument, you can write down the suggestions you get from your classmates here. If you are listening to a presentation and you see a good idea from another group, you can write down that idea here as well. Once we are done with the presentations, I will give you a chance to use these suggestions or ideas to improve your arguments."

    - *If using the gallery walk format,* point to the box and tell the students, "If you see a good idea from another group, you can write it down here. Once we are done reviewing the different arguments, I will give you a chance to use these ideas to improve your own arguments. It is important to share ideas like this."

    - *If using the modified gallery walk format,* point to the box and tell the students, "If you are a presenter, you can write down the suggestions you get from the travelers here. If you are a traveler and you see a good idea from another group, you can write down that idea here. Once we are done with the presentations, I will give you a chance to use these suggestions or ideas to improve your arguments."

5. Ask the students, "Do you have any questions about what you need to do?"

6. Answer any questions that come up.

Teacher Notes

7. Give the following instructions:

    - *If using the whole-class presentation format,* tell the students, "Okay. Let's get started."

    - *If using the gallery walk format,* tell the students, "Okay, I'm now going to tell you which argument to go to and review."

    - *If using the modified gallery walk format,* tell the students, "Okay, I'm now going to assign you to be a presenter or a traveler." Assign one or two students from each group to be presenters and one or two students from each group to be travelers.

8. Give the students an opportunity to review the arguments.

    - *If using the whole-class presentation format,* have each group present their argument one at a time. Give each group only two to three minutes to present their argument. Then give the class two to three minutes to ask them questions and offer suggestions. Encourage as much participation from the students as possible.

    - *If using the gallery walk format,* tell the students, "Okay. Let's get started. Each group, move one argument to the left. Don't move to the next argument until I tell you to move. Once you get there, read the argument and then offer suggestions about how to make it better. I will put some sticky notes next to each argument. You can use the sticky notes to leave your suggestions." Give each group about three to four minutes to read the arguments, talk, and offer suggestions.

        a. After three to four minutes, tell the students, "Okay. Let's move on to the next argument. Please move one group to the left."

        b. Again, give each group three to four minutes to read, talk, and offer suggestions.

        c. Repeat this process until each group has had their argument read and critiqued three times.

    - *If using the modified gallery walk format,* tell the students, "Okay. Let's get started. Reviewers, move one group to the left. Don't move to the next group until I tell you to move. Presenters, go ahead and share your argument with the travelers when they get there." Give each group of presenters and travelers about three to four minutes to talk.

        a. Tell the students, "Okay. Let's move on to the next argument. Travelers, move one group to the left."

        b. Again, give each group of presenters and travelers about three to four minutes to talk.

        c. Repeat this process until each group has had their argument read and critiqued three times.

9. Tell the students to return to their workstations.

Investigation 11. Patterns in Shadows:
How Does the Location of the Sun in the Sky Affect the Direction and Length of Shadows?

10. Give the following instructions about revising the argument:
    - *If using the whole-class presentation format,* tell the students, "I'm now going to give you all about 10 minutes to revise your argument. Take a few minutes to talk in your groups and determine what you want to change to make your argument better. Once you have decided what to change, go ahead and make the changes to your whiteboard."
    - *If using the gallery walk format,* tell the students, "I'm now going to give you all about 10 minutes to revise your argument. Take a few minutes to read the suggestions that were left at your argument. Then talk in your groups and determine what you want to change to make your argument better. Once you have decided what to change, go ahead and make the changes to your whiteboard."
    - *If using the modified gallery walk format,* tell the students, "I'm now going to give you all about 10 minutes to revise your argument. Please return to your original groups." Wait for the students to move back into their original groups and then tell the students, "Okay, take a few minutes to talk in your groups and determine what you want to change to make your argument better. Once you have decided what to change, go ahead and make the changes to your whiteboard."
11. Ask the students, "Do you have any questions about what you need to do?"
12. Answer any questions that come up.
13. Tell the students, "Okay. Let's get started."
14. Give the students 10 minutes to work in their groups on their arguments. As they work, move from group to group to check in, ask probing questions, and offer a suggestion if a group gets stuck.

Stage 5: Reflective Discussion (15 minutes)

1. Tell the students, "We are now going to take a minute to talk about some of the core ideas and crosscutting concepts that we have used during our investigation."
2. Show Figure 11.3 (p. 452) on the screen.
3. Ask the students, "What do you all see going on here?"
4. Allow the students to share their ideas.
5. Show Figure 11.4 (p. 452) on the screen.
6. Ask the students, "What do you all see going on here?"
7. Allow the students to share their ideas.
8. Show Figures 11.3 and 11.4 on the screen at the same time.

Teacher Notes

FIGURE 11.3
An action figure casting a long shadow

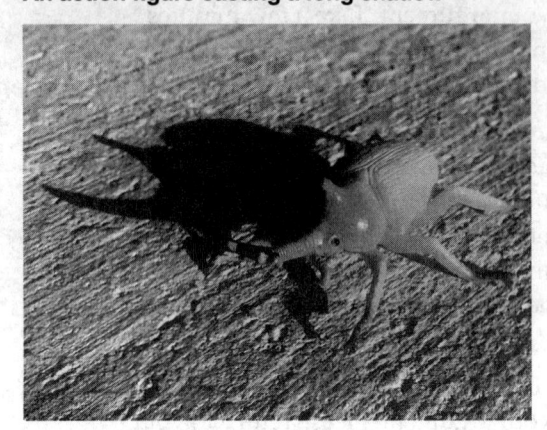

FIGURE 11.4
An action figure casting a small shadow

9. Ask the students, "How could we use what we know about how the Sun changes location in the sky over the course of a day to determine when these pictures were taken?"

10. Allow the students to share their ideas. As they share their ideas, ask different questions to encourage them to expand on their thinking (e.g., "Can you tell me more about that?"), clarify a contribution (e.g., "Can you say that in another way?"), support an idea (e.g., "Why do you think that?"), add to an idea mentioned by a classmate (e.g., "Would anyone like to add to the idea?"), re-voice an idea offered by a classmate (e.g., "Who can explain that to me in another way?"), or critique an idea during the discussion (e.g., "Do you agree or disagree with that idea and why?") until students are able to generate an adequate explanation.

11. Tell the students, "We needed to think about the components of a system, how these components interact, and how to create a model of this system to learn more about shadows." Then ask, "Can anyone tell me why it is useful to create models of systems?"

12. Allow the students to share their ideas.

13. Tell the students, "We also needed to look for patterns during our investigation." Then ask, "Can anyone tell me why it is useful to look for patterns?"

14. Allow the students to share their ideas.

15. Show an image of the question "What do you think are the most important core ideas or crosscutting concepts that we used during this investigation to help us make sense of what we observed?" Tell the students, "Okay, let's make sure we are all on the same page. Please take a moment to discuss this question with the other people in your group." Give them a few minutes to discuss the question.

16. Ask the students, "What do you all think? Who would like to share?"

17. Allow the students to share their ideas.

18. Tell the students, "We are now going to take a minute to talk about what scientists do to figure out answers to their questions."

19. Show an image of the question "Do all scientists follow the same method regardless of what they are trying to figure out?" on the screen. Tell the students, "Take a few minutes to talk about how you would answer this question with the other people in your group. Be ready to share with the rest of the class." Give the students two to three minutes to talk in their group.

20. Ask the students, "What do you all think? Who would like to share an idea?"

21. Allow the students to share their ideas.

22. Tell the students, "I think there is no universal step-by step scientific method that all scientists follow; rather, scientists who work in different scientific disciplines such as biology or physics and fields within a discipline such as genetics or ecology use different types of methods, use different core ideas, and rely on different standards to figure out how the world works."

23. Ask the students, "What might be some different methods that scientists can use?" Allow the students to share their ideas.

24. Tell the students, "There are many examples of different methods, including experiments, systematic observations, creating physical models to test ideas, literature reviews, and analysis of existing data sets. The choice of method depends on what the scientist is trying to accomplish."

25. Ask the students, "What method do you think we used to today?"

26. Allow the students to share their ideas.

27. Tell the students, "I think we created a physical model to test our ideas because we were looking for patterns in the shadows. It wasn't an experiment because we did not have a control group and we couldn't move the Sun to a specific spot to see what happens."

28. Ask the students, "Does anyone have any questions about the different methods that scientists use or why scientists use different methods?"

29. Answer any questions that come up.

30. Tell the students, "We are now going to take a minute to talk about what went well and what didn't go so well during our investigation. We need to talk about this because you all are going to be planning and carrying out your own investigations like this a lot this year, and I want to help you all get better at it."

31. Show an image of the question "What made your investigation scientific?" on the screen. Tell the students, "Take a few minutes to talk about how you would answer this question with the other people in your group. Be ready to share

Teacher Notes

with the rest of the class." Give the students two to three minutes to talk in their group.

32. Ask the students, "What do you all think? Who would like to share an idea?"

33. Allow the students to share their ideas. Be sure to expand on their ideas about what makes an investigation scientific.

34. Show an image of the question "What made your investigation not so scientific?" on the screen. Tell the students, "Take a few minutes to talk about how you would answer this question with the other people in your group. Be ready to share with the rest of the class." Give the students two to three minutes to talk in their group.

35. Ask the students, "What do you all think? Who would like to share an idea?"

36. Allow the students to share their ideas. Be sure to expand on their ideas about what makes an investigation less scientific.

37. Show an image of the question "What rules can we put into place to help us make sure our next investigation is more scientific?" on the screen. Tell the students, "Take a few minutes to talk about how you would answer this question with the other people in your group. Be ready to share with the rest of the class." Give the students two to three minutes to talk in their group.

38. Ask the students, "What do you all think? Who would like to share an idea?"

39. Allow the students to share their ideas. Once they have shared their ideas, offer a suggestion for a possible class rule.

40. Ask the students, "What do you all think? Should we make this a rule?"

41. If the students agree, write the rule on the board or on a class "Rules for Scientific Investigation" chart so you can refer to it during the next investigation.

Stage 6: Write a Draft Report (30 minutes)

Your students will use either the Investigation Handout or the Investigation Log in the student workbook when writing the draft report. When you give the directions shown in quotes in the following steps, substitute "Investigation Log in your workbook" or just "Investigation Log" (as shown in brackets) for "handout" if they are using the workbook.

1. Tell the students, "You are now going to write an investigation report to share what you have learned. Please take out a pencil and turn to the 'Draft Report' section of your handout [Investigation Log in your workbook]."

2. If possible, use a document camera to project the "Introduction" section of the draft report from the Investigation Handout (or the Investigation Log in their workbook) on a screen or board (or take a picture of it and project the picture on a screen or board).

Investigation 11. Patterns in Shadows:
How Does the Location of the Sun in the Sky Affect the Direction and Length of Shadows?

3. Tell the students, "The first part of the report is called the 'Introduction.' In this section of the report you want to explain to the reader what you were investigating, why you were investigating it, and what question you were trying to answer. All this information can be found in the text at the beginning of your handout [Investigation Log]." Point to the image. "Here are some sentence starters to help you begin writing."

4. Ask the students, "Do you have any questions about what you need to do?"

5. Answer any questions that come up.

6. Tell the students, "Okay, let's write."

7. Give the students 10 minutes to write the "Introduction" section of the report. As they work, move from student to student to check in, ask probing questions, and offer a suggestion if a student gets stuck.

8. If possible, use a document camera to project the "Method" section of the draft report from the Investigation Handout (or the Investigation Log in their workbook) on a screen or board (or take a picture of it and project the picture on a screen or board).

9. Tell the students, "The second part of the report is called the 'Method.' In this section of the report you want to explain to the reader what you did during the investigation, what data you collected and why, and how you went about analyzing your data. All this information can be found in the 'Plan Your Investigation' section of the handout [Investigation Log]. Remember that you all planned and carried out different investigations, so do not assume that the reader will know what you did." Point to the image. "Here are some sentence starters to help you begin writing."

10. Ask the students, "Do you have any questions about what you need to do?"

11. Answer any questions that come up.

12. Tell the students, "Okay, let's write."

13. Give the students 10 minutes to write the "Method" section of the report. As they work, move from student to student to check in, ask probing questions, and offer a suggestion if a student gets stuck.

14. If possible, use a document camera to project the "Argument" section of the draft report from the Investigation Handout (or the Investigation Log in their workbook) on a screen or board (or take a picture of it and project the picture on a screen or board).

15. Tell the students, "The last part of the report is called the 'Argument.' In this section of the report you want to share your claim, evidence, and justification of the evidence with the reader. All this information can be found on your whiteboard." Point to the image. "Here are some sentence starters to help you begin writing."

16. Ask the students, "Do you have any questions about what you need to do?"
17. Answer any questions that come up.
18. Tell the students, "Okay, let's write."
19. Give the students 10 minutes to write the "Argument" section of the report. As they work, move from student to student to check in, ask probing questions, and offer a suggestion if a student gets stuck.

Stage 7: Peer Review (35 minutes)

Your students will use either the Investigation Handout or their workbook when doing the peer review. Except where noted below, the directions are the same whether using the handout or the workbook.

1. Tell the students, "We are now going to review our reports to find ways to make them better. I'm going to come around and collect your draft reports. While I do that, please take out a pencil."
2. Collect the handouts or the workbooks with the draft reports from the students.
3. If possible, use a document camera to project the peer-review guide (see Appendix 4) on a screen or board (or take a picture of it and project the picture on a screen or board).
4. Tell the students, "We are going to use this peer-review guide to give each other feedback." Point to the image.
5. Tell the students, "I'm going to ask you to work with a partner to do this. I'm going to give you and your partner a draft report to read. You two will then read the report together. Once you are done reading the report, I want you to answer each of the questions on the peer-review guide." Point to the review questions on the image of the peer-review guide.
6. Tell the students, "You can check 'no,' 'almost,' or 'yes' after each question." Point to the checkboxes on the image of the peer-review guide.
7. Tell the students, "This will be your rating for this part of the report. Make sure you agree on the rating you give the author. If you mark 'no' or 'almost,' then you need to tell the author what he or she needs to do to get a 'yes.'" Point to the space for the reviewer feedback on the image of the peer-review guide.
8. Tell the students, "It is really important for you to give the authors feedback that is helpful. That means you need to tell them exactly what they need to do to make their report better."
9. Ask the students, "Do you have any questions about what you need to do?"
10. Answer any questions that come up.

Investigation 11. Patterns in Shadows:
How Does the Location of the Sun in the Sky Affect the Direction and Length of Shadows?

11. Tell the students, "Please sit with a partner who is not in your current group." Allow the students time to sit with a partner.

12. Tell the students, "Okay, I'm now going to give you one report to read." Pass out one Investigation Handout with a draft report or one workbook to each pair. Make sure that the report you give a pair was not written by one of the students in that pair. Give each pair one peer-review guide to fill out. If the students are using workbooks, the peer-review guide is included right after the draft report so you do not need to pass out copies of the peer-review guide.

13. Tell the students, "Okay, I'm going to give you 15 minutes to read the report I gave you and to fill out the peer-review guide. Go ahead and get started."

14. Give the students 15 minutes to work. As they work, move around from pair to pair to check in and see how things are going, answer questions, and offer advice.

15. After 15 minutes pass, tell the students, "Okay, time is up. Please give me the report and the peer-review guide that you filled out."

16. Collect the Investigation Handouts and the peer-review guides, or collect the workbooks if students are using them. If the students are using the Investigation Handouts and separate peer-review guides, be sure you keep each handout with its corresponding peer-review guide.

17. Tell the students, "Okay, I am now going to give you a different report to read and a new peer-review guide to fill out." Pass out one more report to each pair. Make sure that the report you give a pair was not written by one of the students in that pair. Give each pair a new peer-review guide to fill out as a group.

18. Tell the students, "Okay, I'm going to give you 15 minutes to read this new report and to fill out the peer-review guide. Go ahead and get started."

19. Give the students 15 minutes to work. As they work, move around from pair to pair to check in and see how things are going, answer questions, and offer advice.

20. After 15 minutes pass, tell the students, "Okay, time is up. Please give me the report and the peer-review guide that you filled out."

21. Collect the Investigation Handouts and the peer-review guides, or collect the workbooks if students are using them. If the students are using the Investigation Handouts and separate peer-review guides, be sure you keep each handout with its corresponding peer-review guide.

Stage 8: Revise the Report (30 minutes)

Your students will use either the Investigation Handout or their workbook when revising the report. Except where noted below, the directions are the same whether using the handout or the workbook.

Teacher Notes

1. Tell the students, "You are now going to revise your draft report based on the feedback you get from your classmates. Please take out a pencil."

2. Return the reports to the students.
   - *If the students used the Investigation Handout and a copy of the peer-review guide,* pass back the handout and the peer-review guide to each student.
   - *If the students used the workbook,* pass that back to each student.

3. Tell the students, "Please take a few minutes to read over the peer-review guide. You should use it to figure out what you need to change in your report and how you will change it."

4. Allow the students to read the peer-review guide.

5. *If the students used the workbook,* if possible use a document camera to project the "Write Your Final Report" section from the Investigation Log on a screen or board (or take a picture of it and project the picture on a screen or board).

6. Give the following directions about how to revise their reports:
   - *If the students used the Investigation Handout and a copy of the peer-review guide,* tell them, "Okay, let's revise our reports. Please take out a piece of paper. I would like you to rewrite your report. You can use your draft report as a starting point, but you also need to change it to make it better. Use the feedback on the peer-review guide to make it better."
   - *If the students used the workbook,* tell them, "Okay, let's revise our reports. I would like you to rewrite your report in the section of the Investigation Log called "Write Your Final Report." You can use your draft report as a starting point, but you also need to change it to make it better. Use the feedback on the peer-review guide to make it better."

7. Ask the students, "Do you have any questions about what you need to do?"

8. Answer any questions that come up.

9. Tell the students, "Okay, let's write." Allow about 20 minutes for the students to revise their reports.

10. After about 20 minutes, give the following directions:
    - *If the students used the Investigation Handout,* tell them, "Okay, time's up. I will now come around and collect your Investigation Handout, the peer-review guide, and your final report."
    - *If the students used the workbook,* tell them, "Okay, time's up. I will now come around and collect your workbooks."

11. *If the students used the Investigation Handout,* collect all the Investigation Handouts, peer-review guides, and final reports. *If the students used the workbook,* collect all the workbooks.

Investigation 11. Patterns in Shadows:
How Does the Location of the Sun in the Sky Affect the Direction and Length of Shadows?

12. *If the students used the Investigation Handout,* use the "Teacher Score" column in the peer-review guide to grade the final report. *If the students used the workbook,* use the "Investigation Report Grading Rubric" in the Investigation Log to grade the final report. Whether you are using the handout or the log, you can give the students feedback about their writing in the "Teacher Comments" section.

How to Use the Checkout Questions

The Checkout Questions are an optional assessment. We recommend giving them to students at the start of the next class period after the students finish stage 8 of the investigation. You can then look over the student answers to determine if you need to reteach the core idea from the investigation. Appendix 6 gives the answers to the Checkout Questions that should be given by a student who can apply the core idea correctly and can explain the pattern.

Alignment With Standards

Table 11.2 highlights how the investigation can be used to address specific performance expectations from the *Next Generation Science Standards, Common Core State Standards for English Language Arts (CCSS ELA)* and *Common Core State Standards for Mathematics (CCSS Mathematics),* and *English Language Proficiency (ELP) Standards.*

TABLE 11.2

Investigation 11 alignment with standards

NGSS performance expectation	5-ESS1-2: Represent data in graphical displays to reveal patterns of daily changes in length and direction of shadows, day and night, and the seasonal appearance of some stars in the night sky.
CCSS ELA—Reading: Informational Text	Key ideas and details • CCSS.ELA-LITERACY.RI.5.1: Quote accurately from a text when explaining what the text says explicitly and when drawing inferences from the text. • CCSS.ELA-LITERACY.RI.5.2: Determine two or more main ideas of a text and explain how they are supported by key details; summarize the text. • CCSS.ELA-LITERACY.RI.5.3: Explain the relationships or interactions between two or more individuals, events, ideas, or concepts in a historical, scientific, or technical text based on specific information in the text. Craft and structure • CCSS.ELA-LITERACY.RI.5.4: Determine the meaning of general academic and domain-specific words and phrases in a text relevant to a *grade 5 topic or subject area*. • CCSS.ELA-LITERACY.RI.5.5: Compare and contrast the overall structure (e.g., chronology, comparison, cause/effect, problem/solution) of events, ideas, concepts, or information in two or more texts.

Continued

Teacher Notes

Table 11.2 (continued)

CCSS ELA—Reading: Informational Text (continued)	• CCSS.ELA-LITERACY.RI.5.6: Analyze multiple accounts of the same event or topic, noting important similarities and differences in the point of view they represent. Integration of knowledge and ideas • CCSS.ELA-LITERACY.RI.5.7: Draw on information from multiple print or digital sources, demonstrating the ability to locate an answer to a question quickly or to solve a problem efficiently. • CCSS.ELA-LITERACY.RI.5.8: Explain how an author uses reasons and evidence to support particular points in a text, identifying which reasons and evidence support which point(s). Range of reading and level of text complexity • CCSS.ELA-LITERACY.RI.5.10: By the end of the year, read and comprehend informational texts, including history/social studies, science, and technical texts, at the high end of the grades 4–5 text complexity band independently and proficiently.

CCSS ELA—Writing	Text types and purposes
	
	• CCSS.ELA-LITERACY.W.5.1: Write opinion pieces on topics or texts, supporting a point of view with reasons.
	   ○ CCSS.ELA-LITERACY.W.5.1.A: Introduce a topic or text clearly, state an opinion, and create an organizational structure in which ideas are logically grouped to support the writer's purpose.
	   ○ CCSS.ELA-LITERACY.W.5.1.B: Provide logically ordered reasons that are supported by facts and details.
	   ○ CCSS.ELA-LITERACY.W.5.1.C: Link opinion and reasons using words, phrases, and clauses (e.g., *consequently*, *specifically*).
	   ○ CCSS.ELA-LITERACY.W.5.1.D: Provide a concluding statement or section related to the opinion presented.
	
	• CCSS.ELA-LITERACY.W.5.2: Write informative or explanatory texts to examine a topic and convey ideas and information clearly.
	   ○ CCSS.ELA-LITERACY.W.5.2.A: Introduce a topic clearly, provide a general observation and focus, and group related information logically; include formatting (e.g., headings), illustrations, and multimedia when useful to aiding comprehension.
	   ○ CCSS.ELA-LITERACY.W.5.2.B: Develop the topic with facts, definitions, concrete details, quotations, or other information and examples related to the topic.
	   ○ CCSS.ELA-LITERACY.W.5.2.C: Link ideas within and across categories of information using words, phrases, and clauses (e.g., *in contrast*, *especially*).
	   ○ CCSS.ELA-LITERACY.W.5.2.D: Use precise language and domain-specific vocabulary to inform about or explain the topic.
	   ○ CCSS.ELA-LITERACY.W.5.2.E: Provide a concluding statement or section related to the information or explanation presented.

Continued

Investigation 11. Patterns in Shadows: How Does the Location of the Sun in the Sky Affect the Direction and Length of Shadows?

Table 11.2 (*continued*)

CCSS ELA—Writing (*continued*)	Production and distribution of writing
	• CCSS.ELA-LITERACY.W.5.4: Produce clear and coherent writing in which the development and organization are appropriate to task, purpose, and audience.
	• CCSS.ELA-LITERACY.W.5.5: With guidance and support from peers and adults, develop and strengthen writing as needed by planning, revising, editing, rewriting, or trying a new approach.
	• CCSS.ELA-LITERACY.W.5.6: With some guidance and support from adults, use technology, including the internet, to produce and publish writing as well as to interact and collaborate with others; demonstrate sufficient command of keyboarding skills to type a minimum of two pages in a single sitting.
	Research to build and present knowledge
	• CCSS.ELA-LITERACY.W.5.8: Recall relevant information from experiences or gather relevant information from print and digital sources; summarize or paraphrase information in notes and finished work, and provide a list of sources.
	• CCSS.ELA-LITERACY.W.5.9: Draw evidence from literary or informational texts to support analysis, reflection, and research.
	Range of writing
	• CCSS.ELA-LITERACY.W.5.10: Write routinely over extended time frames (time for research, reflection, and revision) and shorter time frames (a single sitting or a day or two) for a range of discipline-specific tasks, purposes, and audiences.
CCSS ELA— Speaking and Listening	Comprehension and collaboration
	• CCSS.ELA-LITERACY.SL.5.1: Engage effectively in a range of collaborative discussions (one-on-one, in groups, and teacher-led) with diverse partners on *grade 5 topics and texts,* building on others' ideas and expressing their own clearly.
	○ CCSS.ELA-LITERACY.SL.5.1.A: Come to discussions prepared, having read or studied required material; explicitly draw on that preparation and other information known about the topic to explore ideas under discussion.
	○ CCSS.ELA-LITERACY.SL.5.1.B: Follow agreed-upon rules for discussions and carry out assigned roles.
	○ CCSS.ELA-LITERACY.SL.5.1.C: Pose and respond to specific questions by making comments that contribute to the discussion and elaborate on the remarks of others.
	○ CCSS.ELA-LITERACY.SL.5.1.D: Review the key ideas expressed and draw conclusions in light of information and knowledge gained from the discussions.
	• CCSS.ELA-LITERACY.SL.5.2: Summarize a written text read aloud or information presented in diverse media and formats, including visually, quantitatively, and orally.
	• CCSS.ELA-LITERACY.SL.5.3: Summarize the points a speaker makes and explain how each claim is supported by reasons and evidence.

Continued

Teacher Notes

Table 11.2 (*continued*)

CCSS ELA—Speaking and Listening (*continued*)	Presentation of knowledge and ideas • CCSS.ELA-LITERACY.SL.5.4: Report on a topic or text or present an opinion, sequencing ideas logically and using appropriate facts and relevant, descriptive details to support main ideas or themes; speak clearly at an understandable pace. • CCSS.ELA-LITERACY.SL.5.5: Include multimedia components (e.g., graphics, sound) and visual displays in presentations when appropriate to enhance the development of main ideas or themes. • CCSS.ELA-LITERACY.SL.5.6: Adapt speech to a variety of contexts and tasks, using formal English when appropriate to task and situation.
CCSS Mathematics—Measurement and Data	Convert like measurement units within a given measurement system. • CCSS.MATH.CONTENT.5.MD.A.1: Convert among different-sized standard measurement units within a given measurement system (e.g., convert 5 cm to 0.05 m), and use these conversions in solving multi-step, real-world problems.
ELP Standards	Receptive modalities • ELP 1: Construct meaning from oral presentations and literary and informational text through grade-appropriate listening, reading, and viewing. • ELP 8: Determine the meaning of words and phrases in oral presentations and literary and informational text. Productive modalities • ELP 3: Speak and write about grade-appropriate complex literary and informational texts and topics. • ELP 4: Construct grade-appropriate oral and written claims and support them with reasoning and evidence. • ELP 7: Adapt language choices to purpose, task, and audience when speaking and writing. Interactive modalities • ELP 2: Participate in grade-appropriate oral and written exchanges of information, ideas, and analyses, responding to peer, audience, or reader comments and questions. • ELP 5: Conduct research and evaluate and communicate findings to answer questions or solve problems. • ELP 6: Analyze and critique the arguments of others orally and in writing. Linguistic structures of English • ELP 9: Create clear and coherent grade-appropriate speech and text. • ELP 10: Make accurate use of standard English to communicate in grade-appropriate speech and writing.

Investigation Handout

Investigation 11

Patterns in Shadows: How Does the Location of the Sun in the Sky Affect the Direction and Length of Shadows?

Introduction

You will see lots of different shadows on the ground when the Sun is out and the sky is clear of clouds. The shadows that you see will change over time. Your teacher will show you a video of how shadows can change over the course of a day. As you watch the video, keep track of things you notice and things you wonder about in the boxes below.

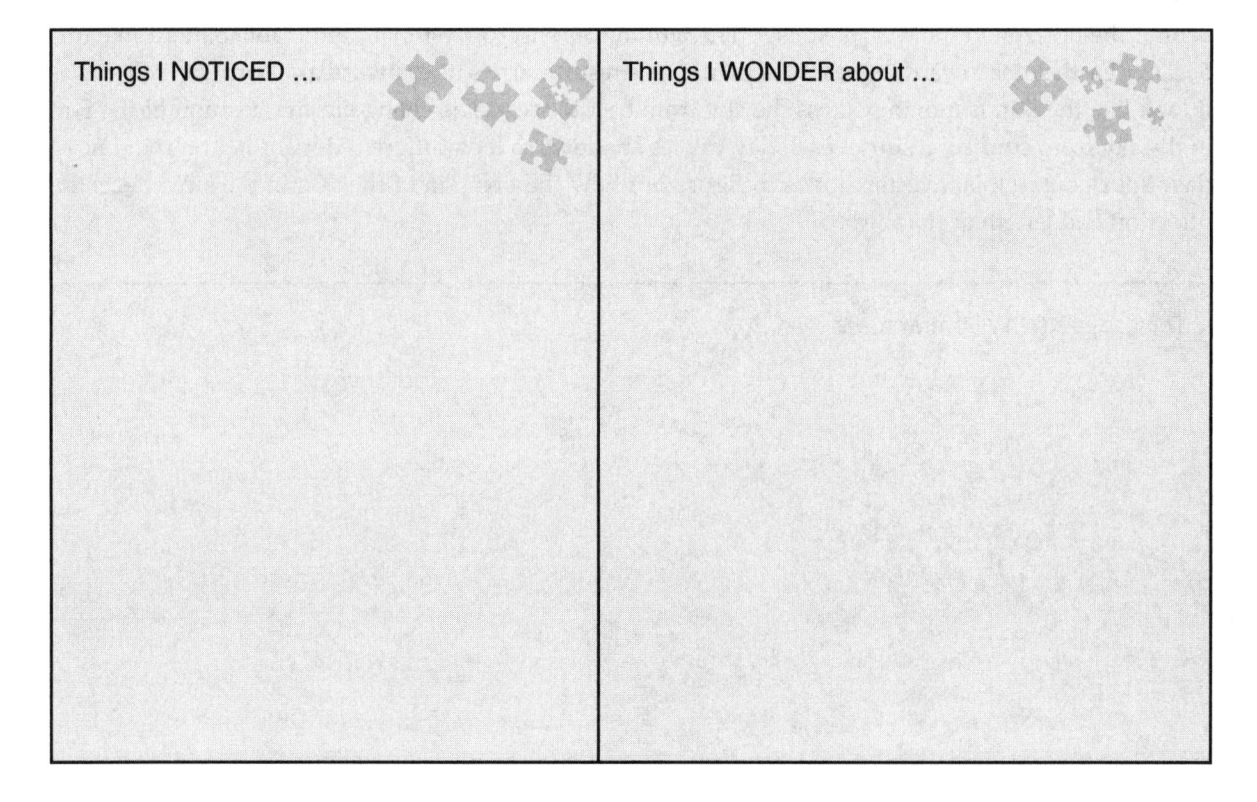

Things I NOTICED …	Things I WONDER about …

A *shadow* is a dark area that is created when an object blocks the light from a light source. A light source is anything that produces light. An object that blocks light is called an *opaque* object. A shadow will appear any time an opaque object is placed in the path of light because the light from the light source is unable to travel through that object.

The light source for most of the shadows that we see outside during the day is the Sun. When light from the Sun strikes an opaque object, such as a tree, a building, a bicycle, or a person, that object will

Investigation Handout

create a shadow. The Sun, however, sets each evening, so the light sources for most of the shadows that we see outside at night are street lamps, ceiling lights, and flashlights.

We have day and night on Earth because only one-half of Earth is facing the Sun at any given time. It is day on the side of Earth that is facing the Sun and night on the side of Earth that is facing away from the Sun. The side of Earth that is facing the Sun changes over time because Earth rotates or spins counterclockwise on an axis once every 24 hours. The axis is a line that runs between the North and South Poles through the center of Earth. The picture on the right shows a model of how Earth rotates on its axis.

The Sun does not stay in the same position in the sky during the daylight hours. The Sun appears to come up from under the horizon in the east at sunrise, moves to its highest point in the sky at noon, and then disappears behind the horizon in the west at sunset. The Sun moves east to west in a predictable pattern like this because Earth spins toward the east. The planet, as a result, carries us eastward as it turns, which makes it look like the Sun is moving across the sky from east to west. The change in the position of the Sun in the sky from sunrise to sunset each day causes shadows to look different depending on the time of day. Your goal in this investigation is to figure out how the position of the Sun in the sky affects the direction and length of shadows.

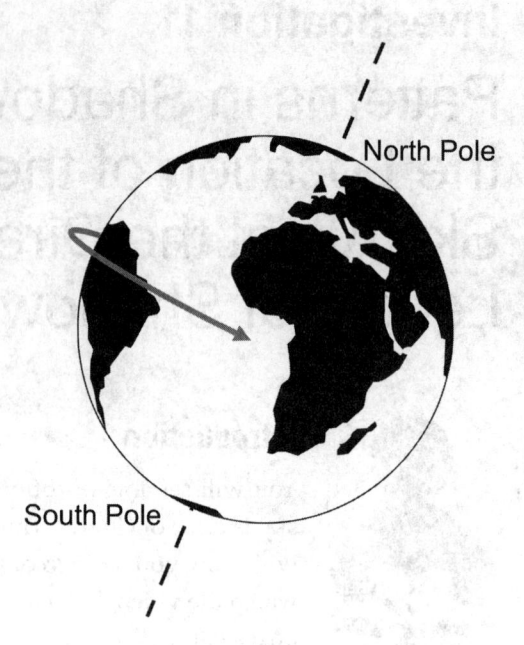

Things we KNOW from what we read ...

Investigation 11. Patterns in Shadows:
How Does the Location of the Sun in the Sky Affect the Direction and Length of Shadows?

Your Task

Use what you know about patterns, systems and system models, and the rotation of Earth about an axis between its North and South Poles to design and carry out an investigation to figure out how the location of the Sun affects the length and position of shadows.

The *guiding question* of this investigation is, **How does the location of the Sun in the sky affect the direction and length of shadows?**

Materials

You may use any of the following materials during your investigation:

- Safety goggles (required)
- Wooden dowel (6 inches)
- One-hole rubber stopper
- Grid paper (1-cm grid)
- String
- Meterstick
- Protractor
- Flashlight
- Compass

Safety Rules

Follow all normal safety rules. In addition, be sure to follow these rules:

- Wear sanitized safety goggles during setup, investigation activity, and cleanup.
- Don't put any of the materials in your mouth.
- Wash your hands with soap and water when you are done cleaning up.

Plan Your Investigation

Prepare a plan for your investigation by filling out the chart on the next page; this plan is called an *investigation proposal*. Before you start developing your plan, be sure to discuss the following questions with the other members of your group:

- How can we use **patterns** to help us determine a **cause-and-effect relationship**?
- What are the components of the **system** we are studying, how do the components interact, and how could we create a **model** of it?

Investigation Handout

Our guiding question:

This is a picture of how we will set up the equipment:

We will collect the following data:

These are the steps we will follow to collect data:

I approve of this investigation proposal.

Teacher's signature   Date

National Science Teaching Association

Investigation 11. Patterns in Shadows:
How Does the Location of the Sun in the Sky Affect the Direction and Length of Shadows?

Collect Your Data

Keep a record of what you measure or observe during your investigation in the space below.

Analyze Your Data

You will need to analyze the data you collected before you can develop an answer to the guiding question. To analyze the data you collected, create a table or graph that shows the relationship between what you changed and what you measured or observed as a result of what you changed.

Argument-Driven Inquiry in **Fifth-Grade Science**: Three-Dimensional Investigations

Investigation Handout

Draft Argument

Develop an argument on a whiteboard. It should include the following:

1. A *claim:* Your answer to the guiding question.
2. *Evidence:* An analysis of the data and an explanation of what the analysis means.
3. A *justification of the evidence:* Why your group thinks the evidence is important.

The Guiding Question:	
Our Claim:	
Our Evidence:	Our Justification of the Evidence:

Argumentation Session

Share your argument with your classmates. Be sure to ask them how to make your draft argument better. Keep track of their suggestions in the space below.

Ways to IMPROVE our argument …

Investigation 11. Patterns in Shadows:
How Does the Location of the Sun in the Sky Affect the Direction and Length of Shadows?

Draft Report

Prepare an *investigation report* to share what you have learned. Use the information in this handout and your group's final argument to write a *draft* of your investigation report.

Introduction

We have been studying _____ in class.

Before we started this investigation, we explored _____

_____

_____

We noticed _____

_____

_____

My goal for this investigation was to figure out _____

_____

_____

The guiding question was _____

_____

_____

Method

To gather the data I needed to answer this question, I _____

_____

_____

_____

_____

I then analyzed the data I collected by _____

Argument-Driven Inquiry in **Fifth-Grade Science:** Three-Dimensional Investigations

Investigation Handout

_____
_____
_____

Argument

My claim is _____
_____
_____
_____

The _____ below shows _____
_____
_____
_____
_____

Investigation 11. Patterns in Shadows:
How Does the Location of the Sun in the Sky Affect the Direction and Length of Shadows?

This analysis of the data I collected suggests _____

_____

_____

_____

_____

This evidence is based on several important scientific concepts. The first one is _____

_____

_____

_____

_____

_____

Review

Your classmates need your help! Review the draft of their investigation reports and give them ideas about how to improve. Use the *peer-review guide* when doing your review.

Submit Your Final Report

Once you have received feedback from your classmates about your draft report, create your final investigation report and hand it in to your teacher.

Checkout Questions

Investigation 11. Patterns in Shadows

1. What does the shadow of the stick look like when the Sun is high in the sky? Draw the length and direction of the stick's shadow in the picture below.

2. What does the shadow of the stick look like when the Sun is near the horizon? Draw the length and direction of the stick's shadow in the picture below.

3. Explain your thinking. What pattern helped you predict how the shadows will look?

_____
_____
_____
_____
_____

Investigation 11. Patterns in Shadows:
How Does the Location of the Sun in the Sky Affect the Direction and Length of Shadows?

Teacher Scoring Rubric for the Checkout Questions

Level	Description
3	The student can apply the core idea correctly and can explain the pattern.
2	The student can apply the core idea correctly but cannot explain the pattern.
1	The student cannot apply the core idea correctly but can explain the pattern.
0	The student cannot apply the core idea correctly and cannot explain the pattern.

Teacher Notes

Investigation 12

Daylight and Location: Why Does the Duration of Daylight Change in Different Locations Throughout the Year, and Why Isn't the Pattern the Same in All Locations on Earth?

Purpose

The purpose of this investigation is to give students an opportunity to use one disciplinary core idea (DCI), two crosscutting concepts (CCs), and eight scientific and engineering practices (SEPs) to figure out why the duration of daylight at a given location on Earth changes over the course of a year and why different locations on Earth have a different duration of daylight on the same day. Students will also learn about how scientists use patterns and models to study and understand systems within nature.

The DCI, CCs, and SEPs That Students Use During This Investigation to Figure Things Out

DCI

- *ESS1.B: Earth and the Solar System:* The orbits of Earth around the Sun and of the Moon around Earth, together with the rotation of Earth about an axis between its North and South Poles, cause observable patterns. These include day and night; daily changes in the length and direction of shadows; and different positions of the Sun, Moon, and stars at different times of the day, month, and year.

CCs

- *CC 1: Patterns:* Similarities and differences in patterns can be used to sort, classify, communicate, and analyze simple rates of change for natural phenomena and designed products. Patterns of change can be used to make predictions. Patterns can be used as evidence to support an explanation.
- *CC 4: Systems and System Models:* A system can be described in terms of its components and their interactions. A system is a group of related parts that make up a whole and can carry out functions its individual parts cannot.

SEPs

- *SEP 1: Asking Questions and Defining Problems:* Ask questions about what would happen if a variable is changed. Ask questions that can be investigated and predict reasonable outcomes based on patterns such as cause-and-effect relationships.

Investigation 12. Daylight and Location:
Why Does the Duration of Daylight Change in Different Locations
Throughout the Year, and Why Isn't the Pattern the Same in All Locations on Earth?

- *SEP 2: Developing and Using Models:* Develop and/or use models to describe and/or predict phenomena.

- *SEP 3: Planning and Carrying Out Investigations:* Plan and conduct an investigation collaboratively to produce data to serve as the basis for evidence, using fair tests in which variables are controlled and the number of trials considered. Evaluate appropriate methods and/or tools for collecting data.

- *SEP 4: Analyzing and Interpreting Data:* Represent data in tables and/or various graphical displays (bar graphs, pictographs, and/or pie charts) to reveal patterns that indicate relationships. Analyze and interpret data to make sense of phenomena, using logical reasoning, mathematics, and/or computation. Compare and contrast data collected by different groups in order to discuss similarities and differences in their findings.

- *SEP 5: Using Mathematics and Computational Thinking:* Organize simple data sets to reveal patterns that suggest relationships. Describe, measure, estimate, and/or graph quantities (e.g., area, volume, weight, time) to address scientific and engineering questions and problems.

- *SEP 6: Constructing Explanations and Designing Solutions:* Construct an explanation of observed relationships. Use evidence to construct or support an explanation. Identify the evidence that supports particular points in an explanation.

- *SEP 7: Engaging in Argument From Evidence:* Compare and refine arguments based on an evaluation of the evidence presented. Distinguish among facts, reasoned judgment based on research findings, and speculation in an explanation. Respectfully provide and receive critiques from peers about a proposed procedure, explanation, or model by citing relevant evidence and posing specific questions.

- *SEP 8: Obtaining, Evaluating, and Communicating Information:* Read and comprehend grade-appropriate complex texts and/or other reliable media to summarize and obtain scientific and technical ideas. Combine information in written text with that contained in corresponding tables, diagrams, and/or charts to support the engagement in other scientific and/or engineering practices. Communicate scientific and/or technical information orally and/or in written formats, including various forms of media as well as tables, diagrams, and charts.

Other Concepts That Students May Use During This Investigation

Students might also use some of the following concepts:

- Earth revolves around or orbits the Sun.
- Earth rotates on an axis.
- One revolution around the Sun is equal to one year or 365 days.
- One rotation of Earth is equal to one day or 24 hours.

Teacher Notes

- Sunrise marks the transition from night to day.
- Sunset marks the transition from day to night.
- The axis of Earth is tilted.
- The side of Earth facing the Sun is illuminated, while the side of Earth facing away from the Sun is not.

What Students Figure Out

Cities that are located farther from the equator experience greater changes in the duration of daylight during the year than cities located near the equator. When a hemisphere of Earth is tilted toward the Sun, cities located in that hemisphere will have a longer duration of daylight during each day compared with cities located at the equator or cities located in the other hemisphere. These phenomena occur because the Earth's axis of rotation is tilted at a 23.5° angle from vertical, and that angle does not change during Earth's orbit around the Sun.

Background Information About This Investigation for the Teacher

Earth and the Sun form a system in which Earth orbits, or revolves around, the Sun once every 365.25 days, or once a year. At the same time, Earth rotates around an imaginary axis. This rotation takes 24 hours, or one day. The duration of daylight at a given location on Earth, however, depends on how far north or south the location is away from the equator. The equator is an imaginary line that divides the Earth in half and creates a Northern Hemisphere and a Southern Hemisphere.

Figure 12.1 depicts Earth being illuminated by the Sun. In this model, half of Earth is illuminated by the Sun (the white side) and the other half (the gray side) is not. It is day

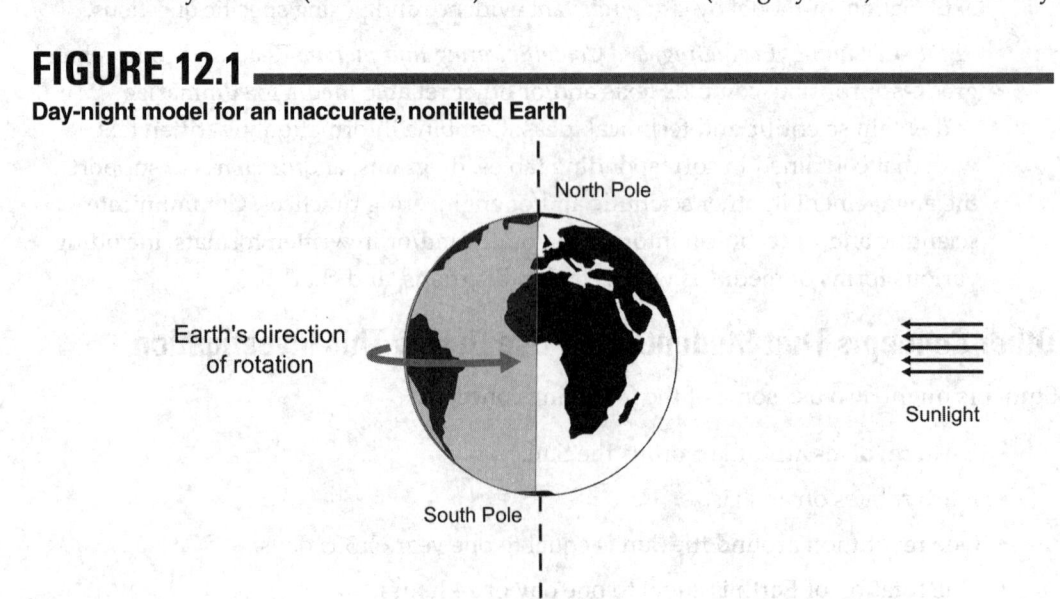

FIGURE 12.1
Day-night model for an inaccurate, nontilted Earth

Investigation 12. Daylight and Location:
Why Does the Duration of Daylight Change in Different Locations Throughout the Year, and Why Isn't the Pattern the Same in All Locations on Earth?

for people who live in cities on the side of Earth that is facing the Sun at a given point in time and night for the people who live in cities on the side that is facing away from the Sun at the same point in time. Notice that in this model the axis of rotation for Earth is perpendicular to the incoming rays of sunlight and that the line dividing the daytime and nighttime halves of Earth is also perpendicular to the incoming rays of sunlight. If this model were accurate, then all cities would experience 12 hours of daylight each day of the year. However, we know this is not the case, so this model is inaccurate.

A revised model for the Earth-Sun relationship is shown in Figure 12.2. In this revised model, Earth's axis of rotation is tilted at a 23.5° angle from vertical. In this model (as in Figure 12.1), half of Earth is illuminated by the Sun (the white side) and the other half (the gray side) is not. The line dividing day and night for people living on Earth is perpendicular to the incoming rays of sunlight in both models; however, in Figure 12.2 that line no longer follows the north-south axis of rotation for Earth.

FIGURE 12.2
Day-night model for an accurate, tilted Earth

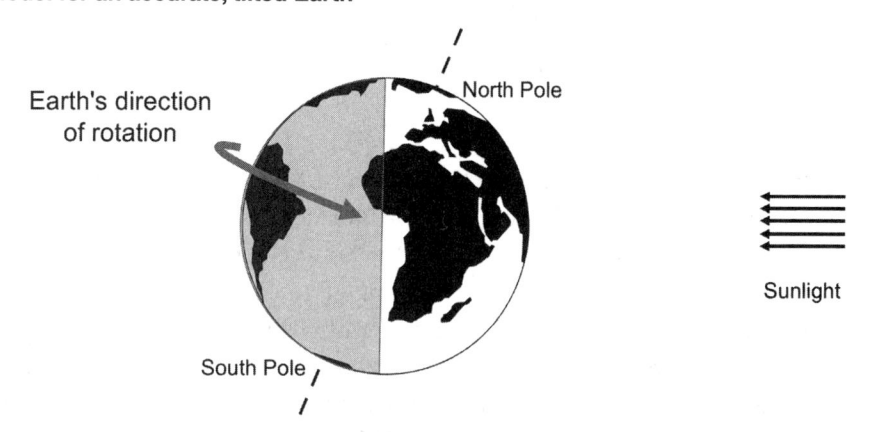

Modeling is an important scientific practice. By creating models, scientists are better able to study and explain various phenomena. In this investigation, the Earth-Sun relationship is difficult to observe because we are in fact standing on Earth, so our view of the orientation of Earth to the Sun and the Sun to Earth is skewed. Using models helps better illustrate these relationships and study them. The two models presented in Figures 12.1 and 12.2 are helpful, but they still lack important information. Figure 12.3 continues to refine the model by illustrating lines of longitude and latitude.

Lines of latitude circle Earth from east to west and run parallel to the equator. Lines of longitude circle Earth from north to south and run roughly perpendicular to the equator. Figure 12.3 (p. 478) shows two cities, City A and City B, that are located along the same line of longitude but along different lines of latitude: City B is located along the equator, and City A

Teacher Notes

is many degrees of latitude north of the equator. Based on this figure, we can determine that people who live in City A see the Sun rise over the horizon in the east, whereas people who live in City B have already seen the Sun rise and have been in the daylight for a few hours. As time passes and Earth continues to rotate on its axis, the people in City A will see the Sun set beneath the horizon in the west before the people in City B will see it set. Using these two cities as examples, it is clear how a city's position relative to the equator dictates the duration of daylight for people who live in that city.

FIGURE 12.3

Day-night model for an accurate, tilted Earth with lines of latitude and longitude

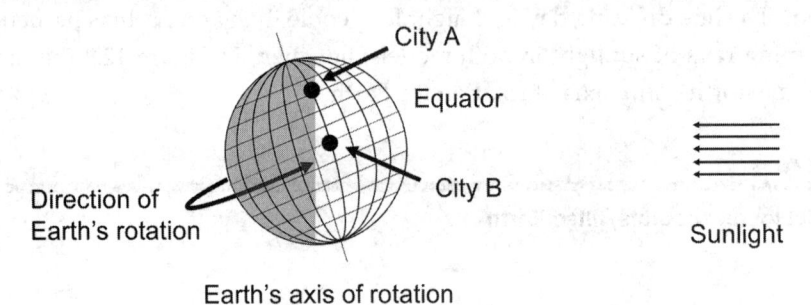

Finally, in the Figure 12.3 model, the Northern Hemisphere of Earth is experiencing the winter season. We know this because the Northern Hemisphere is angled away from the incoming rays of sunlight. In contrast, the Southern Hemisphere is experiencing the summer season because that hemisphere is angled toward the incoming rays of sunlight. Therefore, during the winter months for either hemisphere, the farther a city is from the equator (greater degree of latitude either north or south) the shorter the duration of daylight. Conversely, during the summer months for either hemisphere, the farther a city is from the equator (greater degree of latitude either north or south) the longer the duration of daylight. The extremes of this phenomena occur near the North and South Poles. For example, during winter months at the North Pole or extreme northern latitudes, cities experience nearly 24 hours of darkness, but during summer months those same cities experience nearly 24 hours of daylight. Cities located near the equator experience nearly 12 hours of daylight throughout the year.

Timeline

The time needed to complete this investigation is 270 minutes (4 hours and 30 minutes). The amount of instructional time needed for each stage of the investigation is as follows:

- *Stage 1.* Introduce the task and the guiding question: 35 minutes

Investigation 12. Daylight and Location:
Why Does the Duration of Daylight Change in Different Locations Throughout the Year, and Why Isn't the Pattern the Same in All Locations on Earth?

- *Stage 2.* Design a method and collect data: 50 minutes
- *Stage 3.* Create a draft argument: 45 minutes
- *Stage 4.* Argumentation session: 30 minutes
- *Stage 5.* Reflective discussion: 15 minutes
- *Stage 6.* Write a draft report: 30 minutes
- *Stage 7.* Peer review: 35 minutes
- *Stage 8.* Revise the report: 30 minutes

Materials and Preparation

The materials needed for this investigation are listed in Table 12.1. The items can be purchased from a craft store such as Michaels, a big-box retail store such as Walmart or Target, or an online retailer such as Amazon. The materials for this investigation can also be purchased as a complete kit (which includes enough materials for 24 students, or six groups of four students) at *www.argumentdriveninquiry.com*.

TABLE 12.1
Materials for Investigation 12

Item	Quantity
Safety goggles	1 per student
Light source (flashlight or lamp)	1 per group
Foam ball for Earth model (5–6 inches diameter)	1 per group
Wooden dowel (5/16 inches diameter, ~14 inches long)	1 per group
String (1 yard)	1 per group
Ruler or meterstick	1 per group
Scissors	1 per group
Markers	1 per group
Pushpins	5 per group
Handout with data on Buffalo and Quito (for opening activity)	1 per student
Data cards A–E with information on various cities	1 set per group
Whiteboard, 2' × 3'*	1 per group
Investigation Handout	1 per student
Peer-review guide and teacher scoring rubric	1 per student
Checkout Questions (optional)	1 per student

*As an alternative, students can use computer and presentation software such as Microsoft PowerPoint or Apple Keynote to create their arguments.

Teacher Notes

A sample setup is shown in Figure 12.4. In this figure, the equator of Earth is shown with a dashed line. You can use markers to draw this line on the foam balls. The dowel rod is used for holding the model, but it also represents Earth's axis of rotation—hence it is tilted in Figure 12.4. We recommend using the smoothest foam balls available for this investigation. A smooth surface will allow students to better see the shadow line on the foam ball. Be sure to insert the wooden dowel into the foam ball and secure it in place using hot glue before the investigation begins because it can be difficult for students to insert the wooden dowel into the foam ball.

It is important to provide plenty of space between groups and to angle the light sources away from each other during the investigation. If there is too much light contamination from other groups' light sources, overhead lights, or classroom windows, it can be difficult for students to clearly see day and night on their foam ball Earth model.

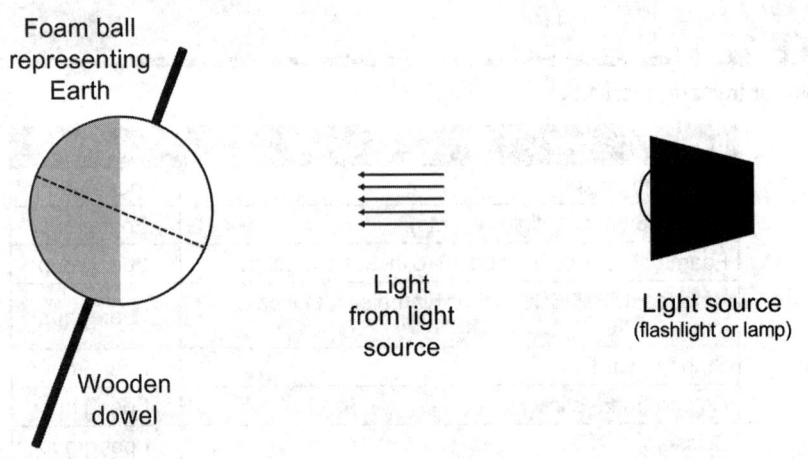

FIGURE 12.4
Sample setup for the modeling activity

This investigation also makes use of five supplementary data cards, which can be downloaded from the book's Extras page at *www.nsta.org/adi-5th*. These cards include information about the duration of daylight and darkness on specific dates in March, June, September, and December for Buffalo, New York; Miami, Florida; Quito, Ecuador; Lima, Peru; and Santiago, Chile. The information on the data cards was obtained from the U.S. Naval Observatory website: *www.usno.navy.mil/USNO*. This website allows you to create data tables related to sunrise and sunset times, duration of daylight and darkness, and other interesting data for specific U.S. cities and cities in other countries by using their latitude and longitude coordinates. You can use this website to generate data cards for other cities or even have your students look up cities on their own. Consider having students

Investigation 12. Daylight and Location:
Why Does the Duration of Daylight Change in Different Locations Throughout the Year, and Why Isn't the Pattern the Same in All Locations on Earth?

generate a list of cities they are interested in to help increase student engagement and ownership of the investigation.

Safety Precautions

Remind students to follow all normal safety rules. In addition, tell the students to take the following safety precautions:

- Wear sanitized safety goggles during setup, investigation activity, and cleanup.
- Be careful with the light source, because it can get hot and burn skin.
- Do not shine the light into their eyes or the eyes of their classmates.
- Be careful when working with sharp objects (such as pushpins and scissors), because they can cut or puncture skin.
- Make sure all materials are put away after completing the activity.
- Wash their hands with soap and water when they are done cleaning up.

Lesson Plan by Stage

This lesson plan is only a suggestion. It is included here to illustrate what you can say and do during each stage of ADI for this specific investigation. We encourage you to modify this lesson plan by asking different questions, using different examples, and providing different scaffolds as needed to better meet the needs of students in your class.

Stage 1: Introduce the Task and the Guiding Question (35 minutes)

1. Ask the students to sit in six groups, with three or four students in each group.
2. Ask the students to clear off their desks except for a pencil (and their *Student Workbook for Argument-Driven Inquiry in Fifth-Grade Science* if they have one).
3. Pass out an Investigation Handout to each student (or ask students to turn to the Investigation Log for Investigation 12 in their workbook).
4. Read the first paragraph of the "Introduction" aloud to the class. Ask the students to follow along as you read.
5. Use the foam ball Earth model and a light source to model day and night on Earth. Model the counter-clockwise rotation and point out how half the Earth is always illuminated (day) and the other half is in darkness (night). However, do not show them that the Earth is tilted at this point in time because it will spoil the investigation. Have the students complete the first "NOTICED/WONDER" chart in the "Introduction."
6. Ask the students to share what they observed.

Teacher Notes

7. Ask the students to share what questions they have about the day and night on Earth.

8. Tell the students, "Some of your questions might be answered by reading the next paragraph of the 'Introduction.'"

9. Ask the students to read the next paragraph of the "Introduction" on their own *or* ask them to follow along as you read it aloud.

10. Provide the students with the handout containing data about hours of darkness and daylight for Buffalo, New York, and Quito, Ecuador, and ask them to make observations about the data provided. This handout is available at *www.nsta.org/adi-5th*.

11. Ask the students to record their observations and questions about the data in the second "NOTICED/WONDER" chart in the "Introduction."

12. Ask the students to share what they observed about the data.

13. Ask the students to share what questions they have about the data or what they are now wondering after reviewing the data.

14. Tell the students, "Some of your questions might be answered by reading the rest of the 'Introduction.'"

15. Ask the students to read the rest of the "Introduction" on their own *or* ask them to follow along as you read it aloud.

16. Once the students have read the rest of the "Introduction," ask them to fill out the "Things we KNOW" chart on their Investigation Handout (or in their Investigation Log) as a group.

17. Ask the students to share what they learned from the reading. Add these ideas to a class "Things we KNOW" chart.

18. Tell the students, "Let's see what we will need to figure out during our investigation."

19. Read the task and the guiding question aloud.

20. Tell the students, "I have lots of materials here that you can use." Note that the students should already be familiar with the foam ball Earth model and the light source.

21. Introduce the students to the other materials available for them to use during the investigation by holding each one up and asking how it might be used.

Stage 2: Design a Method and Collect Data (50 minutes)

1. Tell the students, "I am now going to give you and the other members of your group about 15 minutes to plan your investigation. Before you begin, I want

Investigation 12. Daylight and Location:
Why Does the Duration of Daylight Change in Different Locations
Throughout the Year, and Why Isn't the Pattern the Same in All Locations on Earth?

you to take a couple of minutes to discuss the following questions with the rest of your group."

2. Show the following questions on the screen or board:
   - What components of the *system* should be included in our *model*?
   - What *patterns* should we look for or use to test our model?

3. Tell the students, "Please take a few minutes to come up with an answer to these questions."

4. Give the students two or three minutes to discuss these two questions.

5. Ask two or three different groups to share their answers. Highlight or write down any important ideas on the board so students can refer to them later.

6. If possible, use a document camera to project an image of the graphic organizer for this investigation on a screen or board (or take a picture of it and project the picture on a screen or board). Tell the students, "I now want you all to plan out your investigation. To do that, you will need to fill out this investigation proposal."

7. Point to the box labeled "Our guiding question:" and tell the students, "You can put the question we are trying to answer in this box." Then ask, "Where can we find the guiding question?"

8. Wait for a student to answer.

9. Point to the box labeled "This is a picture of how we will set up the equipment:" and tell the students, "You can draw a picture in this box of how you will set up the equipment in order to carry out this investigation."

10. Point to the box labeled "We will collect the following data:" and tell the students, "You can list the measurements or observations that you will need to collect during the investigation in this box."

11. Point to the box labeled "These are the steps we will follow to collect data:" and tell the students, "You can list what you are going to do to collect the data you need and what you will do with your data once you have it. Be sure to give enough detail that I could do your investigation for you."

12. Ask the students, "Do you have any questions about what you need to do?"

13. Answer any questions that come up.

14. Tell the students, "Once you are done, raise your hand and let me know. I'll then come by and look over your proposal and give you some feedback. You may not begin collecting data until I have approved your proposal by signing it. You need to have your proposal done in the next 15 minutes."

15. Give the students 15 minutes to work in their groups on their investigation proposal. As they work, move from group to group to check in, ask probing questions, and offer a suggestion if a group gets stuck.

Teacher Notes

16. As each group finishes its investigation proposal, read it over and determine if it will be productive or not. If you feel the investigation will be productive (not necessarily what you would do or what the other groups are doing), sign your name on the proposal and let the group start collecting data. If the plan needs to be changed, offer some suggestions or ask some probing questions, and have the group make the changes before you approve it.

17. Pass out the materials or have one student from each group collect the materials they need from a central supply table or cart for the groups that have an approved proposal. Included in these materials should be a set of data cards A–E for each group. Alternatively, recall that you can generate data handouts for other cities your students are interested in or have students research cities on their own using the U.S. Naval Observatory website.

18. Remind students of the safety rules and precautions for this investigation.

19. Tell the students to collect their data and record their observations or measurements in the "Collect Your Data" box in their Investigation Handout (or the Investigation Log in their workbook).

20. Give the students 30 minutes to collect their data. Collect the materials from each group before asking them to analyze their data.

What should a student-designed investigation look like?

There are a number of different investigations that students can design to generate models to answer the question "Why does the duration of daylight change in different locations throughout the year, and why isn't the pattern the same in all locations on Earth?" For example, one method might include the following steps:

1. Review the city data on data cards A–E.
2. Estimate where those cities would be located on the foam ball Earth model and mark them with pushpins.
3. Move the foam ball Earth model in front of the light source (representing the Sun) to try to mimic the data on the data cards.

If students use this method, they will need to collect the following data:

1. Location of cities (distance from equator in terms of latitude or general description)
2. Duration of daylight for each location they choose
3. Relative position of Earth and the Sun to mimic the daylight duration data

Investigation 12. Daylight and Location:
Why Does the Duration of Daylight Change in Different Locations
Throughout the Year, and Why Isn't the Pattern the Same in All Locations on Earth?

Stage 3: Create a Draft Argument (45 minutes)

1. Tell the students, "Now that we have all this data, we need to analyze the data so we can figure out an answer to the guiding question."

2. If possible, project an image of the "Analyze Your Data" section for this investigation on a screen or board using a document camera (or take a picture of it and project the picture on a screen or board). Point to the section and tell the students, "You can create a table, graph, or picture as a way to analyze your data. You can draw your graph or model in this section."

3. Ask the students, "What information do we need to include in a graph, table, or picture?"

4. Tell the students, "Please take a few minutes to discuss this question with your group and be ready to share."

5. Give the students five minutes to discuss.

6. Ask two or three different groups to share their answers. Highlight or write down any important ideas on the board so students can refer to them later.

7. Tell the students, "I am now going to give you and the other members of your group about 10 minutes to create your graph, table, or picture." If the students are having trouble making a graph, table, or picture, you can take a few minutes to provide a mini-lesson about how to create a graph, table, or picture that illustrates the patterns and trends they noticed from analyzing the data and manipulating their physical model of the Earth-Sun system (this strategy is called just-in-time instruction because it is offered only when students get stuck).

8. Give the students 10 minutes to analyze their data and make their graph, table, or picture. As they work, move from group to group to check in, ask probing questions, and offer suggestions.

9. Tell the students, "I am now going to give you and the other members of your group about 15 minutes to create an argument to share what you have learned and convince others that they should believe you. Before you do that, we need to take a few minutes to discuss what you need to include in your argument."

10. If possible, use a document camera to project the "Argument Presentation on a Whiteboard" image from the "Draft Argument" section of the Investigation Handout (or the Investigation Log in their workbook) on a screen or board (or take a picture of it and project the picture on a screen or board).

11. Point to the box labeled "The Guiding Question:" and tell the students, "You can put the question we are trying to answer here on your whiteboard."

12. Point to the box labeled "Our Claim:" and tell the students, "You can put your claim here on your whiteboard. The claim is your answer to the guiding question." For this investigation, students' claims will likely be the model they have

Teacher Notes

created that shows the Earth-Sun relationship that illustrates why duration of daylight is different for different cities.

13. Point to the box labeled "Our Evidence:" and tell the students, "You can put the evidence that you are using to support your claim here on your whiteboard. Your evidence will need to include the analysis you just did and an explanation of what your analysis means or shows. Scientists always need to support their claims with evidence."

14. Point to the box labeled "Our Justification of the Evidence:" and tell the students, "You can put your justification of your evidence here on your whiteboard. Your justification needs to explain why your evidence is important. Scientists often use core ideas to explain why the evidence they are using matters. Core ideas are important concepts that scientists use to help them make sense of what happens during an investigation."

15. Ask the students, "What are some core ideas that we read about earlier that might help us explain why the evidence we are using is important?"

16. Ask the students to share some of the core ideas from the "Introduction" section of the Investigation Handout (or the Investigation Log in the workbook). List these core ideas on the board.

17. Tell the students, "That is great. I would like to see everyone try to include these core ideas in your justification of the evidence. Your goal is to use these core ideas to help explain why your evidence matters and why the rest of us should pay attention to it."

18. Ask the students, "Do you have any questions about what you need to do?"

19. Answer any questions that come up.

20. Tell the students, "Okay, go ahead and start working on your arguments. You need to have your argument done in the next 15 minutes. It doesn't need to be perfect. We just need something down on the whiteboards so we can share our ideas."

21. Give the students 15 minutes to work in their groups on their arguments. As they work, move from group to group to check in, ask probing questions, and offer a suggestion if a group gets stuck. Figure 12.5 shows an example of an argument created by students for this investigation.

Investigation 12. Daylight and Location: Why Does the Duration of Daylight Change in Different Locations Throughout the Year, and Why Isn't the Pattern the Same in All Locations on Earth?

What should the table, graph, or picture for this investigation look like?

There are a number of different ways that students can analyze the data and observations they collected during this investigation. Students often include a table, a graph, and several pictures. For example, students might use a clustered column graph to show the relationship between latitude and hours of daylight by month. The horizontal axis, or *x*-axis, should include the degree latitude with clusters of bars that are labeled by month (such as June and December); the vertical axis, or *y*-axis, should show duration of daylight (measured in hours). Students can also draw pictures to show the tilt of the Earth and the location of different cities. An example of a graph and picture can be seen in Figure 12.5. There are other options for analyzing the collected data. Students often come up with some unique ways of analyzing their data, so be sure to give them some voice and choice during this stage.

FIGURE 12.5
Example of an argument

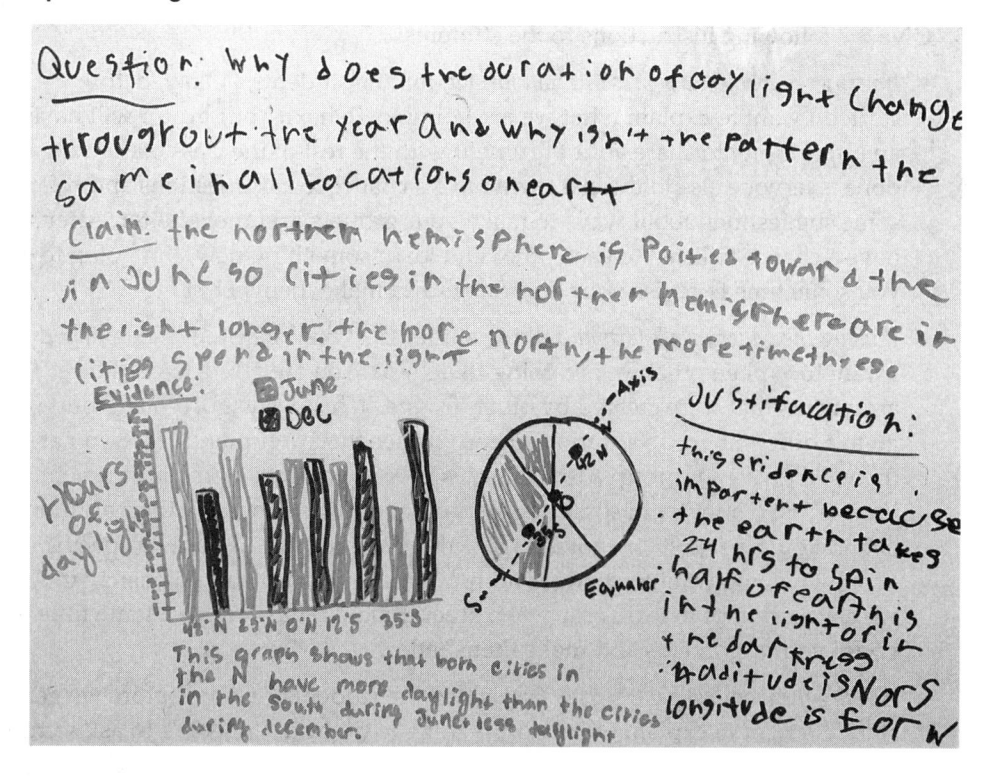

Question: Why does the duration of daylight change throughout the year and why isn't the pattern the same in all locations on earth

Claim: the northern hemisphere is pointed toward the sun in June so cities in the northern hemisphere are in the light longer. the more north, the more time these cities spend in the light

Evidence: [bar graph: June, Dec; y-axis: Hours of daylight; x-axis: 48°N, 23°N, 0°N, 12°S, 35°S] This graph shows that both cities in the N have more daylight than the cities in the south during June. less daylight during december.

Justification: this evidence is important because the earth takes 24 hrs to spin. half of earth is in the light or in the darkness. latitude is N or S longitude is E or W

Argument-Driven Inquiry in **Fifth-Grade Science**: Three-Dimensional Investigations

Teacher Notes

Stage 4: Argumentation Session (30 minutes)

The argumentation session can be conducted in a whole-class presentation format, a gallery walk format, or a modified gallery walk format. We recommend using a whole-class presentation format for the first investigation, but try to transition to either the gallery walk or modified gallery walk format as soon as possible because that will maximize student voice and choice inside the classroom. The following list shows the steps for the three formats; unless otherwise noted, the steps are the same for all three formats.

1. Begin by introducing the use of the whiteboard.

   - *If using the whole-class presentation format,* tell the students, "We are now going to share our arguments. Please set up your whiteboards so everyone can see them."

   - *If using the gallery walk or modified gallery walk format,* tell the students, "We are now going to share our arguments. Please set up your whiteboards so they are facing the walls."

2. Allow the students to set up their whiteboards.

   - *If using the whole-class presentation format,* the whiteboards should be set up on stands or chairs so they are facing toward the center of the room.

   - *If using the gallery walk or modified gallery walk format,* the whiteboards should be set up on stands or chairs so they are facing toward the outside of the room.

3. Give the following instructions to the students:

   - *If using the whole-class presentation format,* tell the students, "Okay, before we get started I want to explain what we are going to do next. Your group will have an opportunity to share your argument with the rest of the class. After you are done, everyone else in the class will have a chance to ask questions and offer some suggestions about ways to make your group's argument better. After we have a chance to listen to each other and learn something new, I'm going to give you some time to revise your arguments and make them better."

   - *If using the gallery walk format,* tell the students, "Okay, before we get started I want to explain what we are going to do next. You are going to read the arguments that were created by other groups. When I say 'go,' your group will go to a different group's station so you can see their argument. Once you are there, I'll give your group a few minutes to read and review their argument. Your job is to offer them some suggestions about ways to make their argument better. You can use sticky notes to give them suggestions. Please be specific about what you want to change and how you think they should change it. After we have a chance to learn from each other, I'm going to give you some time to revise your arguments and make them better."

   - *If using the modified gallery walk format,* tell the students, "Okay, before we get started I want to explain what we are going to do next. I'm going to ask some of

Investigation 12. Daylight and Location:
Why Does the Duration of Daylight Change in Different Locations Throughout the Year, and Why Isn't the Pattern the Same in All Locations on Earth?

you to present your arguments to your classmates. If you are presenting your argument, your job is to share your group's claim, evidence, and justification of the evidence. The rest of you will be travelers. If you are a traveler, your job is to listen to the presenters, ask the presenters questions if you do not understand something, and then offer them some suggestions about ways to make their argument better. After we have a chance to learn from each other, I'm going to give you some time to revise your arguments and make them better."

4. Use a document camera to project the "Ways to IMPROVE our argument …" box from the Investigation Handout (or the Investigation Log in their workbook) on a screen or board (or take a picture of it and project the picture on a screen or board).

    - *If using the whole-class presentation format,* point to the box and tell the students, "After your group presents your argument, you can write down the suggestions you get from your classmates here. If you are listening to a presentation and you see a good idea from another group, you can write down that idea here as well. Once we are done with the presentations, I will give you a chance to use these suggestions or ideas to improve your arguments."

    - *If using the gallery walk format,* point to the box and tell the students, "If you see a good idea from another group, you can write it down here. Once we are done reviewing the different arguments, I will give you a chance to use these ideas to improve your own arguments. It is important to share ideas like this."

    - *If using the modified gallery walk format,* point to the box and tell the students, "If you are a presenter, you can write down the suggestions you get from the travelers here. If you are a traveler and you see a good idea from another group, you can write down that idea here. Once we are done with the presentations, I will give you a chance to use these suggestions or ideas to improve your arguments."

5. Ask the students, "Do you have any questions about what you need to do?"

6. Answer any questions that come up.

7. Give the following instructions:

    - *If using the whole-class presentation format,* tell the students, "Okay. Let's get started."

    - *If using the gallery walk format,* tell the students, "Okay, I'm now going to tell you which argument to go to and review."

    - *If using the modified gallery walk format,* tell the students, "Okay, I'm now going to assign you to be a presenter or a traveler." Assign one or two students from each group to be presenters and one or two students from each group to be travelers.

8. Give the students an opportunity to review the arguments.

Teacher Notes

- *If using the whole-class presentation format,* have each group present their argument one at a time. Give each group only two to three minutes to present their argument. Then give the class two to three minutes to ask them questions and offer suggestions. Encourage as much participation from the students as possible.

- *If using the gallery walk format,* tell the students, "Okay. Let's get started. Each group, move one argument to the left. Don't move to the next argument until I tell you to move. Once you get there, read the argument and then offer suggestions about how to make it better. I will put some sticky notes next to each argument. You can use the sticky notes to leave your suggestions." Give each group about three to four minutes to read the arguments, talk, and offer suggestions.

    a. After three to four minutes, tell the students, "Okay. Let's move on to the next argument. Please move one group to the left."

    b. Again, give each group three to four minutes to read, talk, and offer suggestions.

    c. Repeat this process until each group has had their argument read and critiqued three times.

- *If using the modified gallery walk format,* tell the students, "Okay. Let's get started. Reviewers, move one group to the left. Don't move to the next group until I tell you to move. Presenters, go ahead and share your argument with the travelers when they get there." Give each group of presenters and travelers about three to four minutes to talk.

    a. Tell the students, "Okay. Let's move on to the next argument. Travelers, move one group to the left."

    b. Again, give each group of presenters and travelers about three to four minutes to talk.

    c. Repeat this process until each group has had their argument read and critiqued three times.

9. Tell the students to return to their workstations.

10. Give the following instructions about revising the argument:

    - *If using the whole-class presentation format,* tell the students, "I'm now going to give you all about 10 minutes to revise your argument. Take a few minutes to talk in your groups and determine what you want to change to make your argument better. Once you have decided what to change, go ahead and make the changes to your whiteboard."

    - *If using the gallery walk format,* tell the students, "I'm now going to give you all about 10 minutes to revise your argument. Take a few minutes to read the suggestions that were left at your argument. Then talk in your groups and determine what you want to change to make your argument better. Once

Investigation 12. Daylight and Location:
Why Does the Duration of Daylight Change in Different Locations
Throughout the Year, and Why Isn't the Pattern the Same in All Locations on Earth?

you have decided what to change, go ahead and make the changes to your whiteboard."

- *If using the modified gallery walk format,* tell the students, "I'm now going to give you all about 10 minutes to revise your argument. Please return to your original groups." Wait for the students to move back into their original groups and then tell the students, "Okay, take a few minutes to talk in your groups and determine what you want to change to make your argument better. Once you have decided what to change, go ahead and make the changes to your whiteboard."

11. Ask the students, "Do you have any questions about what you need to do?"

12. Answer any questions that come up.

13. Tell the students, "Okay. Let's get started."

14. Give the students 10 minutes to work in their groups on their arguments. As they work, move from group to group to check in, ask probing questions, and offer a suggestion if a group gets stuck.

Stage 5: Reflective Discussion (15 minutes)

1. Tell the students, "We are now going to take a minute to talk about some of the core ideas and crosscutting concepts that we have used during our investigation."

2. Show Figure 12.6 on the screen.

FIGURE 12.6

Information about city A (Spokane, WA)

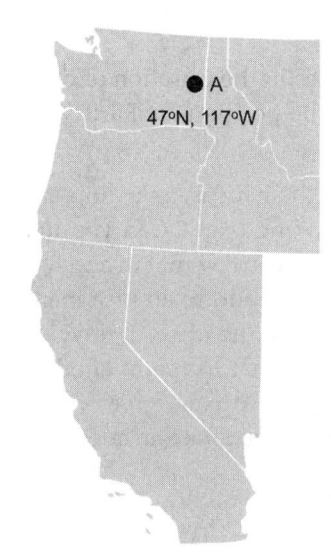

Date	Sunrise	Sunset	Duration of Daylight
3/20	6:51 am	7:03 pm	12 hours 12 minutes
6/20	4:51 am	8:51 pm	15 hours 59 minutes
9/20	6:34 am	6:50 pm	12 hours 8 minutes
12/20	7:35 am	4:00 pm	8 hours 25 minutes

A: 47°N, 117°W

Teacher Notes

3. Ask the students, "What do you all see going on here?"
4. Allow the students to share their ideas.
5. Show Figure 12.7 on the screen.

FIGURE 12.7
Information about city B (San Diego, CA)

32°N, 117°W

Date	Sunrise	Sunset	Duration of Daylight
3/20	6:51 am	7:00 pm	12 hours 9 minutes
6/20	5:41 am	7:59 pm	14 hours 18 minutes
9/20	6:35 am	6:47 pm	12 hours 11 minutes
12/20	6:46 am	4:46 pm	9 hours 59 minutes

6. Ask the students, "What do you all see going on here?"
7. Allow the students to share their ideas.
8. Show Figure 12.8 on the screen.
9. Ask the students, "How can we explain the differences in the duration of daylight in these two cities using what we know about the positions of Earth and the Sun and the movement of Earth?"
10. Allow students to share their ideas. As they share their ideas, ask different questions to encourage them to expand on their thinking (e.g., "Can you tell me more about that?"), clarify a contribution (e.g., "Can you say that in another way?"), support an idea (e.g., "Why do you think that?"), add to an idea mentioned by a classmate (e.g., "Would anyone like to add to the idea?"), re-voice an idea offered by a classmate (e.g., "Who can explain that to me in another way?"), or critique an idea during the discussion (e.g., "Do you agree or disagree with that idea and why?") until students are able to generate an adequate explanation.

Investigation 12. Daylight and Location:
Why Does the Duration of Daylight Change in Different Locations
Throughout the Year, and Why Isn't the Pattern the Same in All Locations on Earth?

FIGURE 12.8

Information about city A (Spokane, WA) and city B (San Diego, CA)

City A — 47°N, 117°W

Date	Sunrise	Sunset	Duration of Daylight
3/20	6:51 am	7:03 pm	12 hours 12 minutes
6/20	4:51 am	8:51 pm	15 hours 59 minutes
9/20	6:34 am	6:50 pm	12 hours 8 minutes
12/20	7:35 am	4:00 pm	8 hours 25 minutes

City B — 32°N, 117°W

Date	Sunrise	Sunset	Duration of Daylight
3/20	6:51 am	7:00 pm	12 hours 9 minutes
6/20	5:41 am	7:59 pm	14 hours 18 minutes
9/20	6:35 am	6:47 pm	12 hours 11 minutes
12/20	6:46 am	4:46 pm	9 hours 59 minutes

11. Tell the students, "We had to look for patterns during this investigation." Then ask, "Can anyone tell me why this was important?"

12. Allow the students to share their ideas.

13. Tell the students, "I think noticing patterns is an important skill for scientists, because patterns of change can be used to make predictions of what might happen next or predict how a change in the system might change an observed pattern. Patterns can also be used as evidence to support an explanation."

14. Ask the students, "What patterns did you uncover and use today?"

15. Allow the students to share their ideas.

16. Show an image of the question "What do you think are the most important core ideas or crosscutting concepts that we used during this investigation to help us make sense of what we observed?" Tell the students, "Okay, let's make sure we are all on the same page. Please take a moment to discuss this question with the other people in your group." Give them a few minutes to discuss the question.

17. Ask the students, "What do you all think? Who would like to share?"

18. Allow the students to share their ideas.

19. Tell the students, "We are now going to take a minute to talk about the use of models as tools for reasoning about natural phenomena."

Teacher Notes

20. Show an image of the question "What is a model?" on the screen. Tell the students, "Take a few minutes to talk about how you would answer this question with the other people in your group. Be ready to share with the rest of the class." Give the students two to three minutes to talk in their group.

21. Ask the students, "What do you all think? Who would like to share an idea?"

22. Allow the students to share their ideas.

23. Tell the students, "Okay, let's make sure we are all using the same definition. I think a model is something scientists use to help them understand how systems work. Models can be pictures. Or they can be graphs. Or they can even be physical models of things like a skeleton."

24. Show Figure 12.9 (without the caption and the footnote) on a screen with the statement "Is this a model?".

FIGURE 12.9
Model drawing of the solar system

Note: A full-color version of this figure is available on the book's Extras page at *www.nsta.org/adi-5th*.

25. Ask the students, "Do you think this is a model? Why or why not?"

26. Allow the students to share their ideas.

27. Tell the students, "I think this is a model because it allows us to understand what happened during the investigation."

28. Ask the students, "What is the system being modeled?"

29. Allow the students to share their ideas.

30. Ask the students, "What aspects of this model do you think are accurate?" Also ask, "What aspects of this model do you think are inaccurate?" Students should recognize that not all models are 100% accurate; for example, the model in Figure 12.9 is not to scale, nor do the planets necessarily have those colors, nor are the orbits of planets "drawn" in space. Nevertheless, models are useful so long as we keep their constraints or inaccuracies in mind while using them.

31. Ask the students, "What questions do you have about how scientists use models?"

32. Answer any questions that come up.

33. Tell the students, "We are now going to take a minute to talk about what went well and what didn't go so well during our investigation. We need to talk about this because you all are going to be planning and carrying out your own investigations like this a lot this year, and I want to help you all get better at it."

34. Show an image of the question "What made your investigation scientific?" on the screen. Tell the students, "Take a few minutes to talk about how you would answer this question with the other people in your group. Be ready to share with the rest of the class." Give the students two to three minutes to talk in their group.

35. Ask the students, "What do you all think? Who would like to share an idea?"

36. Allow the students to share their ideas. Be sure to expand on their ideas about what makes an investigation scientific.

37. Show an image of the question "What made your investigation not so scientific?" on the screen. Tell the students, "Take a few minutes to talk about how you would answer this question with the other people in your group. Be ready to share with the rest of the class." Give the students two to three minutes to talk in their group.

38. Ask the students, "What do you all think? Who would like to share an idea?"

39. Allow the students to share their ideas. Be sure to expand on their ideas about what makes an investigation less scientific.

40. Show an image of the question "What rules can we put into place to help us make sure our next investigation is more scientific?" on the screen. Tell the students, "Take a few minutes to talk about how you would answer this question with the other people in your group. Be ready to share with the rest of the class." Give the students two to three minutes to talk in their group.

41. Ask the students, "What do you all think? Who would like to share an idea?"

42. Allow the students to share their ideas. Once they have shared their ideas, offer a suggestion for a possible class rule.

Teacher Notes

43. Ask the students, "What do you all think? Should we make this a rule?"
44. If the students agree, write the rule on the board or on a class "Rules for Scientific Investigation" chart so you can refer to it during the next investigation.

Stage 6: Write a Draft Report (30 minutes)

Your students will use either the Investigation Handout or the Investigation Log in the student workbook when writing the draft report. When you give the directions shown in quotes in the following steps, substitute "Investigation Log in your workbook" or just "Investigation Log" (as shown in brackets) for "handout" if they are using the workbook.

1. Tell the students, "You are now going to write an investigation report to share what you have learned. Please take out a pencil and turn to the 'Draft Report' section of your handout [Investigation Log in your workbook]."

2. If possible, use a document camera to project the "Introduction" section of the draft report from the Investigation Handout (or the Investigation Log in their workbook) on a screen or board (or take a picture of it and project the picture on a screen or board).

3. Tell the students, "The first part of the report is called the 'Introduction.' In this section of the report you want to explain to the reader what you were investigating, why you were investigating it, and what question you were trying to answer. All this information can be found in the text at the beginning of your handout [Investigation Log]." Point to the image. "Here are some sentence starters to help you begin writing."

4. Ask the students, "Do you have any questions about what you need to do?"

5. Answer any questions that come up.

6. Tell the students, "Okay, let's write."

7. Give the students 10 minutes to write the "Introduction" section of the report. As they work, move from student to student to check in, ask probing questions, and offer a suggestion if a student gets stuck.

8. If possible, use a document camera to project the "Method" section of the draft report from the Investigation Handout (or the Investigation Log in their workbook) on a screen or board (or take a picture of it and project the picture on a screen or board).

9. Tell the students, "The second part of the report is called the 'Method.' In this section of the report you want to explain to the reader what you did during the investigation, what data you collected and why, and how you went about analyzing your data. All this information can be found in the 'Plan Your Investigation' section of the handout [Investigation Log]. Remember that you all planned and carried out different investigations, so do not assume that the

reader will know what you did." Point to the image. "Here are some sentence starters to help you begin writing."

10. Ask the students, "Do you have any questions about what you need to do?"
11. Answer any questions that come up.
12. Tell the students, "Okay, let's write."
13. Give the students 10 minutes to write the "Method" section of the report. As they work, move from student to student to check in, ask probing questions, and offer a suggestion if a student gets stuck.
14. If possible, use a document camera to project the "Argument" section of the draft report from the Investigation Handout (or the Investigation Log in their workbook) on a screen or board (or take a picture of it and project the picture on a screen or board).
15. Tell the students, "The last part of the report is called the 'Argument.' In this section of the report you want to share your claim, evidence, and justification of the evidence with the reader. All this information can be found on your whiteboard." Point to the image. "Here are some sentence starters to help you begin writing."
16. Ask the students, "Do you have any questions about what you need to do?"
17. Answer any questions that come up.
18. Tell the students, "Okay, let's write."
19. Give the students 10 minutes to write the "Argument" section of the report. As they work, move from student to student to check in, ask probing questions, and offer a suggestion if a student gets stuck.

Stage 7: Peer Review (35 minutes)

Your students will use either the Investigation Handout or their workbook when doing the peer review. Except where noted below, the directions are the same whether using the handout or the workbook.

1. Tell the students, "We are now going to review our reports to find ways to make them better. I'm going to come around and collect your draft reports. While I do that, please take out a pencil."
2. Collect the handouts or the workbooks with the draft reports from the students.
3. If possible, use a document camera to project the peer-review guide (see Appendix 4) on a screen or board (or take a picture of it and project the picture on a screen or board).
4. Tell the students, "We are going to use this peer-review guide to give each other feedback." Point to the image.

Teacher Notes

5. Tell the students, "I'm going to ask you to work with a partner to do this. I'm going to give you and your partner a draft report to read. You two will then read the report together. Once you are done reading the report, I want you to answer each of the questions on the peer-review guide." Point to the review questions on the image of the peer-review guide.

6. Tell the students, "You can check 'no,' 'almost,' or 'yes' after each question." Point to the checkboxes on the image of the peer-review guide.

7. Tell the students, "This will be your rating for this part of the report. Make sure you agree on the rating you give the author. If you mark 'no' or 'almost,' then you need to tell the author what he or she needs to do to get a 'yes.'" Point to the space for the reviewer feedback on the image of the peer-review guide.

8. Tell the students, "It is really important for you to give the authors feedback that is helpful. That means you need to tell them exactly what they need to do to make their report better."

9. Ask the students, "Do you have any questions about what you need to do?"

10. Answer any questions that come up.

11. Tell the students, "Please sit with a partner who is not in your current group." Allow the students time to sit with a partner.

12. Tell the students, "Okay, I'm now going to give you one report to read." Pass out one Investigation Handout with a draft report or one workbook to each pair. Make sure that the report you give a pair was not written by one of the students in that pair. Give each pair one peer-review guide to fill out. If the students are using workbooks, the peer-review guide is included right after the draft report so you do not need to pass out copies of the peer-review guide.

13. Tell the students, "Okay, I'm going to give you 15 minutes to read the report I gave you and to fill out the peer-review guide. Go ahead and get started."

14. Give the students 15 minutes to work. As they work, move around from pair to pair to check in and see how things are going, answer questions, and offer advice.

15. After 15 minutes pass, tell the students, "Okay, time is up. Please give me the report and the peer-review guide that you filled out."

16. Collect the Investigation Handouts and the peer-review guides, or collect the workbooks if students are using them. If the students are using the Investigation Handouts and separate peer-review guides, be sure you keep each handout with its corresponding peer-review guide.

17. Tell the students, "Okay, I am now going to give you a different report to read and a new peer-review guide to fill out." Pass out one more report to each pair. Make sure that the report you give a pair was not written by one of the students in that pair. Give each pair a new peer-review guide to fill out as a group.

Investigation 12. Daylight and Location:
Why Does the Duration of Daylight Change in Different Locations
Throughout the Year, and Why Isn't the Pattern the Same in All Locations on Earth?

18. Tell the students, "Okay, I'm going to give you 15 minutes to read this new report and to fill out the peer-review guide. Go ahead and get started."

19. Give the students 15 minutes to work. As they work, move around from pair to pair to check in and see how things are going, answer questions, and offer advice.

20. After 15 minutes pass, tell the students, "Okay, time is up. Please give me the report and the peer-review guide that you filled out."

21. Collect the Investigation Handouts and the peer-review guides, or collect the workbooks if students are using them. If the students are using the Investigation Handouts and separate peer-review guides, be sure you keep each handout with its corresponding peer-review guide.

Stage 8: Revise the Report (30 minutes)

Your students will use either the Investigation Handout or their workbook when revising the report. Except where noted below, the directions are the same whether using the handout or the workbook.

1. Tell the students, "You are now going to revise your draft report based on the feedback you get from your classmates. Please take out a pencil."

2. Return the reports to the students.
   - *If the students used the Investigation Handout and a copy of the peer-review guide,* pass back the handout and the peer-review guide to each student.
   - *If the students used the workbook,* pass that back to each student.

3. Tell the students, "Please take a few minutes to read over the peer-review guide. You should use it to figure out what you need to change in your report and how you will change it."

4. Allow the students to read the peer-review guide.

5. *If the students used the workbook,* if possible use a document camera to project the "Write Your Final Report" section from the Investigation Log on a screen or board (or take a picture of it and project the picture on a screen or board).

6. Give the following directions about how to revise their reports:
   - *If the students used the Investigation Handout and a copy of the peer-review guide,* tell them, "Okay, let's revise our reports. Please take out a piece of paper. I would like you to rewrite your report. You can use your draft report as a starting point, but you also need to change it to make it better. Use the feedback on the peer-review guide to make it better."
   - *If the students used the workbook,* tell them, "Okay, let's revise our reports. I would like you to rewrite your report in the section of the Investigation Log called "Write Your Final Report." You can use your draft report as a starting

Teacher Notes

point, but you also need to change it to make it better. Use the feedback on the peer-review guide to make it better."

7. Ask the students, "Do you have any questions about what you need to do?"
8. Answer any questions that come up.
9. Tell the students, "Okay, let's write." Allow about 20 minutes for the students to revise their reports.
10. After about 20 minutes, give the following directions:
    - *If the students used the Investigation Handout,* tell them, "Okay, time's up. I will now come around and collect your Investigation Handout, the peer-review guide, and your final report."
    - *If the students used the workbook,* tell them, "Okay, time's up. I will now come around and collect your workbooks."
11. *If the students used the Investigation Handout,* collect all the Investigation Handouts, peer-review guides, and final reports. *If the students used the workbook,* collect all the workbooks.
12. *If the students used the Investigation Handout,* use the "Teacher Score" column in the peer-review guide to grade the final report. *If the students used the workbook,* use the "Investigation Report Grading Rubric" in the Investigation Log to grade the final report. Whether you are using the handout or the log, you can give the students feedback about their writing in the "Teacher Comments" section.

How to Use the Checkout Questions

The Checkout Questions are an optional assessment. We recommend giving them to students at the start of the next class period after the students finish stage 8 of the investigation. You can then look over the student answers to determine if you need to reteach the core idea from the investigation. Appendix 6 gives the answers to the Checkout Questions that should be given by a student who can apply the core idea correctly in all cases and can explain (1) the importance of models and (2) the pattern in changes in duration of daylight.

Alignment With Standards

Table 12.2 highlights how the investigation can be used to address specific performance expectations from the *Next Generation Science Standards, Common Core State Standards for English Language Arts (CCSS ELA),* and *English Language Proficiency (ELP) Standards.*

Investigation 12. Daylight and Location:
Why Does the Duration of Daylight Change in Different Locations
Throughout the Year, and Why Isn't the Pattern the Same in All Locations on Earth?

TABLE 12.2

Investigation 12 alignment with standards

***NGSS* performance expectation**	5-ESS1-2: Represent data in graphical displays to reveal patterns of daily changes in length of direction of shadows, day and night, and the seasonal appearance of some stars in the night sky.
***CCSS ELA*—Reading: Informational Text**	Key ideas and details ○ CCSS.ELA-LITERACY.RI.5.1: Quote accurately from a text when explaining what the text says explicitly and when drawing inferences from the text. ○ CCSS.ELA-LITERACY.RI.5.2: Determine two or more main ideas of a text and explain how they are supported by key details; summarize the text. ○ CCSS.ELA-LITERACY.RI.5.3: Explain the relationships or interactions between two or more individuals, events, ideas, or concepts in a historical, scientific, or technical text based on specific information in the text. Craft and structure • CCSS.ELA-LITERACY.RI.5.4: Determine the meaning of general academic and domain-specific words and phrases in a text relevant to a *grade 5 topic or subject area*. • CCSS.ELA-LITERACY.RI.5.5: Compare and contrast the overall structure (e.g., chronology, comparison, cause/effect, problem/solution) of events, ideas, concepts, or information in two or more texts. • CCSS.ELA-LITERACY.RI.5.6: Analyze multiple accounts of the same event or topic, noting important similarities and differences in the point of view they represent. Integration of knowledge and ideas • CCSS.ELA-LITERACY.RI.5.7: Draw on information from multiple print or digital sources, demonstrating the ability to locate an answer to a question quickly or to solve a problem efficiently. • CCSS.ELA-LITERACY.RI.5.8: Explain how an author uses reasons and evidence to support particular points in a text, identifying which reasons and evidence support which point(s). Range of reading and level of text complexity • CCSS.ELA-LITERACY.RI.5.10: By the end of the year, read and comprehend informational texts, including history/social studies, science, and technical texts, at the high end of the grades 4–5 text complexity band independently and proficiently.
***CCSS ELA*—Writing**	Text types and purposes • CCSS.ELA-LITERACY.W.5.1: Write opinion pieces on topics or texts, supporting a point of view with reasons. ○ CCSS.ELA-LITERACY.W.5.1.A: Introduce a topic or text clearly, state an opinion, and create an organizational structure in which ideas are logically grouped to support the writer's purpose. ○ CCSS.ELA-LITERACY.W.5.1.B: Provide logically ordered reasons that are supported by facts and details.

Continued

Teacher Notes

Table 12.2 (*continued*)

***CCSS ELA*—Writing** (*continued*)	○ CCSS.ELA-LITERACY.W.5.1.C: Link opinion and reasons using words, phrases, and clauses (e.g., *consequently*, *specifically*).
	○ CCSS.ELA-LITERACY.W.5.1.D: Provide a concluding statement or section related to the opinion presented.
	• CCSS.ELA-LITERACY.W.5.2: Write informative or explanatory texts to examine a topic and convey ideas and information clearly.
	○ CCSS.ELA-LITERACY.W.5.2.A: Introduce a topic clearly, provide a general observation and focus, and group related information logically; include formatting (e.g., headings), illustrations, and multimedia when useful to aiding comprehension.
	○ CCSS.ELA-LITERACY.W.5.2.B: Develop the topic with facts, definitions, concrete details, quotations, or other information and examples related to the topic.
	○ CCSS.ELA-LITERACY.W.5.2.C: Link ideas within and across categories of information using words, phrases, and clauses (e.g., *in contrast*, *especially*).
	○ CCSS.ELA-LITERACY.W.5.2.D: Use precise language and domain-specific vocabulary to inform about or explain the topic.
	○ CCSS.ELA-LITERACY.W.5.2.E: Provide a concluding statement or section related to the information or explanation presented.
	Production and distribution of writing
	• CCSS.ELA-LITERACY.W.5.4: Produce clear and coherent writing in which the development and organization are appropriate to task, purpose, and audience.
	• CCSS.ELA-LITERACY.W.5.5: With guidance and support from peers and adults, develop and strengthen writing as needed by planning, revising, editing, rewriting, or trying a new approach.
	• CCSS.ELA-LITERACY.W.5.6: With some guidance and support from adults, use technology, including the internet, to produce and publish writing as well as to interact and collaborate with others; demonstrate sufficient command of keyboarding skills to type a minimum of two pages in a single sitting.
	Research to build and present knowledge
	• CCSS.ELA-LITERACY.W.5.8: Recall relevant information from experiences or gather relevant information from print and digital sources; summarize or paraphrase information in notes and finished work, and provide a list of sources.
	• CCSS.ELA-LITERACY.W.5.9: Draw evidence from literary or informational texts to support analysis, reflection, and research.
	Range of writing
	• CCSS.ELA-LITERACY.W.5.10: Write routinely over extended time frames (time for research, reflection, and revision) and shorter time frames (a single sitting or a day or two) for a range of discipline-specific tasks, purposes, and audiences.

Continued

Investigation 12. Daylight and Location:
Why Does the Duration of Daylight Change in Different Locations Throughout the Year, and Why Isn't the Pattern the Same in All Locations on Earth?

Table 12.2 (*continued*)

CCSS ELA—Speaking and Listening	Comprehension and collaboration
	
	• CCSS.ELA-LITERACY.SL.5.1: Engage effectively in a range of collaborative discussions (one-on-one, in groups, and teacher-led) with diverse partners on *grade 5 topics and texts,* building on others' ideas and expressing their own clearly.
	
	    o CCSS.ELA-LITERACY.SL.5.1.A: Come to discussions prepared, having read or studied required material; explicitly draw on that preparation and other information known about the topic to explore ideas under discussion.
	
	    o CCSS.ELA-LITERACY.SL.5.1.B: Follow agreed-upon rules for discussions and carry out assigned roles.
	
	    o CCSS.ELA-LITERACY.SL.5.1.C: Pose and respond to specific questions by making comments that contribute to the discussion and elaborate on the remarks of others.
	
	    o CCSS.ELA-LITERACY.SL.5.1.D: Review the key ideas expressed and draw conclusions in light of information and knowledge gained from the discussions.
	
	• CCSS.ELA-LITERACY.SL.5.2: Summarize a written text read aloud or information presented in diverse media and formats, including visually, quantitatively, and orally.
	
	• CCSS.ELA-LITERACY.SL.5.3: Summarize the points a speaker makes and explain how each claim is supported by reasons and evidence.
	
	Presentation of knowledge and ideas
	
	• CCSS.ELA-LITERACY.SL.5.4: Report on a topic or text or present an opinion, sequencing ideas logically and using appropriate facts and relevant, descriptive details to support main ideas or themes; speak clearly at an understandable pace.
	
	• CCSS.ELA-LITERACY.SL.5.5: Include multimedia components (e.g., graphics, sound) and visual displays in presentations when appropriate to enhance the development of main ideas or themes.
	
	• CCSS.ELA-LITERACY.SL.5.6: Adapt speech to a variety of contexts and tasks, using formal English when appropriate to task and situation.

ELP Standards	Receptive modalities • ELP 1: Construct meaning from oral presentations and literary and informational text through grade-appropriate listening, reading, and viewing. • ELP 8: Determine the meaning of words and phrases in oral presentations and literary and informational text. Productive modalities • ELP 3: Speak and write about grade-appropriate complex literary and informational texts and topics. • ELP 4: Construct grade-appropriate oral and written claims and support them with reasoning and evidence.

Continued

Teacher Notes

Table 12.2 *(continued)*

ELP Standards *(continued)*	• ELP 7: Adapt language choices to purpose, task, and audience when speaking and writing. Interactive modalities • ELP 2: Participate in grade-appropriate oral and written exchanges of information, ideas, and analyses, responding to peer, audience, or reader comments and questions. • ELP 5: Conduct research and evaluate and communicate findings to answer questions or solve problems. • ELP 6: Analyze and critique the arguments of others orally and in writing. Linguistic structures of English • ELP 9: Create clear and coherent grade-appropriate speech and text. • ELP 10: Make accurate use of standard English to communicate in grade-appropriate speech and writing.

Investigation Handout

Investigation 12

Daylight and Location: Why Does the Duration of Daylight Change in Different Locations Throughout the Year, and Why Isn't the Pattern the Same in All Locations on Earth?

Introduction

People experience daylight and darkness on Earth because only half of Earth is facing the Sun at any point in time and Earth rotates (spins) on its axis. Earth's axis is an imaginary line that passes through Earth from the North Pole to the South Pole. Take a minute to watch your teacher model day and night on Earth with a lamp and a foam ball. As you watch your teacher use these materials, keep track of things you notice and things you wonder about in the boxes below.

Things I NOTICED …	Things I WONDER about …

Argument-Driven Inquiry in **Fifth-Grade Science**: Three-Dimensional Investigations

Investigation Handout

The sun rises and sets at different times of the day depending on where you live. The duration of daylight can therefore be different for different cities on the same day of the year. Your teacher will give you a handout with two graphs showing the hours of daylight and darkness on the first day of each month in Buffalo, New York, and Quito, Ecuador. Take some time to look for patterns in these data. Keep track of things you notice and things you wonder about in the boxes below.

Things I NOTICED ...	Things I WONDER about ...

It takes Earth about 365 days to orbit the Sun. Each year on Earth is therefore 365 days long. Earth takes 24 hours to make one complete rotation on its axis. Each day on Earth is therefore 24 hours. The duration of daylight that people experience during a 24-hour day is different in different cities on Earth. People who live in Quito, for example, experience about 12 hours of daylight every day of the year, but people who live in Buffalo experience 15 hours of daylight each day during the month of June and about 9 hours of daylight each day during the month of January.

Buffalo and Quito are located on the same side of Earth at a longitude of roughly 78° West. The *longitude* of a city is the distance east or west of the *prime meridian* (an imaginary line that runs from the North Pole to the South Pole through Greenwich, England). Buffalo and Quito, however, are located at different latitudes. The *latitude* of a city describes how far north or south the location is relative

Investigation 12. Daylight and Location:
Why Does the Duration of Daylight Change in Different Locations Throughout the Year, and Why Isn't the Pattern the Same in All Locations on Earth?

to the equator. The equator "divides" Earth into two halves: the Northern Hemisphere, which is the half above the equator; and the Southern Hemisphere, which is the half below the equator. Buffalo is located in the Northern Hemisphere and Quito is located on the equator.

In this investigation, you will need to figure out why different cities have different durations of daylight and why the duration of daylight changes in some cities over the course of a year. Some cities you many want to explore are Buffalo, New York; Miami, Florida; Quito, Ecuador; Lima, Peru; and Santiago, Chile, because they are all located at a similar longitude. You can explore other cities or locations as well. You will also be able to create a physical model of the Earth-Sun system like your teacher did earlier so you can test your different ideas about how Earth moves in relation to the Sun and how this affects the duration of daylight in different cities.

Things we KNOW from what we read …

Your Task

Use what you know about the movement of Earth relative to the Sun and patterns to develop a model that you can use to explain why the duration of daylight in a given

Investigation Handout

location changes during the year and why different locations on Earth have a different duration of daylight on the same day.

The *guiding question* of this investigation is, **Why does the duration of daylight change in different locations throughout the year, and why isn't the pattern the same in all locations on Earth?**

Materials

You may use any of the following materials during your investigation:

- Safety goggles (required)
- Light source (flashlight or lamp)
- Earth model (foam ball with a wooden dowel through it)
- String
- Ruler or meterstick
- Scissors
- Markers
- Pushpins
- Data cards A–E

Safety Rules

Follow all normal safety rules. In addition, be sure to follow these rules:

- Wear sanitized safety goggles during setup, investigation activity, and cleanup.
- Be careful with the light source, because it can get hot and burn your skin.
- Do not shine the light into your eyes or your classmates' eyes.
- Be careful when working with sharp objects, because they can cut or puncture skin.
- Wash your hands with soap and water when you are done cleaning up.

Plan Your Investigation

Prepare a plan for your investigation by filling out the chart on the next page; this plan is called an *investigation proposal*. Before you start developing your plan, be sure to discuss the following questions with the other members of your group:

- What components of the **system** should be included in our **model**?
- What **patterns** should we look for or use to test our model?

Investigation 12. Daylight and Location:
Why Does the Duration of Daylight Change in Different Locations
Throughout the Year, and Why Isn't the Pattern the Same in All Locations on Earth?

Our guiding question:

This is a picture of how we will set up the equipment:

We will collect the following data:

These are the steps we will follow to collect data:

I approve of this investigation proposal.

_____   _____
Teacher's signature   Date

Investigation Handout

Collect Your Data
Keep a record of what you measure or observe during your investigation in the space below.

Analyze Your Data
You will need to analyze the data you collected before you can develop an answer to the guiding question. To analyze the data you collected, create a table, graph, or picture that will help illustrate important patterns.

Investigation 12. Daylight and Location:
Why Does the Duration of Daylight Change in Different Locations
Throughout the Year, and Why Isn't the Pattern the Same in All Locations on Earth?

Draft Argument

Develop an argument on a whiteboard. It should include the following:

1. A *claim*: Your answer to the guiding question.
2. *Evidence*: An analysis of the data and an explanation of what the analysis means.
3. A justification of the evidence: Why your group thinks the evidence is important.

The Guiding Question:	
Our Claim:	
Our Evidence:	Our Justification of the Evidence:

Argumentation Session

Share your argument with your classmates. Be sure to ask them how to make your draft argument better. Keep track of their suggestions in the space below.

Ways to IMPROVE our argument …

Argument-Driven Inquiry in **Fifth-Grade Science**: Three-Dimensional Investigations

Investigation Handout

Draft Report

Prepare an *investigation report* to share what you have learned. Use the information in this handout and your group's final argument to write a *draft* of your investigation report.

Introduction

We have been studying _____ in class.

Before we started this investigation, we explored _____

_____

_____

We noticed _____

_____

_____

My goal for this investigation was to figure out _____

_____

_____

The guiding question was _____

_____

_____

Method

To gather the data I needed to answer this question, I _____

_____

_____

_____

_____

Investigation 12. Daylight and Location:
Why Does the Duration of Daylight Change in Different Locations
Throughout the Year, and Why Isn't the Pattern the Same in All Locations on Earth?

I then analyzed the data I collected by _____

_____

_____

_____

Argument

My claim is _____

_____

_____

_____

The _____ below shows _____

_____

_____

_____

_____

_____

Argument-Driven Inquiry in **Fifth-Grade Science:** Three-Dimensional Investigations

Investigation Handout

This analysis of the data I collected suggests _____
_____
_____
_____
_____

This evidence is based on several important scientific concepts. The first one is _____
_____
_____
_____
_____

Review

Your classmates need your help! Review the draft of their investigation reports and give them ideas about how to improve. Use the peer-review guide when doing your review.

Submit Your Final Report

Once you have received feedback from your classmates about your draft report, create your final investigation report and hand it in to your teacher.

Checkout Questions

Investigation 12. Daylight and Location

The pictures below show two different models of the Earth-Sun system. These two models were made by two different people who were trying to explain why they see a sunrise in their city each morning. Use these models to answer questions 1 and 2.

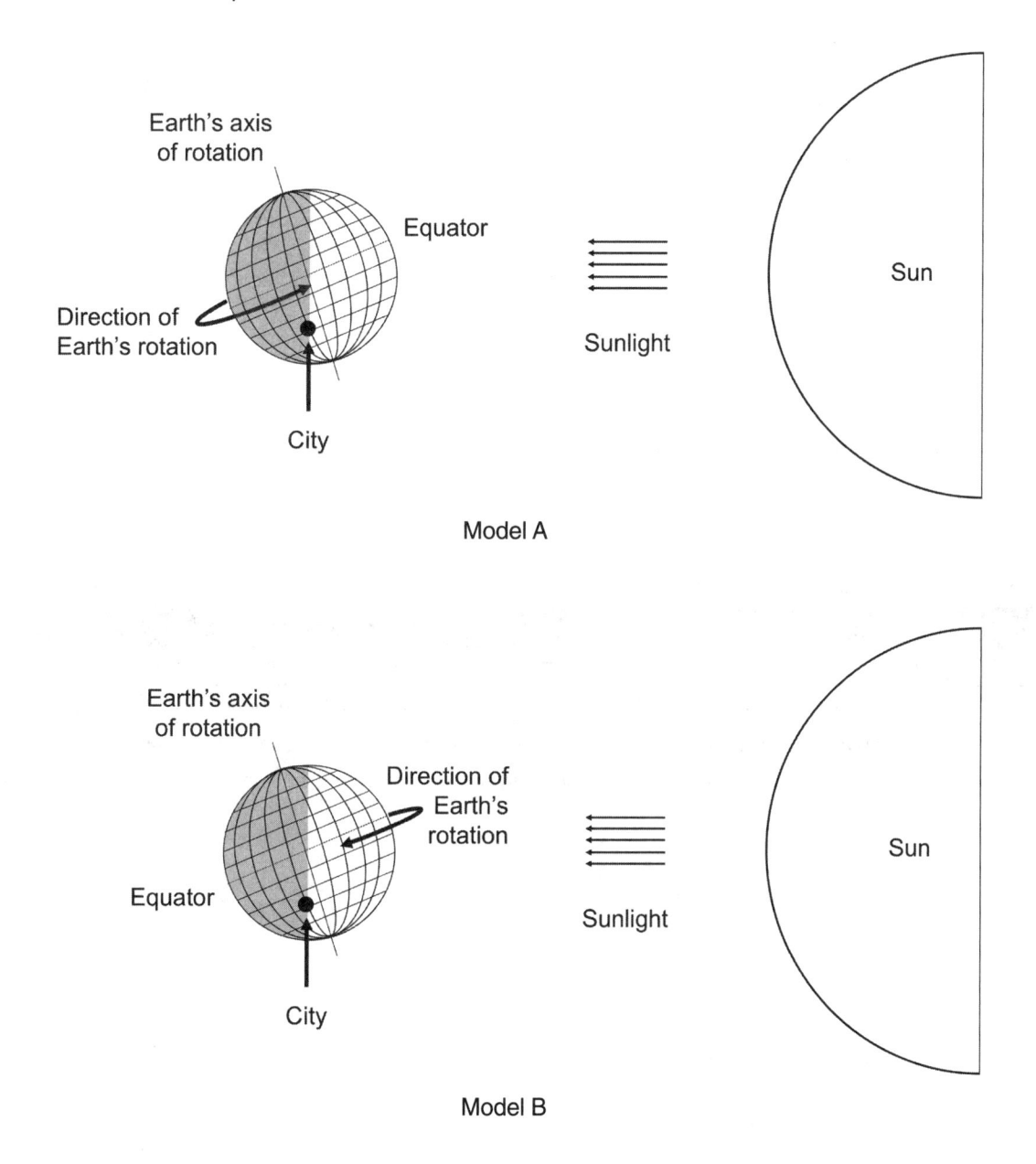

Model A

Model B

1. Which model best explains why they see a sunrise in their city each morning?

    a. Model A

    b. Model B

Checkout Questions

2. Explain your thinking. Why are models used for explaining observations?

___

Teacher Scoring Rubric for Checkout Questions 1 and 2

Level	Description
3	The student can apply the core idea correctly and can explain the importance of models.
2	The student can apply the core idea correctly but cannot explain the importance of models.
1	The student cannot apply the core idea correctly but can explain the importance of models.
0	The student cannot apply the core idea correctly and cannot explain the importance of models.

Investigation 12. Daylight and Location:
Why Does the Duration of Daylight Change in Different Locations
Throughout the Year, and Why Isn't the Pattern the Same in All Locations on Earth?

The picture below shows a model of Earth and three different cities. Use this picture to answer questions 3–5.

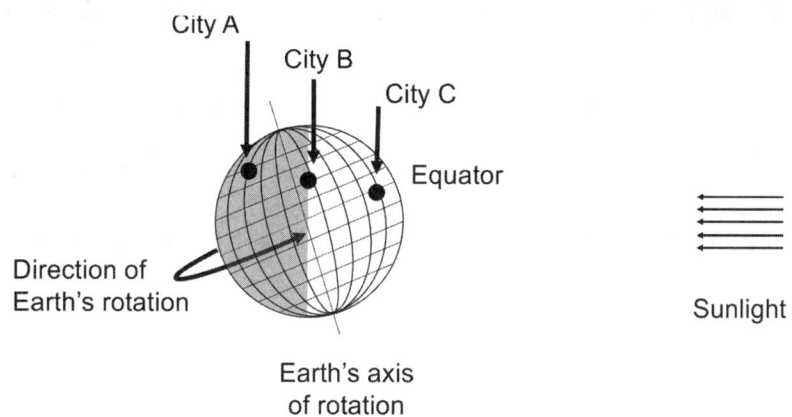

3. Which of the cities would have the longest duration of daylight?

   a. City A   c. City C
   b. City B   d. Unable to tell

4. Which of the cities would have the shortest duration of daylight?

   a. City A   c. City C
   b. City B   d. Unable to tell

5. Explain your thinking. What pattern did you use to answer questions 3 and 4?

_____
_____
_____
_____
_____
_____
_____
_____
_____
_____

Checkout Questions

Teacher Scoring Rubric for Checkout Questions 3–5

Level	Description
3	The student can apply the core idea correctly in all cases and can explain the pattern in changes in duration of daylight.
2	The student can apply the core idea correctly in all cases but cannot explain the pattern in changes in duration of daylight.
1	The student cannot apply the core idea correctly in all cases but can explain the pattern in changes in duration of daylight.
0	The student cannot apply the core idea correctly and cannot explain the pattern in changes in duration of daylight.

Teacher Notes

Investigation 13

Stars in the Night Sky: How Do the Number and Location of the Constellations That We Can See in the Night Sky Change Based on the Time of Year?

Purpose

The purpose of this investigation is to give students an opportunity to use one disciplinary core idea (DCI), one crosscutting concept (CC), and seven scientific and engineering practices (SEPs) to figure out how visible constellations change over the course of a year. Students will also learn about the assumptions that scientists make about order and consistency in nature.

The DCI, CC, and SEPs That Students Use During This Investigation to Figure Things Out

DCI

- *ESS1.B: Earth and the Solar System:* The orbits of Earth around the Sun and of the Moon around Earth, together with the rotation of Earth about an axis between its North and South Poles, cause observable patterns. These include day and night; daily changes in the length and direction of shadows; and different positions of the Sun, Moon, and stars at different times of the day, month, and year.

CC

- *CC 1: Patterns:* Similarities and differences in patterns can be used to sort, classify, communicate, and analyze simple rates of change for natural phenomena and designed products. Patterns of change can be used to make predictions. Patterns can be used as evidence to support an explanation.

SEPs

- *SEP 1: Asking Questions and Defining Problems:* Ask questions about what would happen if a variable is changed. Ask questions that can be investigated and predict reasonable outcomes based on patterns such as cause-and-effect relationships.
- *SEP 2: Developing and Using Models:* Develop and/or use models to describe and/or predict phenomena.

Teacher Notes

- *SEP 3: Planning and Carrying Out Investigations:* Plan and conduct an investigation collaboratively to produce data to serve as the basis for evidence, using fair tests in which variables are controlled and the number of trials considered. Evaluate appropriate methods and/or tools for collecting data.

- *SEP 4: Analyzing and Interpreting Data:* Represent data in tables and/or various graphical displays (bar graphs, pictographs, and/or pie charts) to reveal patterns that indicate relationships. Analyze and interpret data to make sense of phenomena, using logical reasoning, mathematics, and/or computation. Compare and contrast data collected by different groups in order to discuss similarities and differences in their findings.

- *SEP 6: Constructing Explanations and Designing Solutions:* Construct an explanation of observed relationships. Use evidence to construct or support an explanation. Identify the evidence that supports particular points in an explanation.

- *SEP 7: Engaging in Argument From Evidence:* Compare and refine arguments based on an evaluation of the evidence presented. Distinguish among facts, reasoned judgment based on research findings, and speculation in an explanation. Respectfully provide and receive critiques from peers about a proposed procedure, explanation, or model by citing relevant evidence and posing specific questions.

- *SEP 8: Obtaining, Evaluating, and Communicating Information:* Read and comprehend grade-appropriate complex texts and/or other reliable media to summarize and obtain scientific and technical ideas. Combine information in written text with that contained in corresponding tables, diagrams, and/or charts to support the engagement in other scientific and/or engineering practices. Communicate scientific and/or technical information orally and/or in written formats, including various forms of media as well as tables, diagrams, and charts.

Other Concepts That Students May Use During This Investigation

Students might also use some of the following concepts:

- A *constellation* is a group of stars that form a recognizable shape.
- The *north celestial pole* is a point in the sky directly above the North Pole. It is only visible to people who live in the Northern Hemisphere.
- The *south celestial pole* is a point in the sky directly above the South Pole. It is only visible to people who live in the Southern Hemisphere.
- As Earth spins on its axis, the north and south celestial poles do not move in the sky. All the stars that we can see, however, appear to rotate around these poles.
- Stars appear to move counterclockwise around the north celestial pole and clockwise around the south celestial pole.

Investigation 13. Stars in the Night Sky:
How Do the Number and Location of the Constellations That We Can See in
the Night Sky Change Based on the Time of Year?

What Students Figure Out

The constellations that are located near the celestial poles are visible all year, but the constellations that are located farther away are only visible on a seasonal basis. We can see between 10 and 12 major constellations during a season: 12 in winter and summer, 11 in spring, and 11 in fall.

Background Information About This Investigation for the Teacher

The stars outside our solar system are very distant objects. The distance of these stars from Earth differs, but even the closest stars to us are very far away. Proxima Centauri, for example, is the star outside our solar system that is closest to us, and it takes about four years for the light from this star to reach us (Byrd 2018). The stars do not stay in the same place in the sky when we look at them from Earth. Instead, they appear to move from east to west across the sky. The stars appear to move in the sky because Earth spins counterclockwise around its axis.

Earth's axis is an imaginary line that runs through the center of Earth between the North Pole and the South Pole. The north celestial pole is a point in the sky that is directly above the North Pole. A star named Polaris is located very close to the north celestial pole in the sky. The south celestial pole is a point in the sky directly above the South Pole. The north and south celestial poles do not move in the sky as Earth spins on its axis. All the stars that we can see, however, appear to rotate around these poles. Stars appear to move counterclockwise around the north celestial pole and clockwise around the south celestial pole.

Stars that are located close to a celestial pole, like Polaris, have a very small circle of spin. The further a star is located from a celestial pole, the wider the circle the star will trace during a night. Stars that make a full circle around a celestial pole are called circumpolar stars (McClure 2017). These stars are visible in the sky all night and do not set. For people who live at the equator, there are no circumpolar stars because the celestial poles are located at the horizon. All stars observed by people who live at the equator, as a result, rise in the east and set in the west.

Earth completes a single turn on its axis every 23 hours and 56 minutes if you use the location of the stars as a reference. The stars that rise and set in the sky therefore appear to rise and set 4 minutes earlier each night. These extra 4 minutes each night add up over weeks and months, and, as a result, the stars rise and set an hour earlier every two weeks and two hours earlier each month. After 12 months, the stars will be in the same position as they started.

A constellation is a group of stars that form a recognizable pattern in the sky. Figure 13.1 (p. 522) is a portion of a sky chart that includes some examples of constellations. This chart shows the constellations Ursa Minor (sometimes called the Little Dipper), Ursa Major, Gemini, and Orion. The last star that makes up the tail of Ursa Minor is Polaris.

Teacher Notes

This constellation, as a result, appears to spin around this star in the night sky because of its location near the celestial north pole. This constellation also does not set when viewed from the Northern Hemisphere because it is made up of circumpolar stars. Constellations such as Gemini and Orion, however, will rise and set, but the rise and set times for these constellations change throughout the year.

FIGURE 13.1
A section of a sky chart and some constellations

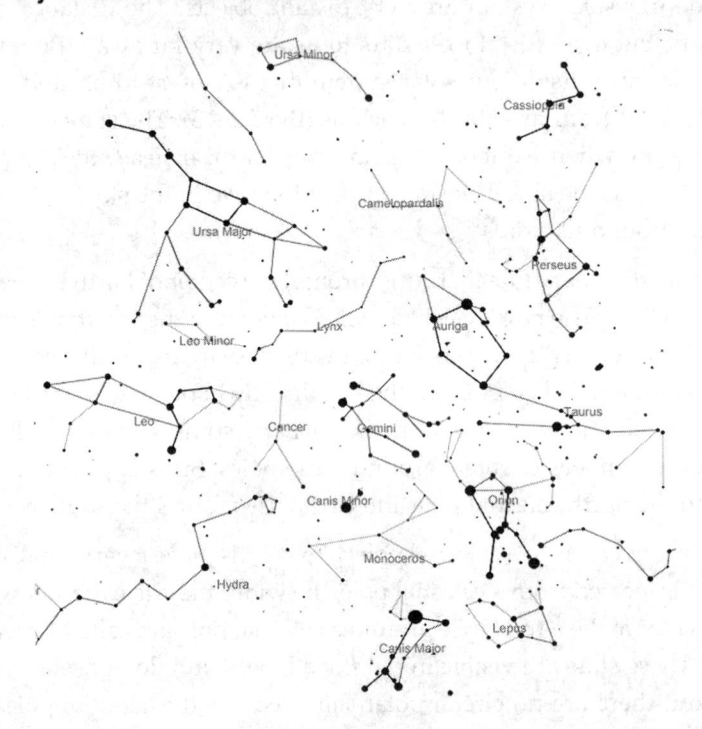

The constellations that are visible at night change with the seasons (Martin 2020). The constellations we can see each season differ because Earth also orbits around our Sun as it spins on its axis. As Earth moves around the Sun, we see different constellations in the sky depending on which constellations are overhead during the night and which constellations are overhead during the day (see Figure 13.2); the constellations that are overhead at night will be visible, but the constellations that are overhead during the day will not be visible because the brightness of day prevents people from seeing stars. In the Northern Hemisphere in winter, for example, the constellation Gemini is overhead at night and is visible, but the constellation Sagittarius is not visible because it is overhead during the day. A different part of the sky is visible at night during the spring. The constellation Cancer is overhead at night, so it is visible to people who live in the Northern Hemisphere, but the constellation Aries is overhead during the day and so it is not visible. The stars move by about 90° in the sky from one season to the next.

FIGURE 13.2
Earth's orbit around the Sun as viewed from above

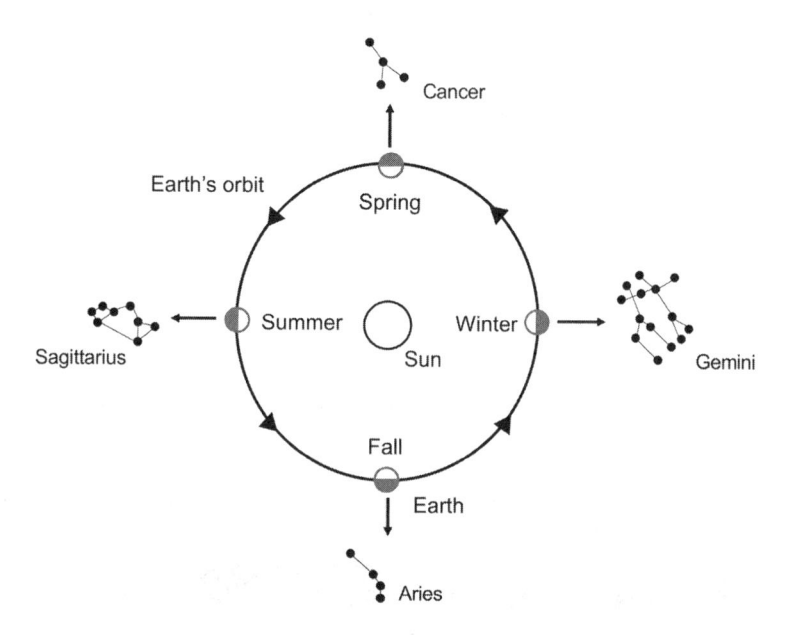

There are five major constellations that are visible all year long in the Northern Hemisphere. These constellations, which are all located near the north celestial pole, are Cassiopeia, Cepheus, Draco, Ursa Major, and Ursa Minor. There are another 26 constellations that are visible during specific seasons:

- Canis Major, Cetus, Eridanus, Gemini, Orion, Perseus, and Taurus are visible during the winter months.
- Boötes, Cancer, Crater, Hydra, Leo, and Virgo are visible during the spring.
- Aquila, Cygnus, Hercules, Lyra, Ophiuchus, Sagittarius, and Scorpius are visible during the summer.
- Andromeda, Aquarius, Aries, Capricornus, Pegasus, and Pisces are visible in the fall.

People have long used the appearance of these 26 constellations in the night sky as a way to track the passage of time and the change of seasons.

Teacher Notes

Timeline

The time needed to complete this investigation is 270 minutes (4 hours and 30 minutes). The amount of instructional time needed for each stage of the investigation is as follows:

- *Stage 1.* Introduce the task and the guiding question: 35 minutes
- *Stage 2.* Design a method and collect data: 50 minutes
- *Stage 3.* Create a draft argument: 45 minutes
- *Stage 4.* Argumentation session: 30 minutes
- *Stage 5.* Reflective discussion: 15 minutes
- *Stage 6.* Write a draft report: 30 minutes
- *Stage 7.* Peer review: 35 minutes
- *Stage 8.* Revise the report: 30 minutes

Materials and Preparation

The materials needed to implement this investigation are listed in Table 13.1.

TABLE 13.1
Materials for Investigation 13

Item	Quantity
Computer or tablet with internet access	1 per group
Whiteboard, 2' × 3'*	1 per group
Investigation Handout	1 per student
Peer-review guide and teacher scoring rubric	1 per student
Checkout Questions (optional)	1 per student

*As an alternative, students can use computer and presentation software such as Microsoft PowerPoint or Apple Keynote to create their arguments.

The students will use an online interactive sky chart during this investigation to collect the data they need to answer the guiding question. The sky chart, which was created by Sky & Telescope and is free to use, is available at *https://skyandtelescope.org/interactive-sky-chart*. Figure 13.3 is a screenshot of the website.

This interactive sky chart shows the stars that are visible in the sky from any location on Earth at any point in time. To set the location, students will need to enter their zip code. They can change the date by typing in the year and then using the drop-down menus to set the month and day. The time defaults to a 24-hour setting, but students can change it to a.m./p.m. if that will be easier. After they have set the date and time, they simply have to press the "Submit" button to update the chart. They can also manipulate

Investigation 13. Stars in the Night Sky:
How Do the Number and Location of the Constellations That We Can See in
the Night Sky Change Based on the Time of Year?

FIGURE 13.3

Screenshot of the interactive sky chart from Sky & Telescope

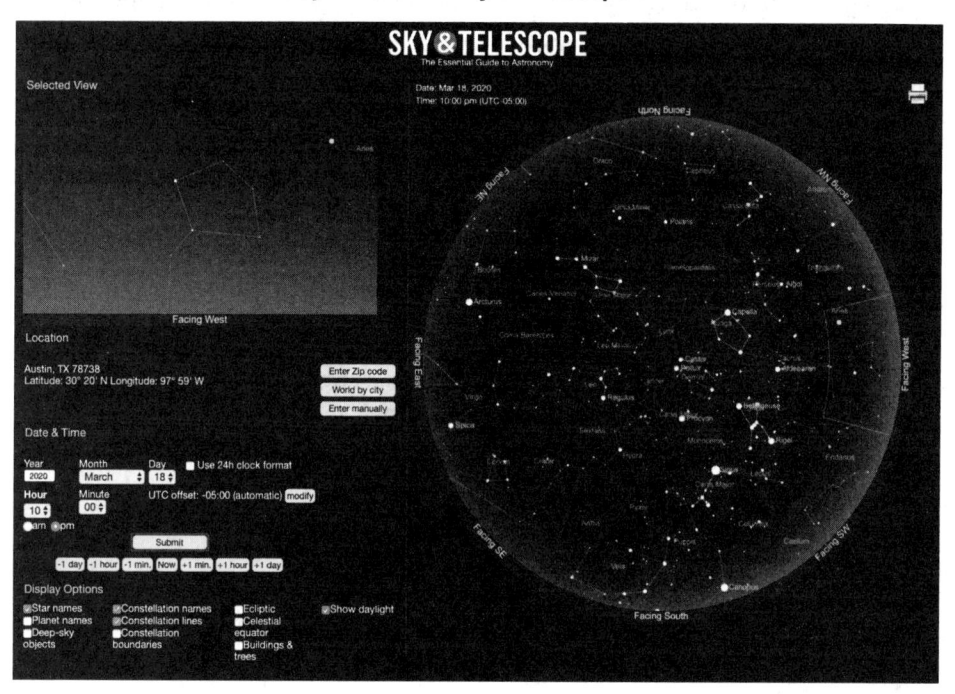

time more granularly by selecting +1 or –1 minute, hour, or day. In terms of the display options, we recommend having "Constellation names," "Constellation lines," and "Show daylight" selected and leaving the options "Star names," "Planet names," "Deep-sky objects," "Constellation boundaries," "Ecliptic," "Celestial equator," and "Buildings and trees" unselected to keep the sky chart from becoming too cluttered and confusing for the students. Finally, there is a green box on the full sky view that can be dragged around to change the smaller field-of-view box on the left to zoom in on certain parts of the sky. It defaults to "Facing West," but that can be changed by clicking and dragging.

Be sure to access the interactive sky chart and learn how it works before beginning this investigation with your students. In addition, it is important to check if students can access and use the website from a school computer, because some schools have set up firewalls and other restrictions on web browsing.

Safety Precautions

Remind students to follow all normal safety rules.

Teacher Notes

Lesson Plan by Stage

This lesson plan is only a suggestion. It is included here to illustrate what you can say and do during each stage of ADI for this specific investigation. We encourage you to modify this lesson plan by asking different questions, using different examples, and providing different scaffolds as needed to better meet the needs of students in your class.

Stage 1: Introduce the Task and the Guiding Question (35 minutes)

1. Ask the students to sit in six groups, with three or four students in each group.

2. Ask the students to clear off their desks except for a pencil (and their *Student Workbook for Argument-Driven Inquiry in Fifth-Grade Science* if they have one).

3. Pass out an Investigation Handout to each student (or ask students to turn to the Investigation Log for Investigation 13 in their workbook).

4. Read the first paragraph of the "Introduction" aloud to the class. Ask the students to follow along as you read.

5. Show a time-lapse video of stars moving in the night sky. Some good examples include the following:

    - *www.youtube.com/watch?v=HsJxGpDmJrQ* (Northern Hemisphere)
    - *www.youtube.com/watch?v=zXLqETYWb2o* (Northern Hemisphere)
    - *www.youtube.com/watch?v=3V3rmDG5J8A* (Northern Hemisphere)
    - *www.youtube.com/watch?v=huysYcz-AiQ&t=42s* (Southern Hemisphere)
    - *www.youtube.com/watch?v=wdpQk1qLrjo* (Southern Hemisphere)

6. Tell the students to record their observations and questions on the first "NOTICED/WONDER" chart in the "Introduction."

7. Ask the students to share what they observed about the video.

8. Ask the students to share what questions they have about what happened in the video.

9. Read the next two paragraphs of the "Introduction" aloud to the class. Ask the students to follow along as you read.

10. Project the web page *www.skyandtelescope.com/interactive-sky-chart* on a screen so all the students can see the interactive sky chart.

11. Tell the students, "I am now going to show you what the stars will look like tonight where we live once the Sun goes down." Set the location of the sky chart to your current location by entering the school's zip code. Set the date and time to the current date and 5:00 p.m. The sky chart should show a blue sky. Explain to the students what you are doing as you change the settings in the interactive sky chart.

Investigation 13. Stars in the Night Sky:
How Do the Number and Location of the Constellations That We Can See in
the Night Sky Change Based on the Time of Year?

12. Tell the students, "I am now going to change the time so the sky chart will show how the stars will look over time tonight. As I change the time, be sure to record your observations and questions in the second "NOTICED/WONDER" chart in the "'Introduction.'"

13. Click the "+1 hour" button every 10 seconds or so. This will move the time in the interactive sky chart by one hour. After a few clicks, the stars will appear (because the Sun will have set). Keep clicking this button until all the stars disappear (because of sunrise). You may want to repeat this process to give students a second or third chance to make observations.

14. Ask the students to share what they observed while they were watching the interactive sky chart.

15. Ask the students to share what questions they have about what happened as they were watching the interactive sky chart.

16. Tell the students, "Some of your questions might be answered by reading the rest of the "'Introduction.'"

17. Ask the students to read the rest of the "Introduction" on their own *or* ask them to follow along as you read it aloud.

18. Once the students have read the rest of the "Introduction," ask them to fill out the "Things we KNOW" chart on their Investigation Handout (or in their Investigation Log) as a group.

19. Ask the students to share what they learned from the reading. Add these ideas to a class "Things we KNOW" chart.

20. Tell the students, "Let's see what we will need to figure out during our investigation."

21. Read the task and the guiding question aloud.

22. Tell the students, "In this investigation you are going to use the same website that I just showed you."

23. Pull up the website once again and remind the students of the basics of how to use it, including how to set a location, how to manipulate time by clicking +1 or −1 minutes, hours, or days, and which settings you'd like them to use as default. Give them 5–10 minutes to play with the simulation on their own computers to figure out some different kinds of things they can see.

24. Have the students put the computers away before moving on to the next stage.

Stage 2: Design a Method and Collect Data (50 minutes)

1. Tell the students, "I am now going to give you and the other members of your group about 15 minutes to plan your investigation. Before you begin, I want

Teacher Notes

you all to take a couple of minutes to discuss the following questions with the rest of your group."

2. Show the following questions on the screen or board:
   - How can a *pattern* help us understand or predict a change over the course of a year?
   - What information do we need in order to find a *pattern*?

3. Tell the students, "Please take a few minutes to come up with an answer to these questions."

4. Give the students two or three minutes to discuss these two questions.

5. Ask two or three different groups to share their answers. Highlight or write down any important ideas on the board so students can refer to them later.

6. If possible, use a document camera to project an image of the graphic organizer for this investigation on a screen or board (or take a picture of it and project the picture on a screen or board). Tell the students, "I now want you all to plan out your investigation. To do that, you will need to fill out this investigation proposal."

7. Point to the box labeled "Our guiding question:" and tell the students, "You can put the question we are trying to answer in this box." Then ask, "Where can we find the guiding question?"

8. Wait for a student to answer.

9. Point to the box labeled "We will collect the following data:" and tell the students, "You can list the measurements or observations that you will need to collect during the investigation in this box."

10. Point to the box labeled "These are the steps we will follow to collect data:" and tell the students, "You can list what you are going to do to collect the data you need and what you will do with your data once you have it. Be sure to give enough detail that I could do your investigation for you."

11. Ask the students, "Do you have any questions about what you need to do?"

12. Answer any questions that come up.

13. Tell the students, "Once you are done, raise your hand and let me know. I'll then come by and look over your proposal and give you some feedback. You may not begin collecting data until I have approved your proposal by signing it. You need to have your proposal done in the next 15 minutes."

14. Give the students 15 minutes to work in their groups on their investigation proposal. As they work, move from group to group to check in, ask probing questions, and offer a suggestion if a group gets stuck.

15. As each group finishes its investigation proposal, read it over and determine if it will be productive or not. If you feel the investigation will be productive (not

necessarily what you would do or what the other groups are doing), sign your name on the proposal and let the group start collecting data. If the plan needs to be changed, offer some suggestions or ask some probing questions, and have the group make the changes before you approve it.

16. Tell the students to collect their data and record their observations in the "Collect Your Data" box in their Investigation Handout (or the Investigation Log in their workbook).

17. Give the students 30 minutes to collect their data.

What should a student-designed investigation look like?

There are a number of different investigations that students can design to answer the question "How do the number and location of constellations that we can see in the night sky change based on the time of year?" One of the ways they could answer this is by choosing a single location on Earth and then making a list of (1) which constellations are visible each month at a particular time of night and (2) the locations of those constellations. Some students may choose to collect data for all 12 months, while others may only collect data for a particular month within each season.

Stage 3: Create a Draft Argument (45 minutes)

1. Tell the students, "Now that we have all this data, we need to analyze the data so we can figure out an answer to the guiding question."

2. If possible, project an image of the "Analyze Your Data" section for this investigation on a screen or board using a document camera (or take a picture of it and project the picture on a screen or board). Point to the section and tell the students, "You can create a table, graph, or other representation as a way to analyze your data. You can make your table, graph, or other representation in this section."

3. Ask the students, "What information do we need to include in this analysis?"

4. Tell the students, "Please take a few minutes to discuss this question with your group and be ready to share."

5. Give the students five minutes to discuss.

Teacher Notes

6. Ask two or three different groups to share their answers. Highlight or write down any important ideas on the board so students can refer to them later.

7. Tell the groups of students, "I am now going to give you and the other members of your group about 10 minutes to analyze your data." If the students are having trouble analyzing their data, you can take a few minutes to provide a mini-lesson about possible ways to analyze the data they collected (this strategy is called just-in-time instruction because it is offered only when students get stuck).

8. Give the students 10 minutes to analyze their data. As they work, move from group to group to check in, ask probing questions, and offer suggestions.

9. Tell the students, "I am now going to give you and the other members of your group about 15 minutes to create an argument to share what you have learned and convince others that they should believe you. Before you do that, we need to take a few minutes to discuss what you need to include in your argument."

10. If possible, use a document camera to project the "Argument Presentation on a Whiteboard" image from the "Draft Argument" section of the Investigation Handout (or the Investigation Log in their workbook) on a screen or board (or take a picture of it and project the picture on a screen or board).

11. Point to the box labeled "The Guiding Question:" and tell the students, "You can put the question we are trying to answer here on your whiteboard."

12. Point to the box labeled "Our Claim:" and tell the students, "You can put your claim here on your whiteboard. The claim is your answer to the guiding question."

13. Point to the box labeled "Our Evidence:" and tell the students, "You can put the evidence that you are using to support your claim here on your whiteboard. Your evidence will need to include the analysis you just did and an explanation of what your analysis means or shows. Scientists always need to support their claims with evidence."

14. Point to the box labeled "Our Justification of the Evidence:" and tell the students, "You can put your justification of your evidence here on your whiteboard. Your justification needs to explain why your evidence is important. Scientists often use core ideas to explain why the evidence they are using matters. Core ideas are important concepts that scientists use to help them make sense of what happens during an investigation."

15. Ask the students, "What are some core ideas that we read about earlier that might help us explain why the evidence we are using is important?"

16. Ask the students to share some of the core ideas from the "Introduction" section of the Investigation Handout (or the Investigation Log in the workbook). List these core ideas on the board.

Investigation 13. Stars in the Night Sky:
How Do the Number and Location of the Constellations That We Can See in
the Night Sky Change Based on the Time of Year?

17. Tell the students, "That is great. I would like to see everyone try to include these core ideas in your justification of the evidence. Your goal is to use these core ideas to help explain why your evidence matters and why the rest of us should pay attention to it."

18. Ask the students, "Do you have any questions about what you need to do?"

19. Answer any questions that come up.

20. Tell the students, "Okay, go ahead and start working on your arguments. You need to have your argument done in the next 15 minutes. It doesn't need to be perfect. We just need something down on the whiteboards so we can share our ideas."

21. Give the students 15 minutes to work in their groups on their arguments. As they work, move from group to group to check in, ask probing questions, and offer a suggestion if a group gets stuck. Figure 13.4 shows an example of an argument created by students for this investigation.

FIGURE 13.4

Example of an argument

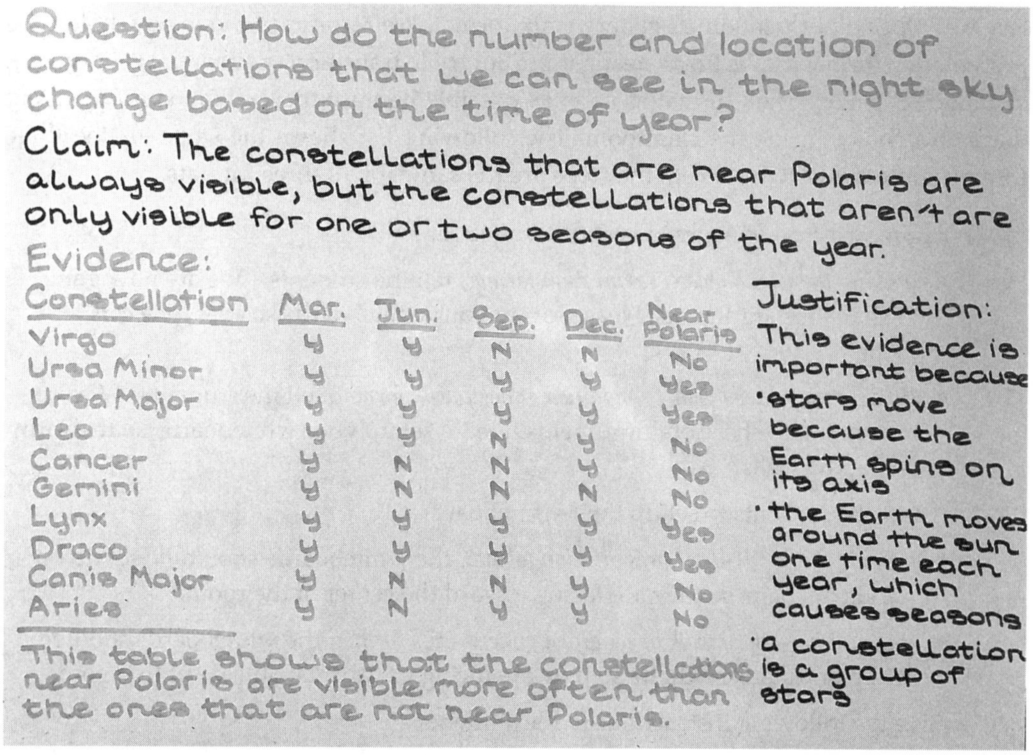

Question: How do the number and location of constellations that we can see in the night sky change based on the time of year?

Claim: The constellations that are near Polaris are always visible, but the constellations that aren't are only visible for one or two seasons of the year.

Evidence:

Constellation	Mar.	Jun.	Sep.	Dec.	Near Polaris
Virgo	Y	Y	N	N	No
Ursa Minor	Y	Y	Y	Y	Yes
Ursa Major	Y	Y	Y	Y	Yes
Leo	Y	Y	N	Y	No
Cancer	Y	N	N	Y	No
Gemini	Y	N	N	Y	No
Lynx	Y	Y	Y	Y	Yes
Draco	Y	Y	Y	Y	Yes
Canis Major	Y	N	N	Y	No
Aries	Y	N	Y	Y	No

This table shows that the constellations near Polaris are visible more often than the ones that are not near Polaris.

Justification: This evidence is important because
• stars move because the Earth spins on its axis
• the Earth moves around the sun one time each year, which causes seasons
• a constellation is a group of stars

Teacher Notes

> ## What should the table, graph, or other representation for this investigation look like?
>
> There are a number of different ways that students can analyze the data they collect during this investigation. One option would be to create a table with the names of constellations and the months or seasons in which they are visible. See Figure 13.4 for an example of this type of table. Students might also create a graph with seasons as the independent variable and the number of constellations as the dependent variable. Some students may choose to focus on a few of the constellations and track their visibility and location over time, while others may attempt to represent all or almost all the constellations. There are other options for analyzing the collected data. Students often come up with some unique ways of analyzing their data, so be sure to give them some voice and choice during this stage.

Stage 4: Argumentation Session (30 minutes)

The argumentation session can be conducted in a whole-class presentation format, a gallery walk format, or a modified gallery walk format. We recommend using a whole-class presentation format for the first investigation, but try to transition to either the gallery walk or modified gallery walk format as soon as possible because that will maximize student voice and choice inside the classroom. The following list shows the steps for the three formats; unless otherwise noted, the steps are the same for all three formats.

1. Begin by introducing the use of the whiteboard.
   - *If using the whole-class presentation format*, tell the students, "We are now going to share our arguments. Please set up your whiteboards so everyone can see them."
   - *If using the gallery walk or modified gallery walk format*, tell the students, "We are now going to share our arguments. Please set up your whiteboards so they are facing the walls."
2. Allow the students to set up their whiteboards.
   - *If using the whole-class presentation format*, the whiteboards should be set up on stands or chairs so they are facing toward the center of the room.
   - *If using the gallery walk or modified gallery walk format*, the whiteboards should be set up on stands or chairs so they are facing toward the outside of the room.
3. Give the following instructions to the students:
   - *If using the whole-class presentation format*, tell the students, "Okay, before we get started I want to explain what we are going to do next. Your group will have

Investigation 13. Stars in the Night Sky:
How Do the Number and Location of the Constellations That We Can See in the Night Sky Change Based on the Time of Year?

an opportunity to share your argument with the rest of the class. After you are done, everyone else in the class will have a chance to ask questions and offer some suggestions about ways to make your group's argument better. After we have a chance to listen to each other and learn something new, I'm going to give you some time to revise your arguments and make them better."

- *If using the gallery walk format,* tell the students, "Okay, before we get started I want to explain what we are going to do next. You are going to read the arguments that were created by other groups. When I say 'go,' your group will go to a different group's station so you can see their argument. Once you are there, I'll give your group a few minutes to read and review their argument. Your job is to offer them some suggestions about ways to make their argument better. You can use sticky notes to give them suggestions. Please be specific about what you want to change and how you think they should change it. After we have a chance to learn from each other, I'm going to give you some time to revise your arguments and make them better."

- *If using the modified gallery walk format,* tell the students, "Okay, before we get started I want to explain what we are going to do next. I'm going to ask some of you to present your arguments to your classmates. If you are presenting your argument, your job is to share your group's claim, evidence, and justification of the evidence. The rest of you will be travelers. If you are a traveler, your job is to listen to the presenters, ask the presenters questions if you do not understand something, and then offer them some suggestions about ways to make their argument better. After we have a chance to learn from each other, I'm going to give you some time to revise your arguments and make them better."

4. Use a document camera to project the "Ways to IMPROVE our argument …" box from the Investigation Handout (or the Investigation Log in their workbook) on a screen or board (or take a picture of it and project the picture on a screen or board).

- *If using the whole-class presentation format,* point to the box and tell the students, "After your group presents your argument, you can write down the suggestions you get from your classmates here. If you are listening to a presentation and you see a good idea from another group, you can write down that idea here as well. Once we are done with the presentations, I will give you a chance to use these suggestions or ideas to improve your arguments."

- *If using the gallery walk format,* point to the box and tell the students, "If you see a good idea from another group, you can write it down here. Once we are done reviewing the different arguments, I will give you a chance to use these ideas to improve your own arguments. It is important to share ideas like this."

- *If using the modified gallery walk format,* point to the box and tell the students, "If you are a presenter, you can write down the suggestions you get from the travelers here. If you are a traveler and you see a good idea from another group,

Teacher Notes

you can write down that idea here. Once we are done with the presentations, I will give you a chance to use these suggestions or ideas to improve your arguments."

5. Ask the students, "Do you have any questions about what you need to do?"
6. Answer any questions that come up.
7. Give the following instructions:
   - *If using the whole-class presentation format,* tell the students, "Okay. Let's get started."
   - *If using the gallery walk format,* tell the students, "Okay, I'm now going to tell you which argument to go to and review."
   - *If using the modified gallery walk format,* tell the students, "Okay, I'm now going to assign you to be a presenter or a traveler." Assign one or two students from each group to be presenters and one or two students from each group to be travelers.
8. Give the students an opportunity to review the arguments.
   - *If using the whole-class presentation format,* have each group present their argument one at a time. Give each group only two to three minutes to present their argument. Then give the class two to three minutes to ask them questions and offer suggestions. Encourage as much participation from the students as possible.
   - *If using the gallery walk format,* tell the students, "Okay. Let's get started. Each group, move one argument to the left. Don't move to the next argument until I tell you to move. Once you get there, read the argument and then offer suggestions about how to make it better. I will put some sticky notes next to each argument. You can use the sticky notes to leave your suggestions." Give each group about three to four minutes to read the arguments, talk, and offer suggestions.

     a. After three to four minutes, tell the students, "Okay. Let's move on to the next argument. Please move one group to the left."

     b. Again, give each group three to four minutes to read, talk, and offer suggestions.

     c. Repeat this process until each group has had their argument read and critiqued three times.
   - *If using the modified gallery walk format,* tell the students, "Okay. Let's get started. Reviewers, move one group to the left. Don't move to the next group until I tell you to move. Presenters, go ahead and share your argument with the travelers when they get there." Give each group of presenters and travelers about three to four minutes to talk.

Investigation 13. Stars in the Night Sky: How Do the Number and Location of the Constellations That We Can See in the Night Sky Change Based on the Time of Year?

   a. Tell the students, "Okay. Let's move on to the next argument. Travelers, move one group to the left."

   b. Again, give each group of presenters and travelers about three to four minutes to talk.

   c. Repeat this process until each group has had their argument read and critiqued three times.

9. Tell the students to return to their workstations.

10. Give the following instructions about revising the argument:

    - *If using the whole-class presentation format,* tell the students, "I'm now going to give you all about 10 minutes to revise your argument. Take a few minutes to talk in your groups and determine what you want to change to make your argument better. Once you have decided what to change, go ahead and make the changes to your whiteboard."

    - *If using the gallery walk format,* tell the students, "I'm now going to give you all about 10 minutes to revise your argument. Take a few minutes to read the suggestions that were left at your argument. Then talk in your groups and determine what you want to change to make your argument better. Once you have decided what to change, go ahead and make the changes to your whiteboard."

    - *If using the modified gallery walk format,* tell the students, " I'm now going to give you all about 10 minutes to revise your argument. Please return to your original groups." Wait for the students to move back into their original groups and then tell the students, "Okay, take a few minutes to talk in your groups and determine what you want to change to make your argument better. Once you have decided what to change, go ahead and make the changes to your whiteboard."

11. Ask the students, "Do you have any questions about what you need to do?"

12. Answer any questions that come up.

13. Tell the students, "Okay. Let's get started."

14. Give the students 10 minutes to work in their groups on their arguments. As they work, move from group to group to check in, ask probing questions, and offer a suggestion if a group gets stuck.

Stage 5: Reflective Discussion (15 minutes)

1. Tell the students, "We are now going to take a minute to talk about some of the core ideas and crosscutting concepts that we have used during our investigation."

2. Show Figure 13.5 on page 536 (without the caption) on the screen.

Teacher Notes

FIGURE 13.5
Trails of circumpolar stars in the sky

Note: A full-color version of this figure is available on the book's Extras page at *www.nsta.org/adi-5th*.

3. Ask the students, "What do all you see going on here?"
4. Allow the students to share their ideas.
5. Ask the students, "Why do you think the paths of the stars follow this pattern?
6. Allow the students to share their ideas. As they share their ideas, ask different questions to encourage them to expand on their thinking (e.g., "Can you tell me more about that?"), clarify a contribution (e.g., "Can you say that in another way?"), support an idea (e.g., "Why do you think that?"), add to an idea mentioned by a classmate (e.g., "Would anyone like to add to the idea?"), re-voice an idea offered by a classmate (e.g., "Who can explain that to me in another way?"), or critique an idea during the discussion (e.g., "Do you agree or disagree with that idea and why?") until students are able to generate an adequate explanation.
7. Show Figure 13.6 (without the caption) on the screen.
8. Ask the students, "What constellations do you see?"
9. Allow the students to share their ideas.

Investigation 13. Stars in the Night Sky:
How Do the Number and Location of the Constellations That We Can See in
the Night Sky Change Based on the Time of Year?

FIGURE 13.6

Constellations during winter in Austin, Texas (not labeled)

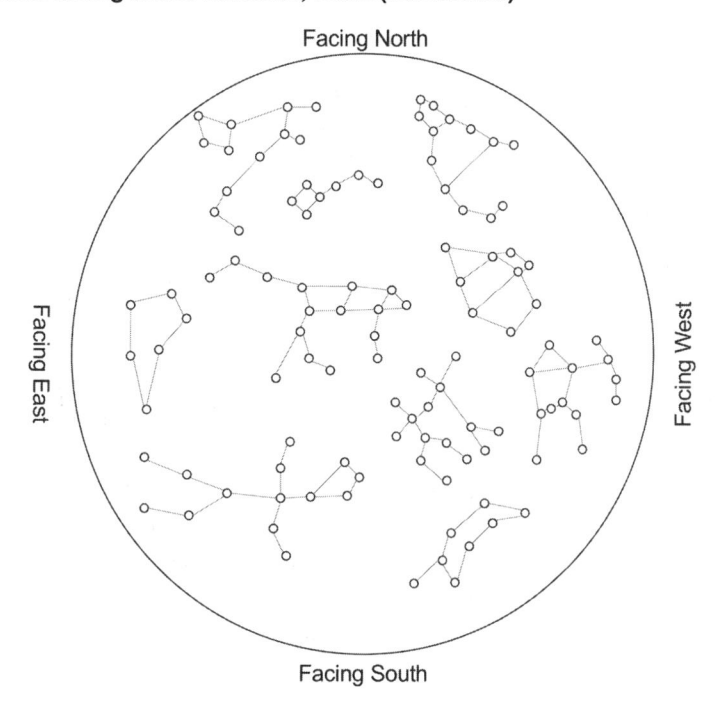

10. Show Figure 13.7 (without the caption) on the screen.

11. Ask the students, "Which of these constellations did you see during your investigation?"

12. Allow the students to share their ideas.

13. Ask the students, "Which of these constellations do you think you would be able to see in the sky all year, and why do you think that?"

14. Allow the students to share their ideas. As they share their ideas, ask different questions to encourage them to expand on their thinking (e.g., "Can you tell me more about that?"), clarify a contribution (e.g., "Can you say that in another way?"), support an idea (e.g., "Why do you think that?"), add to an idea mentioned by a classmate (e.g., "Would anyone like to add to the idea?"), re-voice an idea offered by a classmate (e.g., "Who can explain that to me in another way?"), or critique an idea during the discussion (e.g., "Do you agree or disagree with that idea and why?") until students are able to generate an adequate explanation.

15. Show an image of the question "What do you think are the most important core ideas or crosscutting concepts that we used during this investigation to help us make sense of what we observed?" Tell the students, "Okay, let's make sure we

Teacher Notes

FIGURE 13.7
Constellations during winter in Austin, Texas (labeled)

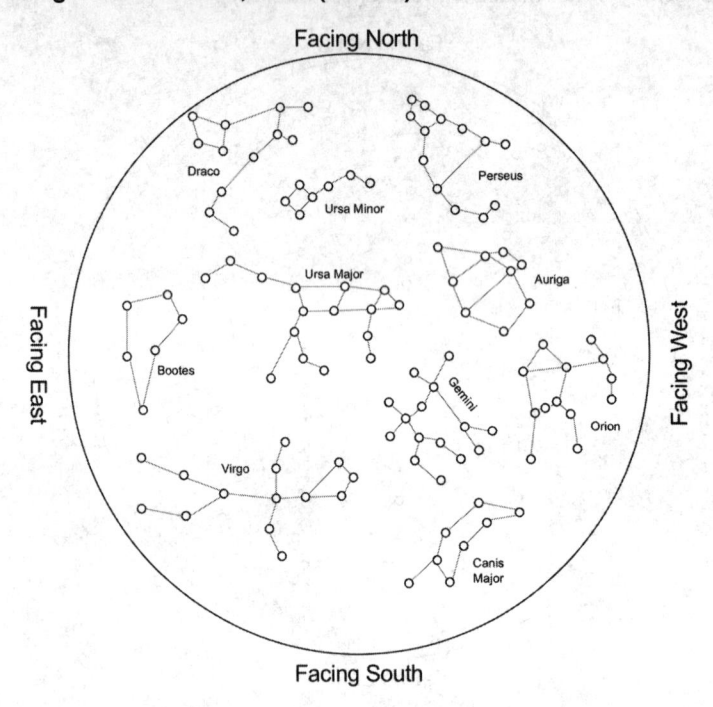

Facing North / *Facing East* / *Facing West* / *Facing South*

Constellations labeled: Draco, Ursa Minor, Perseus, Ursa Major, Auriga, Bootes, Gemini, Orion, Virgo, Canis Major

are all on the same page. Please take a moment to discuss this question with the other people in your group." Give them a few minutes to discuss the question.

16. Ask the students, "What do you all think? Who would like to share?"

17. Allow the students to share their ideas.

18. Tell the students, "We are now going to take a minute to talk about how scientists think about the world."

19. Show an image of the question "Are the laws of nature the same everywhere?" on the screen. Tell the students, "Take a few minutes to talk about how you would answer this question with the other people in your group. Be ready to share with the rest of the class." Give the students two to three minutes to talk in their group.

20. Ask the students, "What do you all think? Who would like to share an idea?"

21. Allow the students to share their ideas.

22. Ask the students, "Why would it be important for scientists to assume that the laws of nature stay the same over time?"

23. Allow the students to share their ideas.

24. Tell the students, "I think it is important for scientists to assume that the laws of nature stay the same over time because it allows them to explain the past and the present."

25. Ask the students, "Suppose I wanted to know where the stars were in the night sky 50 years ago, would I be able to do that if I understand how the stars move over time?"

26. Allow the students to share their ideas.

27. Ask the students, "Suppose I wanted to know where the stars would be in the night sky on my birthday in 20 years, would I be able to do that if I understand how the stars move over time?"

28. Allow the students to share their ideas.

29. Ask the students, "Why can I figure out where the stars will be in the sky in the past or in the future?"

30. Ask the students, "Does anyone have any questions about why scientists assume that the laws of nature stay the same over time?"

31. Answer any questions that come up.

32. Tell the students, "We are now going to take a minute to talk about what went well and what didn't go so well during our investigation. We need to talk about this because you all are going to be planning and carrying out your own investigations like this a lot this year, and I want to help you all get better at it."

33. Show an image of the question "What made your investigation scientific?" on the screen. Tell the students, "Take a few minutes to talk about how you would answer this question with the other people in your group. Be ready to share with the rest of the class." Give the students two to three minutes to talk in their group.

34. Ask the students, "What do you all think? Who would like to share an idea?"

35. Allow the students to share their ideas. Be sure to expand on their ideas about what makes an investigation scientific.

36. Show an image of the question "What made your investigation not so scientific?" on the screen. Tell the students, "Take a few minutes to talk about how you would answer this question with the other people in your group. Be ready to share with the rest of the class." Give the students two to three minutes to talk in their group.

37. Ask the students, "What do you all think? Who would like to share an idea?"

38. Allow the students to share their ideas. Be sure to expand on their ideas about what makes an investigation less scientific.

39. Show an image of the question "What rules can we put into place to help us make sure our next investigation is more scientific?" on the screen. Tell the

Teacher Notes

students, "Take a few minutes to talk about how you would answer this question with the other people in your group. Be ready to share with the rest of the class." Give the students two to three minutes to talk in their group.

40. Ask the students, "What do you all think? Who would like to share an idea?"
41. Allow the students to share their ideas. Once they have shared their ideas, offer a suggestion for a possible class rule.
42. Ask the students, "What do you all think? Should we make this a rule?"
43. If the students agree, write the rule on the board or on a class "Rules for Scientific Investigation" chart so you can refer to it during the next investigation.

Stage 6: Write a Draft Report (30 minutes)

Your students will use either the Investigation Handout or the Investigation Log in the student workbook when writing the draft report. When you give the directions shown in quotes in the following steps, substitute "Investigation Log in your workbook" or just "Investigation Log" (as shown in brackets) for "handout" if they are using the workbook.

1. Tell the students, "You are now going to write an investigation report to share what you have learned. Please take out a pencil and turn to the 'Draft Report' section of your handout [Investigation Log in your workbook]."
2. If possible, use a document camera to project the "Introduction" section of the draft report from the Investigation Handout (or the Investigation Log in their workbook) on a screen or board (or take a picture of it and project the picture on a screen or board).
3. Tell the students, "The first part of the report is called the 'Introduction.' In this section of the report you want to explain to the reader what you were investigating, why you were investigating it, and what question you were trying to answer. All this information can be found in the text at the beginning of your handout [Investigation Log]." Point to the image. "Here are some sentence starters to help you begin writing."
4. Ask the students, "Do you have any questions about what you need to do?"
5. Answer any questions that come up.
6. Tell the students, "Okay, let's write."
7. Give the students 10 minutes to write the "Introduction" section of the report. As they work, move from student to student to check in, ask probing questions, and offer a suggestion if a student gets stuck.
8. If possible, use a document camera to project the "Method" section of the draft report from the Investigation Handout (or the Investigation Log in their workbook) on a screen or board (or take a picture of it and project the picture on a screen or board).

Investigation 13. Stars in the Night Sky:
How Do the Number and Location of the Constellations That We Can See in the Night Sky Change Based on the Time of Year?

9. Tell the students, "The second part of the report is called the 'Method.' In this section of the report you want to explain to the reader what you did during the investigation, what data you collected and why, and how you went about analyzing your data. All this information can be found in the 'Plan Your Investigation' section of the handout [Investigation Log]. Remember that you all planned and carried out different investigations, so do not assume that the reader will know what you did." Point to the image. "Here are some sentence starters to help you begin writing."

10. Ask the students, "Do you have any questions about what you need to do?"

11. Answer any questions that come up.

12. Tell the students, "Okay, let's write."

13. Give the students 10 minutes to write the "Method" section of the report. As they work, move from student to student to check in, ask probing questions, and offer a suggestion if a student gets stuck.

14. If possible, use a document camera to project the "Argument" section of the draft report from the Investigation Handout (or the Investigation Log in their workbook) on a screen or board (or take a picture of it and project the picture on a screen or board).

15. Tell the students, "The last part of the report is called the 'Argument.' In this section of the report you want to share your claim, evidence, and justification of the evidence with the reader. All this information can be found on your whiteboard." Point to the image. "Here are some sentence starters to help you begin writing."

16. Ask the students, "Do you have any questions about what you need to do?"

17. Answer any questions that come up.

18. Tell the students, "Okay, let's write."

19. Give the students 10 minutes to write the "Argument" section of the report. As they work, move from student to student to check in, ask probing questions, and offer a suggestion if a student gets stuck.

Stage 7: Peer Review (35 minutes)

Your students will use either the Investigation Handout or their workbook when doing the peer review. Except where noted below, the directions are the same whether using the handout or the workbook.

1. Tell the students, "We are now going to review our reports to find ways to make them better. I'm going to come around and collect your draft reports. While I do that, please take out a pencil."

2. Collect the handouts or the workbooks with the draft reports from the students.

Teacher Notes

3. If possible, use a document camera to project the peer-review guide (see Appendix 4) on a screen or board (or take a picture of it and project the picture on a screen or board).

4. Tell the students, "We are going to use this peer-review guide to give each other feedback." Point to the image.

5. Tell the students, "I'm going to ask you to work with a partner to do this. I'm going to give you and your partner a draft report to read. You two will then read the report together. Once you are done reading the report, I want you to answer each of the questions on the peer-review guide." Point to the review questions on the image of the peer-review guide.

6. Tell the students, "You can check 'no,' 'almost,' or 'yes' after each question." Point to the checkboxes on the image of the peer-review guide.

7. Tell the students, "This will be your rating for this part of the report. Make sure you agree on the rating you give the author. If you mark 'no' or 'almost,' then you need to tell the author what he or she needs to do to get a 'yes.'" Point to the space for the reviewer feedback on the image of the peer-review guide.

8. Tell the students, "It is really important for you to give the authors feedback that is helpful. That means you need to tell them exactly what they need to do to make their report better."

9. Ask the students, "Do you have any questions about what you need to do?"

10. Answer any questions that come up.

11. Tell the students, "Please sit with a partner who is not in your current group." Allow the students time to sit with a partner.

12. Tell the students, "Okay, I'm now going to give you one report to read." Pass out one Investigation Handout with a draft report or one workbook to each pair. Make sure that the report you give a pair was not written by one of the students in that pair. Give each pair one peer-review guide to fill out. If the students are using workbooks, the peer-review guide is included right after the draft report so you do not need to pass out copies of the peer-review guide.

13. Tell the students, "Okay, I'm going to give you 15 minutes to read the report I gave you and to fill out the peer-review guide. Go ahead and get started."

14. Give the students 15 minutes to work. As they work, move around from pair to pair to check in and see how things are going, answer questions, and offer advice.

15. After 15 minutes pass, tell the students, "Okay, time is up. Please give me the report and the peer-review guide that you filled out."

16. Collect the Investigation Handouts and the peer-review guides, or collect the workbooks if students are using them. If the students are using the Investigation

Handouts and separate peer-review guides, be sure you keep each handout with its corresponding peer-review guide.

17. Tell the students, "Okay, I am now going to give you a different report to read and a new peer-review guide to fill out." Pass out one more report to each pair. Make sure that the report you give a pair was not written by one of the students in that pair. Give each pair a new peer-review guide to fill out as a group.

18. Tell the students, "Okay, I'm going to give you 15 minutes to read this new report and to fill out the peer-review guide. Go ahead and get started."

19. Give the students 15 minutes to work. As they work, move around from pair to pair to check in and see how things are going, answer questions, and offer advice.

20. After 15 minutes pass, tell the students, "Okay, time is up. Please give me the report and the peer-review guide that you filled out."

21. Collect the Investigation Handouts and the peer-review guides, or collect the workbooks if students are using them. If the students are using the Investigation Handouts and separate peer-review guides, be sure you keep each handout with its corresponding peer-review guide.

Stage 8: Revise the Report (30 minutes)

Your students will use either the Investigation Handout or their workbook when revising the report. Except where noted below, the directions are the same whether using the handout or the workbook.

1. Tell the students, "You are now going to revise your draft report based on the feedback you get from your classmates. Please take out a pencil."

2. Return the reports to the students.
   - *If the students used the Investigation Handout and a copy of the peer-review guide,* pass back the handout and the peer-review guide to each student.
   - *If the students used the workbook,* pass that back to each student.

3. Tell the students, "Please take a few minutes to read over the peer-review guide. You should use it to figure out what you need to change in your report and how you will change it."

4. Allow the students to read the peer-review guide.

5. *If the students used the workbook,* if possible use a document camera to project the "Write Your Final Report" section from the Investigation Log on a screen or board (or take a picture of it and project the picture on a screen or board).

6. Give the following directions about how to revise their reports:
   - *If the students used the Investigation Handout and a copy of the peer-review guide,* tell them, "Okay, let's revise our reports. Please take out a piece of paper. I would

Teacher Notes

like you to rewrite your report. You can use your draft report as a starting point, but you also need to change it to make it better. Use the feedback on the peer-review guide to make it better."

- *If the students used the workbook,* tell them, "Okay, let's revise our reports. I would like you to rewrite your report in the section of the Investigation Log called "Write Your Final Report." You can use your draft report as a starting point, but you also need to change it to make it better. Use the feedback on the peer-review guide to make it better."

7. Ask the students, "Do you have any questions about what you need to do?"

8. Answer any questions that come up.

9. Tell the students, "Okay, let's write." Allow about 20 minutes for the students to revise their reports.

10. After about 20 minutes, give the following directions:
    - *If the students used the Investigation Handout,* tell them, "Okay, time's up. I will now come around and collect your Investigation Handout, the peer-review guide, and your final report."
    - *If the students used the workbook,* tell them, "Okay, time's up. I will now come around and collect your workbooks."

11. *If the students used the Investigation Handout,* collect all the Investigation Handouts, peer-review guides, and final reports. *If the students used the workbook,* collect all the workbooks.

12. *If the students used the Investigation Handout,* use the "Teacher Score" column in the peer-review guide to grade the final report. *If the students used the workbook,* use the "Investigation Report Grading Rubric" in the Investigation Log to grade the final report. Whether you are using the handout or the log, you can give the students feedback about their writing in the "Teacher Comments" section.

How to Use the Checkout Questions

The Checkout Questions are an optional assessment. We recommend giving them to students at the start of the next class period after the students finish stage 8 of the investigation. You can then look over the student answers to determine if you need to reteach the core idea from the investigation. Appendix 6 gives the answers to the Checkout Questions that should be given by a student who can apply the core idea correctly and can explain the importance of patterns.

Investigation 13. Stars in the Night Sky: How Do the Number and Location of the Constellations That We Can See in the Night Sky Change Based on the Time of Year?

Alignment With Standards

Table 13.2 highlights how the investigation can be used to address specific performance expectations from the *Next Generation Science Standards, Common Core State Standards for English Language Arts (CCSS ELA),* and *English Language Proficiency (ELP) Standards.*

TABLE 13.2

Investigation 13 alignment with standards

NGSS performance expectation	5-ESS1-2: Represent data in graphical displays to reveal patterns of daily changes in length and direction of shadows, day and night, and the seasonal appearance of some stars in the night sky.
CCSS ELA—Reading: Informational Text	Key ideas and details • CCSS.ELA-LITERACY.RI.5.1: Quote accurately from a text when explaining what the text says explicitly and when drawing inferences from the text. • CCSS.ELA-LITERACY.RI.5.2: Determine two or more main ideas of a text and explain how they are supported by key details; summarize the text. • CCSS.ELA-LITERACY.RI.5.3: Explain the relationships or interactions between two or more individuals, events, ideas, or concepts in a historical, scientific, or technical text based on specific information in the text. Craft and structure • CCSS.ELA-LITERACY.RI.5.4: Determine the meaning of general academic and domain-specific words and phrases in a text relevant to a *grade 5 topic or subject area*. • CCSS.ELA-LITERACY.RI.5.5: Compare and contrast the overall structure (e.g., chronology, comparison, cause/effect, problem/solution) of events, ideas, concepts, or information in two or more texts. • CCSS.ELA-LITERACY.RI.5.6: Analyze multiple accounts of the same event or topic, noting important similarities and differences in the point of view they represent. Integration of knowledge and ideas • CCSS.ELA-LITERACY.RI.5.7: Draw on information from multiple print or digital sources, demonstrating the ability to locate an answer to a question quickly or to solve a problem efficiently. • CCSS.ELA-LITERACY.RI.5.8: Explain how an author uses reasons and evidence to support particular points in a text, identifying which reasons and evidence support which point(s). Range of reading and level of text complexity • CCSS.ELA-LITERACY.RI.5.10: By the end of the year, read and comprehend informational texts, including history/social studies, science, and technical texts, at the high end of the grades 4–5 text complexity band independently and proficiently.

Continued

Table 13.2 *(continued)*

CCSS ELA—Writing	Text types and purposes
	• CCSS.ELA-LITERACY.W.5.1: Write opinion pieces on topics or texts, supporting a point of view with reasons.
	○ CCSS.ELA-LITERACY.W.5.1.A: Introduce a topic or text clearly, state an opinion, and create an organizational structure in which ideas are logically grouped to support the writer's purpose.
	○ CCSS.ELA-LITERACY.W.5.1.B: Provide logically ordered reasons that are supported by facts and details.
	○ CCSS.ELA-LITERACY.W.5.1.C: Link opinion and reasons using words, phrases, and clauses (e.g., *consequently*, *specifically*).
	○ CCSS.ELA-LITERACY.W.5.1.D: Provide a concluding statement or section related to the opinion presented.
	• CCSS.ELA-LITERACY.W.5.2: Write informative or explanatory texts to examine a topic and convey ideas and information clearly.
	○ CCSS.ELA-LITERACY.W.5.2.A: Introduce a topic clearly, provide a general observation and focus, and group related information logically; include formatting (e.g., headings), illustrations, and multimedia when useful to aiding comprehension.
	○ CCSS.ELA-LITERACY.W.5.2.B: Develop the topic with facts, definitions, concrete details, quotations, or other information and examples related to the topic.
	○ CCSS.ELA-LITERACY.W.5.2.C: Link ideas within and across categories of information using words, phrases, and clauses (e.g., *in contrast*, *especially*).
	○ CCSS.ELA-LITERACY.W.5.2.D: Use precise language and domain-specific vocabulary to inform about or explain the topic.
	○ CCSS.ELA-LITERACY.W.5.2.E: Provide a concluding statement or section related to the information or explanation presented.
	Production and distribution of writing
	• CCSS.ELA-LITERACY.W.5.4: Produce clear and coherent writing in which the development and organization are appropriate to task, purpose, and audience.
	• CCSS.ELA-LITERACY.W.5.5: With guidance and support from peers and adults, develop and strengthen writing as needed by planning, revising, editing, rewriting, or trying a new approach.
	• CCSS.ELA-LITERACY.W.5.6: With some guidance and support from adults, use technology, including the internet, to produce and publish writing as well as to interact and collaborate with others; demonstrate sufficient command of keyboarding skills to type a minimum of two pages in a single sitting.
	Research to build and present knowledge
	• CCSS.ELA-LITERACY.W.5.8: Recall relevant information from experiences or gather relevant information from print and digital sources; summarize or paraphrase information in notes and finished work, and provide a list of sources.
	• CCSS.ELA-LITERACY.W.5.9: Draw evidence from literary or informational texts to support analysis, reflection, and research.

Continued

Table 13.2 (*continued*)

CCSS ELA—Writing (*continued*)	Range of writing • CCSS.ELA-LITERACY.W.5.10: Write routinely over extended time frames (time for research, reflection, and revision) and shorter time frames (a single sitting or a day or two) for a range of discipline-specific tasks, purposes, and audiences.

CCSS ELA—Speaking and Listening	Comprehension and collaboration
	• CCSS.ELA-LITERACY.SL.5.1: Engage effectively in a range of collaborative discussions (one-on-one, in groups, and teacher-led) with diverse partners on *grade 5 topics and texts,* building on others' ideas and expressing their own clearly.
	  ○ CCSS.ELA-LITERACY.SL.5.1.A: Come to discussions prepared, having read or studied required material; explicitly draw on that preparation and other information known about the topic to explore ideas under discussion.
	  ○ CCSS.ELA-LITERACY.SL.5.1.B: Follow agreed-upon rules for discussions and carry out assigned roles.
	  ○ CCSS.ELA-LITERACY.SL.5.1.C: Pose and respond to specific questions by making comments that contribute to the discussion and elaborate on the remarks of others.
	  ○ CCSS.ELA-LITERACY.SL.5.1.D: Review the key ideas expressed and draw conclusions in light of information and knowledge gained from the discussions.
	• CCSS.ELA-LITERACY.SL.5.2: Summarize a written text read aloud or information presented in diverse media and formats, including visually, quantitatively, and orally.
	• CCSS.ELA-LITERACY.SL.5.3: Summarize the points a speaker makes and explain how each claim is supported by reasons and evidence.
	Presentation of knowledge and ideas
	• CCSS.ELA-LITERACY.SL.5.4: Report on a topic or text or present an opinion, sequencing ideas logically and using appropriate facts and relevant, descriptive details to support main ideas or themes; speak clearly at an understandable pace.
	• CCSS.ELA-LITERACY.SL.5.5: Include multimedia components (e.g., graphics, sound) and visual displays in presentations when appropriate to enhance the development of main ideas or themes.
	• CCSS.ELA-LITERACY.SL.5.6: Adapt speech to a variety of contexts and tasks, using formal English when appropriate to task and situation.

ELP Standards	Receptive modalities • ELP 1: Construct meaning from oral presentations and literary and informational text through grade-appropriate listening, reading, and viewing. • ELP 8: Determine the meaning of words and phrases in oral presentations and literary and informational text.

Continued

Teacher Notes

Table 13.2 *(continued)*

ELP Standards *(continued)*	Productive modalities • ELP 3: Speak and write about grade-appropriate complex literary and informational texts and topics. • ELP 4: Construct grade-appropriate oral and written claims and support them with reasoning and evidence. • ELP 7: Adapt language choices to purpose, task, and audience when speaking and writing. Interactive modalities • ELP 2: Participate in grade-appropriate oral and written exchanges of information, ideas, and analyses, responding to peer, audience, or reader comments and questions. • ELP 5: Conduct research and evaluate and communicate findings to answer questions or solve problems. • ELP 6: Analyze and critique the arguments of others orally and in writing. Linguistic structures of English • ELP 9: Create clear and coherent grade-appropriate speech and text. • ELP 10: Make accurate use of standard English to communicate in grade-appropriate speech and writing.

References

Byrd, D. 2018. Meet Proxima Centauri, closest star to Sun. EarthSky. *https://earthsky.org/space/proxima-centauri-our-suns-nearest-neighbor*.

Martin, L. 2020. A list of constellations visible seasonally. Sciencing. *https://sciencing.com/list-constellations-visible-seasonally-7789783.html*.

McClure, B. 2017. Circumpolar stars never rise or set. EarthSky. *https://earthsky.org/space/what-are-circumpolar-stars*.

Investigation Handout

Investigation 13

Stars in the Night Sky: How Do the Number and Location of the Constellations That We Can See in the Night Sky Change Based on the Time of Year?

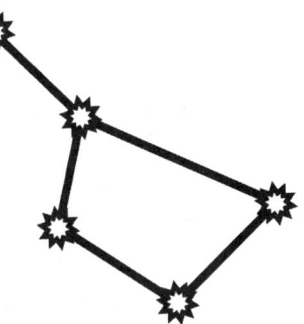

Introduction

When you look up at the sky on a dark and clear night you can see many different stars. Your teacher will show you a time-lapse video of the night sky. As you watch the video, keep track of things you notice and things you wonder about in the boxes below.

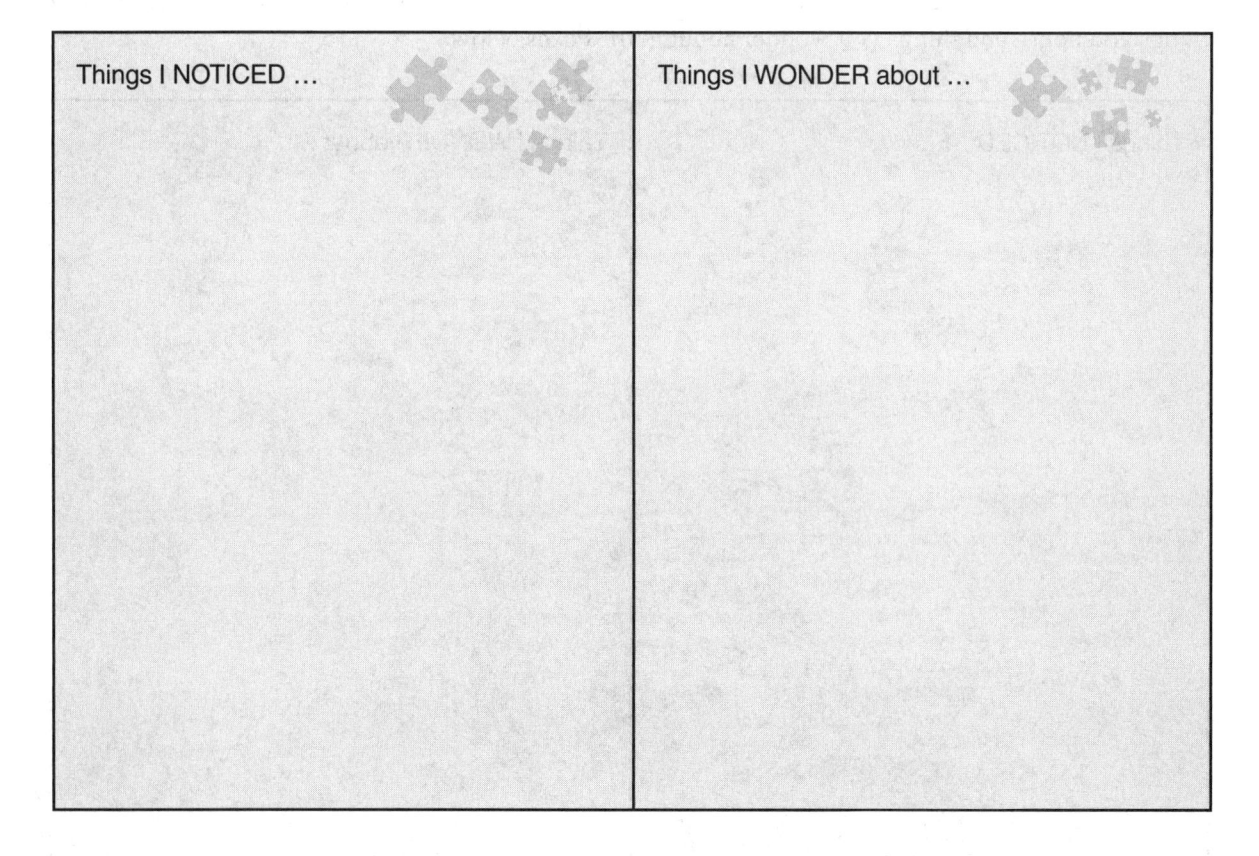

Things I NOTICED ...	Things I WONDER about ...

The stars appear to move in the sky because Earth spins counterclockwise around its axis. Earth's axis is an imaginary line that runs through the center of Earth between the North Pole and the South Pole (see the picture on the next page). The north celestial pole is a point in the sky that is directly above

Investigation Handout

the North Pole. A star named Polaris is located very close to the north celestial pole in the sky. The south celestial pole is a point in the sky directly above the South Pole. The north and south celestial poles do not move in the sky as Earth spins on its axis. All the stars that we can see, however, appear to rotate around these poles. Stars appear to move counterclockwise around the north celestial pole and appear to move clockwise around the south celestial pole.

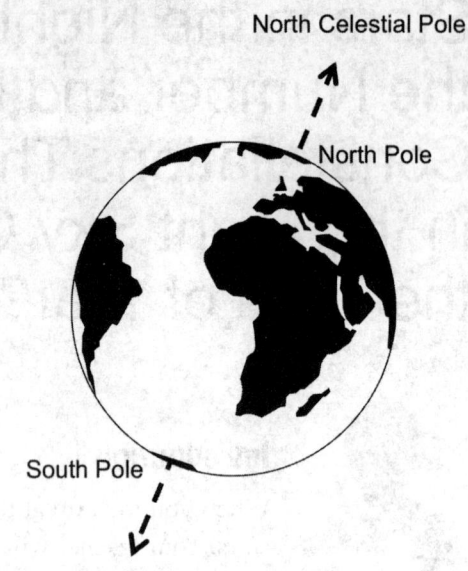

A constellation is a group of stars that form a recognizable shape. Which constellations we can see in the night sky depends on where we are on Earth. People who live in the Northern Hemisphere see different constellations than people who live in the Southern Hemisphere. Which constellations people see in the night sky also depends on how far north or south they live from the equator. Your teacher will use a computer simulation to show you the constellations that will be in the sky tonight where you live and how these constellations will move across the sky over time. As your watch, keep track of things you notice and things you wonder about in the boxes below.

The celestial poles

Things I NOTICED …	Things I WONDER about …

Investigation 13. Stars in the Night Sky: How Do the Number and Location of the Constellations That We Can See in the Night Sky Change Based on the Time of Year?

Constellations that are visible in the Northern Hemisphere

You may have noticed that some of the constellations that you can see in the sky from where you live are visible for the whole night and some of the constellations are only visible for a few hours. Constellations that are located near a celestial pole are visible all night. For example, people who live in the Northern Hemisphere can see a constellation called Ursa Minor from sunset to sunrise (see the picture on the right). People can see this constellation all night because it is located near the north celestial pole. Constellations that are farther away from a celestial pole, such as Capricornus, Sagittarius, and Libra, will only be visible at certain times during the night because they rise and set like the Moon or the Sun. Unlike Ursa Minor, these constellations are not visible all night because they may be beneath the horizon when it is dark enough outside to see them.

There is another reason why you may or may not be able to see a constellation in the night sky besides where you live and the time of night. The constellations that you can see in the sky also change depending on the season. For example, if you live in Texas and you look at the sky at 10:00 p.m. in June, you will be able to see a constellation called Virgo, but you will not be able to see Virgo if you look at the sky at the same time of night in January. You will, however, be able to see a constellation called Orion. Orion will be in about the same location in the sky in December that Virgo was in June. In this investigation you will have a chance to figure out how the number and location of the constellations that you can see where you live change based on the time of year.

Things we KNOW from what we read …

Investigation Handout

Your Task

Use what you know about Earth and solar system, the rotation of Earth around its axis, and patterns to determine how the visible constellations change during the year. To complete this task, you will be using a computer simulation of the night sky to monitor changes and collect data.

The *guiding question* of this investigation is, **How do the number and location of the constellations that we can see in the night sky change based on the time of year?**

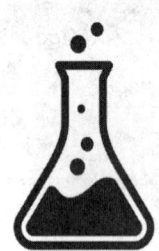

Materials

You will use a computer or tablet to access a website to collect data. The URL for the website is *www.skyandtelescope.com/interactive-sky-chart*.

Safety Rules

Follow all normal safety rules.

Plan Your Investigation

Prepare a plan for your investigation by filling out the chart on the next page; this plan is called an *investigation proposal*. Before you start developing your plan, be sure to discuss the following questions with the other members of your group:

- How can a **pattern** help us understand or predict a change over the course of a year?
- What information do we need in order to find a **pattern**?

Investigation 13. Stars in the Night Sky: How Do the Number and Location of the Constellations That We Can See in the Night Sky Change Based on the Time of Year?

Our guiding question:

We will collect the following data:

These are the steps we will follow to collect data:

I approve of this investigation proposal.

_____   _____
Teacher's signature   Date

Argument-Driven Inquiry in **Fifth-Grade Science**: Three-Dimensional Investigations

Investigation Handout

Collect Your Data

Keep a record of what you measure or observe during your investigation in the space below.

Analyze Your Data

You will need to analyze the data you collected before you can develop an answer to the guiding question. To analyze the data you collected, create a table, graph, or other representation to compare and contrast the constellations that are visible in the night sky over time.

Investigation 13. Stars in the Night Sky: How Do the Number and Location of the Constellations That We Can See in the Night Sky Change Based on the Time of Year?

Draft Argument

Develop an argument on a whiteboard. It should include the following:

1. *A claim:* Your answer to the guiding question.
2. *Evidence:* An analysis of the data and an explanation of what the analysis means.
3. A *justification of the evidence:* Why your group thinks the evidence is important.

The Guiding Question:	
Our Claim:	
Our Evidence:	Our Justification of the Evidence:

Argumentation Session

Share your argument with your classmates. Be sure to ask them how to make your draft argument better. Keep track of their suggestions in the space below.

Ways to IMPROVE our argument …

Investigation Handout

Draft Report

Prepare an *investigation report* to share what you have learned. Use the information in this handout and your group's final argument to write a *draft* of your investigation report.

Introduction

We have been studying _____ in class.

Before we started this investigation, we explored _____

_____

_____

We noticed _____

_____

_____

My goal for this investigation was to figure out _____

_____

_____

The guiding question was _____

_____

_____

Method

To gather the data I needed to answer this question, I _____

_____

_____

_____

_____

Investigation 13. Stars in the Night Sky: How Do the Number and Location of the Constellations That We Can See in the Night Sky Change Based on the Time of Year?

I then analyzed the data I collected by _____

_____

_____

_____

Argument

My claim is _____

_____

_____

_____

The _____ below shows _____

_____

_____

_____

_____

Investigation Handout

This analysis of the data I collected suggests _____

_____

_____

_____

_____

This evidence is based on several important scientific concepts. The first one is _____

_____

_____

_____

_____

Review

Your classmates need your help! Review the draft of their investigation reports and give them ideas about how to improve. Use the peer-review guide when doing your review.

Submit Your Final Report

Once you have received feedback from your classmates about your draft report, create your final investigation report and hand it in to your teacher.

Checkout Questions

Investigation 13. Stars in the Night Sky

1. If this is the view of the night sky in the Northern Hemisphere, draw arrows showing the path that you would expect the stars to take if you watched them all night.

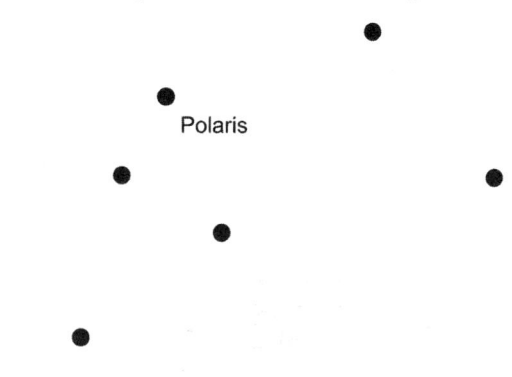

2. If this is the view of the night sky in the Northern Hemisphere in June, which constellations would you expect to see in December? Circle the letters of all the constellations that you think you would still be able to see.

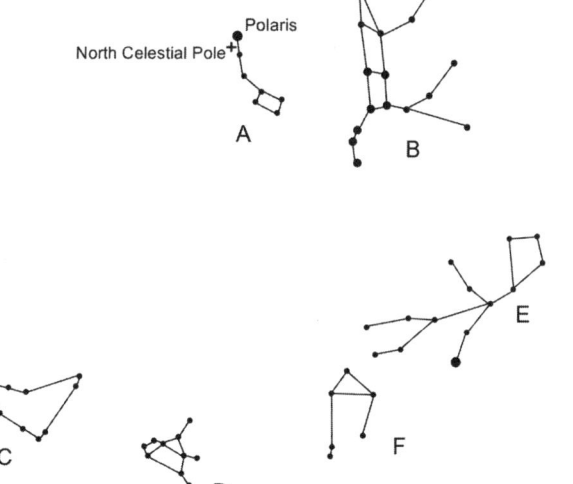

3. How do you know? Use what you know about patterns to help explain your thinking.

_____

_____

_____

_____

_____

Argument-Driven Inquiry in **Fifth-Grade Science:** *Three-Dimensional Investigations*

Checkout Questions

Teacher Scoring Rubric for the Checkout Questions

Level	Description
3	The student can apply the core idea correctly in all cases and can explain the importance of patterns.
2	The student cannot apply the core idea correctly in all cases but can explain the importance of patterns.
1	The student can apply the core idea correctly in all cases but cannot explain the importance of patterns.
0	The student cannot apply the core idea correctly and cannot explain the importance of patterns.

Teacher Notes

Investigation 14
Star Brightness: How Does Distance Affect the Apparent Brightness of a Star?

Purpose

The purpose of this investigation is to give students an opportunity to use two disciplinary core ideas (DCIs), two crosscutting concepts (CCs), and eight scientific and engineering practices (SEPs) to figure out how distance affects the apparent brightness of a star. Students will also learn about the use of models as tools for reasoning about natural phenomena during the reflective discussion.

The DCIs, CCs, and SEPs That Students Use During This Investigation to Figure Things Out

DCIs

- *PS3.B: Conservation of Energy and Energy Transfer:* Light transfers energy from place to place
- *ESS1.A: The Universe and Its Stars:* Stars range greatly in their distance from Earth.

CCs

- *CC2: Cause and Effect*: Cause-and-effect relationships are routinely identified, tested, and used to explain change. Events that occur together with regularity might or might not be a cause-and-effect relationship.
- *CC3: Scale, Proportion, and Quantity*: Standard units are used to measure and describe physical quantities such as weight, time, temperature, and volume.

SEPs

- *SEP 1: Asking Questions and Defining Problems:* Ask questions about what would happen if a variable is changed. Ask questions that can be investigated and predict reasonable outcomes based on patterns such as cause-and-effect relationships.
- *SEP 2: Developing and Using Models:* Develop and/or use models to describe and/or predict phenomena.
- *SEP 3: Planning and Carrying Out Investigations:* Plan and conduct an investigation collaboratively to produce data to serve as the basis for evidence, using fair tests in which variables are controlled and the number of trials considered. Evaluate appropriate methods and/or tools for collecting data.
- *SEP 4: Analyzing and Interpreting Data:* Represent data in tables and/or various graphical displays (bar graphs, pictographs, and/or pie charts) to reveal

Teacher Notes

patterns that indicate relationships. Analyze and interpret data to make sense of phenomena, using logical reasoning, mathematics, and/or computation. Compare and contrast data collected by different groups in order to discuss similarities and differences in their findings.

- *SEP 5: Using Mathematics and Computational Thinking:* Organize simple data sets to reveal patterns that suggest relationships. Describe, measure, estimate, and/or graph quantities (e.g., area, volume, weight, time) to address scientific and engineering questions and problems.

- *SEP 6: Constructing Explanations and Designing Solutions:* Construct an explanation of observed relationships. Use evidence to construct or support an explanation. Identify the evidence that supports particular points in an explanation.

- *SEP 7: Engaging in Argument From Evidence:* Compare and refine arguments based on an evaluation of the evidence presented. Distinguish among facts, reasoned judgment based on research findings, and speculation in an explanation. Respectfully provide and receive critiques from peers about a proposed procedure, explanation, or model by citing relevant evidence and posing specific questions.

- *SEP 8: Obtaining, Evaluating, and Communicating Information:* Read and comprehend grade-appropriate complex texts and/or other reliable media to summarize and obtain scientific and technical ideas. Combine information in written text with that contained in corresponding tables, diagrams, and/or charts to support the engagement in other scientific and/or engineering practices. Communicate scientific and/or technical information orally and/or in written formats, including various forms of media as well as tables, diagrams, and charts.

Other Concepts That Students May Use During This Investigation

Students might also use some of the following concepts:

- *Luminosity* is the total amount of energy emitted by a star in a unit of time. The luminosity of a star depends on its size and temperature.
- *Apparent brightness* is how bright a star looks to us from Earth.
- *Lux* is a unit of measurement used to describe the amount of light that falls on an object.
- *Lumens* is a unit of measurement used to describe the amount of light produced by a bulb.

What Students Figure Out

Stars that are closer to Earth will appear brighter in the sky than stars that produce more energy but are farther away.

Investigation 14. Star Brightness:
How Does Distance Affect the Apparent Brightness of a Star?

Background Information About This Investigation for the Teacher

There are a vast number of stars in the universe. Astronomers classify stars using different characteristics such as color, size, temperature, and luminosity. Luminosity is the total amount of energy emitted by a star in a unit of time. This energy then transfers from the star to other objects around it by light. Stars that produce more energy in a given amount of time have a greater luminosity and emit more light than stars that produce less energy in the same amount of time. The Sun, for example, has a white-yellow color, a diameter of about 864,000 miles, a surface temperature of about 5500°C, and a luminosity of 3.846 × 1026 watts, an amount equal to 1 *solar luminosity* (Gregersen 2009, NASA 2017). Astronomers use solar luminosity as the baseline for measuring the luminosity of other stars. Rigel is another example of a star that we can see in the sky. Rigel has a blue-white color, a diameter of about 68,250,000 miles, a surface temperature of about 11600°C, and a luminosity that is 120,000 times greater than the Sun (Sessions 2020, Star Facts 2019).

Astronomers use the term *apparent brightness* to describe how bright a star looks to us on Earth. Apparent brightness is not the same as luminosity. Stars that produce the same amount of energy each second may have a different apparent brightness when viewed from Earth, and stars with the same apparent brightness when viewed from Earth may have different luminosities. The Sun is a good example of this phenomenon. The Sun produces less energy per second than many other stars in the universe (such as Rigel). It is also smaller and cooler than many of these stars. The Sun, however, appears much brighter than any of the other stars that we can see in the sky because it is closer to Earth than any other star. The Sun is about 93 million miles away from us, and it takes the light that is produced by the Sun about 8 minutes to travel this distance through space. The next closest star to Earth is called Proxima Centauri. This star is about 25 trillion miles away from Earth, and light from Proxima Centauri takes about 4.24 years to reach Earth (Temming 2014). Rigel is even farther away than Proxima Centauri. It takes 863 years for energy from Rigel to travel through space and reach Earth (Sessions 2020). There are other stars that are even farther away from the Earth.

Timeline

The time needed to complete this investigation is 260 minutes (4 hours and 20 minutes). The amount of instructional time needed for each stage of the investigation is as follows:

- *Stage 1.* Introduce the task and the guiding question: 35 minutes
- *Stage 2.* Design a method and collect data: 40 minutes
- *Stage 3.* Create a draft argument: 45 minutes
- *Stage 4.* Argumentation session: 30 minutes
- *Stage 5.* Reflective discussion: 15 minutes

Teacher Notes

- *Stage 6.* Write a draft report: 30 minutes
- *Stage 7.* Peer review: 35 minutes
- *Stage 8.* Revise the report: 30 minutes

Materials and Preparation

The materials needed for this investigation are listed in Table 14.1. The items can be purchased from a big-box retail store such as Walmart or Target or through an online retailer such as Amazon. The materials for this investigation can also be purchased as a complete kit (which includes enough materials for 24 students, or six groups of four students) at *www.argumentdriveninquiry.com*.

TABLE 14.1
Materials for Investigation 14

Item	Quantity
Safety goggles	1 per student
Lamp with 40 W bulb	1 per class
Lamp with 60 W bulb	1 per class
Lamp with 100 W bulb	1 per class
Masking tape	1 roll per class
Meterstick or soft tape measure	1 per group
Tablet with a light meter app	1 per group
10 strips (1" × 8.5") of paper stapled together	1 per group
Whiteboard, 2' × 3'*	1 per group
Investigation Handout	1 per student
Peer-review guide and teacher scoring rubric	1 per student
Checkout Questions (optional)	1 per student

*As an alternative, students can use computer and presentation software such as Microsoft PowerPoint or Apple Keynote to create their arguments.

Students will need to create physical models of stars during this investigation. The 40 W, 60 W, and 100 W bulbs produce different amounts of light, so these bulbs can be used to create a physical model of three different stars with different luminosities. The amount of light produced by a bulb is reported in a unit of measurement called lumens. The amount of light produced by a bulb, in lumens, is labeled on the top of each bulb. Students can measure the apparent brightness of each bulb at different distances using a light meter app loaded onto a tablet with a camera. A light meter app measures the intensity of light using a unit of measurement called lux. There are several free light meter apps that can be used to measure light intensity available to download at the Apple App Store or Google Play. Some examples follow.

> **Investigation 14.** Star Brightness:
> How Does Distance Affect the Apparent Brightness of a Star?

- Lux Light Meter Pro at *https://apps.apple.com/us/app/lux-light-meter-pro/id1292598866*
- Lux Light Meter Free at *https://apps.apple.com/us/app/lux-light-meter-free/id1171685960*
- Lux Light Meter Free at *https://play.google.com/store/apps/details?id=com.doggoapps.luxlight&hl=en_US*

You may need to ask someone from your school or district IT department to load one of these apps on the tablets that the students will use. If you do not have tablets available for students to use in your school or district, set up a meeting with someone from your IT department to discuss other options. You should learn how to use the light meter app that you decide to use before starting the investigation so you can help students when they get stuck.

Students can also measure the apparent brightness of the bulbs at different distances by counting the number of pieces of paper that the light shines through. Figure 14.1 shows how you can make a basic tool for measuring apparent brightness by stapling 10 strips of paper together and labeling the number of strips with a pen or pencil. Brighter or more intense lights will shine through more pieces of paper than dimmer lights.

FIGURE 14.1

How to make a tool to measure light intensity with 10 strips of paper

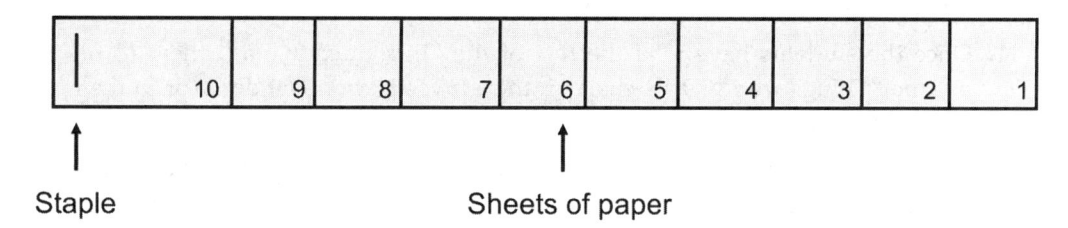

| | 10 | 9 | 8 | 7 | 6 | 5 | 4 | 3 | 2 | 1 |

↑ Staple   ↑ Sheets of paper

Safety Precautions

Remind students to follow all normal safety rules. In addition, tell the students to take the following safety precautions:

- Wear sanitized safety goggles during setup, investigation activity, and cleanup.
- Lightbulbs can get very hot. Do not touch a lightbulb when it is on or for several minutes after turning it off.
- Keep electrical equipment away from water sources to prevent shock.
- Make sure all materials are put away after completing the activity.
- Wash their hands with soap and water when they are done cleaning up.

Teacher Notes

Lesson Plan by Stage

Stage 1: Introduce the Task and the Guiding Question (35 minutes)

1. Ask the students to sit in six groups, with three or four students in each group.
2. Ask the students to clear off their desks except for a pencil (and their *Student Workbook for Argument-Driven Inquiry in Fifth-Grade Science* if they have one).
3. Pass out an Investigation Handout to each student (or ask students to turn to the Investigation Log for Investigation 14 in their workbook).
4. Read the first paragraph of the "Introduction" aloud to the class. Ask the students to follow along as you read.
5. Show the following video: *www.youtube.com/watch?v=0FXJUP6_O1w*.
6. Tell the students to record their observations and questions about what they see in the video on the "NOTICED/WONDER" chart in the "Introduction."
7. Ask the students to share what they observed.
8. Ask the students to share what questions they have.
9. Tell the students, "Some of your questions might be answered by reading the rest of the 'Introduction.'"
10. Ask the students to read the rest of the "Introduction" on their own *or* ask them to follow along as you read it aloud.
11. Once the students have read the rest of the "Introduction," ask them to fill out the "Things we KNOW" chart on their Investigation Handout (or in their Investigation Log) as a group.
12. Ask the students to share what they learned from the reading. Add these ideas to a class "Things we KNOW" chart.
13. Tell the students, "Let's see what we will need to figure out during our investigation."
14. Read the task and the guiding question aloud.
15. Tell the students, "I have lots of materials here that you can use."
16. Introduce the students to the materials available for them to use during the investigation by holding each one up and then asking how it might be used to collect data. We also recommend that you show the students how to use the light meter app or the strips of paper to measure light intensity and then give them a chance to play with the app or the strips of paper so they can see how they work before moving on to stage 2.

Investigation 14. Star Brightness:
How Does Distance Affect the Apparent Brightness of a Star?

Stage 2: Design a Method and Collect Data (40 minutes)

1. Tell the students, "I am now going to give you and the other members of your group about 15 minutes to plan your investigation. Before you begin, I want you all to take a couple of minutes to discuss the following questions with the rest of your group."

2. Show the following questions on the screen or board:
   - What information do we need to find a relationship between a *cause* and an *effect*?
   - What measurement *scale* and *units* might we use as we collect data?

3. Tell the students, "Please take a few minutes to come up with an answer to these questions."

4. Give the students two or three minutes to discuss these two questions.

5. Ask two or three different groups to share their answers. Highlight or write down any important ideas on the board so students can refer to them later.

6. If possible, use a document camera to project an image of the graphic organizer for this investigation on a screen or board (or take a picture of it and project the picture on a screen or board). Tell the groups of students, "I now want you all to plan out your investigation. To do that, you will need to fill out this proposal."

7. Point to the box labeled "Our guiding question:" and tell the students, "You can put the question we are trying to answer in this box." Then ask, "Where can we find the guiding question?"

8. Wait for a student to answer.

9. Point to the box labeled "This is a picture of how we will set up the equipment:" and tell the students, "You can draw a picture in this box of how you will set up the equipment in order to carry out this investigation."

10. Point to the box labeled "We will collect the following data:" and tell the students, "You can list the measurements or observations that you will need to collect during the investigation in this box."

11. Point to the box labeled "These are the steps we will follow to collect data:" and tell the students, "You can list what you are going to do to collect the data you need and what you will do with your data once you have it. Be sure to give enough detail that I could do your investigation for you."

12. Ask the students, "Do you have any questions about what you need to do?"

13. Wait for questions. Answer any questions that come up.

14. Tell the students, "Once you are done, raise your hand and let me know. I'll then come by and look over your proposal and give you some feedback. You may not begin collecting data until I have approved your proposal by signing it. You need to have your proposal done in the next 15 minutes."

Teacher Notes

15. Give the students 15 minutes to work in their groups on their investigation proposal. As they work, move from group to group to check in, ask probing questions, and offer a suggestion if a group gets stuck.

16. As each group finishes its investigation proposal, read it over and determine if it will be productive or not. If you feel the investigation will be productive (not necessarily what you would do or what the other groups are doing), sign your name on the proposal and let the group start collecting data. If the plan needs to be changed, offer some suggestions or ask some probing questions, and have the group make the changes before you approve it.

What should a student-designed investigation look like?

There are a number of different investigations that students can design to answer the question "How does distance affect the apparent brightness of a star?" For example, one method might include the following steps:

1. Go to the lamp with the 40 W lightbulb and turn it on.
2. Use the light meter app to measure the apparent brightness of the bulb (in lux) when the tablet is positioned 100 cm, 200 cm, 300 cm, 400 cm, and 500 cm from the bulb.
3. Go to the lamp with the 60 W lightbulb and turn it on.
4. Repeat step 2.
5. Go to the lamp with a 100 W lightbulb and turn it on.
6. Repeat step 2.

If students use this method, they will need to collect data on (1) distance from the light source and (2) apparent brightness of the light in lux.

17. Pass out the materials, or have one student from each group collect the materials they need from a central supply table or cart for the groups that have an approved proposal.

18. Remind students of the safety rules and precautions for this investigation.

19. Tell the students to collect their data and record their observations or measurements in the "Collect Your Data" box in their Investigation Handout (or the Investigation Log in their workbook).

20. Give the students 30 minutes to collect their data. Collect the materials from each group before asking them to analyze their data.

Investigation 14. Star Brightness:
How Does Distance Affect the Apparent Brightness of a Star?

Stage 3: Create a Draft Argument (45 minutes)

1. Tell the students, "Now that we have all this data, we need to analyze the data so we can figure out an answer to the guiding question."

2. If possible, project an image of the "Analyze Your Data" section for this investigation on a screen or board using a document camera (or take a picture of it and project the picture on a screen or board). Point to the section and tell the students, "You can create a graph as a way to analyze your data. You can make your graph in this section."

3. Ask the students, "What information do we need to include in a graph?"

4. Tell the students, "Please take a few minutes to discuss this question with your group and be ready to share."

5. Give the students five minutes to discuss.

6. Ask two or three different groups to share their answers. Highlight or write down any important ideas on the board so students can refer to them later.

7. Tell the students, "I am now going to give you and the other members of your group about 10 minutes to create your graph." The graph they create should include the distance from the lightbulb and the amount of light falling on the tablet. If the students are having trouble making a graph, you can take a few minutes to provide a mini-lesson about how to create a graph from a bunch of measurements (this strategy is called just-in-time instruction because it is offered only when students get stuck).

8. Give the students 10 minutes to analyze their data by creating a graph. As they work, move from group to group to check in, ask probing questions, and offer suggestions.

What should the graph for this investigation look like?

There are a number of different ways that students can analyze the measurements they collect during this investigation. One of the most straightforward ways is to create a grouped bar graph with the distance from the bulb on the horizontal axis, or *x*-axis, and the amount of light falling on the tablet on the vertical axis, or *y*-axis. The groups of bars can be labeled by type of bulb (40 W, 60 W, or 100 W). There are other options for analyzing the collected data. Students often come up with some unique ways of analyzing their data, so be sure to give them some voice and choice during this stage.

Teacher Notes

9. Tell the students, "I am now going to give you and the other members of your group about 15 minutes to create an argument to share what you have learned and convince others that they should believe you. Before you do that, we need to take a few minutes to discuss what you need to include in your argument."

10. If possible, use a document camera to project the "Argument Presentation on a Whiteboard" image from the "Draft Argument" section of the Investigation Handout (or the Investigation Log in their workbook) on a screen or board (or take a picture of it and project the picture on a screen or board).

11. Point to the box labeled "The Guiding Question:" and tell the students, "You can put the question we are trying to answer here on your whiteboard."

12. Point to the box labeled "Our Claim:" and tell the students, "You can put your claim here on your whiteboard. The claim is your answer to the guiding question."

13. Point to the box labeled "Our Evidence:" and tell the students, "You can put the evidence that you are using to support your claim here on your whiteboard. Your evidence will need to include the analysis you just did and an explanation of what your analysis means or shows. Scientists always need to support their claims with evidence."

14. Point to the box labeled "Our Justification of the Evidence:" and tell the students, "You can put your justification of your evidence here on your whiteboard. Your justification needs to explain why your evidence is important. Scientists often use core ideas to explain why the evidence they are using matters. Core ideas are important concepts that scientists use to help them make sense of what happens during an investigation."

15. Ask the students, "What are some core ideas that we read about earlier that might help us explain why the evidence we are using is important?"

16. Ask the students to share some of the core ideas from the "Introduction" section of the Investigation Handout (or the Investigation Log in the workbook). List these core ideas on the board.

17. Tell the students, "That is great. I would like to see everyone try to include these core ideas in your justification of the evidence. Your goal is to use these core ideas to help explain why your evidence matters and why the rest of us should pay attention to it."

18. Ask the students, "Do you have any questions about what you need to do?"

19. Answer any questions that come up.

20. Tell the students, "Okay, go ahead and start working on your arguments. You need to have your argument done in the next 15 minutes. It doesn't need to be perfect. We just need something down on the whiteboards so we can share our ideas."

Investigation 14. Star Brightness:
How Does Distance Affect the Apparent Brightness of a Star?

21. Give the students 15 minutes to work in their groups on their arguments. As they work, move from group to group to check in, ask probing questions, and offer a suggestion if a group gets stuck. Figure 14.2 shows an example of an argument created by students for this investigation.

FIGURE 14.2
Example of an argument

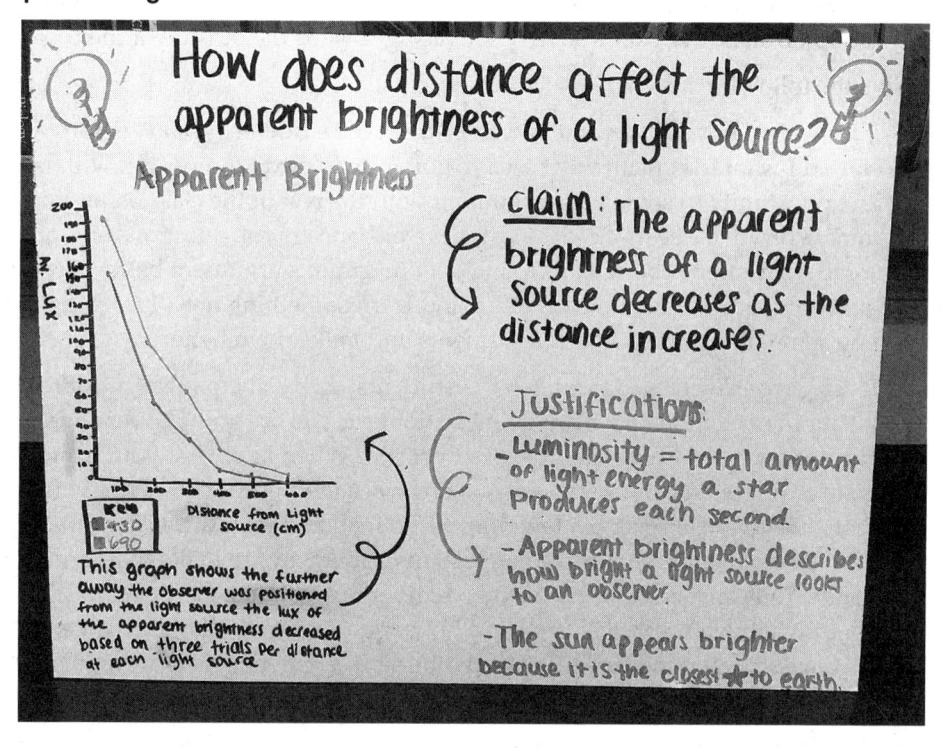

[Figure shows a student-created whiteboard argument with the question "How does distance affect the apparent brightness of a light source?" It includes a graph of Apparent Brightness (N. Lux) vs. Distance from Light Source (cm) with keys for 430 and 690. The claim states: "The apparent brightness of a light source decreases as the distance increases." Justifications include: "Luminosity = total amount of light energy a star produces each second," "Apparent brightness describes how bright a light source looks to an observer," and "The sun appears brighter because it is the closest to earth." A caption on the graph reads: "This graph shows the further away the observer was positioned from the light source the lux of the apparent brightness decreased based on three trials per distance at each light source."]

Stage 4: Argumentation Session (30 minutes)

The argumentation session can be conducted in a whole-class presentation format, a gallery walk format, or a modified gallery walk format. We recommend using a whole-class presentation format for the first investigation, but try to transition to either the gallery walk or modified gallery walk format as soon as possible because that will maximize student voice and choice inside the classroom. The following list shows the steps for the three formats; unless otherwise noted, the steps are the same for all three formats.

1. Begin by introducing the use of the whiteboard.
   - *If using the whole-class presentation format,* tell the students, "We are now going to share our arguments. Please set up your whiteboards so everyone can see them."

Teacher Notes

- *If using the gallery walk or modified gallery walk format,* tell the students, "We are now going to share our arguments. Please set up your whiteboards so they are facing the walls."

2. Allow the students to set up their whiteboards.

    - *If using the whole-class presentation format,* the whiteboards should be set up on stands or chairs so they are facing toward the center of the room.

    - *If using the gallery walk or modified gallery walk format,* the whiteboards should be set up on stands or chairs so they are facing toward the outside of the room.

3. Give the following instructions to the students:

    - *If using the whole-class presentation format,* tell the students, "Okay, before we get started I want to explain what we are going to do next. Your group will have an opportunity to share your argument with the rest of the class. After you are done, everyone else in the class will have a chance to ask questions and offer some suggestions about ways to make your group's argument better. After we have a chance to listen to each other and learn something new, I'm going to give you some time to revise your arguments and make them better."

    - *If using the gallery walk format,* tell the students, "Okay, before we get started I want to explain what we are going to do next. You are going to read the arguments that were created by other groups. When I say 'go,' your group will go to a different group's station so you can see their argument. Once you are there, I'll give your group a few minutes to read and review their argument. Your job is to offer them some suggestions about ways to make their argument better. You can use sticky notes to give them suggestions. Please be specific about what you want to change and how you think they should change it. After we have a chance to learn from each other, I'm going to give you some time to revise your arguments and make them better."

    - *If using the modified gallery walk format,* tell the students, "Okay, before we get started I want to explain what we are going to do next. I'm going to ask some of you to present your arguments to your classmates. If you are presenting your argument, your job is to share your group's claim, evidence, and justification of the evidence. The rest of you will be travelers. If you are a traveler, your job is to listen to the presenters, ask the presenters questions if you do not understand something, and then offer them some suggestions about ways to make their argument better. After we have a chance to learn from each other, I'm going to give you some time to revise your arguments and make them better."

4. Use a document camera to project the "Ways to IMPROVE our argument …" box from the Investigation Handout (or the Investigation Log in their workbook) on a screen or board (or take a picture of it and project the picture on a screen or board).

Investigation 14. Star Brightness:
How Does Distance Affect the Apparent Brightness of a Star?

- *If using the whole-class presentation format,* point to the box and tell the students, "After your group presents your argument, you can write down the suggestions you get from your classmates here. If you are listening to a presentation and you see a good idea from another group, you can write down that idea here as well. Once we are done with the presentations, I will give you a chance to use these suggestions or ideas to improve your arguments."

- *If using the gallery walk format,* point to the box and tell the students, "If you see a good idea from another group, you can write it down here. Once we are done reviewing the different arguments, I will give you a chance to use these ideas to improve your own arguments. It is important to share ideas like this."

- *If using the modified gallery walk format,* point to the box and tell the students, "If you are a presenter, you can write down the suggestions you get from the travelers here. If you are a traveler and you see a good idea from another group, you can write down that idea here. Once we are done with the presentations, I will give you a chance to use these suggestions or ideas to improve your arguments."

5. Ask the students, "Do you have any questions about what you need to do?"

6. Answer any questions that come up.

7. Give the following instructions:

    - *If using the whole-class presentation format,* tell the students, "Okay. Let's get started."

    - *If using the gallery walk format,* tell the students, "Okay, I'm now going to tell you which argument to go to and review."

    - *If using the modified gallery walk format,* tell the students, "Okay, I'm now going to assign you to be a presenter or a traveler." Assign one or two students from each group to be presenters and one or two students from each group to be travelers.

8. Give the students an opportunity to review the arguments.

    - *If using the whole-class presentation format,* have each group present their argument one at a time. Give each group only two to three minutes to present their argument. Then give the class two to three minutes to ask them questions and offer suggestions. Encourage as much participation from the students as possible.

    - *If using the gallery walk format,* tell the students, "Okay. Let's get started. Each group, move one argument to the left. Don't move to the next argument until I tell you to move. Once you get there, read the argument and then offer suggestions about how to make it better. I will put some sticky notes next to each argument. You can use the sticky notes to leave your suggestions." Give each group about three to four minutes to read the arguments, talk, and offer suggestions.

Teacher Notes

   a. After three to four minutes, tell the students, "Okay. Let's move on to the next argument. Please move one group to the left."
   b. Again, give each group three to four minutes to read, talk, and offer suggestions.
   c. Repeat this process until each group has had their argument read and critiqued three times.

- *If using the modified gallery walk format,* tell the students, "Okay. Let's get started. Reviewers, move one group to the left. Don't move to the next group until I tell you to move. Presenters, go ahead and share your argument with the travelers when they get there." Give each group of presenters and travelers about three to four minutes to talk.

   a. Tell the students, "Okay. Let's move on to the next argument. Travelers, move one group to the left."
   b. Again, give each group of presenters and travelers about three to four minutes to talk.
   c. Repeat this process until each group has had their argument read and critiqued three times.

9. Tell the students to return to their workstations.
10. Give the following instructions about revising the argument:

    - *If using the whole-class presentation format,* tell the students, "I'm now going to give you all about 10 minutes to revise your argument. Take a few minutes to talk in your groups and determine what you want to change to make your argument better. Once you have decided what to change, go ahead and make the changes to your whiteboard."

    - *If using the gallery walk format,* tell the students, "I'm now going to give you all about 10 minutes to revise your argument. Take a few minutes to read the suggestions that were left at your argument. Then talk in your groups and determine what you want to change to make your argument better. Once you have decided what to change, go ahead and make the changes to your whiteboard."

    - *If using the modified gallery walk format,* tell the students, "I'm now going to give you all about 10 minutes to revise your argument. Please return to your original groups." Wait for the students to move back into their original groups and then tell the students, "Okay, take a few minutes to talk in your groups and determine what you want to change to make your argument better. Once you have decided what to change, go ahead and make the changes to your whiteboard."

11. Ask the students, "Do you have any questions about what you need to do?"
12. Answer any questions that come up.
13. Tell the students, "Okay. Let's get started."

Investigation 14. Star Brightness:
How Does Distance Affect the Apparent Brightness of a Star?

14. Give the students 10 minutes to work in their groups on their arguments. As they work, move from group to group to check in, ask probing questions, and offer a suggestion if a group gets stuck.

Stage 5: Reflective Discussion (15 minutes)

1. Tell the students, "We are now going to take a minute to talk about some of the core ideas and crosscutting concepts that we have used during our investigation."
2. Show Figure 14.3 (without the caption) on the screen.
3. Ask the students, "What do you notice?"
4. Allow the students to share their ideas.

FIGURE 14.3
Stars of different apparent brightness

5. Ask the students, "What are some different properties or characteristics of stars that we can use to help us describe or classify all these stars?"
6. Allow the students to share their ideas. As they share their ideas, encourage them to build off each contribution and keep asking inviting questions (e.g.,

Teacher Notes

"Would anyone else like to add to that idea?") and probing questions (e.g., "Could you tell me more about that?") until students bring up the characteristics of color, size, temperature, and luminosity.

7. Ask the students, "How can we use what we know about the characteristics of stars to help explain what we see in this picture?"

8. Allow students to share their ideas. As they share their ideas, ask different questions to encourage them to expand on their thinking (e.g., "Can you tell me more about that?"), clarify a contribution (e.g., "Can you say that in another way?"), support an idea (e.g., "Why do you think that?"), add to an idea mentioned by a classmate (e.g., "Would anyone like to add to the idea?"), re-voice an idea offered by a classmate (e.g., "Who can explain that to me in another way?"), or critique an idea during the discussion (e.g., "Do you agree or disagree with that idea and why?") until students are able to generate an adequate explanation.

9. Tell the students, "We also needed to think about cause and effect during our investigations." Then ask, "Can anyone tell me why it is important to think about cause-and-effect relationships during an investigation?"

10. Allow the students to share their ideas.

11. Tell the students, "I think cause-and-effect relationships are really important in science because they can be used to explain things we see happening."

12. Ask the students, "What cause-and-effect relationship was important to keep in mind during this investigation?"

13. Allow the students to share their ideas.

14. Tell the students, "Scientists often need to think about scales during an investigation." Then ask, "Can anyone tell me why this is important?"

15. Allow the students to share their ideas.

16. Tell the students, "I think natural objects exist from the very small, such as atoms, to the immensely large, such as stars, and from very short to very long distances. Scientists therefore need to think about what measurement scales to use when collecting data."

17. Ask the students, "How did you need to think about measurement scales today?"

18. Show an image of the question "What do you think are the most important core ideas or crosscutting concepts that we used during this investigation to help us make sense of what we observed?" Tell the students, "Okay, let's make sure we are all on the same page. Please take a moment to discuss this question with the other people in your group." Give them a few minutes to discuss the question.

19. Ask the students, "What do you all think? Who would like to share?"

Investigation 14. Star Brightness:
How Does Distance Affect the Apparent Brightness of a Star?

20. Allow the students to share their ideas.

21. Tell the students, "We are now going to take a minute to talk about what went well and what didn't go so well during our investigation. We need to talk about this because you all are going to be planning and carrying out your own investigations like this a lot this year, and I want to help you all get better at it."

22. Show an image of the question "What made your investigation scientific?" on the screen. Tell the students, "Take a few minutes to talk about how you would answer this question with the other people in your group. Be ready to share with the rest of the class." Give the students two to three minutes to talk in their group.

23. Ask the students, "What do you all think? Who would like to share an idea?"

24. Allow the students to share their ideas. Be sure to expand on their ideas about what makes an investigation scientific.

25. Show an image of the question "What made your investigation not so scientific?" on the screen. Tell the students, "Take a few minutes to talk about how you would answer this question with the other people in your group. Be ready to share with the rest of the class." Give the students two to three minutes to talk in their group.

26. Ask the students, "What do you all think? Who would like to share an idea?"

27. Allow the students to share their ideas. Be sure to expand on their ideas about what makes an investigation less scientific.

28. Show an image of the question "What rules can we put into place to help us make sure our next investigation is more scientific?" on the screen. Tell the students, "Take a few minutes to talk about how you would answer this question with the other people in your group. Be ready to share with the rest of the class." Give the students two to three minutes to talk in their group.

29. Ask the students, "What do you all think? Who would like to share an idea?"

30. Allow the students to share their ideas. Once they have shared their ideas, offer a suggestion for a possible class rule.

31. Ask the students, "What do you all think? Should we make this a rule?"

32. If the students agree, write the rule on the board or on a class "Rules for Scientific Investigation" chart so you can refer to it during the next investigation.

Stage 6: Write a Draft Report (30 minutes)

Your students will use either the Investigation Handout or the Investigation Log in the student workbook when writing the draft report. When you give the directions shown in quotes in the following steps, substitute "Investigation Log in your workbook" or just "Investigation Log" (as shown in brackets) for "handout" if they are using the workbook.

Teacher Notes

1. Tell the students, "You are now going to write an investigation report to share what you have learned. Please take out a pencil and turn to the 'Draft Report' section of your handout [Investigation Log in your workbook]."

2. If possible, use a document camera to project the "Introduction" section of the draft report from the Investigation Handout (or the Investigation Log in their workbook) on a screen or board (or take a picture of it and project the picture on a screen or board).

3. Tell the students, "The first part of the report is called the 'Introduction.' In this section of the report you want to explain to the reader what you were investigating, why you were investigating it, and what question you were trying to answer. All this information can be found in the text at the beginning of your handout [Investigation Log]." Point to the image. "Here are some sentence starters to help you begin writing."

4. Ask the students, "Do you have any questions about what you need to do?"

5. Answer any questions that come up.

6. Tell the students, "Okay, let's write."

7. Give the students 10 minutes to write the "Introduction" section of the report. As they work, move from student to student to check in, ask probing questions, and offer a suggestion if a student gets stuck.

8. If possible, use a document camera to project the "Method" section of the draft report from the Investigation Handout (or the Investigation Log in their workbook) on a screen or board (or take a picture of it and project the picture on a screen or board).

9. Tell the students, "The second part of the report is called the 'Method.' In this section of the report you want to explain to the reader what you did during the investigation, what data you collected and why, and how you went about analyzing your data. All this information can be found in the 'Plan Your Investigation' section of the handout [Investigation Log]. Remember that you all planned and carried out different investigations, so do not assume that the reader will know what you did." Point to the image. "Here are some sentence starters to help you begin writing."

10. Ask the students, "Do you have any questions about what you need to do?"

11. Answer any questions that come up.

12. Tell the students, "Okay, let's write."

13. Give the students 10 minutes to write the "Method" section of the report. As they work, move from student to student to check in, ask probing questions, and offer a suggestion if a student gets stuck.

14. If possible, use a document camera to project the "Argument" section of the draft report from the Investigation Handout (or the Investigation Log in their

workbook) on a screen or board (or take a picture of it and project the picture on a screen or board).

15. Tell the students, "The last part of the report is called the 'Argument.' In this section of the report you want to share your claim, evidence, and justification of the evidence with the reader. All this information can be found on your whiteboard." Point to the image. "Here are some sentence starters to help you begin writing."

16. Ask the students, "Do you have any questions about what you need to do?"

17. Answer any questions that come up.

18. Tell the students, "Okay, let's write."

19. Give the students 10 minutes to write the "Argument" section of the report. As they work, move from student to student to check in, ask probing questions, and offer a suggestion if a student gets stuck.

Stage 7: Peer Review (35 minutes)

Your students will use either the Investigation Handout or their workbook when doing the peer review. Except where noted below, the directions are the same whether using the handout or the workbook.

1. Tell the students, "We are now going to review our reports to find ways to make them better. I'm going to come around and collect your draft reports. While I do that, please take out a pencil."

2. Collect the handouts or the workbooks with the draft reports from the students.

3. If possible, use a document camera to project the peer-review guide (see Appendix 4) on a screen or board (or take a picture of it and project the picture on a screen or board).

4. Tell the students, "We are going to use this peer-review guide to give each other feedback." Point to the image.

5. Tell the students, "I'm going to ask you to work with a partner to do this. I'm going to give you and your partner a draft report to read. You two will then read the report together. Once you are done reading the report, I want you to answer each of the questions on the peer-review guide." Point to the review questions on the image of the peer-review guide.

6. Tell the students, "You can check 'no,' 'almost,' or 'yes' after each question." Point to the checkboxes on the image of the peer-review guide.

7. Tell the students, "This will be your rating for this part of the report. Make sure you agree on the rating you give the author. If you mark 'no' or 'almost,' then you need to tell the author what he or she needs to do to get a 'yes.'" Point to the space for the reviewer feedback on the image of the peer-review guide.

Teacher Notes

8. Tell the students, "It is really important for you to give the authors feedback that is helpful. That means you need to tell them exactly what they need to do to make their report better."
9. Ask the students, "Do you have any questions about what you need to do?"
10. Answer any questions that come up.
11. Tell the students, "Please sit with a partner who is not in your current group." Allow the students time to sit with a partner.
12. Tell the students, "Okay, I'm now going to give you one report to read." Pass out one Investigation Handout with a draft report or one workbook to each pair. Make sure that the report you give a pair was not written by one of the students in that pair. Give each pair one peer-review guide to fill out. If the students are using workbooks, the peer-review guide is included right after the draft report so you do not need to pass out copies of the peer-review guide.
13. Tell the students, "Okay, I'm going to give you 15 minutes to read the report I gave you and to fill out the peer-review guide. Go ahead and get started."
14. Give the students 15 minutes to work. As they work, move around from pair to pair to check in and see how things are going, answer questions, and offer advice.
15. After 15 minutes pass, tell the students, "Okay, time is up. Please give me the report and the peer-review guide that you filled out."
16. Collect the Investigation Handouts and the peer-review guides, or collect the workbooks if students are using them. If the students are using the Investigation Handouts and separate peer-review guides, be sure you keep each handout with its corresponding peer-review guide.
17. Tell the students, "Okay, I am now going to give you a different report to read and a new peer-review guide to fill out." Pass out one more report to each pair. Make sure that the report you give a pair was not written by one of the students in that pair. Give each pair a new peer-review guide to fill out as a group.
18. Tell the students, "Okay, I'm going to give you 15 minutes to read this new report and to fill out the peer-review guide. Go ahead and get started."
19. Give the students 15 minutes to work. As they work, move around from pair to pair to check in and see how things are going, answer questions, and offer advice.
20. After 15 minutes pass, tell the students, "Okay, time is up. Please give me the report and the peer-review guide that you filled out."
21. Collect the Investigation Handouts and the peer-review guides, or collect the workbooks if students are using them. If the students are using the Investigation Handouts and separate peer-review guides, be sure you keep each handout with its corresponding peer-review guide.

Investigation 14. Star Brightness:
How Does Distance Affect the Apparent Brightness of a Star?

Stage 8: Revise the Report (30 minutes)

Your students will use either the Investigation Handout or their workbook when revising the report. Except where noted below, the directions are the same whether using the handout or the workbook.

1. Tell the students, "You are now going to revise your draft report based on the feedback you get from your classmates. Please take out a pencil."

2. Return the reports to the students.
   - *If the students used the Investigation Handout and a copy of the peer-review guide,* pass back the handout and the peer-review guide to each student.
   - *If the students used the workbook,* pass that back to each student.

3. Tell the students, "Please take a few minutes to read over the peer-review guide. You should use it to figure out what you need to change in your report and how you will change it."

4. Allow the students to read the peer-review guide.

5. *If the students used the workbook,* if possible use a document camera to project the "Write Your Final Report" section from the Investigation Log on a screen or board (or take a picture of it and project the picture on a screen or board).

6. Give the following directions about how to revise their reports:
   - *If the students used the Investigation Handout and a copy of the peer-review guide,* tell them, "Okay, let's revise our reports. Please take out a piece of paper. I would like you to rewrite your report. You can use your draft report as a starting point, but you also need to change it to make it better. Use the feedback on the peer-review guide to make it better."
   - *If the students used the workbook,* tell them, "Okay, let's revise our reports. I would like you to rewrite your report in the section of the Investigation Log called "Write Your Final Report." You can use your draft report as a starting point, but you also need to change it to make it better. Use the feedback on the peer-review guide to make it better."

7. Ask the students, "Do you have any questions about what you need to do?"

8. Answer any questions that come up.

9. Tell the students, "Okay, let's write." Allow about 20 minutes for the students to revise their reports.

10. After about 20 minutes, give the following directions:
    - *If the students used the Investigation Handout,* tell them, "Okay, time's up. I will now come around and collect your Investigation Handout, the peer-review guide, and your final report."
    - *If the students used the workbook,* tell them, "Okay, time's up. I will now come around and collect your workbooks."

Teacher Notes

11. *If the students used the Investigation Handout,* collect all the Investigation Handouts, peer-review guides, and final reports. *If the students used the workbook,* collect all the workbooks.

12. *If the students used the Investigation Handout,* use the "Teacher Score" column in the peer-review guide to grade the final report. *If the students used the workbook,* use the "Investigation Report Grading Rubric" in the Investigation Log to grade the final report. Whether you are using the handout or the log, you can give the students feedback about their writing in the "Teacher Comments" section.

How to Use the Checkout Questions

The Checkout Questions are an optional assessment. We recommend giving them to students at the start of the next class period after the students finish stage 8 of the investigation. You can then look over the student answers to determine if you need to reteach the core idea from the investigation. Appendix 6 gives the answers to the Checkout Questions that should be given by a student who can apply the core idea correctly in all cases and can explain the cause-and-effect relationship.

Alignment With Standards

Table 14.2 highlights how the investigation can be used to address specific performance expectations from the *Next Generation Science Standards, Common Core State Standards for English Language Arts* (*CCSS ELA*) and *Common Core State Standards for Mathematics* (*CCSS Mathematics*), and *English Language Proficiency* (*ELP*) *Standards*.

TABLE 14.2

Investigation 14 alignment with standards

NGSS performance expectation	5-ESS1-1: Support an argument that differences in the apparent brightness of the Sun compared to other stars is due to their relative distances from the Earth.
CCSS ELA—Reading: Informational Text	Key ideas and details • CCSS.ELA-LITERACY.RI.5.1: Quote accurately from a text when explaining what the text says explicitly and when drawing inferences from the text. • CCSS.ELA-LITERACY.RI.5.2: Determine two or more main ideas of a text and explain how they are supported by key details; summarize the text. • CCSS.ELA-LITERACY.RI.5.3: Explain the relationships or interactions between two or more individuals, events, ideas, or concepts in a historical, scientific, or technical text based on specific information in the text.

Continued

Investigation 14. Star Brightness:
How Does Distance Affect the Apparent Brightness of a Star?

Table 14.2 (*continued*)

***CCSS ELA**—Reading:* **Informational Text** (*continued*)	Craft and structure • CCSS.ELA-LITERACY.RI.5.4: Determine the meaning of general academic and domain-specific words and phrases in a text relevant to a *grade 5 topic or subject area*. • CCSS.ELA-LITERACY.RI.5.5: Compare and contrast the overall structure (e.g., chronology, comparison, cause/effect, problem/solution) of events, ideas, concepts, or information in two or more texts. • CCSS.ELA-LITERACY.RI.5.6: Analyze multiple accounts of the same event or topic, noting important similarities and differences in the point of view they represent. Integration of knowledge and ideas • CCSS.ELA-LITERACY.RI.5.7: Draw on information from multiple print or digital sources, demonstrating the ability to locate an answer to a question quickly or to solve a problem efficiently. • CCSS.ELA-LITERACY.RI.5.8: Explain how an author uses reasons and evidence to support particular points in a text, identifying which reasons and evidence support which point(s). Range of reading and level of text complexity • CCSS.ELA-LITERACY.RI.5.10: By the end of the year, read and comprehend informational texts, including history/social studies, science, and technical texts, at the high end of the grades 4–5 text complexity band independently and proficiently.

***CCSS ELA**—Writing*	Text types and purposes
	• CCSS.ELA-LITERACY.W.5.1: Write opinion pieces on topics or texts, supporting a point of view with reasons.
	   ○ CCSS.ELA-LITERACY.W.5.1.A: Introduce a topic or text clearly, state an opinion, and create an organizational structure in which ideas are logically grouped to support the writer's purpose.
	   ○ CCSS.ELA-LITERACY.W.5.1.B: Provide logically ordered reasons that are supported by facts and details.
	   ○ CCSS.ELA-LITERACY.W.5.1.C: Link opinion and reasons using words, phrases, and clauses (e.g., *consequently*, *specifically*).
	   ○ CCSS.ELA-LITERACY.W.5.1.D: Provide a concluding statement or section related to the opinion presented.
	• CCSS.ELA-LITERACY.W.5.2: Write informative or explanatory texts to examine a topic and convey ideas and information clearly.
	   ○ CCSS.ELA-LITERACY.W.5.2.A: Introduce a topic clearly, provide a general observation and focus, and group related information logically; include formatting (e.g., headings), illustrations, and multimedia when useful to aiding comprehension.
	   ○ CCSS.ELA-LITERACY.W.5.2.B: Develop the topic with facts, definitions, concrete details, quotations, or other information and examples related to the topic.
	   ○ CCSS.ELA-LITERACY.W.5.2.C: Link ideas within and across categories of information using words, phrases, and clauses (e.g., *in contrast*, *especially*).

Continued

Table 14.2 (*continued*)

***CCSS ELA*—Writing** (*continued*)	○ CCSS.ELA-LITERACY.W.5.2.D: Use precise language and domain-specific vocabulary to inform about or explain the topic. ○ CCSS.ELA-LITERACY.W.5.2.E: Provide a concluding statement or section related to the information or explanation presented. Production and distribution of writing • CCSS.ELA-LITERACY.W.5.4: Produce clear and coherent writing in which the development and organization are appropriate to task, purpose, and audience. • CCSS.ELA-LITERACY.W.5.5: With guidance and support from peers and adults, develop and strengthen writing as needed by planning, revising, editing, rewriting, or trying a new approach. • CCSS.ELA-LITERACY.W.5.6: With some guidance and support from adults, use technology, including the internet, to produce and publish writing as well as to interact and collaborate with others; demonstrate sufficient command of keyboarding skills to type a minimum of two pages in a single sitting. Research to build and present knowledge • CCSS.ELA-LITERACY.W.5.8: Recall relevant information from experiences or gather relevant information from print and digital sources; summarize or paraphrase information in notes and finished work, and provide a list of sources. • CCSS.ELA-LITERACY.W.5.9: Draw evidence from literary or informational texts to support analysis, reflection, and research. Range of writing • CCSS.ELA-LITERACY.W.5.10: Write routinely over extended time frames (time for research, reflection, and revision) and shorter time frames (a single sitting or a day or two) for a range of discipline-specific tasks, purposes, and audiences.
***CCSS ELA*—Speaking and Listening**	Comprehension and collaboration • CCSS.ELA-LITERACY.SL.5.1: Engage effectively in a range of collaborative discussions (one-on-one, in groups, and teacher-led) with diverse partners on *grade 5 topics and texts,* building on others' ideas and expressing their own clearly. ○ CCSS.ELA-LITERACY.SL.5.1.A: Come to discussions prepared, having read or studied required material; explicitly draw on that preparation and other information known about the topic to explore ideas under discussion. ○ CCSS.ELA-LITERACY.SL.5.1.B: Follow agreed-upon rules for discussions and carry out assigned roles. ○ CCSS.ELA-LITERACY.SL.5.1.C: Pose and respond to specific questions by making comments that contribute to the discussion and elaborate on the remarks of others. ○ CCSS.ELA-LITERACY.SL.5.1.D: Review the key ideas expressed and draw conclusions in light of information and knowledge gained from the discussions.

Continued

Investigation 14. Star Brightness:
How Does Distance Affect the Apparent Brightness of a Star?

Table 14.2 (*continued*)

CCSS ELA—Speaking and Listening (*continued*)	• CCSS.ELA-LITERACY.SL.5.2: Summarize a written text read aloud or information presented in diverse media and formats, including visually, quantitatively, and orally. • CCSS.ELA-LITERACY.SL.5.3: Summarize the points a speaker makes and explain how each claim is supported by reasons and evidence. Presentation of knowledge and ideas • CCSS.ELA-LITERACY.SL.5.4: Report on a topic or text or present an opinion, sequencing ideas logically and using appropriate facts and relevant, descriptive details to support main ideas or themes; speak clearly at an understandable pace. • CCSS.ELA-LITERACY.SL.5.5: Include multimedia components (e.g., graphics, sound) and visual displays in presentations when appropriate to enhance the development of main ideas or themes. • CCSS.ELA-LITERACY.SL.5.6: Adapt speech to a variety of contexts and tasks, using formal English when appropriate to task and situation.
CCSS Mathematics—Numbers and Operations in Base Ten	Perform operations with multi-digit whole numbers and with decimals to hundredths. • CCSS.MATH.CONTENT.5.NBT.B.7: Add, subtract, multiply, and divide decimals to hundredths.
CCSS Mathematics—Measurement and Data	Convert like measurement units within a given measurement system. • CCSS.MATH.CONTENT.5.MD.A.1: Convert among different-sized standard measurement units within a given measurement system (e.g., convert 5 cm to 0.05 m), and use these conversions in solving multi-step, real-world problems.
ELP Standards (*continued*)	Receptive modalities • ELP 1: Construct meaning from oral presentations and literary and informational text through grade-appropriate listening, reading, and viewing. • ELP 8: Determine the meaning of words and phrases in oral presentations and literary and informational text. Productive modalities • ELP 3: Speak and write about grade-appropriate complex literary and informational texts and topics. • ELP 4: Construct grade-appropriate oral and written claims and support them with reasoning and evidence. • ELP 7: Adapt language choices to purpose, task, and audience when speaking and writing.

Continued

Teacher Notes

Table 14.2 (*continued*)

ELP Standards (continued)	Interactive modalities
	- ELP 2: Participate in grade-appropriate oral and written exchanges of information, ideas, and analyses, responding to peer, audience, or reader comments and questions.
	- ELP 5: Conduct research and evaluate and communicate findings to answer questions or solve problems.
	- ELP 6: Analyze and critique the arguments of others orally and in writing.
	Linguistic structures of English
	- ELP 9: Create clear and coherent grade-appropriate speech and text.
	- ELP 10: Make accurate use of standard English to communicate in grade-appropriate speech and writing.

References

Gregersen, E. 2009. Luminosity. Encyclopaedia Brittanica. *www.britannica.com/science/luminosity*.

NASA. 2017. The Sun. *www.nasa.gov/sun*.

Sessions, L. 2020. Rigel in Orion is blue-white. EarthSky. *https://earthsky.org/brightest-stars/blue-white-rigel-is-orions-brightest-star*.

Star Facts. 2019. Rigel. *www.star-facts.com/rigel*.

Temming, M. 2014. How far is the closest star? Sky & Telescope. *https://skyandtelescope.org/astronomy-resources/far-closest-star*.

Investigation Handout

Investigation 14
Star Brightness: How Does Distance Affect the Apparent Brightness of a Star?

Introduction

We can see many different objects in the night sky. Take a few minutes to watch a video of the night sky that was taken from different locations in California and Oregon. As you watch the video, keep track of things you notice and things you wonder about in the boxes below.

Things I NOTICED ...	Things I WONDER about ...

We can see many different objects in the night sky because these objects either produce light or reflect light. A star is an example of an object that produces light. Stars produce light because they are very hot. The light that a star produces travels through space and transfers energy from the star to other objects around it. Stars, like the Sun, are therefore an important source of energy.

Argument-Driven Inquiry in **Fifth-Grade Science:** Three-Dimensional Investigations

Investigation Handout

There are a vast number of stars in the universe. Stars differ in color, size, temperature, and luminosity. *Luminosity* is the total amount of energy that a star produces in a unit of time. The luminosity of a star depends on its size and temperature. Stars that are large and hot have a greater luminosity than stars that are small and cool. Astronomers use the color, size, temperature, and luminosity of a star to describe it or to compare it with other stars. The Sun, for example, has a white-yellow color, a diameter of about 864,000 miles, a surface temperature of about 5500°C, and a luminosity of 384.6 septillion watts per second (Gregersen 2009, NASA 2017). Rigel is another example of a star that we can see. Rigel has a blue-white color, a diameter of about 68,250,000 miles, a surface temperature of about 11600°C, and a luminosity that is 120,000 times greater than the Sun (Sessions 2020, Star Facts 2019).

Not all stars are the same distance from Earth. The Sun is the closest star to us. It is about 93 million miles away from Earth. It takes the light that is produced by the Sun about 8 minutes to travel this distance through space. The next closest star to Earth is called Proxima Centauri. This star is about 25 trillion miles away from Earth, and light from Proxima Centauri takes about 4.24 years to reach Earth (Temming 2014). Rigel is even farther away. It takes 863 years for light from Rigel to travel though space and reach us (Sessions 2020). There are other stars that are even farther away from Earth.

Astronomers use the term *apparent brightness* to describe how bright a star looks to us on Earth. Apparent brightness is not the same as luminosity. Stars that produce the same amount of energy each second may have a different apparent brightness when viewed from Earth, and stars with the same apparent brightness when viewed from Earth may have different luminosities. In this investigation, you will have a chance to explore how distance affects the apparent brightness of a star. Unfortunately, you cannot change how much energy a star produces or the distance between Earth and a star, so you will need to create a *physical model* to collect the data you need.

Things we KNOW from what we read …

Investigation 14. Star Brightness:
How Does Distance Affect the Apparent Brightness of a Star?

Your Task

Use what you know about stars, light, physical models, scale, and quantity to carry out an investigation to figure out how changing the distance between a light source and an observer (a cause) affects its apparent brightness (the effect). Your teacher will also show you how to measure the apparent brightness of the bulb using a light meter or 10 strips of paper that are stapled together. The light meter measures apparent brightness in a unit of measurement called lux. The unit of measurement for the strips of paper will be number of pieces of paper.

The *guiding question* of this investigation is, *How does distance affect the apparent brightness of a star?*

Materials

You may use any of the following materials during your investigation:

- Safety goggles (required)
- Tablet with a light meter app
- 10 strips of paper stapled together.
- Meterstick or soft tape measure
- Lamp with a 40 W bulb
- Lamp with a 60 W bulb
- Lamp with a 100 W bulb

Safety Rules

Follow all normal lab safety rules. In addition, be sure to follow these rules:

- Wear sanitized safety goggles during setup, investigation activity, and cleanup.
- Lightbulbs can get very hot. Do not touch a lightbulb when it is on or for several minutes after turning it off.
- Keep electrical equipment away from water sources to prevent shock.
- Wash your hands with soap and water when you are done cleaning up.

Plan Your Investigation

Prepare a plan for your investigation by filling out the chart on the next page; this plan is called an *investigation proposal*. Before you start developing your plan, be sure to discuss the following questions with the other members of your group:

- What information do we need to find a relationship between a **cause** and an **effect**?
- What measurement **scale** and **units** might we use as we collect data?

Investigation Handout

Our guiding question:

This is a picture of how we will set up the equipment:

We will collect the following data:

These are the steps we will follow to collect data:

I approve of this investigation proposal.

_____   _____
Teacher's signature   Date

National Science Teaching Association

Investigation 14. Star Brightness:
How Does Distance Affect the Apparent Brightness of a Star?

Collect Your Data

Keep a record of what you measure or observe during your investigation in the space below.

Analyze Your Data

You will need to analyze the data you collected before you can develop an answer to the guiding question. To analyze the data you collected, create a graph that shows the relationship between the cause and the effect.

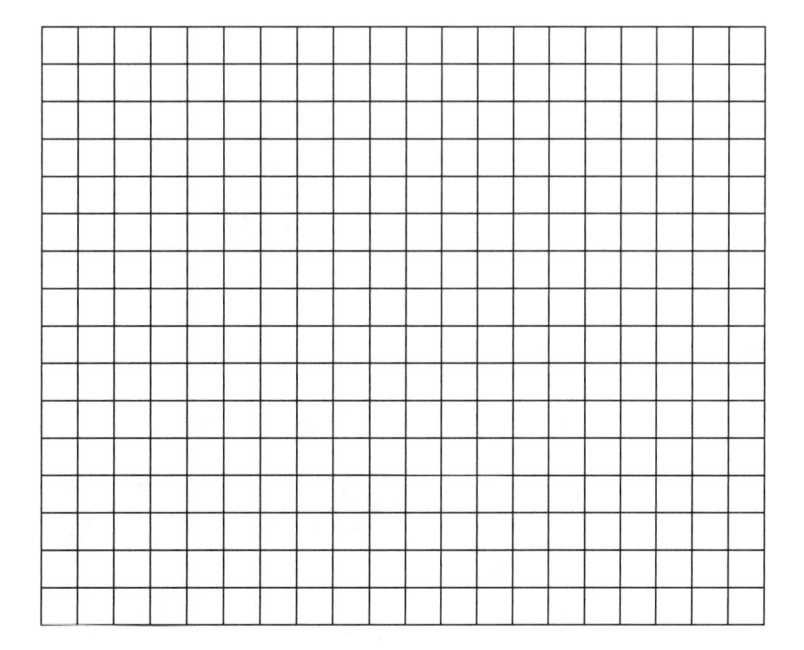

Investigation Handout

Draft Argument

Develop an argument on a whiteboard. It should include the following:

1. A *claim*: Your answer to the guiding question.
2. *Evidence*: An analysis of the data and an explanation of what the analysis means.
3. A *justification of the evidence*: Why your group thinks the evidence is important.

The Guiding Question:	
Our Claim:	
Our Evidence:	Our Justification of the Evidence:

Argumentation Session

Share your argument with your classmates. Be sure to ask them how to make your draft argument better. Keep track of their suggestions in the space below.

Ways to IMPROVE our argument …

Investigation 14. Star Brightness:
How Does Distance Affect the Apparent Brightness of a Star?

Draft Report

Prepare an *investigation report* to share what you have learned. Use the information in this handout and your group's final argument to write a *draft* of your investigation report.

Introduction

We have been studying _____ in class.

Before we started this investigation, we explored _____

_____

_____

We noticed _____

_____

_____

My goal for this investigation was to figure out _____

_____

_____

The guiding question was _____

_____

_____

Method

To gather the data I needed to answer this question, I _____

_____

_____

_____

_____

Investigation Handout

I then analyzed the data I collected by _____

_____

_____

_____

Argument

My claim is _____

_____

_____

_____

The graph below shows _____

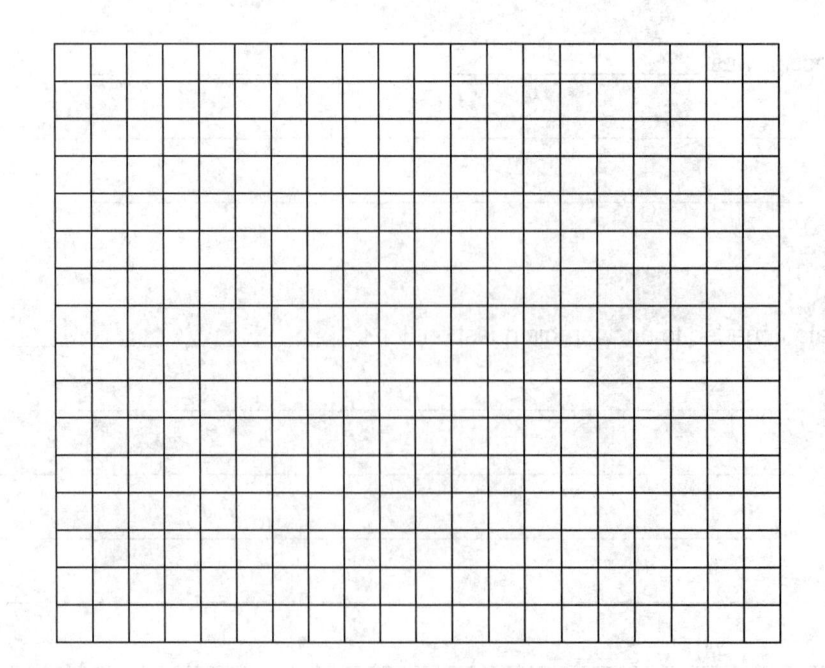

_____

_____

_____

_____

Investigation 14. Star Brightness:
How Does Distance Affect the Apparent Brightness of a Star?

This analysis of the data I collected suggests _____

_____

_____

_____

_____

This evidence is based on several important scientific concepts. The first one is _____

_____

_____

_____

_____

_____

Review

Your classmates need your help! Review the draft of their investigation reports and give them ideas about how to improve. Use the *peer-review guide* when doing your review.

Submit Your Final Report

Once you have received feedback from your classmates about your draft report, create your final investigation report and hand it in to your teacher.

References

Gregersen, E. 2009. Luminosity. Encyclopaedia Brittanica. *www.britannica.com/science/luminosity*.

NASA. 2017. The Sun. *www.nasa.gov/sun*.

Sessions, L. 2020. Rigel in Orion is blue-white. EarthSky. *https://earthsky.org/brightest-stars/blue-white-rigel-is-orions-brightest-star*.

Star Facts. 2019. Rigel. *www.star-facts.com/rigel*.

Temming, M. 2014. How far is the closest star? Sky & Telescope. *https://skyandtelescope.org/astronomy-resources/far-closest-star*.

Checkout Questions

Investigation 14. Star Brightness

The picture below shows a night sky. Use this picture to answer questions 1–4.

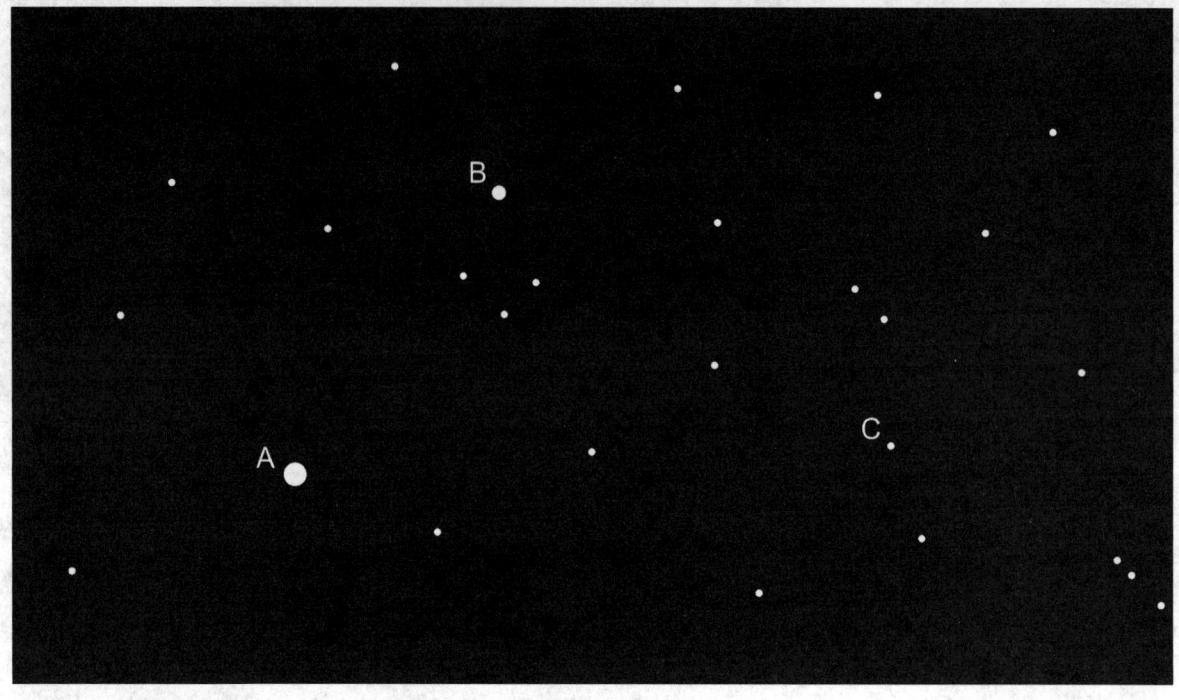

1. Which star in the picture appears to be the *brightest*?

    a. Star A

    b. Star B

    c. Star C

    d. Unable to tell

2. Which star in the picture is the *largest*?

    a. Star A

    b. Star B

    c. Star C

    d. Unable to tell

3. Which star in the picture produces the most energy in a hour?

    a. Star A

    b. Star B

    c. Star C

    d. Unable to tell

Investigation 14. Star Brightness:
How Does Distance Affect the Apparent Brightness of a Star?

4. Explain your thinking. What cause-and-effect relationship did you use to answer questions 1–3?

___

Teacher Scoring Rubric for the Checkout Questions

Level	Description
3	The student can apply the core idea correctly in all cases and can explain the cause-and-effect relationship.
2	The student can apply the core idea correctly in all cases but cannot explain the cause-and-effect relationship.
1	The student cannot apply the core idea correctly in all cases but can explain the cause-and-effect relationship.
0	The student cannot apply the core idea correctly and cannot explain the cause-and-effect relationship.

Section 6
Earth's Systems

Teacher Notes

Investigation 15
Geographic Position and Climate: Why Is the Climate in Western and Eastern Washington State So Different?

Purpose

The purpose of this investigation is to give students an opportunity to use one disciplinary core idea (DCI), two crosscutting concepts (CCs), and seven scientific and engineering practices (SEPs) to figure out why the climate on the western side of Washington State is so different from the eastern side. Students will also learn about how scientists use different methods to answer different types of questions and how scientists use models as tools for reasoning about natural phenomena.

The DCI, CCs, and SEPs That Students Use During This Investigation to Figure Things Out

DCI

- *ESS2.A: Earth Materials and Systems:* Earth's major systems are the geosphere (solid and molten rock, soil, and sediments), the hydrosphere (water and ice), the atmosphere (air), and the biosphere (living things, including humans). These systems interact in multiple ways to affect Earth's surface materials and processes. The ocean supports a variety of ecosystems and organisms, shapes landforms, and influences climate. Winds and clouds in the atmosphere interact with the landforms to determine patterns of weather.

CCs

- *CC 1: Patterns:* Similarities and differences in patterns can be used to sort, classify, communicate, and analyze simple rates of change for natural phenomena and designed products. Patterns of change can be used to make predictions. Patterns can be used as evidence to support an explanation.

- *CC 4: Systems and System Models:* A system can be described in terms of its components and their interactions.

SEPs

- *SEP 1: Asking Questions and Defining Problems:* Ask questions about what would happen if a variable is changed. Ask questions that can be investigated and predict reasonable outcomes based on patterns such as cause-and-effect relationships.

Investigation 15. Geographic Position and Climate:
Why Is the Climate in Western and Eastern Washington State So Different?

- *SEP 2: Developing and Using Models:* Develop and/or use models to describe and/or predict phenomena.
- *SEP 3: Planning and Carrying Out Investigations:* Plan and conduct an investigation collaboratively to produce data to serve as the basis for evidence, using fair tests in which variables are controlled and the number of trials considered. Evaluate appropriate methods and/or tools for collecting data.
- *SEP 4: Analyzing and Interpreting Data:* Represent data in tables and/or various graphical displays (bar graphs, pictographs, and/or pie charts) to reveal patterns that indicate relationships. Analyze and interpret data to make sense of phenomena, using logical reasoning, mathematics, and/or computation. Compare and contrast data collected by different groups in order to discuss similarities and differences in their findings.
- *SEP 6: Constructing Explanations and Designing Solutions:* Construct an explanation of observed relationships. Use evidence to construct or support an explanation. Identify the evidence that supports particular points in an explanation.
- *SEP 7: Engaging in Argument From Evidence:* Compare and refine arguments based on an evaluation of the evidence presented. Distinguish among facts, reasoned judgment based on research findings, and speculation in an explanation. Respectfully provide and receive critiques from peers about a proposed procedure, explanation, or model by citing relevant evidence and posing specific questions.
- *SEP 8: Obtaining, Evaluating, and Communicating Information:* Read and comprehend grade-appropriate complex texts and/or other reliable media to summarize and obtain scientific and technical ideas. Combine information in written text with that contained in corresponding tables, diagrams, and/or charts to support the engagement in other scientific and/or engineering practices. Communicate scientific and/or technical information orally and/or in written formats, including various forms of media as well as tables, diagrams, and charts.

Other Concepts That Students May Use During This Investigation

Students might also use some of the following concepts:

- *Weather* is the current condition of the atmosphere at a specific place.
- *Climate* is a pattern of weather in a particular region over a long period of time.

What Students Figure Out

The different climate in western and eastern Washington is caused by rain shadows created by the Olympic mountain range and the Cascade mountain range.

Teacher Notes

Background Information About This Investigation for the Teacher

People often want to know what the weather will be like in a city before making plans to travel to that city. *Weather* is the current condition of the atmosphere at a specific place. We describe the weather by measuring the air temperature, humidity, wind speed, precipitation, and cloud cover in a city. *Climate* is a pattern of weather in a particular region over a long period of time.

The climate of western Washington is different from the climate in eastern Washington. In western Washington (which includes the city of Seattle), daytime temperatures in summer are rarely above 79°F (26°C), and daytime temperatures in winter are rarely below 45°F (7°C). However, winter temperatures can easily dip into the 20s and 30s Fahrenheit (−6°C to 4°C) at night. Snow is rare in western Washington, but annual rainfall in the greater Seattle area is about 37 inches (94 cm). July and August are the driest months in western Washington, and January and February are the wettest months.

Compared with western Washington, summers are much hotter and winters are much colder in eastern Washington. The average summer daytime temperatures are in the upper 80s to mid-90s Fahrenheit (27° to 35°C); in winter, average daytime temperatures can range from the upper 30s to just above 0°F (3°C to −17°C). Eastern Washington is much drier than western Washington, with most cities in eastern Washington having close to 300 days of sunshine a year. Eastern Washington has far less annual rainfall than western Washington. In the central part of the state, cities such as Wenatchee, Chelan, and Kennewick average only 7–9 inches (18–23 cm) of rainfall annually. In Spokane, which marks the eastern edge of the state, rainfall averages 15–30 inches (38–76 cm) a year. July and August are the driest months in eastern Washington, and November and December are the wettest.

Western and eastern Washington State have different climates because of their locations in relation to two large bodies of water and two mountain ranges. The western edge of Washington State borders the Pacific Ocean, and a large body of water called Puget Sound cuts through the western part of the state. The cities in western Washington are therefore located near two large bodies of water, but the cities in eastern Washington are not. Additionally, the cities in western Washington are located near the Olympic mountain range, which is in the western part of the state, and are also located near the Cascade mountain range, which separates the western and eastern parts of the state. The cities in eastern Washington are located near the Cascade mountain range but not near the Olympic mountain range.

In Washington State there is a great deal of rain on the western sides of the Cascade and Olympic mountain ranges and less rain and snow on the eastern sides of these mountain ranges. This difference in rainfall is due to rain shadows. A *rain shadow* is a dry region of land on the side of a mountain range that is opposite a large body of water and protected from prevailing winds. A rain shadow is produced under the following conditions:

Investigation 15. Geographic Position and Climate:
Why Is the Climate in Western and Eastern Washington State So Different?

1. Warm and humid air masses form over water and then move over land due to prevailing winds that blow in the direction of the land.
2. Clouds form in the air masses as they travel over land with the prevailing winds.
3. When the warm and humid air masses reach a mountain range, they are forced to increase in altitude as they move over the range.
4. As the air masses increase in altitude, they cool.
5. As the clouds in the air masses move up the side of the mountain, they drop water in the form of rain or snow as they gain altitude, because cool air cannot hold as much moisture as warm air.
6. After the air masses cross over the peak of the mountain range and start down the other side, they decrease in altitude and the air warms up again.
7. The clouds dissipate due to lack of moisture in the air (see Figure 15.1).

Rain shadows are located on the eastern side of both the Olympic and Cascade mountain ranges because warm and moist air masses form over the Pacific Ocean and Puget Sound. These air masses then move west to east with the prevailing winds.

FIGURE 15.1
Model of a rain shadow

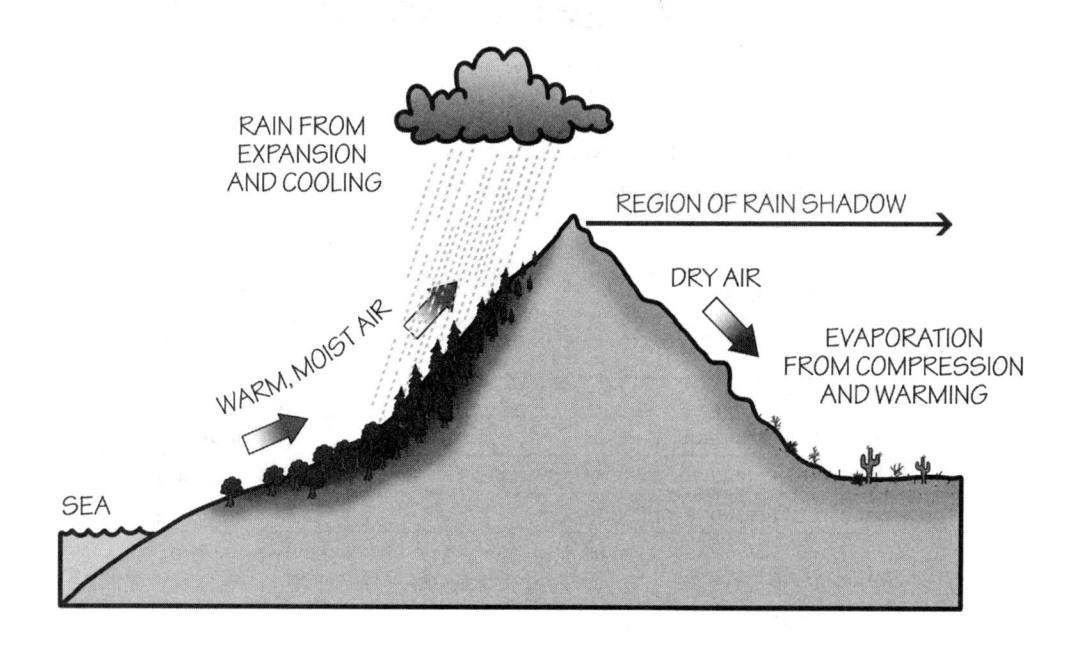

Teacher Notes

Timeline

The time needed to complete this investigation is 270 minutes (4 hours and 30 minutes). The amount of instructional time needed for each stage of the investigation is as follows:

- *Stage 1.* Introduce the task and the guiding question: 35 minutes
- *Stage 2.* Design a method and collect data: 50 minutes
- *Stage 3.* Create a draft argument: 45 minutes
- *Stage 4.* Argumentation session: 30 minutes
- *Stage 5.* Reflective discussion: 15 minutes
- *Stage 6.* Write a draft report: 30 minutes
- *Stage 7.* Peer review: 35 minutes
- *Stage 8.* Revise the report: 30 minutes

Materials and Preparation

The materials needed for this investigation are listed in Table 15.1.

TABLE 15.1
Materials for Investigation 15

Item	Quantity
Computer or tablet with internet access	1 per group
Map of Washington State cities and precipitation (in color)	1 per group
Map of Washington State elevations (in color)	1 per group
Map of Washington State (blank, in black and white)	1 per student
Colored pencils	1 set per group
Wet-erase markers	1 set per group
Transparency sheets	3 per group
Whiteboard, 2' × 3'*	1 per group
Investigation Handout	1 per student
Peer-review guide and teacher scoring rubric	1 per student
Checkout Questions (optional)	1 per student

*As an alternative, students can use computer and presentation software such as Microsoft PowerPoint or Apple Keynote to create their arguments.

The three maps can be downloaded from the book's Extras page at *www.nsta.org/adi-5th*. Students will also need to access the following web pages during this investigation:

Investigation 15. Geographic Position and Climate:
Why Is the Climate in Western and Eastern Washington State So Different?

- U.S. Climate Data: *www.usclimatedata.com*
- WA Cloud Cover: *www.climate.washington.edu/cloudcover*
- Ventusky (a visualization of current wind direction and speed): *www.ventusky.com*
- Water Cycle: *www.nationalgeographic.org/encyclopedia/water-cycle*
- Cloud: *www.nationalgeographic.org/encyclopedia/cloud*

We recommend bookmarking each web page in the internet browser that the students will use or adding links to a shared document that students can access to make it easier and faster for them to visit the web page during the investigation. In addition, it is important to check if students can access and use the websites from a school computer, because some schools have set up firewalls and other restrictions on web browsing.

Safety Precautions

Remind students to follow all normal safety rules.

Lesson Plan by Stage

This lesson plan is only a suggestion. It is included here to illustrate what you can say and do during each stage of ADI for this specific investigation. We encourage you to modify this lesson plan by asking different questions, using different examples, and providing different scaffolds as needed to better meet the needs of students in your class.

Stage 1: Introduce the Task and the Guiding Question (35 minutes)

1. Ask the students to sit in six groups, with three or four students in each group.
2. Ask students to clear off their desks except for a pencil (and their *Student Workbook for Argument-Driven Inquiry in Fifth-Grade Science* if they have one).
3. Pass out an Investigation Handout to each student (or ask students to turn to the Investigation Log for Investigation 15 in their workbook).
4. Read the first paragraph of the "Introduction" aloud to the class. Ask the students to follow along as you read.
5. Give each group a color copy of the map of Washington State cities and precipitation.
6. Tell the students to record their observations and questions about the map in the "NOTICED/WONDER" chart in the "Introduction."
7. Ask the students to share what they noticed about the map.
8. Ask the students to share what questions they have about the map.
9. Tell the students, "Some of your questions might be answered by reading the rest of the 'Introduction.'"

Teacher Notes

10. Ask the students to read the rest of the "Introduction" on their own *or* ask them to follow along as you read it aloud.

11. Once the students have read the rest of the "Introduction," ask them to fill out the "Things we KNOW" chart on their Investigation Handout (or in their Investigation Log) as a group.

12. Ask the students to share what they learned from the reading. Add these ideas to a class "Things we KNOW" chart.

13. Tell the students, "Let's see what we will need to figure out during our investigation."

14. Read the task and the guiding question aloud.

15. Tell the students, "I have some materials here that you can use."

16. Introduce the students to the materials available for them to use during the investigation. We also recommend showing the students each website and what information they can find there.

Stage 2: Design a Method and Collect Data (50 minutes)

1. Tell the students, "I am now going to give you and the other members of your group about 15 minutes to plan your investigation. Before you begin, I want you all to take a couple of minutes to discuss the following questions with the rest of your group."

2. Show the following questions on the screen or board:
    - What information should we collect so we can *describe* the climate of a city?
    - What types of *patterns* might we look for to help answer the guiding question?

3. Tell the students, "Please take a few minutes to come up with an answer to these questions."

4. Give the students two or three minutes to discuss these two questions.

5. Ask two or three different groups to share their answers. Highlight or write down any important ideas on the board so students can refer to them later.

6. If possible, use a document camera to project an image of the graphic organizer for this investigation on a screen or board (or take a picture of it and project the picture on a screen or board). Tell the groups of students, "I now want you all to plan out your investigation. To do that, you will need to fill out this proposal."

7. Point to the box labeled "Our guiding question:" and tell the students, "You can put the question we are trying to answer in this box." Then ask, "Where can we find the guiding question?"

8. Wait for a student to answer.

Investigation 15. Geographic Position and Climate:
Why Is the Climate in Western and Eastern Washington State So Different?

9. Point to the box labeled "We will collect the following data:" and tell the students, "You can list the measurements or observations that you will need to collect during the investigation in this box."

10. Point to the box labeled "These are the steps we will follow to collect data:" and tell the students, "You can list what you are going to do to collect the data you need and what you will do with your data once you have it. Be sure to give enough detail that I could do your investigation for you."

11. Ask the students, "Do you have any questions about what you need to do?"

12. Wait for questions. Answer any questions that come up.

13. Tell the students, "Once you are done, raise your hand and let me know. I'll then come by and look over your proposal and give you some feedback. You may not begin collecting data until I have approved your proposal by signing it. You need to have your proposal done in the next 15 minutes."

14. Give the students 15 minutes to work in their groups on their investigation proposal. As they work, move from group to group to check in, ask probing questions, and offer a suggestion if a group gets stuck.

What should a student-designed investigation look like?

There are a number of different investigations that students can design to answer the question "Why is the climate in western and eastern Washington State so different?" For example, one method might include the following steps:

1. Visit the U.S. Climate Data website and compare and contrast the average temperature and precipitation for three cities in western Washington and three cities in eastern Washington.

2. Visit the WA Cloud Cover website and compare and contrast the number of days cloudy for three cities in western Washington and three cities in eastern Washington.

3. Visit the Ventusky website and compare and contrast the wind patterns in western and eastern Washington.

4. Determine the location of mountain ranges in Washington State.

If students use this method, they will need to collect the following data:

1. Annual high temperature
2. Annual low temperature
3. Average temperature
4. Average annual precipitation
5. Days cloudy
6. Direction of prevailing winds
7. Locations of mountain ranges

Teacher Notes

15. As each group finishes its investigation proposal, read it over and determine if it will be productive or not. If you feel the investigation will be productive (not necessarily what you would do or what the other groups are doing), sign your name on the proposal and let the group start collecting data. If the plan needs to be changed, offer some suggestions or ask some probing questions, and have the group make the changes before you approve it.

16. Pass out the materials, or have one student from each group collect the materials they need from a central supply table or cart for the groups that have an approved proposal.

17. Tell the students to collect their data and record their observations or measurements in the "Collect Your Data" box in their Investigation Handout (or the Investigation Log in their workbook).

18. Give the students 30 minutes to collect their data. Collect the materials from each group before asking them to analyze their data.

Stage 3: Create a Draft Argument (45 minutes)

1. Tell the students, "Now that we have all this data, we need to analyze the data so we can figure out an answer to the guiding question."

2. If possible, project an image of the "Analyze Your Data" section for this investigation on a screen or board using a document camera (or take a picture of it and project the picture on a screen or board). Point to the section and tell the students, "You can create a table, graph, or other representation as a way to analyze your data. You can make your table, graph, or other representation in this section."

3. Ask the students, "What information do we need to include in this analysis?"

4. Tell the students, "Please take a few minutes to discuss this question with your group and be ready to share."

5. Give the students five minutes to discuss.

6. Ask two or three different groups to share their answers. Highlight or write down any important ideas on the board so students can refer to them later.

7. Tell the groups of students, "I am now going to give you and the other members of your group about 10 minutes to analyze your data." If the students are having trouble analyzing their data, you can take a few minutes to provide a mini-lesson about possible ways to analyze the data they collected (this strategy is called just-in-time instruction because it is offered only when students get stuck).

8. Give the students 10 minutes to analyze their data. As they work, move from group to group to check in, ask probing questions, and offer suggestions.

9. Tell the students, "I am now going to give you and the other members of your group about 15 minutes to create an argument to share what you have learned

Investigation 15. Geographic Position and Climate:
Why Is the Climate in Western and Eastern Washington State So Different?

and convince others that they should believe you. Before you do that, we need to take a few minutes to discuss what you need to include in your argument."

10. If possible, use a document camera to project the "Argument Presentation on a Whiteboard" image from the "Draft Argument" section of the Investigation Handout (or the Investigation Log in their workbook) on a screen or board (or take a picture of it and project the picture on a screen or board).

11. Point to the box labeled "The Guiding Question:" and tell the students, "You can put the question we are trying to answer here on your whiteboard."

12. Point to the box labeled "Our Claim:" and tell the students, "You can put your claim here on your whiteboard. The claim is your answer to the guiding question."

13. Point to the box labeled "Our Evidence:" and tell the students, "You can put the evidence that you are using to support your claim here on your whiteboard. Your evidence will need to include the analysis you just did and an explanation of what your analysis means or shows. Scientists always need to support their claims with evidence."

14. Point to the box labeled "Our Justification of the Evidence:" and tell the students, "You can put your justification of your evidence here on your whiteboard. Your justification needs to explain why your evidence is important. Scientists often use core ideas to explain why the evidence they are using matters. Core ideas are important concepts that scientists use to help them make sense of what happens during an investigation."

15. Ask the students, "What are some core ideas that we read about earlier that might help us explain why the evidence we are using is important?"

16. Ask the students to share some of the core ideas from the "Introduction" section of the Investigation Handout (or the Investigation Log in the workbook). List these core ideas on the board.

17. Tell the students, "That is great. I would like to see everyone try to include these core ideas in your justification of the evidence. Your goal is to use these core ideas to help explain why your evidence matters and why the rest of us should pay attention to it."

18. Ask the students, "Do you have any questions about what you need to do?"

19. Answer any questions that come up.

20. Tell the students, "Okay, go ahead and start working on your arguments. You need to have your argument done in the next 15 minutes. It doesn't need to be perfect. We just need something down on the whiteboards so we can share our ideas."

21. Give the students 15 minutes to work in their groups on their arguments. As they work, move from group to group to check in, ask probing questions, and offer a suggestion if a group gets stuck.

Teacher Notes

> ## What should the table, graph, or other representation for this investigation look like?
>
> There are a number of different ways that students can analyze the data they collect during this investigation. One of the most straightforward ways is to create a map that shows the location of mountain ranges, prevailing wind direction, average high or low temperature, average cloudy days, and amount of precipitation. See Figure 15.2 for an example of a map. There are many other options for analyzing the collected data. Students often come up with some unique ways of analyzing their data, so be sure to give them some voice and choice during this stage.

FIGURE 15.2
Example of an argument

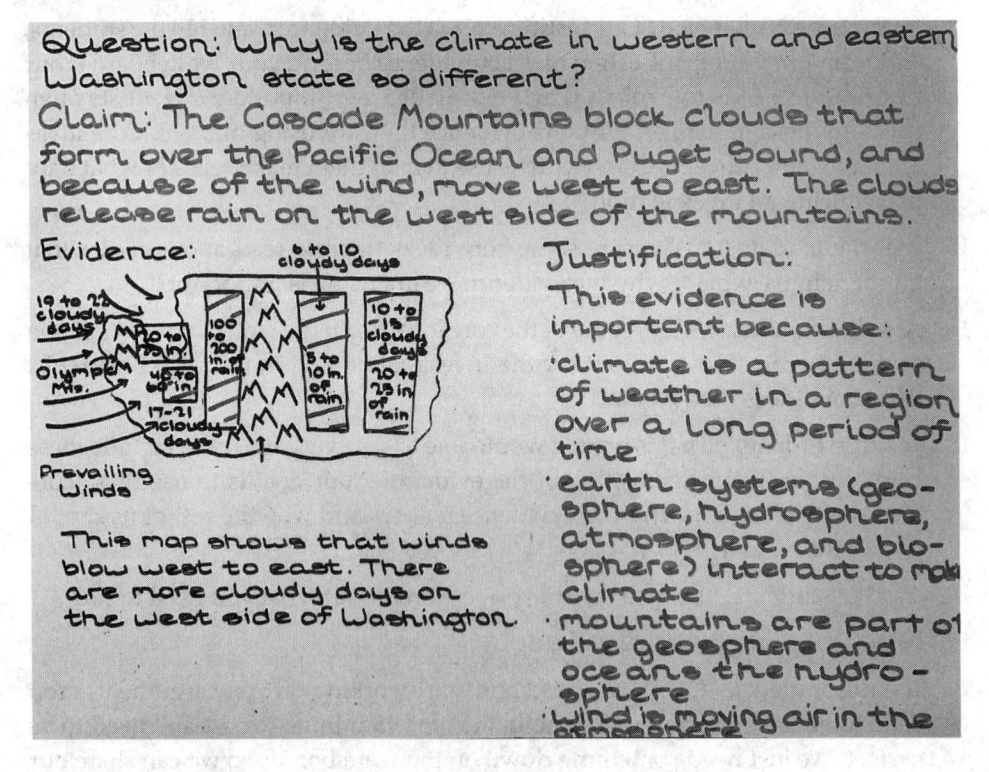

Question: Why is the climate in western and eastern Washington state so different?

Claim: The Cascade Mountains block clouds that form over the Pacific Ocean and Puget Sound, and because of the wind, move west to east. The clouds release rain on the west side of the mountains.

Evidence: [map showing prevailing winds west to east, Olympic Mts, Cascade Mountains, with annotations: 19 to 22 cloudy days, 70 to 100 in. of rain, 6 to 10 cloudy days, 5 to 10 in. of rain, 10 to 13 cloudy days, 20 to 25 in. of rain, 17–21 cloudy days, 45 to 60 in.]

This map shows that winds blow west to east. There are more cloudy days on the west side of Washington.

Justification: This evidence is important because:
- climate is a pattern of weather in a region over a long period of time
- earth systems (geosphere, hydrosphere, atmosphere, and biosphere) interact to make climate
- mountains are part of the geosphere and oceans the hydrosphere
- Wind is moving air in the atmosphere

Stage 4: Argumentation Session (30 minutes)

The argumentation session can be conducted in a whole-class presentation format, a gallery walk format, or a modified gallery walk format. We recommend using a whole-class presentation format for the first investigation, but try to transition to either the gallery walk

Investigation 15. Geographic Position and Climate:
Why Is the Climate in Western and Eastern Washington State So Different?

or modified gallery walk format as soon as possible because that will maximize student voice and choice inside the classroom. The following list shows the steps for the three formats; unless otherwise noted, the steps are the same for all three formats.

1. Begin by introducing the use of the whiteboard.

    - *If using the whole-class presentation format,* tell the students, "We are now going to share our arguments. Please set up your whiteboards so everyone can see them."

    - *If using the gallery walk or modified gallery walk format,* tell the students, "We are now going to share our arguments. Please set up your whiteboards so they are facing the walls."

2. Allow the students to set up their whiteboards.

    - *If using the whole-class presentation format,* the whiteboards should be set up on stands or chairs so they are facing toward the center of the room.

    - *If using the gallery walk or modified gallery walk format,* the whiteboards should be set up on stands or chairs so they are facing toward the outside of the room.

3. Give the following instructions to the students:

    - *If using the whole-class presentation format,* tell the students, "Okay, before we get started I want to explain what we are going to do next. Your group will have an opportunity to share your argument with the rest of the class. After you are done, everyone else in the class will have a chance to ask questions and offer some suggestions about ways to make your group's argument better. After we have a chance to listen to each other and learn something new, I'm going to give you some time to revise your arguments and make them better."

    - *If using the gallery walk format,* tell the students, "Okay, before we get started I want to explain what we are going to do next. You are going to read the arguments that were created by other groups. When I say 'go,' your group will go to a different group's station so you can see their argument. Once you are there, I'll give your group a few minutes to read and review their argument. Your job is to offer them some suggestions about ways to make their argument better. You can use sticky notes to give them suggestions. Please be specific about what you want to change and how you think they should change it. After we have a chance to learn from each other, I'm going to give you some time to revise your arguments and make them better."

    - *If using the modified gallery walk format,* tell the students, "Okay, before we get started I want to explain what we are going to do next. I'm going to ask some of you to present your arguments to your classmates. If you are presenting your argument, your job is to share your group's claim, evidence, and justification of the evidence. The rest of you will be travelers. If you are a traveler, your job is to listen to the presenters, ask the presenters questions if you do not understand

Teacher Notes

something, and then offer them some suggestions about ways to make their argument better. After we have a chance to learn from each other, I'm going to give you some time to revise your arguments and make them better."

4. Use a document camera to project the "Ways to IMPROVE our argument ..." box from the Investigation Handout (or the Investigation Log in their workbook) on a screen or board (or take a picture of it and project the picture on a screen or board).

   - *If using the whole-class presentation format,* point to the box and tell the students, "After your group presents your argument, you can write down the suggestions you get from your classmates here. If you are listening to a presentation and you see a good idea from another group, you can write down that idea here as well. Once we are done with the presentations, I will give you a chance to use these suggestions or ideas to improve your arguments."

   - *If using the gallery walk format,* point to the box and tell the students, "If you see a good idea from another group, you can write it down here. Once we are done reviewing the different arguments, I will give you a chance to use these ideas to improve your own arguments. It is important to share ideas like this."

   - *If using the modified gallery walk format,* point to the box and tell the students, "If you are a presenter, you can write down the suggestions you get from the travelers here. If you are a traveler and you see a good idea from another group, you can write down that idea here. Once we are done with the presentations, I will give you a chance to use these suggestions or ideas to improve your arguments."

5. Ask the students, "Do you have any questions about what you need to do?"

6. Answer any questions that come up.

7. Give the following instructions:

   - *If using the whole-class presentation format,* tell the students, "Okay. Let's get started."

   - *If using the gallery walk format,* tell the students, "Okay, I'm now going to tell you which argument to go to and review."

   - *If using the modified gallery walk format,* tell the students, "Okay, I'm now going to assign you to be a presenter or a traveler." Assign one or two students from each group to be presenters and one or two students from each group to be travelers.

8. Give the students an opportunity to review the arguments.

   - *If using the whole-class presentation format,* have each group present their argument one at a time. Give each group only two to three minutes to present their argument. Then give the class two to three minutes to ask them questions

Investigation 15. Geographic Position and Climate: Why Is the Climate in Western and Eastern Washington State So Different?

and offer suggestions. Encourage as much participation from the students as possible.

- *If using the gallery walk format,* tell the students, "Okay. Let's get started. Each group, move one argument to the left. Don't move to the next argument until I tell you to move. Once you get there, read the argument and then offer suggestions about how to make it better. I will put some sticky notes next to each argument. You can use the sticky notes to leave your suggestions." Give each group about three to four minutes to read the arguments, talk, and offer suggestions.

   a. After three to four minutes, tell the students, "Okay. Let's move on to the next argument. Please move one group to the left."

   b. Again, give each group three to four minutes to read, talk, and offer suggestions.

   c. Repeat this process until each group has had their argument read and critiqued three times.

- *If using the modified gallery walk format,* tell the students, "Okay. Let's get started. Reviewers, move one group to the left. Don't move to the next group until I tell you to move. Presenters, go ahead and share your argument with the travelers when they get there." Give each group of presenters and travelers about three to four minutes to talk.

   a. Tell the students, "Okay. Let's move on to the next argument. Travelers, move one group to the left."

   b. Again, give each group of presenters and travelers about three to four minutes to talk.

   c. Repeat this process until each group has had their argument read and critiqued three times.

9. Tell the students to return to their workstations.

10. Give the following instructions about revising the argument:

    - *If using the whole-class presentation format,* tell the students, "I'm now going to give you all about 10 minutes to revise your argument. Take a few minutes to talk in your groups and determine what you want to change to make your argument better. Once you have decided what to change, go ahead and make the changes to your whiteboard."

    - *If using the gallery walk format,* tell the students, "I'm now going to give you all about 10 minutes to revise your argument. Take a few minutes to read the suggestions that were left at your argument. Then talk in your groups and determine what you want to change to make your argument better. Once you have decided what to change, go ahead and make the changes to your whiteboard."

Teacher Notes

- *If using the modified gallery walk format*, tell the students, "I'm now going to give you all about 10 minutes to revise your argument. Please return to your original groups." Wait for the students to move back into their original groups and then tell the students, "Okay, take a few minutes to talk in your groups and determine what you want to change to make your argument better. Once you have decided what to change, go ahead and make the changes to your whiteboard."

11. Ask the students, "Do you have any questions about what you need to do?"
12. Answer any questions that come up.
13. Tell the students, "Okay. Let's get started."
14. Give the students 10 minutes to work in their groups on their arguments. As they work, move from group to group to check in, ask probing questions, and offer a suggestion if a group gets stuck.

Stage 5: Reflective Discussion (15 minutes)

1. Tell the students, "We are now going to take a minute to talk about some of the core ideas and crosscutting concepts that we have used during our investigation."
2. Show Figure 15.3 on the screen.
3. Ask the students, "What do you all see going on here?"
4. Allow the students to share their ideas.
5. Tell the students, "This is the Cascade mountains as viewed from Seattle in western Washington."
6. Show Figure 15.4 on the screen.
7. Ask the students, "What do you all see going on here?"
8. Allow the students to share their ideas.
9. Tell the students, "This is the Cascade mountains as viewed from Wenatchee in eastern Washington."
10. Show Figures 15.3 and 15.4 on the screen at the same time. Ask the students, "How can we use what we know about Earth's systems to help explain why the view of these mountains looks so different from these two cities?"

FIGURE 15.3
The Cascade mountains as viewed from Seattle

Note: A full-color version of this figure is available on the book's Extras page at *www.nsta.org/adi-5th*.

Investigation 15. Geographic Position and Climate:
Why Is the Climate in Western and Eastern Washington State So Different?

FIGURE 15.4

The Cascade mountains as viewed from Wenatchee

Note: A full-color version of this figure is available on the book's Extras page at *www.nsta.org/adi-5th*.

11. Allow students to share their ideas. As they share their ideas, ask different questions to encourage them to expand on their thinking (e.g., "Can you tell me more about that?"), clarify a contribution (e.g., "Can you say that in another way?"), support an idea (e.g., "Why do you think that?"), add to an idea mentioned by a classmate (e.g., "Would anyone like to add to the idea?"), re-voice an idea offered by a classmate (e.g., "Who can explain that to me in another way?"), or critique an idea during the discussion (e.g., "Do you agree or disagree with that idea and why?") until students are able to generate an adequate explanation.

12. Tell the students, "We had to look for patterns during our investigation." Then ask, "Can anyone tell me why we needed to look for patterns?"

13. Allow the students to share their ideas.

14. Tell the students, "I think patterns are really important in science because differences or similarities in patterns can be used to identify potential explanations for why something happens. Patterns can also be used as evidence to support an explanation."

Teacher Notes

15. Ask the students, "What are some of the patterns you found during your investigation?"

16. Allow the students to share their ideas.

17. Tell the students, "We also had to think about Washington State as a system during our investigation." Then ask, "Can anyone tell me why this was important?"

18. Allow the students to share their ideas.

19. Tell the students, "I think that thinking about Washington State as a system can help us describe it in terms of its components and their interactions. A system is a just a group of related parts that make up a whole and can carry out functions that its individual parts cannot."

20. Ask the students, "What were some of the components and interaction of the Washington State system that you examined today?"

21. Show an image of the question "What do you think are the most important core ideas or crosscutting concepts that we used during this investigation to help us make sense of what we observed?" Tell the students, "Okay, let's make sure we are all on the same page. Please take a moment to discuss this question with the other people in your group." Give them a few minutes to discuss the question.

22. Ask the students, "What do you all think? Who would like to share?"

23. Allow the students to share their ideas.

24. Tell the students, "We are now going to take a minute to talk about what scientists do to figure out answers to their questions."

25. Show an image of the question "Do all scientists follow the same method regardless of what they are trying to figure out?" on the screen. Tell the students, "Take a few minutes to talk about how you would answer this question with the other people in your group. Be ready to share with the rest of the class." Give the students two to three minutes to talk in their group.

26. Ask the students, "What do you all think? Who would like to share an idea?"

27. Allow the students to share their ideas.

28. Tell the students, "I think there is no universal step-by step scientific method that all scientists follow; rather, scientists who work in different scientific disciplines such as biology or physics and fields within a discipline such as genetics or ecology use different types of methods, use different core ideas, and rely on different standards to figure out how the world works."

29. Ask the students, "What might be some different methods that scientists can use?"

30. Allow the students to share their ideas.

31. Tell the students, "There are many examples of different methods, including experiments, systematic observations, literature reviews, and analysis of

existing data sets. The choice of method depends on what the scientist is trying to accomplish."

32. Ask the students, "What type of method do you think we used today?"
33. Allow the students to share their ideas.
34. Tell the students, "I think we analyzed existing data sets. These data sets include observations and measurement that were collected by other people. Scientists often share the data they collect by making it available on websites so other people can use it."
35. Ask the students, "Does anyone have any questions about the different methods that scientists use or why scientists use different methods?"
36. Answer any questions that come up.
37. Show an image of the question "Why do scientists develop or use models?" on the screen. Tell the students, "Take a few minutes to talk about how you would answer this question with the other people in your group. Be ready to share with the rest of the class." Give the students two to three minutes to talk in their group.
38. Ask the students, "What do you all think? Who would like to share an idea?"
39. Allow the students to share their ideas.
40. Tell the students, "I think models are tools for reasoning about phenomena. A model is just a representation of a set of ideas about how something works or why something happens."
41. Ask the students, "How did you use a model today?"
42. Allow the students to share their ideas.
43. Ask the students, "Does anyone have any questions about models or how they are used in science?"
44. Answer any questions that come up.
45. Tell the students, "We are now going to take a minute to talk about what went well and what didn't go so well during our investigation. We need to talk about this because you all are going to be planning and carrying out your own investigations like this a lot this year, and I want to help you all get better at it."
46. Show an image of the question "What made your investigation scientific?" on the screen. Tell the students, "Take a few minutes to talk about how you would answer this question with the other people in your group. Be ready to share with the rest of the class." Give the students two to three minutes to talk in their group.
47. Ask the students, "What do you all think? Who would like to share an idea?"

Teacher Notes

48. Allow the students to share their ideas. Be sure to expand on their ideas about what makes an investigation scientific.

49. Show an image of the question "What made your investigation not so scientific?" on the screen. Tell the students, "Take a few minutes to talk about how you would answer this question with the other people in your group. Be ready to share with the rest of the class." Give the students two to three minutes to talk in their group.

50. Ask the students, "What do you all think? Who would like to share an idea?"

51. Allow the students to share their ideas. Be sure to expand on their ideas about what makes an investigation less scientific.

52. Show an image of the question "What rules can we put into place to help us make sure our next investigation is more scientific?" on the screen. Tell the students, "Take a few minutes to talk about how you would answer this question with the other people in your group. Be ready to share with the rest of the class." Give the students two to three minutes to talk in their group.

53. Ask the students, "What do you all think? Who would like to share an idea?"

54. Allow the students to share their ideas. Once they have shared their ideas, offer a suggestion for a possible class rule.

55. Ask the students, "What do you all think? Should we make this a rule?"

56. If the students agree, write the rule on the board or on a class "Rules for Scientific Investigation" chart so you can refer to it during the next investigation.

Stage 6: Write a Draft Report (30 minutes)

Your students will use either the Investigation Handout or the Investigation Log in the student workbook when writing the draft report. When you give the directions shown in quotes in the following steps, substitute "Investigation Log in your workbook" or just "Investigation Log" (as shown in brackets) for "handout" if they are using the workbook.

1. Tell the students, "You are now going to write an investigation report to share what you have learned. Please take out a pencil and turn to the 'Draft Report' section of your handout [Investigation Log in your workbook]."

2. If possible, use a document camera to project the "Introduction" section of the draft report from the Investigation Handout (or the Investigation Log in their workbook) on a screen or board (or take a picture of it and project the picture on a screen or board).

3. Tell the students, "The first part of the report is called the 'Introduction.' In this section of the report you want to explain to the reader what you were investigating, why you were investigating it, and what question you were trying to answer. All this information can be found in the text at the beginning of

Investigation 15. Geographic Position and Climate:
Why Is the Climate in Western and Eastern Washington State So Different?

your handout [Investigation Log]." Point to the image. "Here are some sentence starters to help you begin writing."

4. Ask the students, "Do you have any questions about what you need to do?"
5. Answer any questions that come up.
6. Tell the students, "Okay, let's write."
7. Give the students 10 minutes to write the "Introduction" section of the report. As they work, move from student to student to check in, ask probing questions, and offer a suggestion if a student gets stuck.
8. If possible, use a document camera to project the "Method" section of the draft report from the Investigation Handout (or the Investigation Log in their workbook) on a screen or board (or take a picture of it and project the picture on a screen or board).
9. Tell the students, "The second part of the report is called the 'Method.' In this section of the report you want to explain to the reader what you did during the investigation, what data you collected and why, and how you went about analyzing your data. All this information can be found in the 'Plan Your Investigation' section of the handout [Investigation Log]. Remember that you all planned and carried out different investigations, so do not assume that the reader will know what you did." Point to the image. "Here are some sentence starters to help you begin writing."
10. Ask the students, "Do you have any questions about what you need to do?"
11. Answer any questions that come up.
12. Tell the students, "Okay, let's write."
13. Give the students 10 minutes to write the "Method" section of the report. As they work, move from student to student to check in, ask probing questions, and offer a suggestion if a student gets stuck.
14. If possible, use a document camera to project the "Argument" section of the draft report from the Investigation Handout (or the Investigation Log in their workbook) on a screen or board (or take a picture of it and project the picture on a screen or board).
15. Tell the students, "The last part of the report is called the 'Argument.' In this section of the report you want to share your claim, evidence, and justification of the evidence with the reader. All this information can be found on your whiteboard." Point to the image. "Here are some sentence starters to help you begin writing."
16. Ask the students, "Do you have any questions about what you need to do?"
17. Answer any questions that come up.
18. Tell the students, "Okay, let's write."

Teacher Notes

19. Give the students 10 minutes to write the "Argument" section of the report. As they work, move from student to student to check in, ask probing questions, and offer a suggestion if a student gets stuck.

Stage 7: Peer Review (35 minutes)

Your students will use either the Investigation Handout or their workbook when doing the peer review. Except where noted below, the directions are the same whether using the handout or the workbook.

1. Tell the students, "We are now going to review our reports to find ways to make them better. I'm going to come around and collect your draft reports. While I do that, please take out a pencil."
2. Collect the handouts or the workbooks with the draft reports from the students.
3. If possible, use a document camera to project the peer-review guide (see Appendix 4) on a screen or board (or take a picture of it and project the picture on a screen or board).
4. Tell the students, "We are going to use this peer-review guide to give each other feedback." Point to the image.
5. Tell the students, "I'm going to ask you to work with a partner to do this. I'm going to give you and your partner a draft report to read. You two will then read the report together. Once you are done reading the report, I want you to answer each of the questions on the peer-review guide." Point to the review questions on the image of the peer-review guide.
6. Tell the students, "You can check 'no,' 'almost,' or 'yes' after each question." Point to the checkboxes on the image of the peer-review guide.
7. Tell the students, "This will be your rating for this part of the report. Make sure you agree on the rating you give the author. If you mark 'no' or 'almost,' then you need to tell the author what he or she needs to do to get a 'yes.'" Point to the space for the reviewer feedback on the image of the peer-review guide.
8. Tell the students, "It is really important for you to give the authors feedback that is helpful. That means you need to tell them exactly what they need to do to make their report better."
9. Ask the students, "Do you have any questions about what you need to do?"
10. Answer any questions that come up.
11. Tell the students, "Please sit with a partner who is not in your current group." Allow the students time to sit with a partner.
12. Tell the students, "Okay, I'm now going to give you one report to read." Pass out one Investigation Handout with a draft report or one workbook to each pair. Make sure that the report you give a pair was not written by one of the

students in that pair. Give each pair one peer-review guide to fill out. If the students are using workbooks, the peer-review guide is included right after the draft report so you do not need to pass out copies of the peer-review guide.

13. Tell the students, "Okay, I'm going to give you 15 minutes to read the report I gave you and to fill out the peer-review guide. Go ahead and get started."

14. Give the students 15 minutes to work. As they work, move around from pair to pair to check in and see how things are going, answer questions, and offer advice.

15. After 15 minutes pass, tell the students, "Okay, time is up. Please give me the report and the peer-review guide that you filled out."

16. Collect the Investigation Handouts and the peer-review guides, or collect the workbooks if students are using them. If the students are using the Investigation Handouts and separate peer-review guides, be sure you keep each handout with its corresponding peer-review guide.

17. Tell the students, "Okay, I am now going to give you a different report to read and a new peer-review guide to fill out." Pass out one more report to each pair. Make sure that the report you give a pair was not written by one of the students in that pair. Give each pair a new peer-review guide to fill out as a group.

18. Tell the students, "Okay, I'm going to give you 15 minutes to read this new report and to fill out the peer-review guide. Go ahead and get started."

19. Give the students 15 minutes to work. As they work, move around from pair to pair to check in and see how things are going, answer questions, and offer advice.

20. After 15 minutes pass, tell the students, "Okay, time is up. Please give me the report and the peer-review guide that you filled out."

21. Collect the Investigation Handouts and the peer-review guides, or collect the workbooks if students are using them. If the students are using the Investigation Handouts and separate peer-review guides, be sure you keep each handout with its corresponding peer-review guide.

Stage 8: Revise the Report (30 minutes)

Your students will use either the Investigation Handout or their workbook when revising the report. Except where noted below, the directions are the same whether using the handout or the workbook.

1. Tell the students, "You are now going to revise your draft report based on the feedback you get from your classmates. Please take out a pencil."

2. Return the reports to the students.

Teacher Notes

- *If the students used the Investigation Handout and a copy of the peer-review guide,* pass back the handout and the peer-review guide to each student.
- *If the students used the workbook,* pass that back to each student.

3. Tell the students, "Please take a few minutes to read over the peer-review guide. You should use it to figure out what you need to change in your report and how you will change it."
4. Allow the students to read the peer-review guide.
5. *If the students used the workbook,* if possible use a document camera to project the "Write Your Final Report" section from the Investigation Log on a screen or board (or take a picture of it and project the picture on a screen or board).
6. Give the following directions about how to revise their reports:
   - *If the students used the Investigation Handout and a copy of the peer-review guide,* tell them, "Okay, let's revise our reports. Please take out a piece of paper. I would like you to rewrite your report. You can use your draft report as a starting point, but you also need to change it to make it better. Use the feedback on the peer-review guide to make it better."
   - *If the students used the workbook,* tell them, "Okay, let's revise our reports. I would like you to rewrite your report in the section of the Investigation Log called "Write Your Final Report." You can use your draft report as a starting point, but you also need to change it to make it better. Use the feedback on the peer-review guide to make it better."
7. Ask the students, "Do you have any questions about what you need to do?"
8. Answer any questions that come up.
9. Tell the students, "Okay, let's write." Allow about 20 minutes for the students to revise their reports.
10. After about 20 minutes, give the following directions:
    - *If the students used the Investigation Handout,* tell them, "Okay, time's up. I will now come around and collect your Investigation Handout, the peer-review guide, and your final report."
    - *If the students used the workbook,* tell them, "Okay, time's up. I will now come around and collect your workbooks."
11. *If the students used the Investigation Handout,* collect all the Investigation Handouts, peer-review guides, and final reports. *If the students used the workbook,* collect all the workbooks.
12. *If the students used the Investigation Handout,* use the "Teacher Score" column in the peer-review guide to grade the final report. *If the students used the workbook,* use the "Investigation Report Grading Rubric" in the Investigation Log to grade

Investigation 15. Geographic Position and Climate:
Why Is the Climate in Western and Eastern Washington State So Different?

the final report. Whether you are using the handout or the log, you can give the students feedback about their writing in the "Teacher Comments" section.

How to Use the Checkout Questions

The Checkout Questions are an optional assessment. We recommend giving them to students at the start of the next class period after the students finish stage 8 of the investigation. You can then look over the student answers to determine if you need to reteach the core idea from the investigation. Appendix 6 gives the answers to the Checkout Questions that should be given by a student who can apply the core idea correctly in all cases and can explain the interactions of Earth's systems.

Alignment With Standards

Table 15.2 highlights how the investigation can be used to address specific performance expectations from the *Next Generation Science Standards*, *Common Core State Standards for English Language Arts* (*CCSS ELA*), and *English Language Proficiency (ELP) Standards*.

TABLE 15.2

Investigation 15 alignment with standards

NGSS performance expectation	5-ESS2-1: Develop a model using an example to describe ways the geosphere, biosphere, hydrosphere, and/or atmosphere interact.
CCSS ELA—Reading: Informational Text	Key ideas and details • CCSS.ELA-LITERACY.RI.5.1: Quote accurately from a text when explaining what the text says explicitly and when drawing inferences from the text. • CCSS.ELA-LITERACY.RI.5.2: Determine two or more main ideas of a text and explain how they are supported by key details; summarize the text. • CCSS.ELA-LITERACY.RI.5.3: Explain the relationships or interactions between two or more individuals, events, ideas, or concepts in a historical, scientific, or technical text based on specific information in the text. Craft and structure • CCSS.ELA-LITERACY.RI.5.4: Determine the meaning of general academic and domain-specific words and phrases in a text relevant to a *grade 5 topic or subject area*. • CCSS.ELA-LITERACY.RI.5.5: Compare and contrast the overall structure (e.g., chronology, comparison, cause/effect, problem/solution) of events, ideas, concepts, or information in two or more texts. • CCSS.ELA-LITERACY.RI.5.6: Analyze multiple accounts of the same event or topic, noting important similarities and differences in the point of view they represent.

Continued

Teacher Notes

Table 15.2 (continued)

CCSS ELA**—**Reading: Informational Text (*continued*)	Integration of knowledge and ideas • CCSS.ELA-LITERACY.RI.5.7: Draw on information from multiple print or digital sources, demonstrating the ability to locate an answer to a question quickly or to solve a problem efficiently. • CCSS.ELA-LITERACY.RI.5.8: Explain how an author uses reasons and evidence to support particular points in a text, identifying which reasons and evidence support which point(s). Range of reading and level of text complexity • CCSS.ELA-LITERACY.RI.5.10: By the end of the year, read and comprehend informational texts, including history/social studies, science, and technical texts, at the high end of the grades 4–5 text complexity band independently and proficiently.

CCSS ELA**—**Writing	Text types and purposes
	• CCSS.ELA-LITERACY.W.5.1: Write opinion pieces on topics or texts, supporting a point of view with reasons.
	   ◦ CCSS.ELA-LITERACY.W.5.1.A: Introduce a topic or text clearly, state an opinion, and create an organizational structure in which ideas are logically grouped to support the writer's purpose.
	   ◦ CCSS.ELA-LITERACY.W.5.1.B: Provide logically ordered reasons that are supported by facts and details.
	   ◦ CCSS.ELA-LITERACY.W.5.1.C: Link opinion and reasons using words, phrases, and clauses (e.g., *consequently*, *specifically*).
	   ◦ CCSS.ELA-LITERACY.W.5.1.D: Provide a concluding statement or section related to the opinion presented.
	• CCSS.ELA-LITERACY.W.5.2: Write informative or explanatory texts to examine a topic and convey ideas and information clearly.
	   ◦ CCSS.ELA-LITERACY.W.5.2.A: Introduce a topic clearly, provide a general observation and focus, and group related information logically; include formatting (e.g., headings), illustrations, and multimedia when useful to aiding comprehension.
	   ◦ CCSS.ELA-LITERACY.W.5.2.B: Develop the topic with facts, definitions, concrete details, quotations, or other information and examples related to the topic.
	   ◦ CCSS.ELA-LITERACY.W.5.2.C: Link ideas within and across categories of information using words, phrases, and clauses (e.g., *in contrast*, *especially*).
	   ◦ CCSS.ELA-LITERACY.W.5.2.D: Use precise language and domain-specific vocabulary to inform about or explain the topic.
	   ◦ CCSS.ELA-LITERACY.W.5.2.E: Provide a concluding statement or section related to the information or explanation presented.

Continued

Investigation 15. Geographic Position and Climate: Why Is the Climate in Western and Eastern Washington State So Different?

Table 15.2 (continued)

CCSS ELA—Writing *(continued)*	Production and distribution of writing	
	• CCSS.ELA-LITERACY.W.5.4: Produce clear and coherent writing in which the development and organization are appropriate to task, purpose, and audience.	
	• CCSS.ELA-LITERACY.W.5.5: With guidance and support from peers and adults, develop and strengthen writing as needed by planning, revising, editing, rewriting, or trying a new approach.	
	• CCSS.ELA-LITERACY.W.5.6: With some guidance and support from adults, use technology, including the internet, to produce and publish writing as well as to interact and collaborate with others; demonstrate sufficient command of keyboarding skills to type a minimum of two pages in a single sitting.	
	Research to build and present knowledge	
	• CCSS.ELA-LITERACY.W.5.8: Recall relevant information from experiences or gather relevant information from print and digital sources; summarize or paraphrase information in notes and finished work, and provide a list of sources.	
	• CCSS.ELA-LITERACY.W.5.9: Draw evidence from literary or informational texts to support analysis, reflection, and research.	
	Range of writing	
	• CCSS.ELA-LITERACY.W.5.10: Write routinely over extended time frames (time for research, reflection, and revision) and shorter time frames (a single sitting or a day or two) for a range of discipline-specific tasks, purposes, and audiences.	
CCSS ELA—Speaking and Listening	Comprehension and collaboration	
	• CCSS.ELA-LITERACY.SL.5.1: Engage effectively in a range of collaborative discussions (one-on-one, in groups, and teacher-led) with diverse partners on *grade 5 topics and texts,* building on others' ideas and expressing their own clearly.	
	o CCSS.ELA-LITERACY.SL.5.1.A: Come to discussions prepared, having read or studied required material; explicitly draw on that preparation and other information known about the topic to explore ideas under discussion.	
	o CCSS.ELA-LITERACY.SL.5.1.B: Follow agreed-upon rules for discussions and carry out assigned roles.	
	o CCSS.ELA-LITERACY.SL.5.1.C: Pose and respond to specific questions by making comments that contribute to the discussion and elaborate on the remarks of others.	
	o CCSS.ELA-LITERACY.SL.5.1.D: Review the key ideas expressed and draw conclusions in light of information and knowledge gained from the discussions.	
	• CCSS.ELA-LITERACY.SL.5.2: Summarize a written text read aloud or information presented in diverse media and formats, including visually, quantitatively, and orally.	
	• CCSS.ELA-LITERACY.SL.5.3: Summarize the points a speaker makes and explain how each claim is supported by reasons and evidence.	

Continued

Teacher Notes

Table 15.2 (*continued*)

CCSS ELA—Speaking and Listening (*continued*)	Presentation of knowledge and ideas • CCSS.ELA-LITERACY.SL.5.4: Report on a topic or text or present an opinion, sequencing ideas logically and using appropriate facts and relevant, descriptive details to support main ideas or themes; speak clearly at an understandable pace. • CCSS.ELA-LITERACY.SL.5.5: Include multimedia components (e.g., graphics, sound) and visual displays in presentations when appropriate to enhance the development of main ideas or themes. • CCSS.ELA-LITERACY.SL.5.6: Adapt speech to a variety of contexts and tasks, using formal English when appropriate to task and situation.
ELP Standards	Receptive modalities • ELP 1: Construct meaning from oral presentations and literary and informational text through grade-appropriate listening, reading, and viewing. • ELP 8: Determine the meaning of words and phrases in oral presentations and literary and informational text. Productive modalities • ELP 3: Speak and write about grade-appropriate complex literary and informational texts and topics. • ELP 4: Construct grade-appropriate oral and written claims and support them with reasoning and evidence. • ELP 7: Adapt language choices to purpose, task, and audience when speaking and writing. Interactive modalities • ELP 2: Participate in grade-appropriate oral and written exchanges of information, ideas, and analyses, responding to peer, audience, or reader comments and questions. • ELP 5: Conduct research and evaluate and communicate findings to answer questions or solve problems. • ELP 6: Analyze and critique the arguments of others orally and in writing. Linguistic structures of English • ELP 9: Create clear and coherent grade-appropriate speech and text. • ELP 10: Make accurate use of standard English to communicate in grade-appropriate speech and writing.

Investigation Handout

Investigation 15

Geographic Position and Climate: Why Is the Climate in Western and Eastern Washington State So Different?

Introduction

People often want to know what the weather will be like in a city before making plans to travel to that city. *Weather* is the current condition of the atmosphere at a specific place. We describe the weather by measuring the air temperature, humidity, wind speed, precipitation, and cloud cover in a city. *Climate* is a pattern of weather in a particular region over a long period of time. When we know the climate in a region, we can make better predictions about what the weather will be like in a city. Take a minute to look at the map of some cities that are found in different regions in the state of Washington. Keep track of things you notice and things you wonder about in the boxes below.

Things I NOTICED ...	Things I WONDER about ...

Investigation Handout

The cities included on the map of Washington State are found in two different geographic regions. The first region is called western Washington. This region includes all the cities on the west coast of Washington as well as some cities that are closer to the center of the state such as Darrington, Baring, and North Bend. The second region is called eastern Washington. This region includes all the cities in the eastern half of the state, such as Omak, Chelan, Wenatchee, Ellensburg, and Yakima. The climate of western Washington is different from the climate in eastern Washington. The climate in a specific region is determined, in part, by differences in the components of the four main Earth systems in that region and how these different components interact with each other over time.

The four main Earth systems are the geosphere, the hydrosphere, the atmosphere, and the biosphere. The *geosphere* includes all the rock, soil, sediments, and landforms (such as mountains) found on Earth. The *hydrosphere* includes all the saltwater and freshwater found on, under, and above the surface of Earth. Oceans, rivers, lakes, underground water, and the water in clouds are all part of the hydrosphere. The *atmosphere* is the layer of air that surrounds Earth. The *biosphere* includes all the living things on Earth. Your goal in this investigation is to develop a model that explains how the components of the four main Earth systems are different in eastern and western Washington and how the interaction of these components results in different climates.

You will need to collect more information about Washington State before you can develop your model. To accomplish this task, you will need to first look for patterns in the high and low temperatures, the average number of cloudy days, and the average amount of rainfall in several different cities located throughout the state. Next, you will need to learn more about the location of any large bodies of water and mountains. From there, you will need to learn about the direction that wind tends to blow in different parts of Washington. You might also find it useful to learn about the water cycle and how clouds are associated with different types of weather. You can then use all this information to figure how the components of the four main Earth systems are different in eastern and western Washington and how the interaction of these components results in different climates.

Things we KNOW from what we read …

Investigation 15. Geographic Position and Climate:
Why Is the Climate in Western and Eastern Washington State So Different?

Your Task

Use what you know about weather, climate, and patterns to determine how climate changes as one moves from western to eastern Washington. Then use what you know about Earth systems and models to explain why cities located in western and eastern Washington have different climates.

The *guiding question* of this investigation is, *Why is the climate in western and eastern Washington State so different?*

Materials

You will use a computer to access several web pages to find the information you need to develop a model; the name and URL for each web page are provided below.

- U.S. Climate Data: *www.usclimatedata.com*
- WA Cloud Cover: *www.climate.washington.edu/cloudcover*
- Ventusky (a visualization of current wind direction and speed): *www.ventusky.com*
- Water Cycle: *www.nationalgeographic.org/encyclopedia/water-cycle*
- Cloud: *www.nationalgeographic.org/encyclopedia/cloud*

You may also use the following materials to create your model:

- A map showing elevation changes across the state of Washington
- A blank map of the state of Washington
- Colored pencils
- Wet-erase markers
- Transparency sheets

Safety Rules

Follow all normal safety rules.

Plan Your Investigation

Prepare a plan for your investigation by filling out the chart on the next page; this plan is called an *investigation proposal*. Before you start developing your plan, be sure to discuss the following questions with the other members of your group:

- What information should we collect so we can **describe** the climate of a city?
- What types of **patterns** might we look for to help answer the guiding question?

Investigation Handout

Our guiding question:

We will collect the following data:

These are the steps we will follow to collect data:

I approve of this investigation proposal.

_____   _____
Teacher's signature   Date

Investigation 15. Geographic Position and Climate:
Why Is the Climate in Western and Eastern Washington State So Different?

Collect Your Data

Keep a record of what you measure or observe during your investigation in the space below.

Analyze Your Data

You will need to analyze the data you collected before you can develop an answer to the guiding question. To analyze the data you collected, create a table, graph, or other representation to illustrate any patterns you found during your investigation.

Investigation Handout

Draft Argument

Develop an argument on a whiteboard. It should include the following:

1. A *claim*: Your answer to the guiding question.
2. *Evidence*: An analysis of the data and an explanation of what the analysis means.
3. A *justification of the evidence*: Why your group thinks the evidence is important.

The Guiding Question:	
Our Claim:	
Our Evidence:	Our Justification of the Evidence:

Argumentation Session

Share your argument with your classmates. Be sure to ask them how to make your draft argument better. Keep track of their suggestions in the space below.

Ways to IMPROVE our argument …

Draft Report

Prepare an *investigation report* to share what you have learned. Use the information in this handout and your group's final argument to write a *draft* of your investigation report.

Investigation 15. Geographic Position and Climate:
Why Is the Climate in Western and Eastern Washington State So Different?

Introduction

We have been studying _____ in class.

Before we started this investigation, we explored _____

_____

_____

We noticed _____

_____

_____

My goal for this investigation was to figure out _____

_____

_____

The guiding question was _____

_____

_____

Method

To gather the data I needed to answer this question, I _____

_____

_____

_____

_____

I then analyzed the data I collected by _____

_____

_____

_____

Investigation Handout

Argument

My claim is _____

_____

_____

_____

The _____ below shows _____

_____

_____

_____

_____

This analysis of the data I collected suggests _____

_____

_____

_____

Investigation 15. Geographic Position and Climate:
Why Is the Climate in Western and Eastern Washington State So Different?

This evidence is based on several important scientific concepts. The first one is _____

_____

_____

_____

_____

_____

Review

Your classmates need your help! Review the draft of their investigation reports and give them ideas about how to improve. Use the *peer-review guide* when doing your review.

Submit Your Final Report

Once you have received feedback from your classmates about your draft report, create your final investigation report and hand it in to your teacher.

Checkout Questions

Investigation 15. Geographic Position and Climate

The picture below is a map of the state of Oregon, with triangles representing large mountains. Use this map to answer questions 1–3.

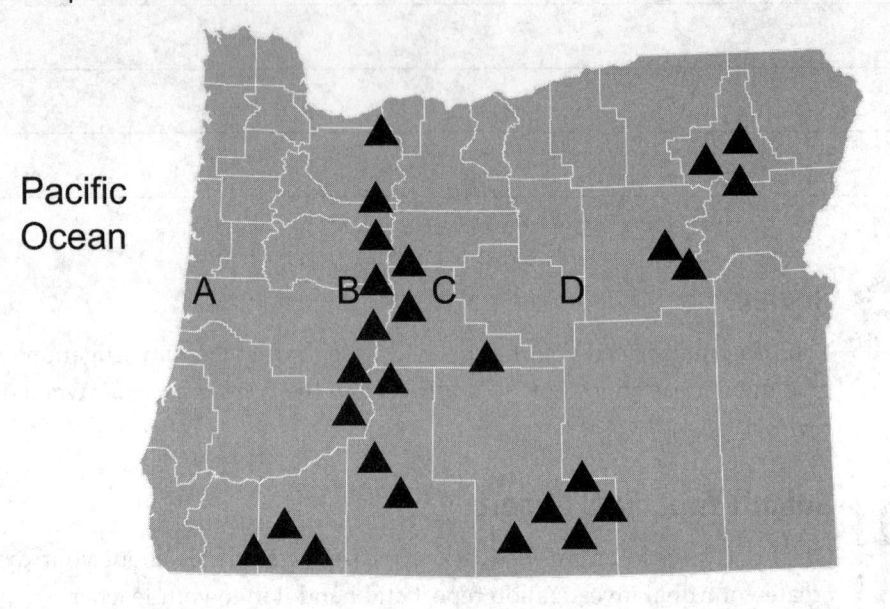

1. Which location in Oregon do you think has the *most* precipitation each year?

    a. Location A

    b. Location B

    c. Location C

    d. Location D

2. Which location in Oregon do you think has the *least* precipitation each year?

    a. Location A

    b. Location B

    c. Location C

    d. Location D

3. Explain your thinking. How do you think Earth's geosphere, hydrosphere, and atmosphere interact to affect the climate in Oregon?

    _____
    _____
    _____
    _____

Investigation 15. Geographic Position and Climate: Why Is the Climate in Western and Eastern Washington State So Different?

Teacher Scoring Rubric for the Checkout Questions

Level	Description
3	The student can apply the core idea correctly in all cases and can explain the interactions of Earth's systems.
2	The student can apply the core idea correctly in all cases but cannot explain the interactions of Earth's systems.
1	The student cannot apply the core idea correctly in all cases but can explain the interactions of Earth's systems.
0	The student cannot apply the core idea correctly and cannot explain the interactions of Earth's systems.

Teacher Notes

Investigation 16

Water Reservoirs: How Are Farmers Able to Grow Crops That Require a Lot of Water When They Live in a State That Does Not Get Much Rain?

Purpose

The purpose of this investigation is to give students an opportunity to use two disciplinary core ideas (DCIs), one crosscutting concept (CC), and seven scientific and engineering practices (SEPs) to figure out where the water needed for irrigation comes from in different states. Students will also learn about the importance of using scientific ideas to protect Earth's resources and how scientific knowledge can change over time.

The DCIs, CC, and SEPs That Students Use During This Investigation to Figure Things Out

DCIs

- *ESS2.C: The Roles of Water in Earth's Surface Processes:* Nearly all of Earth's available water is in the ocean. Most freshwater is in glaciers or underground; only a tiny fraction is in streams, lakes, wetlands, and the atmosphere.
- *ESS3.C: Human Impacts on Earth Systems:* Human activities in agriculture, industry, and everyday life have had major effects on the land, vegetation, streams, ocean, air, and even outer space. But individuals and communities are doing things to help protect Earth's resources and environments.

CC

- *CC 3: Scale, Proportion, and Quantity:* Natural objects and/or observable phenomena exist from the very small to the immensely large or from very short to very long time periods. Standard units are used to measure and describe physical quantities such as weight, time, temperature, and volume.

SEPs

- *SEP 1: Asking Questions and Defining Problems:* Ask questions about what would happen if a variable is changed. Ask questions that can be investigated and predict reasonable outcomes based on patterns such as cause-and-effect relationships.
- *SEP 3: Planning and Carrying Out Investigations:* Plan and conduct an investigation collaboratively to produce data to serve as the basis for evidence, using fair tests

Investigation 16. Water Reservoirs: How Are Farmers Able to Grow Crops That Require a Lot of Water When They Live in a State That Does Not Get Much Rain?

in which variables are controlled and the number of trials considered. Evaluate appropriate methods and/or tools for collecting data.

- *SEP 4: Analyzing and Interpreting Data:* Represent data in tables and/or various graphical displays (bar graphs, pictographs, and/or pie charts) to reveal patterns that indicate relationships. Analyze and interpret data to make sense of phenomena, using logical reasoning, mathematics, and/or computation. Compare and contrast data collected by different groups in order to discuss similarities and differences in their findings.
- *SEP 5: Using Mathematics and Computational Thinking:* Organize simple data sets to reveal patterns that suggest relationships. Describe, measure, estimate, and/or graph quantities (e.g., area, volume, weight, time) to address scientific and engineering questions and problems.
- *SEP 6: Constructing Explanations and Designing Solutions:* Construct an explanation of observed relationships. Use evidence to construct or support an explanation. Identify the evidence that supports particular points in an explanation.
- *SEP 7: Engaging in Argument From Evidence:* Compare and refine arguments based on an evaluation of the evidence presented. Distinguish among facts, reasoned judgment based on research findings, and speculation in an explanation. Respectfully provide and receive critiques from peers about a proposed procedure, explanation, or model by citing relevant evidence and posing specific questions.
- *SEP 8: Obtaining, Evaluating, and Communicating Information:* Read and comprehend grade-appropriate complex texts and/or other reliable media to summarize and obtain scientific and technical ideas. Combine information in written text with that contained in corresponding tables, diagrams, and/or charts to support the engagement in other scientific and/or engineering practices. Communicate scientific and/or technical information orally and/or in written formats, including various forms of media as well as tables, diagrams, and charts.

Other Concepts That Students May Use During This Investigation

Students might also use some of the following concepts:

- Crops need specific soil conditions, warm temperatures, and adequate water to grow well.
- Climate is a pattern of weather in a particular region over a long period of time.
- Water can be used by people for different purposes.
- Water is a limited resource.

Teacher Notes

What Students Figure Out

Farmers who live and work in states that do not get much rain must supply extra water to their crops. This is called *irrigation*. The water for crop irrigation comes from either surface water (e.g., rivers and lakes) or groundwater (underground sources).

Background Information About This Investigation for the Teacher

There are approximately 332.5 million cubic miles of water on Earth (USGS 2020b). About 97.5% of all this water is saline. It is called *saline* or *saltwater* because it contains about 35 grams of salt per liter (USGS 2020a). The remaining 2.5% of all the water on Earth is classified as *freshwater*. Freshwater contains less than 1 gram of salt per liter (USGS 2020a). About 99% of all saltwater is located in oceans, seas, and bays (which is about 96.5% of all the water on Earth). Most freshwater is located in the ice caps or glaciers (68.7% of the freshwater) or in the ground (30.1% of the freshwater); only a tiny fraction is on the surface in lakes, rivers, or wetlands (0.3% of the freshwater) or in the atmosphere (0.04% of the freshwater). Table 16.1 provides one estimate of the total global water distribution (USGS 2020b).

TABLE 16.1

Global water distribution (percentages are rounded so will not add up to 100)

Water source	Water volume (cubic miles)	Percent of total water	Percent of freshwater
Oceans, seas, and bays	321,000,000	96.54	--
Ice caps, glaciers, and permanent snow	5,773,000	1.74	68.7
Saline groundwater	3,088,000	0.93	--
Fresh groundwater	2,526,000	0.76	30.1
Ground ice and permafrost	71,970	0.02	0.86
Fresh lakes	21,830	0.007	0.26
Saline lakes	20,490	0.006	--
Atmosphere	3,095	0.001	0.04
Wetlands	2,752	0.0008	0.03
Rivers	509	0.0002	0.006

Source: Adapted from U.S. Geological Survey. n.d. Where is Earth's water? www.usgs.gov/special-topic/water-science-school/science/where-earths-water?qt-science_center_objects=0#qt-science_center_object.

It is important to understand how people use water and where they get it, because water is a limited resource. The U.S. Geological Survey (USGS), fortunately, publishes a report with this information every five years. According to the USGS report for 2015 (Dieter et al. 2018), people in the United States use about 322 billion gallons of water per day. The majority of the water that we use comes from surface water, such as lakes and rivers (61.5%, or

Investigation 16. Water Reservoirs:
How Are Farmers Able to Grow Crops That Require a Lot of Water
When They Live in a State That Does Not Get Much Rain?

198 billion gallons per day). The next largest source of water comes from groundwater (25.5%, or 82 billion gallons per day). The remaining 13% comes from oceans and bays.

Figure 16.1 shows how the 322 billion gallons of water is used in the United States as a whole (Dieter et al. 2018). The largest amount, 41%, is used in thermoelectric power plants to generate electricity. The states that withdraw the most water to generate electricity are Texas, Florida, Illinois, Michigan, New York, and North Carolina. Surface water withdrawals account for nearly all the water used in this category. Most water used by power plants, however, is returned to the source right after it is used to cool the steam that drives the thermoelectric generators. The second largest use of water in the United States, 37%, is for irrigation. Most of this water is used in the drier parts of the western United States where there is not enough rainfall to meet crop needs. Irrigation withdrawals come from surface water (52%) and groundwater (48%). The next largest use of water is for the public supply, but that is a much smaller amount (12%) than the amount used for thermoelectric power plants or irrigation. The public supply is used for public services such as wastewater treatment, schools, parks, swimming pools, and firefighting. Public supply withdrawals are generally highest in areas with large populations, such as Los Angeles, New York, and Dallas. Industrial use, including manufacturing, processing, and shipping goods, accounts for 5% of total water withdrawals. The two states that use the most water for industrial purposes are Indiana and Louisiana; together these two states account for 30% of industrial water use nationwide. In northwestern Indiana, the water is largely used for steel production, but in Louisiana the water is used for chemical and petroleum industries. The remaining 5% of the water used in the United States is used in homes (domestic), on farms to raise cows, pigs, and chickens (livestock) or fish, crabs, and shrimp (aquaculture), and for mining.

FIGURE 16.1

How water was used in the United States in 2015

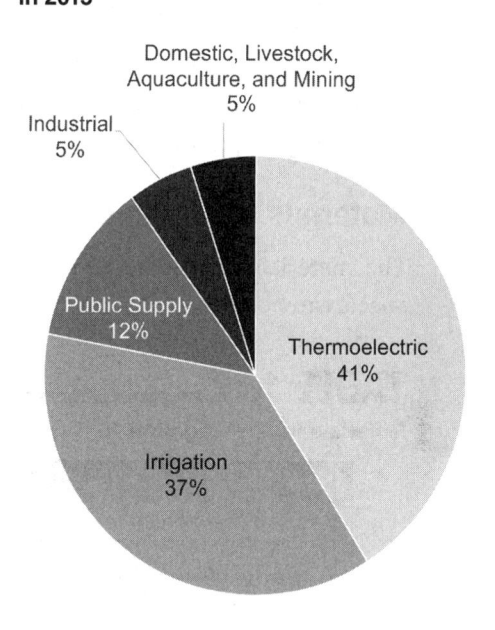

- Domestic, Livestock, Aquaculture, and Mining 5%
- Industrial 5%
- Public Supply 12%
- Thermoelectric 41%
- Irrigation 37%

Most of the water that is used for irrigation in the United States (81 percent) is taken from freshwater sources located in the 17 states west of Minnesota, Iowa, Missouri, Arkansas, and Louisiana (not including Alaska and Hawaii). Agricultural regions in these western states are typically located in areas where average annual precipitation is less than 20 inches per year, which is not enough to support crops such as alfalfa, corn, cotton, and wheat without supplemental water. Surface water is the main source of freshwater for irrigation in Arizona, Colorado, Idaho, Montana, Nevada, North Dakota, Oregon, Utah, Washington, and Wyoming. Groundwater is the main source of freshwater for irrigation in California, Kansas, Nebraska, New Mexico, Oklahoma, Texas, and South Dakota.

Teacher Notes

Timeline

The time needed to complete this investigation is 295 minutes (4 hours and 55 minutes). The amount of instructional time needed for each stage of the investigation is as follows:

- *Stage 1.* Introduce the task and the guiding question: 35 minutes
- *Stage 2.* Design a method and collect data: 60 minutes
- *Stage 3.* Create a draft argument: 60 minutes
- *Stage 4.* Argumentation session: 30 minutes
- *Stage 5.* Reflective discussion: 15 minutes
- *Stage 6.* Write a draft report: 30 minutes
- *Stage 7.* Peer review: 35 minutes
- *Stage 8.* Revise the report: 30 minutes

Materials and Preparation

The materials needed for this investigation are listed in Table 16.2. The maps and fact sheets can be downloaded from the book's Extras page at *www.nsta.org/adi-5th*.

TABLE 16.2
Materials for Investigation 16

Item	Quantity
Map showing annual precipitation in the United States in 2017 (in color)	1 per group
Maps of where crops were grown in the United States in 2017 (in color)	1 per group
Arizona State Fact Sheet	1 per group
California State Fact Sheet	1 per group
Colorado State Fact Sheet	1 per group
Idaho State Fact Sheet	1 per group
Illinois State Fact Sheet	1 per group
Iowa State Fact Sheet	1 per group
Minnesota State Fact Sheet	1 per group
North Dakota State Fact Sheet	1 per group
Texas State Fact Sheet	1 per group
Wisconsin State Fact Sheet	1 per group
Blank map of the United States	1 per student
Colored pencils	1 set per group
Wet-erase markers	1 set per group

Continued

Table 16.2 (*continued*)

Item	Quantity
Transparency sheets	4 per group
Whiteboard, 2' × 3'*	1 per group
Investigation Handout	1 per student
Peer-review guide and teacher scoring rubric	1 per student
Checkout Questions (optional)	1 per student

*As an alternative, students can use computer and presentation software such as Microsoft PowerPoint or Apple Keynote to create their arguments.

Safety Precautions

Remind students to follow all normal safety rules.

Lesson Plan by Stage

This lesson plan is only a suggestion. It is included here to illustrate what you can say and do during each stage of ADI for this specific investigation. We encourage you to modify this lesson plan by asking different questions, using different examples, and providing different scaffolds as needed to better meet the needs of students in your class.

Stage 1: Introduce the Task and the Guiding Question (35 minutes)

1. Ask the students to sit in six groups, with three or four students in each group.

2. Ask the students to clear off their desks except for a pencil (and their *Student Workbook for Argument-Driven Inquiry in Fifth-Grade Science* if they have one).

3. Pass out an Investigation Handout to each student (or ask students to turn to the Investigation Log for Investigation 16 in their workbook).

4. Read the first paragraph of the "Introduction" aloud to the class. Ask the students to follow along as you read.

5. Give each group color copies of the map showing annual precipitation in the United States in 2017 and the maps showing where crops were grown in the United States in 2017.

6. Tell the students to record what they notice about the maps and any questions they have in the "NOTICED/WONDER" chart in the "Introduction."

7. Ask the students to share what they noticed about the maps.

8. Ask the students to share what questions they have about the maps.

9. Tell the students, "Some of your questions might be answered by reading the rest of the 'Introduction.'"

Teacher Notes

10. Ask the students to read the rest of the "Introduction" on their own *or* ask them to follow along as you read it aloud.
11. Once the students have read the rest of the "Introduction," ask them to fill out the "Things we KNOW" chart on their Investigation Handout (or in their Investigation Log) as a group.
12. Ask the students to share what they learned from the reading. Add these ideas to a class "Things we KNOW" chart.
13. Tell the students, "Let's see what we will need to figure out during our investigation."
14. Read the task and the guiding question aloud.
15. Tell the students, "I have some materials here that you can use."
16. Introduce the students to the materials available for them to use during the investigation. We also recommend showing the students what information they can find on the state fact sheets.

Stage 2: Design a Method and Collect Data (60 minutes)

1. Tell the students, "I am now going to give you and the other members of your group about 15 minutes to plan your investigation. Before you begin, I want you all to take a couple of minutes to discuss the following questions with the rest of your group."
2. Show the following questions on the screen or board:
   - What *units* should we use to describe the water use of a state?
   - What types of *patterns* might we look for to help answer the guiding question?
3. Tell the students, "Please take a few minutes to come up with an answer to these questions."
4. Give the students two or three minutes to discuss these two questions.
5. Ask two or three different groups to share their answers. Highlight or write down any important ideas on the board so students can refer to them later.
6. If possible, use a document camera to project an image of the graphic organizer for this investigation on a screen or board (or take a picture of it and project the picture on a screen or board). Tell the groups of students, "I now want you all to plan out your investigation. To do that, you will need to fill out this proposal."
7. Point to the box labeled "Our guiding question:" and tell the students, "You can put the question we are trying to answer in this box." Then ask, "Where can we find the guiding question?"
8. Wait for a student to answer.

Investigation 16. Water Reservoirs: How Are Farmers Able to Grow Crops That Require a Lot of Water When They Live in a State That Does Not Get Much Rain?

9. Point to the box labeled "We will collect the following data:" and tell the students, "You can list the measurements or observations that you will need to collect during the investigation in this box."

10. Point to the box labeled "These are the steps we will follow to collect data:" and tell the students, "You can list what you are going to do to collect the data you need and what you will do with your data once you have it. Be sure to give enough detail that I could do your investigation for you."

11. Ask the students, "Do you have any questions about what you need to do?"

12. Wait for questions. Answer any questions that come up.

13. Tell the students, "Once you are done, raise your hand and let me know. I'll then come by and look over your proposal and give you some feedback. You may not begin collecting data until I have approved your proposal by signing it. You need to have your proposal done in the next 15 minutes."

What should a student-designed investigation look like?

There are a number of different investigations that students can design to answer the question "How are farmers able to grow crops that require a lot of water when they live in a state that does not get much rain?" For example, one method might include the following steps:

1. Record climate information for each state.
2. Record crop production for each state.
3. Record water use for each state
4. Sort the states into two groups: (1) states that get 17 inches or more of rain a year and (2) states that get less than 17 inches of a rain a year.
5. Compare and contrast climate, crop production, and water use across groups.

If students use this method, they will need to collect the following data:

1. Average amount of precipitation
2. Average high and low temperature
3. Types of crops produced
4. Amount of each crop produced
5. Surface water used for irrigation
6. Groundwater used for irrigation

Teacher Notes

14. Give the students 15 minutes to work in their groups on their investigation proposal. As they work, move from group to group to check in, ask probing questions, and offer a suggestion if a group gets stuck.

15. As each group finishes its investigation proposal, read it over and determine if it will be productive or not. If you feel the investigation will be productive (not necessarily what you would do or what the other groups are doing), sign your name on the proposal and let the group start collecting data. If the plan needs to be changed, offer some suggestions or ask some probing questions, and have the group make the changes before you approve it.

16. Pass out the materials, or have one student from each group collect the materials they need from a central supply table or cart for the groups that have an approved proposal.

17. Tell the students to collect their data and record their observations or measurements in the "Collect Your Data" box in their Investigation Handout (or the Investigation Log in their workbook).

18. Give the students 40 minutes to collect their data. Collect the materials from each group before asking them to analyze their data.

Stage 3: Create a Draft Argument (60 minutes)

1. Tell the students, "Now that we have all this data, we need to analyze the data so we can figure out an answer to the guiding question."

2. If possible, project an image of the "Analyze Your Data" section for this investigation on a screen or board using a document camera (or take a picture of it and project the picture on a screen or board). Point to the section and tell the students, "You can create a table, graph, or map as a way to analyze your data. You can make your table, graph, or map in this section."

3. Ask the students, "What information do we need to include in this analysis?"

4. Tell the students, "Please take a few minutes to discuss this question with your group and be ready to share."

5. Give the students five minutes to discuss.

6. Ask two or three different groups to share their answers. Highlight or write down any important ideas on the board so students can refer to them later.

7. Tell the groups of students, "I am now going to give you and the other members of your group about 30 minutes to analyze your data." If the students are having trouble analyzing their data, you can take a few minutes to provide a mini-lesson about possible ways to analyze the data they collected (this strategy is called just-in-time instruction because it is offered only when students get stuck).

8. Give the students 30 minutes to analyze their data. As they work, move from group to group to check in, ask probing questions, and offer suggestions.

Investigation 16. Water Reservoirs:
How Are Farmers Able to Grow Crops That Require a Lot of Water When They Live in a State That Does Not Get Much Rain?

9. Tell the students, "I am now going to give you and the other members of your group about 15 minutes to create an argument to share what you have learned and convince others that they should believe you. Before you do that, we need to take a few minutes to discuss what you need to include in your argument."

10. If possible, use a document camera to project the "Argument Presentation on a Whiteboard" image from the "Draft Argument" section of the Investigation Handout (or the Investigation Log in their workbook) on a screen or board (or take a picture of it and project the picture on a screen or board).

11. Point to the box labeled "The Guiding Question:" and tell the students, "You can put the question we are trying to answer here on your whiteboard."

12. Point to the box labeled "Our Claim:" and tell the students, "You can put your claim here on your whiteboard. The claim is your answer to the guiding question."

13. Point to the box labeled "Our Evidence:" and tell the students, "You can put the evidence that you are using to support your claim here on your whiteboard. Your evidence will need to include the analysis you just did and an explanation of what your analysis means or shows. Scientists always need to support their claims with evidence."

14. Point to the box labeled "Our Justification of the Evidence:" and tell the students, "You can put your justification of your evidence here on your whiteboard. Your justification needs to explain why your evidence is important. Scientists often use core ideas to explain why the evidence they are using matters. Core ideas are important concepts that scientists use to help them make sense of what happens during an investigation."

15. Ask the students, "What are some core ideas that we read about earlier that might help us explain why the evidence we are using is important?"

16. Ask the students to share some of the core ideas from the "Introduction" section of the Investigation Handout (or the Investigation Log in the workbook). List these core ideas on the board.

17. Tell the students, "That is great. I would like to see everyone try to include these core ideas in your justification of the evidence. Your goal is to use these core ideas to help explain why your evidence matters and why the rest of us should pay attention to it."

18. Ask the students, "Do you have any questions about what you need to do?"

19. Answer any questions that come up.

20. Tell the students, "Okay, go ahead and start working on your arguments. You need to have your argument done in the next 15 minutes. It doesn't need to be perfect. We just need something down on the whiteboards so we can share our ideas."

21. Give the students 15 minutes to work in their groups on their arguments. As they work, move from group to group to check in, ask probing questions, and

Teacher Notes

offer a suggestion if a group gets stuck. Figure 16.2 shows an example of an argument created by students for this investigation.

> ### What should the table, graph, or map for this investigation look like?
>
> There are a number of different ways that students can analyze the data they collect during this investigation. One of the most straightforward ways is to create a map that shows the source of irrigation water and how much is used in the states they examined (see Figure 16.2 for an example). Another option is to create a bar graph that shows the category of the states (17 inches or more of rain per year and less than 17 inches of rain per year) on the horizontal axis, or x-axis, and the average amount of irrigation water used on the vertical axis, or y-axis. They could then make color-coded bars to represent the source of the irrigation water (surface water or groundwater). There are many other options for analyzing the collected data. Students often come up with some unique ways of analyzing their data, so be sure to give them some voice and choice during this stage.

FIGURE 16.2

Example of an argument

Question: How are farmers able to grow crops that require a lot of water when they live in a state that does not get much rain?

Claim: They use surface and ground water for irrigation.

Evidence:

ID:
10,400 mil of gal of surface
4,900 of ground

CO:
7,600 of surface
1,310 of ground

CA:
5,130 of surface
13,900 of ground

AZ:
2,560 of surface
1,970 of ground

TX:
1,010 of surface
4,480 of ground

This map shows where irrigation water comes from. States that get more rain, like Illinois, Iowa, and Minnesota, use less surface and ground water for irrigation because they don't need it.

Justification: This evidence is important because:
- plants need freshwater to grow and lots of sunlight and warm temperatures
- many states that are warm do not get much rain
- freshwater is found in lakes, rivers, and underground

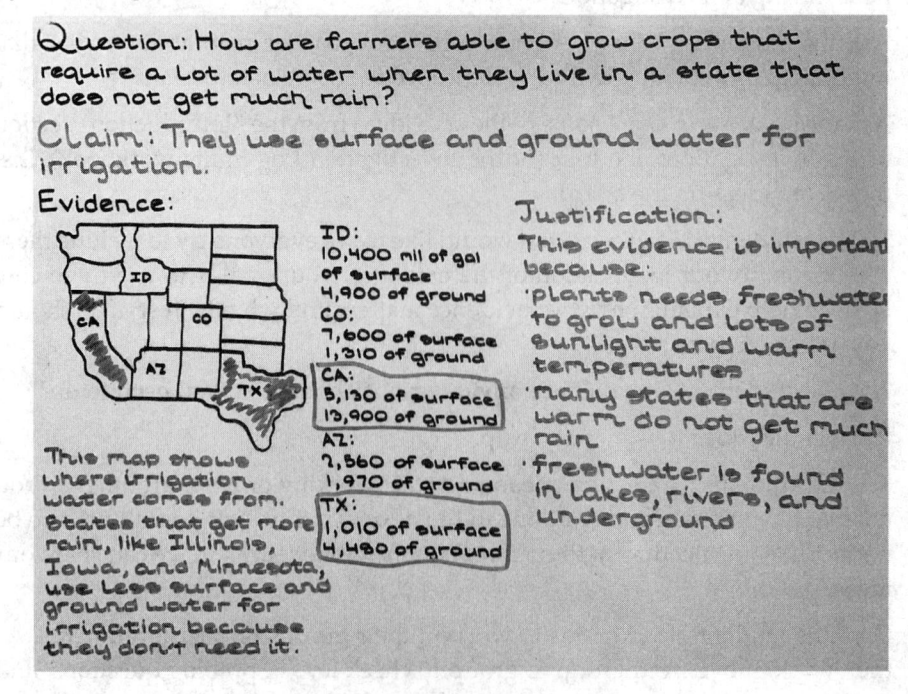

Investigation 16. Water Reservoirs:
How Are Farmers Able to Grow Crops That Require a Lot of Water
When They Live in a State That Does Not Get Much Rain?

Stage 4: Argumentation Session (30 minutes)

The argumentation session can be conducted in a whole-class presentation format, a gallery walk format, or a modified gallery walk format. We recommend using a whole-class presentation format for the first investigation, but try to transition to either the gallery walk or modified gallery walk format as soon as possible because that will maximize student voice and choice inside the classroom. The following list shows the steps for the three formats; unless otherwise noted, the steps are the same for all three formats.

1. Begin by introducing the use of the whiteboard.
   - *If using the whole-class presentation format,* tell the students, "We are now going to share our arguments. Please set up your whiteboards so everyone can see them."
   - *If using the gallery walk or modified gallery walk format,* tell the students, "We are now going to share our arguments. Please set up your whiteboards so they are facing the walls."

2. Allow the students to set up their whiteboards.
   - *If using the whole-class presentation format,* the whiteboards should be set up on stands or chairs so they are facing toward the center of the room.
   - *If using the gallery walk or modified gallery walk format,* the whiteboards should be set up on stands or chairs so they are facing toward the outside of the room.

3. Give the following instructions to the students:
   - *If using the whole-class presentation format,* tell the students, "Okay, before we get started I want to explain what we are going to do next. Your group will have an opportunity to share your argument with the rest of the class. After you are done, everyone else in the class will have a chance to ask questions and offer some suggestions about ways to make your group's argument better. After we have a chance to listen to each other and learn something new, I'm going to give you some time to revise your arguments and make them better."
   - *If using the gallery walk format,* tell the students, "Okay, before we get started I want to explain what we are going to do next. You are going to read the arguments that were created by other groups. When I say 'go,' your group will go to a different group's station so you can see their argument. Once you are there, I'll give your group a few minutes to read and review their argument. Your job is to offer them some suggestions about ways to make their argument better. You can use sticky notes to give them suggestions. Please be specific about what you want to change and how you think they should change it. After we have a chance to learn from each other, I'm going to give you some time to revise your arguments and make them better."
   - *If using the modified gallery walk format,* tell the students, "Okay, before we get started I want to explain what we are going to do next. I'm going to ask some of

Teacher Notes

you to present your arguments to your classmates. If you are presenting your argument, your job is to share your group's claim, evidence, and justification of the evidence. The rest of you will be travelers. If you are a traveler, your job is to listen to the presenters, ask the presenters questions if you do not understand something, and then offer them some suggestions about ways to make their argument better. After we have a chance to learn from each other, I'm going to give you some time to revise your arguments and make them better."

4. Use a document camera to project the "Ways to IMPROVE our argument ..." box from the Investigation Handout (or the Investigation Log in their workbook) on a screen or board (or take a picture of it and project the picture on a screen or board).

   - *If using the whole-class presentation format,* point to the box and tell the students, "After your group presents your argument, you can write down the suggestions you get from your classmates here. If you are listening to a presentation and you see a good idea from another group, you can write down that idea here as well. Once we are done with the presentations, I will give you a chance to use these suggestions or ideas to improve your arguments."

   - *If using the gallery walk format,* point to the box and tell the students, "If you see a good idea from another group, you can write it down here. Once we are done reviewing the different arguments, I will give you a chance to use these ideas to improve your own arguments. It is important to share ideas like this."

   - *If using the modified gallery walk format,* point to the box and tell the students, "If you are a presenter, you can write down the suggestions you get from the travelers here. If you are a traveler and you see a good idea from another group, you can write down that idea here. Once we are done with the presentations, I will give you a chance to use these suggestions or ideas to improve your arguments."

5. Ask the students, "Do you have any questions about what you need to do?"

6. Answer any questions that come up.

7. Give the following instructions:

   - *If using the whole-class presentation format,* tell the students, "Okay. Let's get started."

   - *If using the gallery walk format,* tell the students, "Okay, I'm now going to tell you which argument to go to and review."

   - *If using the modified gallery walk format,* tell the students, "Okay, I'm now going to assign you to be a presenter or a traveler." Assign one or two students from each group to be presenters and one or two students from each group to be travelers.

8. Give the students an opportunity to review the arguments.

Investigation 16. Water Reservoirs:
How Are Farmers Able to Grow Crops That Require a Lot of Water
When They Live in a State That Does Not Get Much Rain?

- *If using the whole-class presentation format,* have each group present their argument one at a time. Give each group only two to three minutes to present their argument. Then give the class two to three minutes to ask them questions and offer suggestions. Encourage as much participation from the students as possible.

- *If using the gallery walk format,* tell the students, "Okay. Let's get started. Each group, move one argument to the left. Don't move to the next argument until I tell you to move. Once you get there, read the argument and then offer suggestions about how to make it better. I will put some sticky notes next to each argument. You can use the sticky notes to leave your suggestions." Give each group about three to four minutes to read the arguments, talk, and offer suggestions.

    a. After three to four minutes, tell the students, "Okay. Let's move on to the next argument. Please move one group to the left."

    b. Again, give each group three to four minutes to read, talk, and offer suggestions.

    c. Repeat this process until each group has had their argument read and critiqued three times.

- *If using the modified gallery walk format,* tell the students, "Okay. Let's get started. Reviewers, move one group to the left. Don't move to the next group until I tell you to move. Presenters, go ahead and share your argument with the travelers when they get there." Give each group of presenters and travelers about three to four minutes to talk.

    a. Tell the students, "Okay. Let's move on to the next argument. Travelers, move one group to the left."

    b. Again, give each group of presenters and travelers about three to four minutes to talk.

    c. Repeat this process until each group has had their argument read and critiqued three times.

9. Tell the students to return to their workstations.

10. Give the following instructions about revising the argument:

    - *If using the whole-class presentation format,* tell the students, "I'm now going to give you all about 10 minutes to revise your argument. Take a few minutes to talk in your groups and determine what you want to change to make your argument better. Once you have decided what to change, go ahead and make the changes to your whiteboard."

    - *If using the gallery walk format,* tell the students, "I'm now going to give you all about 10 minutes to revise your argument. Take a few minutes to read the suggestions that were left at your argument. Then talk in your groups and determine what you want to change to make your argument better. Once

Teacher Notes

you have decided what to change, go ahead and make the changes to your whiteboard."

- *If using the modified gallery walk format*, tell the students, "I'm now going to give you all about 10 minutes to revise your argument. Please return to your original groups." Wait for the students to move back into their original groups and then tell the students, "Okay, take a few minutes to talk in your groups and determine what you want to change to make your argument better. Once you have decided what to change, go ahead and make the changes to your whiteboard."

11. Ask the students, "Do you have any questions about what you need to do?"
12. Answer any questions that come up.
13. Tell the students, "Okay. Let's get started."
14. Give the students 10 minutes to work in their groups on their arguments. As they work, move from group to group to check in, ask probing questions, and offer a suggestion if a group gets stuck.

Stage 5: Reflective Discussion (15 minutes)

1. Tell the students, "We are now going to take a minute to talk about some of the core ideas and crosscutting concepts that we have used during our investigation."
2. Show Figure 16.3 on the screen.

FIGURE 16.3
The distribution of water on Earth

There are approximately **332,507,646 cubic miles** of water on Earth

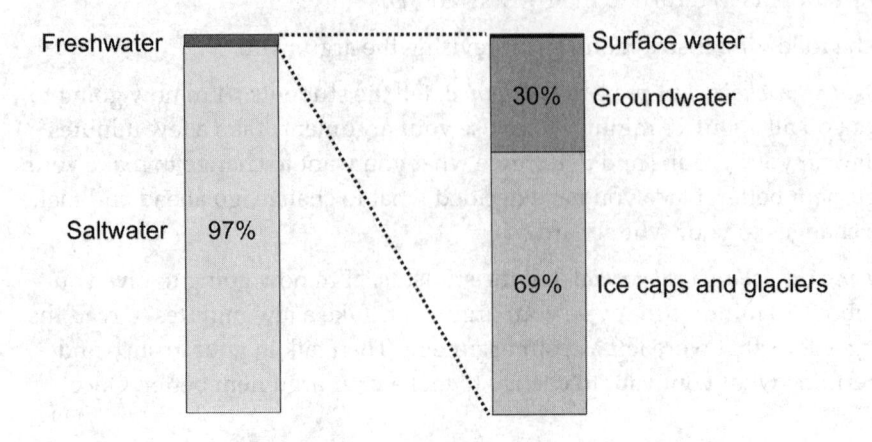

Freshwater | Surface water
Saltwater 97% | 30% Groundwater
 | 69% Ice caps and glaciers

Investigation 16. Water Reservoirs:
How Are Farmers Able to Grow Crops That Require a Lot of Water
When They Live in a State That Does Not Get Much Rain?

3. Ask the students, "What do you all see going on here?"
4. Allow the students to share their ideas.
5. Ask the students, "How can we use what we know about Earth's water to help explain what we see here?"
6. Allow the students to share their ideas. As they share their ideas, ask different questions to encourage them to expand on their thinking (e.g., "Can you tell me more about that?"), clarify a contribution (e.g., "Can you say that in another way?"), support an idea (e.g., "Why do you think that?"), add to an idea mentioned by a classmate (e.g., "Would anyone like to add to the idea?"), re-voice an idea offered by a classmate (e.g., "Who can explain that to me in another way?"), or critique an idea during the discussion (e.g., "Do you agree or disagree with that idea and why?") until students are able to generate an adequate explanation.
7. Tell the students, "We also had to think about the units that we would use to describe how water is used in different states during our investigation." Then ask, "Can anyone tell me why this was important?"
8. Allow the students to share their ideas.
9. Tell the students, "I think it is important to use standard units to measure and describe physical quantities such as volume or other amounts because it allows us to compare results from different groups."
10. Ask the students, "What were some of the units that you used during your investigation?"
11. Show an image of the question "What do you think are the most important core ideas or crosscutting concepts that we used during this investigation to help us make sense of what we observed?" Tell the students, "Okay, let's make sure we are all on the same page. Please take a moment to discuss this question with the other people in your group." Give them a few minutes to discuss the question.
12. Ask the students, "What do you all think? Who would like to share?"
13. Allow the students to share their ideas.
14. Tell the students, "We are now going take a minute to talk about the nature of scientific knowledge."
15. Show an image of the question "Do scientific ideas ever change?" on the screen. Tell the students, "Take a few minutes to talk about how you would answer this question with the other people in your group. Be ready to share with the rest of the class." Give the students two to three minutes to talk in their group.
16. Ask the students, "What do you all think? Who would like to share an idea?"
17. Allow the students to share their ideas.

Teacher Notes

18. Tell the students, "I think scientific ideas may be abandoned or modified in light of new evidence or because existing evidence has been reinterpreted by scientists."

19. Ask the students, "Do you think scientific ideas about where water is found or how much is available have ever changed over time? Why or why not?"

20. Allow the students to share their ideas.

21. Tell the students, "Our understanding of how much water is available and how much is located in different parts of Earth has changed as scientists conduct new studies in new areas and develop new tools for measuring amounts of different types of water."

22. Ask the students, "Does anyone have any questions about the ways scientific knowledge can change?"

23. Answer any questions that come up.

24. Tell the students, "We are now going to take a minute to talk about what went well and what didn't go so well during our investigation. We need to talk about this because you all are going to be planning and carrying out your own investigations like this a lot this year, and I want to help you all get better at it."

25. Show an image of the question "What made your investigation scientific?" on the screen. Tell the students, "Take a few minutes to talk about how you would answer this question with the other people in your group. Be ready to share with the rest of the class." Give the students two to three minutes to talk in their group.

26. Ask the students, "What do you all think? Who would like to share an idea?"

27. Allow the students to share their ideas. Be sure to expand on their ideas about what makes an investigation scientific.

28. Show an image of the question "What made your investigation not so scientific?" on the screen. Tell the students, "Take a few minutes to talk about how you would answer this question with the other people in your group. Be ready to share with the rest of the class." Give the students two to three minutes to talk in their group.

29. Ask the students, "What do you all think? Who would like to share an idea?"

30. Allow the students to share their ideas. Be sure to expand on their ideas about what makes an investigation less scientific.

31. Show an image of the question "What rules can we put into place to help us make sure our next investigation is more scientific?" on the screen. Tell the students, "Take a few minutes to talk about how you would answer this question with the other people in your group. Be ready to share with the rest of the class." Give the students two to three minutes to talk in their group.

Investigation 16. Water Reservoirs:
How Are Farmers Able to Grow Crops That Require a Lot of Water
When They Live in a State That Does Not Get Much Rain?

32. Ask the students, "What do you all think? Who would like to share an idea?"
33. Allow the students to share their ideas. Once they have shared their ideas, offer a suggestion for a possible class rule.
34. Ask the students, "What do you all think? Should we make this a rule?"
35. If the students agree, write the rule on the board or on a class "Rules for Scientific Investigation" chart so you can refer to it during the next investigation.

Stage 6: Write a Draft Report (30 minutes)

Your students will use either the Investigation Handout or the Investigation Log in the student workbook when writing the draft report. When you give the directions shown in quotes in the following steps, substitute "Investigation Log in your workbook" or just "Investigation Log" (as shown in brackets) for "handout" if they are using the workbook.

1. Tell the students, "You are now going to write an investigation report to share what you have learned. Please take out a pencil and turn to the 'Draft Report' section of your handout [Investigation Log in your workbook]."
2. If possible, use a document camera to project the "Introduction" section of the draft report from the Investigation Handout (or the Investigation Log in their workbook) on a screen or board (or take a picture of it and project the picture on a screen or board).
3. Tell the students, "The first part of the report is called the 'Introduction.' In this section of the report you want to explain to the reader what you were investigating, why you were investigating it, and what question you were trying to answer. All this information can be found in the text at the beginning of your handout [Investigation Log]." Point to the image. "Here are some sentence starters to help you begin writing."
4. Ask the students, "Do you have any questions about what you need to do?"
5. Answer any questions that come up.
6. Tell the students, "Okay, let's write."
7. Give the students 10 minutes to write the "Introduction" section of the report. As they work, move from student to student to check in, ask probing questions, and offer a suggestion if a student gets stuck.
8. If possible, use a document camera to project the "Method" section of the draft report from the Investigation Handout (or the Investigation Log in their workbook) on a screen or board (or take a picture of it and project the picture on a screen or board).
9. Tell the students, "The second part of the report is called the 'Method.' In this section of the report you want to explain to the reader what you did during the investigation, what data you collected and why, and how you went

Teacher Notes

about analyzing your data. All this information can be found in the 'Plan Your Investigation' section of the handout [Investigation Log]. Remember that you all planned and carried out different investigations, so do not assume that the reader will know what you did." Point to the image. "Here are some sentence starters to help you begin writing."

10. Ask the students, "Do you have any questions about what you need to do?"
11. Answer any questions that come up.
12. Tell the students, "Okay, let's write."
13. Give the students 10 minutes to write the "Method" section of the report. As they work, move from student to student to check in, ask probing questions, and offer a suggestion if a student gets stuck.
14. If possible, use a document camera to project the "Argument" section of the draft report from the Investigation Handout (or the Investigation Log in their workbook) on a screen or board (or take a picture of it and project the picture on a screen or board).
15. Tell the students, "The last part of the report is called the 'Argument.' In this section of the report you want to share your claim, evidence, and justification of the evidence with the reader. All this information can be found on your whiteboard." Point to the image. "Here are some sentence starters to help you begin writing."
16. Ask the students, "Do you have any questions about what you need to do?"
17. Answer any questions that come up.
18. Tell the students, "Okay, let's write."
19. Give the students 10 minutes to write the "Argument" section of the report. As they work, move from student to student to check in, ask probing questions, and offer a suggestion if a student gets stuck.

Stage 7: Peer Review (35 minutes)

Your students will use either the Investigation Handout or their workbook when doing the peer review. Except where noted below, the directions are the same whether using the handout or the workbook.

1. Tell the students, "We are now going to review our reports to find ways to make them better. I'm going to come around and collect your draft reports. While I do that, please take out a pencil."
2. Collect the handouts or the workbooks with the draft reports from the students.
3. If possible, use a document camera to project the peer-review guide (see Appendix 4) on a screen or board (or take a picture of it and project the picture on a screen or board).

Investigation 16. Water Reservoirs:
How Are Farmers Able to Grow Crops That Require a Lot of Water
When They Live in a State That Does Not Get Much Rain?

4. Tell the students, "We are going to use this peer-review guide to give each other feedback." Point to the image.

5. Tell the students, "I'm going to ask you to work with a partner to do this. I'm going to give you and your partner a draft report to read. You two will then read the report together. Once you are done reading the report, I want you to answer each of the questions on the peer-review guide." Point to the review questions on the image of the peer-review guide.

6. Tell the students, "You can check 'no,' 'almost,' or 'yes' after each question." Point to the checkboxes on the image of the peer-review guide.

7. Tell the students, "This will be your rating for this part of the report. Make sure you agree on the rating you give the author. If you mark 'no' or 'almost,' then you need to tell the author what he or she needs to do to get a 'yes.'" Point to the space for the reviewer feedback on the image of the peer-review guide.

8. Tell the students, "It is really important for you to give the authors feedback that is helpful. That means you need to tell them exactly what they need to do to make their report better."

9. Ask the students, "Do you have any questions about what you need to do?"

10. Answer any questions that come up.

11. Tell the students, "Please sit with a partner who is not in your current group." Allow the students time to sit with a partner.

12. Tell the students, "Okay, I'm now going to give you one report to read." Pass out one Investigation Handout with a draft report or one workbook to each pair. Make sure that the report you give a pair was not written by one of the students in that pair. Give each pair one peer-review guide to fill out. If the students are using workbooks, the peer-review guide is included right after the draft report so you do not need to pass out copies of the peer-review guide.

13. Tell the students, "Okay, I'm going to give you 15 minutes to read the report I gave you and to fill out the peer-review guide. Go ahead and get started."

14. Give the students 15 minutes to work. As they work, move around from pair to pair to check in and see how things are going, answer questions, and offer advice.

15. After 15 minutes pass, tell the students, "Okay, time is up. Please give me the report and the peer-review guide that you filled out."

16. Collect the Investigation Handouts and the peer-review guides, or collect the workbooks if students are using them. If the students are using the Investigation Handouts and separate peer-review guides, be sure you keep each handout with its corresponding peer-review guide.

17. Tell the students, "Okay, I am now going to give you a different report to read and a new peer-review guide to fill out." Pass out one more report to each pair.

Teacher Notes

Make sure that the report you give a pair was not written by one of the students in that pair. Give each pair a new peer-review guide to fill out as a group.

18. Tell the students, "Okay, I'm going to give you 15 minutes to read this new report and to fill out the peer-review guide. Go ahead and get started."

19. Give the students 15 minutes to work. As they work, move around from pair to pair to check in and see how things are going, answer questions, and offer advice.

20. After 15 minutes pass, tell the students, "Okay, time is up. Please give me the report and the peer-review guide that you filled out."

21. Collect the Investigation Handouts and the peer-review guides, or collect the workbooks if students are using them. If the students are using the Investigation Handouts and separate peer-review guides, be sure you keep each handout with its corresponding peer-review guide.

Stage 8: Revise the Report (30 minutes)

Your students will use either the Investigation Handout or their workbook when revising the report. Except where noted below, the directions are the same whether using the handout or the workbook.

1. Tell the students, "You are now going to revise your draft report based on the feedback you get from your classmates. Please take out a pencil."

2. Return the reports to the students.
   - *If the students used the Investigation Handout and a copy of the peer-review guide,* pass back the handout and the peer-review guide to each student.
   - *If the students used the workbook,* pass that back to each student.

3. Tell the students, "Please take a few minutes to read over the peer-review guide. You should use it to figure out what you need to change in your report and how you will change it."

4. Allow the students to read the peer-review guide.

5. *If the students used the workbook,* if possible use a document camera to project the "Write Your Final Report" section from the Investigation Log on a screen or board (or take a picture of it and project the picture on a screen or board).

6. Give the following directions about how to revise their reports:
   - *If the students used the Investigation Handout and a copy of the peer-review guide,* tell them, "Okay, let's revise our reports. Please take out a piece of paper. I would like you to rewrite your report. You can use your draft report as a starting point, but you also need to change it to make it better. Use the feedback on the peer-review guide to make it better."

Investigation 16. Water Reservoirs:
How Are Farmers Able to Grow Crops That Require a Lot of Water
When They Live in a State That Does Not Get Much Rain?

- *If the students used the workbook,* tell them, "Okay, let's revise our reports. I would like you to rewrite your report in the section of the Investigation Log called "Write Your Final Report." You can use your draft report as a starting point, but you also need to change it to make it better. Use the feedback on the peer-review guide to make it better."

7. Ask the students, "Do you have any questions about what you need to do?"

8. Answer any questions that come up.

9. Tell the students, "Okay, let's write." Allow about 20 minutes for the students to revise their reports.

10. After about 20 minutes, give the following directions:
    - *If the students used the Investigation Handout,* tell them, "Okay, time's up. I will now come around and collect your Investigation Handout, the peer-review guide, and your final report."
    - *If the students used the workbook,* tell them, "Okay, time's up. I will now come around and collect your workbooks."

11. *If the students used the Investigation Handout,* collect all the Investigation Handouts, peer-review guides, and final reports. *If the students used the workbook,* collect all the workbooks.

12. *If the students used the Investigation Handout,* use the "Teacher Score" column in the peer-review guide to grade the final report. *If the students used the workbook,* use the "Investigation Report Grading Rubric" in the Investigation Log to grade the final report. Whether you are using the handout or the log, you can give the students feedback about their writing in the "Teacher Comments" section.

How to Use the Checkout Questions

The Checkout Questions are an optional assessment. We recommend giving them to students at the start of the next class period after the students finish stage 8 of the investigation. You can then look over the student answers to determine if you need to reteach the core idea from the investigation. Appendix 6 gives the answers to the Checkout Questions that should be given by a student who can apply the core idea correctly in all cases and can explain the importance of using standard units to measure or describe the physical quantities of matter.

Alignment With Standards

Table 16.3 (p. 660) highlights how the investigation can be used to address specific performance expectations from the *Next Generation Science Standards, Common Core State Standards for English Language Arts* (*CCSS ELA*) and *Common Core State Standards for Mathematics* (*CCSS Mathematics*), and *English Language Proficiency (ELP) Standards.*

Teacher Notes

TABLE 16.3
Investigation 16 alignment with standards

NGSS performance expectations	Strong alignment • 5-ESS2-2: Describe and graph the amounts of water and freshwater in various reservoirs to provide evidence about the distribution of water on Earth. Moderate alignment • 5-ESS3-1: Obtain and combine information about ways individual communities use science ideas to protect the Earth's resources and environment.
CCSS ELA—Reading: Informational Text	Key ideas and details • CCSS.ELA-LITERACY.RI.5.1: Quote accurately from a text when explaining what the text says explicitly and when drawing inferences from the text. • CCSS.ELA-LITERACY.RI.5.2: Determine two or more main ideas of a text and explain how they are supported by key details; summarize the text. • CCSS.ELA-LITERACY.RI.5.3: Explain the relationships or interactions between two or more individuals, events, ideas, or concepts in a historical, scientific, or technical text based on specific information in the text. Craft and structure • CCSS.ELA-LITERACY.RI.5.4: Determine the meaning of general academic and domain-specific words and phrases in a text relevant to a *grade 5 topic or subject area*. • CCSS.ELA-LITERACY.RI.5.5: Compare and contrast the overall structure (e.g., chronology, comparison, cause/effect, problem/solution) of events, ideas, concepts, or information in two or more texts. • CCSS.ELA-LITERACY.RI.5.6: Analyze multiple accounts of the same event or topic, noting important similarities and differences in the point of view they represent. Integration of knowledge and ideas • CCSS.ELA-LITERACY.RI.5.7: Draw on information from multiple print or digital sources, demonstrating the ability to locate an answer to a question quickly or to solve a problem efficiently. • CCSS.ELA-LITERACY.RI.5.8: Explain how an author uses reasons and evidence to support particular points in a text, identifying which reasons and evidence support which point(s). Range of reading and level of text complexity • CCSS.ELA-LITERACY.RI.5.10: By the end of the year, read and comprehend informational texts, including history/social studies, science, and technical texts, at the high end of the grades 4–5 text complexity band independently and proficiently.

Continued

Investigation 16. Water Reservoirs: How Are Farmers Able to Grow Crops That Require a Lot of Water When They Live in a State That Does Not Get Much Rain?

Table 16.3 (*continued*)

***CCSS ELA*—Writing**	Text types and purposes
	• CCSS.ELA-LITERACY.W.5.1: Write opinion pieces on topics or texts, supporting a point of view with reasons.
	○ CCSS.ELA-LITERACY.W.5.1.A: Introduce a topic or text clearly, state an opinion, and create an organizational structure in which ideas are logically grouped to support the writer's purpose.
	○ CCSS.ELA-LITERACY.W.5.1.B: Provide logically ordered reasons that are supported by facts and details.
	○ CCSS.ELA-LITERACY.W.5.1.C: Link opinion and reasons using words, phrases, and clauses (e.g., *consequently*, *specifically*).
	○ CCSS.ELA-LITERACY.W.5.1.D: Provide a concluding statement or section related to the opinion presented.
	• CCSS.ELA-LITERACY.W.5.2: Write informative or explanatory texts to examine a topic and convey ideas and information clearly.
	○ CCSS.ELA-LITERACY.W.5.2.A: Introduce a topic clearly, provide a general observation and focus, and group related information logically; include formatting (e.g., headings), illustrations, and multimedia when useful to aiding comprehension.
	○ CCSS.ELA-LITERACY.W.5.2.B: Develop the topic with facts, definitions, concrete details, quotations, or other information and examples related to the topic.
	○ CCSS.ELA-LITERACY.W.5.2.C: Link ideas within and across categories of information using words, phrases, and clauses (e.g., *in contrast*, *especially*).
	○ CCSS.ELA-LITERACY.W.5.2.D: Use precise language and domain-specific vocabulary to inform about or explain the topic.
	○ CCSS.ELA-LITERACY.W.5.2.E: Provide a concluding statement or section related to the information or explanation presented.
	Production and distribution of writing
	• CCSS.ELA-LITERACY.W.5.4: Produce clear and coherent writing in which the development and organization are appropriate to task, purpose, and audience.
	• CCSS.ELA-LITERACY.W.5.5: With guidance and support from peers and adults, develop and strengthen writing as needed by planning, revising, editing, rewriting, or trying a new approach.
	• CCSS.ELA-LITERACY.W.5.6: With some guidance and support from adults, use technology, including the internet, to produce and publish writing as well as to interact and collaborate with others; demonstrate sufficient command of keyboarding skills to type a minimum of two pages in a single sitting.

Continued

Table 16.3 (continued)

CCSS ELA—Writing (*continued*)	Research to build and present knowledge • CCSS.ELA-LITERACY.W.5.8: Recall relevant information from experiences or gather relevant information from print and digital sources; summarize or paraphrase information in notes and finished work, and provide a list of sources. • CCSS.ELA-LITERACY.W.5.9: Draw evidence from literary or informational texts to support analysis, reflection, and research. Range of writing • CCSS.ELA-LITERACY.W.5.10: Write routinely over extended time frames (time for research, reflection, and revision) and shorter time frames (a single sitting or a day or two) for a range of discipline-specific tasks, purposes, and audiences.

CCSS ELA— Speaking and Listening	Comprehension and collaboration
	• CCSS.ELA-LITERACY.SL.5.1: Engage effectively in a range of collaborative discussions (one-on-one, in groups, and teacher-led) with diverse partners on *grade 5 topics and texts*, building on others' ideas and expressing their own clearly.
	   o CCSS.ELA-LITERACY.SL.5.1.A: Come to discussions prepared, having read or studied required material; explicitly draw on that preparation and other information known about the topic to explore ideas under discussion.
	   o CCSS.ELA-LITERACY.SL.5.1.B: Follow agreed-upon rules for discussions and carry out assigned roles.
	   o CCSS.ELA-LITERACY.SL.5.1.C: Pose and respond to specific questions by making comments that contribute to the discussion and elaborate on the remarks of others.
	   o CCSS.ELA-LITERACY.SL.5.1.D: Review the key ideas expressed and draw conclusions in light of information and knowledge gained from the discussions.
	• CCSS.ELA-LITERACY.SL.5.2: Summarize a written text read aloud or information presented in diverse media and formats, including visually, quantitatively, and orally.
	• CCSS.ELA-LITERACY.SL.5.3: Summarize the points a speaker makes and explain how each claim is supported by reasons and evidence.
	Presentation of knowledge and ideas
	• CCSS.ELA-LITERACY.SL.5.4: Report on a topic or text or present an opinion, sequencing ideas logically and using appropriate facts and relevant, descriptive details to support main ideas or themes; speak clearly at an understandable pace.
	• CCSS.ELA-LITERACY.SL.5.5: Include multimedia components (e.g., graphics, sound) and visual displays in presentations when appropriate to enhance the development of main ideas or themes.
	• CCSS.ELA-LITERACY.SL.5.6: Adapt speech to a variety of contexts and tasks, using formal English when appropriate to task and situation.

Continued

Investigation 16. Water Reservoirs: How Are Farmers Able to Grow Crops That Require a Lot of Water When They Live in a State That Does Not Get Much Rain?

Table 16.3 (*continued*)

CCSS Mathematics—Numbers and Operations in Base Ten	Understand the place value system. • CCSS.MATH.CONTENT.5.NBT.A.1: Recognize that in a multi-digit number, a digit in one place represents 10 times as much as it represents in the place to its right and 1/10 of what it represents in the place to its left • CCSS.MATH.CONTENT.5.NBT.A.4: Use place value understanding to round decimals to any place. Perform operations with multi-digit whole numbers and with decimals to hundredths. • CCSS.MATH.CONTENT.5.NBT.B.5: Fluently multiply multi-digit whole numbers using the standard algorithm. • CCSS.MATH.CONTENT.5.NBT.B.7: Add, subtract, multiply, and divide decimals to hundredths.
CCSS Mathematics—Measurement and Data	Convert like measurement units within a given measurement system. • CCSS.MATH.CONTENT.5.MD.A.1: Convert among different-sized standard measurement units within a given measurement system (e.g., convert 5 cm to 0.05 m), and use these conversions in solving multi-step, real-world problems. Geometric measurement: understand concepts of volume. • CCSS.MATH.CONTENT.5.MD.C.3: Recognize volume as an attribute of solid figures and understand concepts of volume measurement.
ELP Standards	Receptive modalities • ELP 1: Construct meaning from oral presentations and literary and informational text through grade-appropriate listening, reading, and viewing. • ELP 8: Determine the meaning of words and phrases in oral presentations and literary and informational text. Productive modalities • ELP 3: Speak and write about grade-appropriate complex literary and informational texts and topics. • ELP 4: Construct grade-appropriate oral and written claims and support them with reasoning and evidence. • ELP 7: Adapt language choices to purpose, task, and audience when speaking and writing. Interactive modalities • ELP 2: Participate in grade-appropriate oral and written exchanges of information, ideas, and analyses, responding to peer, audience, or reader comments and questions. • ELP 5: Conduct research and evaluate and communicate findings to answer questions or solve problems. • ELP 6: Analyze and critique the arguments of others orally and in writing.

Continued

Teacher Notes

Table 16.3 (continued)

ELP Standards (continued)	Linguistic structures of English • ELP 9: Create clear and coherent grade-appropriate speech and text. • ELP 10: Make accurate use of standard English to communicate in grade-appropriate speech and writing.

References

Dieter, C. A., M. A. Maupin, R. R. Caldwell, M. A. Harris, T. I. Ivahnenko, J. K. Lovelace, N. L. Barber, and K. S. Linsey. 2018. Estimated use of water in the United States in 2015. U.S. Geological Survey Circular 1441. *https://doi.org/10.3133/cir1441*.

U.S. Geological Survey (USGS). 2020a. Saline water and salinity. *www.usgs.gov/special-topic/water-science-school/science/saline-water-and-salinity?qt-science_center_objects=0#qt-science_center_objects*.

U.S. Geological Survey(USGS). 2020b. Where is Earth's water? *www.usgs.gov/special-topic/water-science-school/science/where-earths-water?qt-science_center_objects=0#qt-science_center_object*.

Investigation Handout

Investigation 16

Water Reservoirs: How Are Farmers Able to Grow Crops That Require a Lot of Water When They Live in a State That Does Not Get Much Rain?

Introduction

People depend on farmers to grow the food that we need. Some of this food includes plants, and some of this food includes animals. Your teacher will give you maps showing how much rain falls on different parts of the United States over the course of a year and where different crops are grown in the United States. As you look at the maps, keep track of things you notice and things you wonder about in the boxes below.

Things I NOTICED …	Things I WONDER about …

    Farms that grow crops like alfalfa, corn, cotton, and wheat are only located in certain states because these crops need specific soil conditions and 60–160 warm, frost-free days in a row to grow well. Alfalfa needs a temperature range of 42°F–110°F and a minimum of 400 hours of sunshine before

Investigation Handout

harvest, corn needs a temperature range of 60°F–95°F and a minimum of 800 hours of sunshine before harvest, cotton needs a temperature range of 64°F–86°F and a minimum of 1,000 hours of sunshine before harvest, and wheat needs a temperature range of 60°F–75°F and a minimum of 1,200 hours of sunshine before harvest.

These crops also require a lot of water to grow well. Wheat, for example, needs about 17 inches of water to grow from a seed, flower, and ripen. Corn and cotton plants need about 20 inches of water before they are ready to harvest, and alfalfa plants need about 32 inches of water. Some states that have the right soil conditions for growing crops and have long, warm summers get between 25 and 40 inches of rain a year. Other states have the right soil conditions for growing crops and are warm most of the year but get less than 20 inches of rain a year. Farmers who live and work in states that do not get much rain, such as Arizona, California, Colorado, Idaho, and Texas, therefore need to find other sources of water that they can use to grow crops.

Water is found on the surface of Earth in oceans, lakes, rivers, and wetlands. It can also be found in the ice caps or glaciers. Water can also be found underground, and in the air and clouds. Nearly all of Earth's available water (about 97%) is located in the oceans and seas (USGS 2020). This water is called *saltwater* because it contains about 35 grams of salt per liter. Large amounts of salt are toxic to plants and animals that live on land, so farmers cannot use saltwater to grow crops. They need water without salt in it (or very little salt in it). This type of water is called *freshwater*, and it makes up only about 3% of all the water on Earth. Most freshwater is located in the ice caps or glaciers (69% of the freshwater) or in the ground (30% of the freshwater). Only a tiny fraction of the freshwater found on Earth is on the surface in lakes, rivers, or wetlands (0.3% of the freshwater) or in the atmosphere (0.04% of the freshwater).

You will need to figure out how people who farm in states that do not get much rain, such as Arizona, California, Colorado, Idaho, and Texas, get the freshwater they need to grow crops. To accomplish this task, you will need to learn more about how water is used in at least six different states. You can then use this information to look for *patterns* in the way that water is used in states that get a lot of rain each year and states that do not get a lot of rain. The information on how different states use water each year comes from the U.S. Geological Survey (USGS), which monitors how water is used by people across the United States and where it comes from because human activities can cause a water shortage. People, however, can do things to protect our water supply and prevent a water shortage from happening when they know how water is being used over time.

Things we KNOW from what we read …

Investigation 16. Water Reservoirs:
How Are Farmers Able to Grow Crops That Require a Lot of Water
When They Live in a State That Does Not Get Much Rain?

Your Task

Use what you know about the role of water in Earth's surface processes, human impacts on Earth's systems, and patterns to determine where farmers who live and work in states that do not get much rain get the freshwater they need to grow crops that require a lot of water.

The *guiding question* of this investigation is, How are farmers able to grow crops that require a lot of water when they live in a state that does not get much rain?

Materials

You may use the following materials during this investigation:

- Arizona State Fact Sheet
- California State Fact Sheet
- Colorado State Fact Sheet
- Idaho State Fact Sheet
- Illinois State Fact Sheet
- Iowa State Fact Sheet
- Minnesota State Fact Sheet
- North Dakota State Fact Sheet
- Texas State Fact Sheet
- Wisconsin State Fact Sheet
- Blank map of the United States
- Colored pencils
- Wet-erase markers
- Transparency sheets

Safety Rules

Follow all normal safety rules.

Plan Your Investigation

Prepare a plan for your investigation by filling out the chart on the next page; this plan is called an *investigation proposal*. Before you start developing your plan, be sure to discuss the following questions with the other members of your group:

- What **units** should we use to describe the water use of a state?
- What types of **patterns** might we look for to help answer the guiding question?

Investigation Handout

Our guiding question:

We will collect the following data:

These are the steps we will follow to collect data:

I approve of this investigation proposal.

_____   _____
Teacher's signature   Date

Investigation 16. Water Reservoirs:
How Are Farmers Able to Grow Crops That Require a Lot of Water
When They Live in a State That Does Not Get Much Rain?

Collect Your Data

Keep a record of what you measure or observe during your investigation in the space below.

Analyze Your Data

You will need to analyze the data you collected before you can develop an answer to the guiding question. To analyze the data you collected, create a table, graph, or map to illustrate any patterns you found during your investigation.

Argument-Driven Inquiry in **Fifth-Grade Science:** Three-Dimensional Investigations

Investigation Handout

Draft Argument

Develop an argument on a whiteboard. It should include the following:

1. A *claim*: Your answer to the guiding question.
2. *Evidence*: An analysis of the data and an explanation of what the analysis means.
3. A *justification of the evidence*: Why your group thinks the evidence is important.

The Guiding Question:	
Our Claim:	
Our Evidence:	Our Justification of the Evidence:

Argumentation Session

Share your argument with your classmates. Be sure to ask them how to make your draft argument better. Keep track of their suggestions in the space below.

Ways to IMPROVE our argument …

Investigation 16. Water Reservoirs:
How Are Farmers Able to Grow Crops That Require a Lot of Water
When They Live in a State That Does Not Get Much Rain?

Draft Report

Prepare an *investigation report* to share what you have learned. Use the information in this handout and your group's final argument to write a *draft* of your investigation report.

Introduction

We have been studying _____ in class.

Before we started this investigation, we explored _____

_____

_____

We noticed _____

_____

_____

My goal for this investigation was to figure out _____

_____

_____

The guiding question was _____

_____

_____

Method

To gather the data I needed to answer this question, I _____

_____

_____

_____

_____

Argument-Driven Inquiry in **Fifth-Grade Science:** Three-Dimensional Investigations

Investigation Handout

I then analyzed the data I collected by _____

_____

_____

_____

Argument

My claim is _____

_____

_____

_____

The _____ below shows _____

_____

_____

_____

_____

Investigation 16. Water Reservoirs:
How Are Farmers Able to Grow Crops That Require a Lot of Water
When They Live in a State That Does Not Get Much Rain?

This analysis of the data I collected suggests _____

_____

_____

_____

_____

_____

This evidence is based on several important scientific concepts. The first one is _____

_____

_____

_____

_____

_____

_____

Review

Your classmates need your help! Review the draft of their investigation reports and give them ideas about how to improve. Use the *peer-review guide* when doing your review.

Submit Your Final Report

Once you have received feedback from your classmates about your draft report, create your final investigation report and hand it in to your teacher.

Reference

U.S. Geological Survey (USGS). 2020. Where is Earth's water? *www.usgs.gov/special-topic/water-science-school/science/where-earths-water?qt-science_center_objects=0#qt-science_center_object.*

Checkout Questions

Investigation 16. Water Reservoirs

1. The figure below includes four different bar graphs. Which of the bar graphs shows the correct percentages of saltwater and freshwater on Earth?

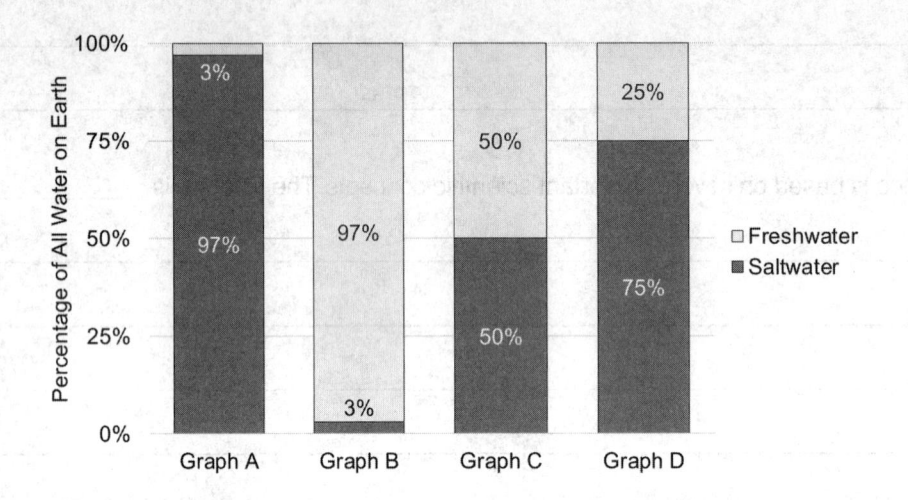

   a. Graph A
   b. Graph B
   c. Graph C
   d. Graph D

2. The figure below includes four different bar graphs. Which of these bar graphs shows the correct percentages of freshwater on Earth that is found underground (groundwater), in the ice caps and glaciers, and in lakes, rivers, or wetlands?

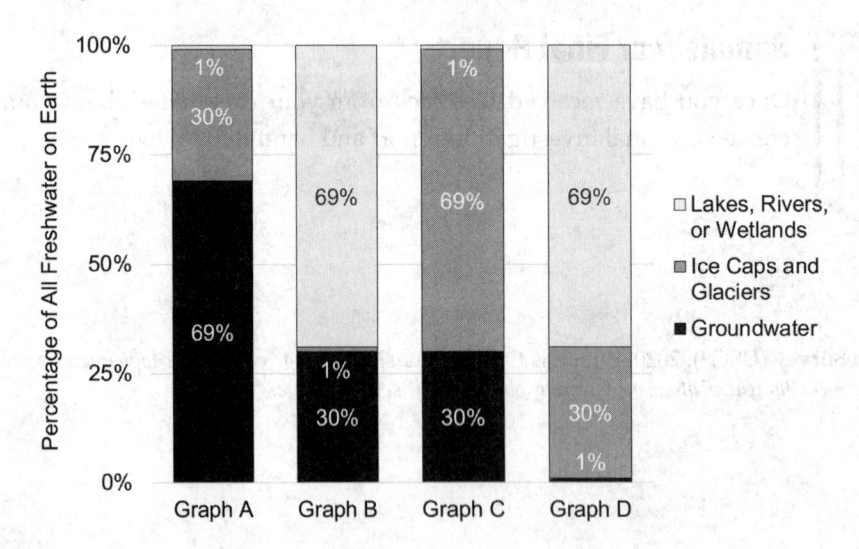

   a. Graph A
   b. Graph B
   c. Graph C
   d. Graph D

Investigation 16. Water Reservoirs:
How Are Farmers Able to Grow Crops That Require a Lot of Water
When They Live in a State That Does Not Get Much Rain?

3. Explain your thinking. Use what you know about the importance of using standard units to measure or describe the physical quantities of matter.

Teacher Scoring Rubric for the Checkout Questions

Level	Description
3	The student can apply the core idea correctly in all cases and can explain the importance of using standard units to measure or describe the physical quantities of matter.
2	The student can apply the core idea correctly in all cases but cannot explain the importance of using standard units to measure or describe the physical quantities of matter.
1	The student cannot apply the core idea correctly in all cases but can explain the importance of using standard units to measure or describe the physical quantities of matter.
0	The student cannot apply the core idea correctly and cannot explain the importance of using standard units to measure or describe the physical quantities of matter.

Section 7
Appendixes

APPENDIX 1
Standards Alignment Matrixes

Standards Matrix A: Alignment of the Argument-Driven Inquiry (ADI) Investigations With the Scientific and Engineering Practices (SEPs), Crosscutting Concepts (CCs), and Disciplinary Core Ideas (DCIs) in *A Framework for K–12 Science Education* (NRC 2012)

SEPs, CCs, and DCIs found in the *Framework*	Inv. 1. Movement of Matter Into and Out of a System	Inv. 2. Movement of Particles in a Liquid	Inv. 3. States of Matter and Weight	Inv. 4. Chemical Reactions	Inv. 5. Reactions and Weight	Inv. 6. Physical and Chemical Properties	Inv. 7. Gravity	Inv. 8. Plant Growth	Inv. 9. Energy in Ecosystems	Inv. 10. Movement of Carbon in Ecosystems	Inv. 11. Patterns in Shadows	Inv. 12. Daylight and Location	Inv. 13. Stars in the Night Sky	Inv. 14. Star Brightness	Inv. 15. Geographic Position and Climate	Inv. 16. Water Reservoirs
SEPs																
SEP 1. Asking Questions and Defining Problems	■	■	■	■	■	■	■	■	■	■	■	■	■	■	■	■
SEP 2. Developing and Using Models	■	■					■		■	■	■	■	■	■	■	
SEP 3. Planning and Carrying Out Investigations	■	■	■	■	■	■	■	■	■	■	■	■	■	■	■	■
SEP 4. Analyzing and Interpreting Data	■	■	■	■	■	■	■			■	■	■	■	■	■	■
SEP 5. Using Mathematics and Computational Thinking	■	■	■		■		■	■		■	■	■		■		■
SEP 6. Constructing Explanations and Designing Solutions	■	■	■	■	■	■	■	■	■	■	■	■	■	■	■	■
SEP 7. Engaging in Argument From Evidence	■	■	■	■	■	■	■	■	■	■	■	■	■	■	■	■
SEP 8. Obtaining, Evaluating, and Communicating Information	■	■	■	■	■	■	■	■	■	■	■	■	■	■	■	■

Key: ■ = strong alignment; ☐ = moderate alignment.

Continued

Appendix 1

Standards Matrix A (*continued*)

SEPs, CCs, and DCIs found in the *Framework*	Investigation 1. Movement of Matter Into and Out of a System	Investigation 2. Movement of Particles in a Liquid	Investigation 3. States of Matter and Weight	Investigation 4. Chemical Reactions	Investigation 5. Reactions and Weight	Investigation 6: Physical and Chemical Properties	Investigation 7. Gravity	Investigation 8. Plant Growth	Investigation 9. Energy in Ecosystems	Investigation 10. Movement of Carbon in Ecosystems	Investigation 11. Patterns in Shadows	Investigation 12. Daylight and Location	Investigation 13. Stars in the Night Sky	Investigation 14. Star Brightness	Investigation 15. Geographic Position and Climate	Investigation 16. Water Reservoirs
CCs																
CC 1. Patterns						■						■	■	■	■	
CC 2. Cause and Effect: Mechanism and Explanation	■	■		■		■								■		
CC 3. Scale, Proportion, and Quantity			■	■	■	■		■						■		■
CC 4. Systems and System Models	■		■		■		■		■	■	■				■	
CC 5. Energy and Matter: Flows, Cycles, and Conservation								■	■							
CC 6. Structure and Function																
CC 7. Stability and Change		■														
DCIs																
PS1. Matter and Its Interactions	■	■	■	■	■	■										
PS2. Motion and Stability: Forces and Interactions							■	■								
PS3. Energy									■					■		
LS1. From Molecules to Organisms: Structures and Process								■	■							
LS2. Ecosystems: Interactions, Energy, and Dynamics										■						
ESS1. Earth's Place in the Universe											■	■	■	■		
ESS2. Earth's Systems															■	■
ESS3. Earth and Human Activity																■

Key: ■ = strong alignment; ☐ = moderate alignment.

Appendix 1

Standards Matrix B: Alignment of the Argument-Driven Inquiry (ADI) Investigations With the *NGSS* Performance Expectations for Fifth-Grade Science (NGSS Lead States 2013)

NGSS performance expectations	Inv. 1. Movement of Matter Into and Out of a System	Inv. 2. Movement of Particles in a Liquid	Inv. 3. States of Matter and Weight	Inv. 4. Chemical Reactions	Inv. 5. Reactions and Weight	Inv. 6. Physical and Chemical Properties	Inv. 7. Gravity	Inv. 8. Plant Growth	Inv. 9. Energy in Ecosystems	Inv. 10. Movement of Carbon in Ecosystems	Inv. 11. Patterns in Shadows	Inv. 12. Daylight and Location	Inv. 13. Stars in the Night Sky	Inv. 14. Star Brightness	Inv. 15. Geographic Position and Climate	Inv. 16. Water Reservoirs
5-PS1-1: Develop a model to describe that matter is made of particles too small to be seen.	■	■														
5-PS1-2: Measure and graph quantities to provide evidence that regardless of the type and change that occurs when heating, cooling, or mixing substances, the total weight of matter is conserved.			■		■											
5-PS1-3: Make observations and measurements to identify materials based on their properties.				■	□	■										
5-PS1-4: Conduct an investigation to determine whether the mixing of two or more substances results in a new substance.				■	□	□										
5-PS2-1: Support an argument that the gravitational force exerted by Earth on objects is directed down.							■									
5-PS3-1: Use models to describe that energy in animals' food was once energy from the Sun.									■							
5-LS1-1: Support an argument that plants get the materials they need for growth chiefly from air and water.								■								

Key: ■ = strong alignment; □ = moderate alignment.

Continued

Appendix 1

Standards Matrix B (continued)

NGSS performance expectations	ADI investigations															
	Investigation 1. Movement of Matter Into and Out of a System	Investigation 2. Movement of Particles in a Liquid	Investigation 3. States of Matter and Weight	Investigation 4. Chemical Reactions	Investigation 5. Reactions and Weight	Investigation 6. Physical and Chemical Properties	Investigation 7. Gravity	Investigation 8. Plant Growth	Investigation 9. Energy in Ecosystems	Investigation 10. Movement of Carbon in Ecosystems	Investigation 11. Patterns in Shadows	Investigation 12. Daylight and Location	Investigation 13. Stars in the Night Sky	Investigation 14. Star Brightness	Investigation 15. Geographic Position and Climate	Investigation 16. Water Reservoirs
5-LS2-1: Develop a model to describe the movement of matter among plants, animals, decomposers, and the environment.									☐	■						
5-ESS1-1: Support an argument that differences in the apparent brightness of the Sun compared to other stars is due to their relative distance from the Earth.														■		
5-ESS1-2: Represent data in graphical displays to reveal patterns of daily changes in length and direction of shadows, day and night, and the seasonal appearance of some stars in the night sky.											■	■	■			
5-ESS2-1: Develop a model using an example to describe ways the geosphere, biosphere, hydrosphere, and/or atmosphere interact.															■	
5-ESS2-2: Describe and graph the amounts of water and freshwater in various reservoirs to provide evidence about the distribution of water on Earth.																■
5-ESS3-1: Obtain and combine information about ways individual communities use science ideas to protect the Earth's resources and the environment.																☐

Key: ■ = strong alignment; ☐ = moderate alignment.

Appendix 1

Standards Matrix C: Alignment of the Argument-Driven Inquiry (ADI) Investigations With the Nature of Scientific Knowledge (NOSK) and Nature of Scientific Inquiry (NOSI) Concepts*

NOSK and NOSI concepts	1. Movement of Matter Into and Out of a System	2. Movement of Particles in a Liquid	3. States of Matter and Weight	4. Chemical Reactions	5. Reactions and Weight	6. Physical and Chemical Properties	7. Gravity	8. Plant Growth	9. Energy in Ecosystems	10. Movement of Carbon in Ecosystems	11. Patterns in Shadows	12. Daylight and Location	13. Stars in the Night Sky	14. Star Brightness	15. Geographic Position and Climate	16. Water Reservoirs
NOSK																
How scientific knowledge changes over time																■
The difference between laws and theories in science																
Models as tools for reasoning about natural phenomena	■						■		■			■		■	■	
The difference between data and evidence in science										■						
The difference between observations and inferences in science		■														
NOSI																
The types of questions that scientists can investigate				■												
How scientists use different methods to answer different types of questions								■			■				■	
The nature and role of experiments in science									■							
The assumptions made by scientists about order and consistency in nature			■		■							■				

Key: ■ = strong alignment

*The NOSK/NOSI concepts listed in this matrix are based on the work of Abd-El-Khalick and Lederman 2000; Akerson, Abd-El-Khalick, and Lederman 2000; Lederman et al. 2002, 2014; Schwartz, Lederman, and Crawford 2004; and NGSS Lead States 2013.

Appendix 1

Standards Matrix D: Alignment of the Argument-Driven Inquiry (ADI) Investigations With the *Common Core State Standards for English Language Arts* (*CCSS ELA*; NGAC and CCSSO 2010)

CCSS ELA for fifth grade	Inv. 1. Movement of Matter Into and Out of a System	Inv. 2. Movement of Particles in a Liquid	Inv. 3. States of Matter and Weight	Inv. 4. Chemical Reactions	Inv. 5. Reactions and Weight	Inv. 6. Physical and Chemical Properties	Inv. 7. Gravity	Inv. 8. Plant Growth	Inv. 9. Energy in Ecosystems	Inv. 10. Movement of Carbon in Ecosystems	Inv. 11. Patterns in Shadows	Inv. 12. Daylight and Location	Inv. 13. Stars in the Night Sky	Inv. 14. Star Brightness	Inv. 15. Geographic Position and Climate	Inv. 16. Water Reservoirs
Reading: Informational Text																
Key ideas and details (CCSS.ELA-LITERACY.RI.5.1-3)	■	■	■	■	■	■	■	■	■	■	■	■	■	■	■	■
Craft and structure (CCSS.ELA-LITERACY.RI.5.4-6)	■	■	■	■	■	■	■	■	■	■	■	■	■	■	■	■
Integration of knowledge and ideas (CCSS.ELA-LITERACY.RI.5.7, 8)	■	■	■	■	■	■	■	■	■	■	■	■	■	■	■	■
Range of reading and level of text complexity (CCSS.ELA-LITERACY.RI.5.10)	■	■	■	■	■	■	■	■	■	■	■	■	■	■	■	■
Writing																
Text types and purposes (CCSS.ELA-LITERACY.W.5.1, 2)	■	■	■	■	■	■	■	■	■	■	■	■	■	■	■	■
Production and distribution of writing (CCSS.ELA-LITERACY.W.5.4-6)	■	■	■	■	■	■	■	■	■	■	■	■	■	■	■	■
Research to build and present knowledge (CCSS.ELA-LITERACY.W.5.8-9)	■	■	■	■	■	■	■	■	■	■	■	■	■	■	■	■
Range of writing (CCSS.ELA-LITERACY.W.5.10)	■	■	■	■	■	■	■	■	■	■	■	■	■	■	■	■
Speaking and Listening																

Key: ■ = strong alignment; □ = moderate alignment.

Continued

Standards Matrix D (*continued*)

CCSS ELA for fifth grade	ADI investigations															
	Investigation 1. Movement of Matter Into and Out of a System	Investigation 2. Movement of Particles in a Liquid	Investigation 3. States of Matter and Weight	Investigation 4. Chemical Reactions	Investigation 5. Reactions and Weight	Investigation 6: Physical and Chemical Properties	Investigation 7. Gravity	Investigation 8. Plant Growth	Investigation 9. Energy in Ecosystems	Investigation 10. Movement of Carbon in Ecosystems	Investigation 11. Patterns in Shadows	Investigation 12. Daylight and Location	Investigation 13. Stars in the Night Sky	Investigation 14. Star Brightness	Investigation 15. Geographic Position and Climate	Investigation 16. Water Reservoirs
Comprehension and collaboration (CCSS.ELA-LITERACY.SL.5.1-3)	■	■	■	■	■	■	■	■	■	■	■	■	■	■	■	■
Presentation of knowledge and ideas (CCSS.ELA-LITERACY.SL.5.4-6)	■	■	■	■	■	■	■	■	■	■	■	■	■	■	■	■
Language																	
Conventions of Standard English (CCSS.ELA-LITERACY.L.5.1-2)	□	□	□	□	□	□	□	□	□	□	□	□	□	□	□	□
Knowledge of Language (CCSS.ELA-LITERACY.L.5.3)	□	□	□	□	□	□	□	□	□	□	□	□	□	□	□	□
Vocabulary Acquisition and Use (CCSS.ELA-LITERACY.L.5.4-6)	□	□	□	□	□	□	□	□	□	□	□	□	□	□	□	□

Key: ■ = strong alignment; □ = moderate alignment.

Appendix 1

Standards Matrix E: Alignment of the Argument-Driven Inquiry (ADI) Investigations With the *Common Core State Standards for Mathematics* (*CCSS Mathematics*; NGAC and CCSSO 2010)

CCSS Mathematics for fifth grade	1. Movement of Matter Into and Out of a System	2. Movement of Particles in a Liquid	3. States of Matter and Weight	4. Chemical Reactions	5. Reactions and Weight	6. Physical and Chemical Properties	7. Gravity	8. Plant Growth	9. Energy in Ecosystems	10. Movement of Carbon in Ecosystems	11. Patterns in Shadows	12. Daylight and Location	13. Stars in the Night Sky	14. Star Brightness	15. Geographic Position and Climate	16. Water Reservoirs
Mathematical Practices																
Make sense of problems and preserver in solving them (CCSS.MATH.PRACTICE.MP1)	■	■	■		■						■		■	■		■
Reason abstractly and quantitatively (CCSS.MATH.PRACTICE.MP2)	■	■	■		■			■	■		■		■	■		■
Construct viable arguments and critique the reasoning of others (CCSS.MATH.PRACTICE.MP3)									■		■		■	■		■
Model with mathematics (CCSS.MATH.PRACTICE.MP4)									■		■		■	■		■
Use appropriate tools strategically (CCSS.MATH.PRACTICE.MP5)	■		■		■			■			■		■	■		■
Attend to precision (CCSS.MATH.PRACTICE.MP6)	■	■	■		■			■			■		■			■
Look for and make use of structure (CCSS.MATH.PRACTICE.MP7)											■		■			
Look for and express regularity in repeated reasoning (CCSS.MATH.PRACTICE.MP8)										■	■		■			

Key: ■ = strong alignment; □ = moderate alignment.

Continued

Standards Matrix E (*continued*)

	ADI investigations															
CCSS Mathematics for fifth grade	Investigation 1. Movement of Matter Into and Out of a System	Investigation 2. Movement of Particles in a Liquid	Investigation 3. States of Matter and Weight	Investigation 4. Chemical Reactions	Investigation 5. Reactions and Weight	Investigation 6: Physical and Chemical Properties	Investigation 7. Gravity	Investigation 8. Plant Growth	Investigation 9. Energy in Ecosystems	Investigation 10. Movement of Carbon in Ecosystems	Investigation 11. Patterns in Shadows	Investigation 12. Daylight and Location	Investigation 13. Stars in the Night Sky	Investigation 14. Star Brightness	Investigation 15. Geographic Position and Climate	Investigation 16. Water Reservoirs
Operations and Algebraic Thinking																
Write and interpret numerical expressions. CCSS.MATH.CONTENT.5.OA.A.2)	■															
Numbers and Operations in Base Ten																
Understand the place value system. (CCSS.MATH.CONTENT.5.NBT.A.1, 3, 4)	□	■	■		■			■								■
Perform operations with multi-digit whole numbers and with decimals to hundredths. (CCSS.MATH.CONTENT.5.NBT.B.5, 7)	■	■	■					■						■		■
Measurement and Data																
Convert like measurement units within a given measurement system. (CCSS.MATH.CONTENT.5.MD.A.1)	■	■	■		■			■				■		■		■
Represent and interpret data. (CCSS.MATH.CONTENT.5.MD.B.2)	■	■	■										■			
Geometric measurement: Understand concepts of volume. (CCSS.MATH.CONTENT.5.MD.C.3)	■															■

Key: ■ = strong alignment; □ = moderate alignment.

Appendix 1

Standards Matrix F: Alignment of the Argument-Driven Inquiry (ADI) Investigations With the *English Language Proficiency* (*ELP*) Standards (CCSSO 2014)

ELP Standards	Investigation 1. Movement of Matter Into and Out of a System	Investigation 2. Movement of Particles in a Liquid	Investigation 3. States of Matter and Weight	Investigation 4. Chemical Reactions	Investigation 5. Reactions and Weight	Investigation 6. Physical and Chemical Properties	Investigation 7. Gravity	Investigation 8. Plant Growth	Investigation 9. Energy in Ecosystems	Investigation 10. Movement of Carbon in Ecosystems	Investigation 11. Patterns in Shadows	Investigation 12. Daylight and Location	Investigation 13. Stars in the Night Sky	Investigation 14. Star Brightness	Investigation 15. Geographic Position and Climate	Investigation 16. Water Reservoirs
Receptive Modalities																
ELP 1: Construct meaning from oral presentations and informational text through grade-appropriate listening, reading, and viewing.	■	■	■	■	■	■	■	■	■	■	■	■	■	■	■	■
ELP 8: Determine the meaning of words and phrases in oral presentations and literary and informational text.	■	■	■	■	■	■	■	■	■	■	■	■	■	■	■	■
Productive Modalities																
ELP 3: Speak and write about grade-appropriate complex literary and informational texts and topics.	■	■	■	■	■	■	■	■	■	■	■	■	■	■	■	■
ELP 4: Construct grade-appropriate oral and written claims and support them with reasoning and evidence.	■	■	■	■	■	■	■	■	■	■	■	■	■	■	■	■
ELP 7: Adapt language choices to purpose, task, and audience when speaking and writing.	■	■	■	■	■	■	■	■	■	■	■	■	■	■	■	■

Key: ■ = strong alignment

Continued

Standards Matrix F (*continued*)

ELP Standards	ADI investigations															
	Investigation 1. Movement of Matter Into and Out of a System	Investigation 2. Movement of Particles in a Liquid	Investigation 3. States of Matter and Weight	Investigation 4. Chemical Reactions	Investigation 5. Reactions and Weight	Investigation 6. Physical and Chemical Properties	Investigation 7. Gravity	Investigation 8. Plant Growth	Investigation 9. Energy in Ecosystems	Investigation 10. Movement of Carbon in Ecosystems	Investigation 11. Patterns in Shadows	Investigation 12. Daylight and Location	Investigation 13. Stars in the Night Sky	Investigation 14. Star Brightness	Investigation 15. Geographic Position and Climate	Investigation 16. Water Reservoirs
Interactive Modalities																
ELP 2: Participate in grade-appropriate oral and written exchanges of information, ideas, and analyses, responding to peer, audience, or reader comments and questions.	■	■	■	■	■	■	■	■	■	■	■	■	■	■	■	■
ELP 5: Conduct research and evaluate and communicate findings to answer questions or solve problems.	■	■	■	■	■	■	■	■	■	■	■	■	■	■	■	■
ELP 6: Analyze and critique the arguments of others orally and in writing.	■	■	■	■	■	■	■	■	■	■	■	■	■	■	■	■
Linguistic Structures of English																
ELP 9: Create clear and coherent grade-appropriate speech and text.	■	■	■	■	■	■	■	■	■	■	■	■	■	■	■	■
ELP 10: Make accurate use of standard English to communicate in grade-appropriate speech and writing.	■	■	■	■	■	■	■	■	■	■	■	■	■	■	■	■

Key: ■ = strong alignment; ☐ = moderate alignment.

Appendix 1

References

Abd-El-Khalick, F., and N. G. Lederman. 2000. Improving science teachers' conceptions of nature of science: A critical review of the literature. *International Journal of Science Education* 22: 665–701.

Akerson, V., F. Abd-El-Khalick, and N. Lederman. 2000. Influence of a reflective explicit activity-based approach on elementary teachers' conception of nature of science. *Journal of Research in Science Teaching* 37 (4): 295–317.

Council of Chief State School Officers (CCSSO). 2014. *English language proficiency (ELP) standards.* Washington, DC: NGAC and CCSSO. *www.ccsso.org/resource-library/english-language-proficiency-elp-standards.*

Lederman, N. G., F. Abd-El-Khalick, R. L. Bell, and R. S. Schwartz. 2002. Views of nature of science questionnaire: Toward a valid and meaningful assessment of learners' conceptions of nature of science. *Journal of Research in Science Teaching* 39 (6): 497–521.

Lederman, J., N. Lederman, S. Bartos, S. Bartels, A. Meyer, and R. Schwartz. 2014. Meaningful assessment of learners' understanding about scientific inquiry: The Views About Scientific Inquiry (VASI) questionnaire. *Journal of Research in Science Teaching* 51 (1): 65–83.

National Governors Association Center for Best Practices and Council of Chief State School Officers (NGAC and CCSSO). 2010. *Common core state standards.* Washington, DC: NGAC and CCSSO.

NGSS Lead States. 2013. *Next Generation Science Standards: For states, by states.* Washington, DC: National Academies Press. *www.nextgenscience.org/next-generation-science-standards.*

National Research Council (NRC). 2012. *A framework for K–12 science education: Practices, crosscutting concepts, and core ideas.* Washington, DC: National Academies Press.

Schwartz, R. S., N. Lederman, and B. Crawford. 2004. Developing views of nature of science in an authentic context: An explicit approach to bridging the gap between nature of science and scientific inquiry. *Science Education* 88: 610–645.

APPENDIX 2

OVERVIEW OF *NGSS* CROSSCUTTING CONCEPTS AND NATURE OF SCIENTIFIC KNOWLEDGE AND SCIENTIFIC INQUIRY CONCEPTS

Overview of *NGSS* Crosscutting Concepts

Patterns

Scientists look for patterns in nature and attempt to understand the underlying cause of these patterns. For example, scientists often collect data and then look for patterns to identify a relationship between two variables, a trend over time, or a difference between groups.

Cause and Effect: Mechanism and Explanation

Natural phenomena have causes, and uncovering causal relationships (e.g., how changes in x affect y) is a major activity of science. Scientists also need to understand that correlation does not imply causation, some effects can have more than one cause, and some cause-and-effect relationships in systems can only be described using probability.

Scale, Proportion, and Quantity

It is critical for scientists to be able to recognize what is relevant at different sizes, times, and scales. An understanding of scale involves not only understanding how systems and processes vary in size, time span, and energy, but also how different mechanisms operate at different scales. Scientists must also be able to recognize proportional relationships between categories, groups, or quantities.

Systems and System Models

Scientists often need to define a system under study, and making a model of the system is a tool for developing a better understanding of natural phenomena in science. Scientists also need to understand that a system may interact with other systems and a system might include several different subsystems. Scientists often describe a system in terms of inputs and outputs or processes and interactions. All models of a system have limitations because they only represent certain aspects of the system under study.

Energy and Matter: Flows, Cycles, and Conservation

It is important to track how energy and matter move into, out of, and within systems during investigations. Scientists understand that the total amount of energy and matter remains the same in a closed system and that energy cannot be created or destroyed; it only moves between objects and/or fields, between one place and another place, or between systems. Energy drives the cycling and transformation of matter within and between systems.

Structure and Function

The way an object or a material is structured or shaped determines how it functions and places limits on what it can and cannot do. Scientists can make inferences about the function of an object or system by making observations about the structure or shape of its component parts and how these components interact with each other.

Appendix 2

Stability and Change

It is critical to understand what makes a system stable or unstable and what controls rates of change in a system. Scientists understand that changes in one part of a system might cause large changes in another part. They also understand that systems in dynamic equilibrium are stable due to a balance of feedback mechanisms, but the stability of these systems can be disturbed by a sudden change in the system or a series of gradual changes that accumulate over time.

Overview of Nature of Scientific Knowledge and Scientific Inquiry Concepts

Nature of Scientific Knowledge Concepts

How scientific knowledge changes over time

A person can have confidence in the validity of scientific knowledge but must also accept that scientific knowledge may be abandoned or modified in light of new evidence or because existing evidence has been reconceptualized by scientists. There are many examples in the history of science of both *evolutionary changes* (i.e., the slow or gradual refinement of ideas) and *revolutionary changes* (i.e., the rapid abandonment of a well-established idea) in scientific knowledge.

The difference between laws and theories in science

A *scientific law* describes the behavior of a natural phenomenon or a generalized relationship under certain conditions; a *scientific theory* is a well-substantiated explanation of some aspect of the natural world. Theories do not become laws even with additional evidence; they explain laws. However, not all scientific laws have an accompanying explanatory theory. It is also important for students to understand that scientists do not discover laws or theories; the scientific community develops them over time.

The use of models as tools for reasoning about natural phenomena

Scientists use conceptual models as tools to understand natural phenomena and to make predictions. A *conceptual model* is a representation of a set of ideas about how something works or why something happens. Models can take the form of diagrams, mathematical relationships, analogies, or simulations. Scientists often develop, use, test, and refine models as part of an investigation. All models are based on a set of assumptions and include approximations that limit how a model can be used and its overall predictive power.

The difference between data and evidence in science

Data are measurements, observations, and findings from other studies that are collected as part of an investigation. *Evidence*, in contrast, is analyzed data and an interpretation of the analysis. Scientists do not collect evidence; they collect data and then transform the data they collect into evidence through a process of analysis and interpretation.

The difference between observations and inferences in science

An *observation* is a descriptive statement about a natural phenomenon, whereas an *inference* is an interpretation of an observation. Students should also understand that current scientific knowledge and the perspectives of individual scientists guide both observations and inferences. Thus, different scientists can have different but equally valid interpretations of the same observations due to differences in their perspectives and background knowledge.

Nature of Scientific Inquiry Concepts

The types of questions that scientists can investigate

Scientists answer questions about the natural or material world, but not all questions can be answered by science. Science and technology may raise ethical issues for which science, by itself, does not provide answers and solutions. Scientists attempt to answer questions about what can happen in natural systems, why things happen, or how things happen. Scientists do not attempt to answer questions about what should happen. To answer questions about what should happen requires consideration of issues related to ethics, morals, values, politics, and economics.

How scientists use different methods to answer different types of questions

Examples of methods include experiments, systematic observations of a phenomenon, literature reviews, and analysis of existing data sets; the choice of method depends on the objectives of the research. There is no universal step-by step scientific method that all scientists follow; rather, different scientific disciplines (e.g., geoscience vs. chemistry) and fields within a discipline (e.g., geophysics vs. paleontology) use different types of methods, use different core theories, and rely on different standards to develop scientific knowledge.

Science as a way of knowing

Science can help us figure out how or why things happen in the world around us. Science is both a body of knowledge (which includes core ideas and crosscutting concepts) and a set of practices (such as asking questions, planning and carrying out investigations, constructing explanations, and arguing from evidence) that people use to add, revise, and refine ideas. It is a way of making sense of the world that is used by many people, not just scientists.

The nature and role of experiments in science

Scientists use experiments to test one or more explanations about how things work or why they happen. When scientists are testing an explanation, they call that explanation a hypothesis because they are not sure if that explanation is the best one. Scientists often make predictions about the expected results of an

Appendix 2

experiment based on the hypothesis they are testing and how they decided to test it. The experiment is then carried out and the predictions are compared with the actual results of the test. If the predictions match the actual results, then the hypothesis they were testing is supported. If the actual results do not match the predicted results, then the hypothesis they were testing is not supported. Scientists also control as many variables as they can when conducting an experiment to help eliminate other possible explanations for the actual results of the test.

The assumptions made by scientists about order and consistency in nature

Scientific investigations are designed based on the assumptions that natural laws operate today as they did in the past and that they will continue to do so in the future. Scientists also assume that the universe is a vast single system in which basic laws are consistent.

APPENDIX 3

SOME FREQUENTLY ASKED QUESTIONS ABOUT ARGUMENT-DRIVEN INQUIRY (ADI)

What are some things I can do to encourage productive talk among my students during ADI?

We suggest that you ...

- Choose a **talk format** that will maximize participation (we recommend small-group or partner format during most stages).
- Establish some **class talk norms** (agreed-upon rules for discussions) or remind your students about the existing ones.
- Make the **goal of the discussion** explicit (e.g., identify important or useful ideas in a text, share ideas, improve an argument, help others improve their argument, reach consensus).
- Post different **conversation starters** in the classroom. Conservation starters are general question stems (e.g., "I disagree with that because ...."; "Could you tell me more about ...?"; "Can you explain what you mean by ...?") that encourage students to think, reason, and collaborate in academically productive ways.
- Make and use a set of **back-pocket question cards**. You can carry a set of cards with you as you move from group to group. Each card has a different question written on it. When you see a group of students no longer talking, hand one of the cards to a student. The student then asks the question written on the card. This is a great way to get a conversation started and to encourage students to engage in more productive talk.
- **Model** how to argue from evidence, critique ideas, or offer useful feedback.

What can I do to help my students comprehend more of what they read?

We suggest that you ...

- **Read the text aloud** so your students can listen to a fluent adult.
- If you are reading the text aloud, stop when you read a sentence that includes an important or useful idea and **tell students to highlight** or "star" that part in the text (e.g., "I think that will be really useful for you later, so you might want to put a star next to that sentence.").
- Encourage your students to **annotate** the text as they read if you ask them to read the text on their own.
- Remind students to **talk with their peers** about what they read.
- Break the text into more **manageable chunks** by telling the students to read only one paragraph at a time and add to their "Things we KNOW" chart before moving on to the next paragraph of text.

Appendix 3

What are some things I can do to support my emerging bilingual students during ADI?

We suggest that you …

- **Group students strategically** to maximize productive interaction between students of varying language proficiencies. For example, you may want to pair a student who is bilingual with one who is learning English (if they share the same first language) so these two students can work together when you ask them to present or critique arguments.
- Encourage students to **use the language they have** to express their ideas (e.g., everyday language, imperfect English, native language).
- Provide a **visual representation** of the key terms and ideas.
- Provide a **copy of the handout** in the students' native language. Ask your school's ELL (English-language learner) support staff for assistance in translation.
- Encourage students to **draw** what they are thinking.
- **Model** language expectations for a task (e.g., demonstrate what critiquing looks like, demonstrate what giving good feedback looks like, highlight how a section of a report is organized, demonstrate how to write a section of a report).
- Allow the students to **work with a peer** as they write.
- Provide **extended time** to complete tasks that require writing.

What are some things that I can do if my students get stuck when they are developing their draft argument?

We suggest that you …

- Provide students with **sentence starters** for each component of the argument. Examples of sentence starters that will help students provide a good interpretation of the data analysis include the following:
    - "Our analysis suggests …"
    - "Our analysis shows …"
    - "This table shows …"

  Examples of sentence starters that will help students provide a good justification of the evidence include the following:
    - "This evidence matters because …"
    - "We included this evidence in our argument because …"
    - "We think this evidence is important because …"

- Provide them with **strong and weak examples** of each component of an argument (claim, evidence, and justification of evidence) and then discuss what makes the component strong or weak.
- Send one or two students from each group to a different group to **learn something from their classmates.** When students visit another group, encourage those students to ask questions about

what the other group is doing and why. They can then return to their own group and share what they have learned.

- Encourage students to **revisit the "Introduction" section of the handout** (or the investigation log) and look for useful ideas to include in the argument.

What are some things that I can do if my students get stuck as they start writing their report?

We suggest that you …

- Provide **strong and weak examples** of each section of the report and then discuss what makes that section strong or weak.
- Allow students to talk to each other as they write so they can **learn something from their classmates**. When the students talk, encourage them to ask questions about what they are writing and why.
- Encourage students to **revisit the "Introduction" section of the handout** (or the investigation log) and look for useful ideas to include in the "Introduction" section of the report.
- Encourage students to **revisit the investigation proposal section of the handout** (or the investigation log) and look for useful ideas to include in the "Method" section of the report.
- Encourage students to **use their group's argument as a foundation** when they write the "Argument" section of their report. All the basic information should be on their whiteboard. All they need to do is write a paragraph or two that includes this information.

What are some things that I can do to help my students give better feedback to each other during the peer-review process?

We suggest that you …

- Provide **strong and weak examples** of each section of the report and review them together as a class. Then **model** how to give good feedback.
- Provide students with **feedback sentence starters** for different types of feedback. Examples of feedback sentence starters include the following:
  - "We suggest adding …"
  - "We suggest making the following changes …"
  - "We think you can make your writing more clear by …"
  - "We think you should change ___ to ___."
- Be sure to establish some **class feedback norms** (agreed-upon rules for critiquing the work of others) or remind them about the existing ones.
- Make the **goal of the activity** explicit (in this case the goal is to help others improve the quality of their report).

APPENDIX 4

ADI INVESTIGATION REPORT PEER-REVIEW GUIDE: ELEMENTARY SCHOOL VERSION

Report By: _____ **Date:** _____
Name

Reviewed By: _____ _____
Name of Reviewer 1   Name of Reviewer 2

Section 1: The Investigation	Reviewer Rating			Teacher Score		
1. Did the author do a good job of explaining what the investigation was about?	☐ No	☐ Almost	☐ Yes	0	1	2
2. Did the author do a good job of making the **guiding question** clear?	☐ No	☐ Almost	☐ Yes	0	1	2
3. Did the author do a good job of describing **what data were collected** during the investigation?	☐ No	☐ Almost	☐ Yes	0	1	2
4. Did the author do a good job describing **how the data were collected** during the investigation?	☐ No	☐ Almost	☐ Yes	0	1	2
5. Did the author do a good job describing **how the data were analyzed** after it was collected?	☐ No	☐ Almost	☐ Yes	0	1	2

Reviewers: If you gave the author any "No" or "Almost" ratings, please give the author some advice about what to do to improve the description of the investigation.

Section 2: The Argument	Reviewer Rating			Teacher Score		
1. Does the author's claim provide a clear and detailed **answer** to the guiding question?	☐ No	☐ Almost	☐ Yes	0	1	2
2. Did the author support the claim with **scientific evidence?** Scientific evidence includes an analysis of the data and an explanation of the analysis.	☐ No	☐ Almost	☐ Yes	0	1	2
3. Did the author include enough **evidence** and does the **evidence support the claim?**	☐ No	☐ Almost	☐ Yes	0	1	2
4. Did the author do a good job of **explaining why the evidence** is important (why it matters)?	☐ No	☐ Almost	☐ Yes	0	1	2
5. Is the content of the argument **correct** based on the science concepts we talked about in class?	☐ No	☐ Almost	☐ Yes	0	1	2

Reviewers: If you gave the author any "No" or "Almost" ratings, please give the author some advice about what to do to improve the argument.

Section 3: Mechanics	Reviewer Rating			Teacher Score
1. **Grammar:** Are the sentences complete? Is there proper subject-verb agreement in each sentence? Is the report written without run-on sentences?	☐ No	☐ Almost	☐ Yes	0 1 2
2. **Conventions:** Did the author use proper spelling, punctuation, and capitalization?	☐ No	☐ Almost	☐ Yes	0 1 2
3. **Word Choice:** Did the author use the right words in each sentence (e.g., *there* vs. *their*, *to* vs. *too*)?	☐ No	☐ Almost	☐ Yes	0 1 2

Reviewers: If you gave the author any "No" or "Almost" ratings, please give the author some advice about what to do to improve the writing mechanics of the investigation report.

General Reviewer Comments	Teacher Comments
We liked … We wonder …	

Total: _____ /26

APPENDIX 5

SAFETY ACKNOWLEDGMENT FORM

I know that it is very important to be as safe as I can during an investigation. My teacher has told me how to be safer in science. I agree to follow these 11 safety rules when I am working with my classmates to figure things out in science:

1. I will act in a responsible manner at all times. I will not run around the classroom, throw things, play jokes on my classmates, or be careless.
2. I will never eat, drink, or chew gum.
3. I will never touch, taste, or smell any materials, tools, or chemicals without permission.
4. I will wear my safety goggles at all times during the activity setup, hands-on work, and cleanup.
5. I will do my best to take care of the materials and tools that my teacher allows me to use.
6. I will always tell my teacher about any accidents as soon as they happen.
7. I will always dress in a way that will help keep me safer. I will wear closed-toed shoes and pants. My clothes will not be loose, baggy, or bulky. I will also use hair ties to keep my hair out of the way while I am working if my hair is long.
8. I will keep my work area clean and neat at all times. I will put my backpack, books, and other personal items where my teacher tells me to put them and I will not get them out unless my teacher tells me that it is okay.
9. I will clean my work area and the materials or tools that I use.
10. I will wash my hands with soap and water at the end of the activity.
11. I will follow my teacher's directions at all times.

_____   _____   _____
Print Name   Signature   Date

I have read and reviewed the 11 investigation safety rules with my child. He or she understands how important it is to follow safety rules in science and has agreed to follow these safety rules at all times. I give my permission for my child to participate in the investigations this year.

_____   _____   _____
Parent or Guardian Name   Parent or Guardian Signature   Date

APPENDIX 6

CHECKOUT QUESTIONS ANSWER GUIDE

The Checkout Questions are specific to each investigation and include a mixture of multiple choice, two-tiered, and free-response items. Except for most multiple-choice questions, there may be various acceptable answers to the questions; the answers given in this appendix include the correct answers to the multiple-choice questions and the key points that should be made by students in their answers to other questions. Answers for multiple choice and two-tiered items, as well as possible responses to free-response items, are provided below for each individual investigation.

Investigation 1: Movement of Matter Into and Out of a System

1. B
2. More air entering the bottle from the balloon will cause more water to be pushed out of the bottle and into the cup.
3. B
4. The balloon, bottle, tube, water, and air make up a system. The water did not move out of the system and into the cup because the air moved out of the system through the hole instead of pushing the water out of the bottle through the tube.

Investigation 2: Movement of Particles in a Liquid

1. A
2. C
3. Increasing the temperature makes the particles that make up tea and water move faster so the tea will spread through the water faster.
4. The model the students draw should include (a) only tea particles in the callout circle on the left and only water particles in the callout circle on the right for the picture labeled "Right before a tea bag is added to the water"; (b) some water particles and many tea particles in the callout circle on the left and only water particles in the callout circle on the right for the picture labeled "Right after a tea bag is added to the water"; and (c) many water particles with a few tea particles in the callout circle for the picture labeled "After the tea bag is removed from the water."
5. Tea particles take time to spread from the tea bag and through the water. The more time that the tea bag is in the water, the more tea particles will break off from the tea leaves and mix with the water particles.

Investigation 3: States of Matter and Weight

1. A
2. The weight of matter stays the same when it changes from a liquid to a solid.
3. C or D
4. The weight of the water stays the same when it changes from a solid to a liquid, but it appears like the water decreased because different units are used to measure weight before and after the water melted.

Appendix 6

Investigation 4: Chemical Reactions

1. A
2. A new substance with different properties will be formed if a chemical reaction take place after two different substances are mixed.
3. B
4. The mass and volume of substances E and G were the same and the mass and volume of substances F and H were the same, so a new substance was not produced when substances E and F were mixed. We can compare the mass and volume of these substances before and after mixing because standard units were used to measure mass and volume.

Investigation 5: Reactions and Weight

1. A
2. The sealed container is a closed system, so matter cannot transfer into or out of it. The closed system will stay the same because the weight of matter stays the same when it changes from one substance to another.
3. C or D
4. The weight of the container stays the same even when the matter inside it seems to disappear because of a chemical reaction, but it looks like the weight of the container decreases because different units are used to measure weight of the container before and after the chemical reaction took place.

Investigation 6: Physical and Chemical Properties

1. C
2. B
3. Powders can be identified based on similarities in their physical and chemical properties. These patterns can be used as evidence to support an explanation.
4. C
5. A
6. The mass and volume of salt and unknown powder D are the same. We can compare the mass and volume of these substances because the same standard units were used to measure these properties.

Investigation 7: Gravity

1. D
2. B
3. The force of gravity causes a ball to fall toward the center of Earth after it is kicked in the air.
4. A
5. Models are used to represent how the components of a system interact with each other. The arrows in the models represent forces. Model A shows that the force of gravity is always acting

on the ball, there is a force from the table acting on the ball while it is on the table, and there is a force acting on the ball when it is pushed but not before or after.

Investigation 8: Plant Growth

1. B

2. Air and water must transfer into a plant in order for it to grow.

Investigation 9: Energy in Ecosystems

1. A possible model is shown below. Students' models should match the description of what each thing eats with the arrows pointing toward the thing doing the eating. Students should correctly describe at least one energy pathway from a producer to the fox.

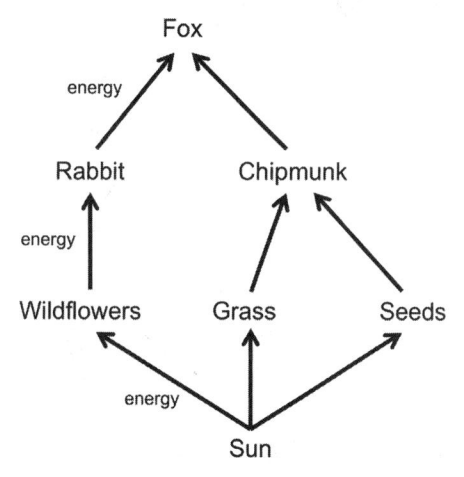

```
                    Fox
                   ↗  ↖
              energy     
           Rabbit      Chipmunk
             ↑          ↑  ↖
          energy        
        Wildflowers  Grass   Seeds
             ↖        ↑     ↗
                  energy
                    Sun
```

2. Students should mention that plants use the energy from sunlight to produce sugar from carbon dioxide and water. They use some of this sugar for the energy they need to survive, and the rest is stored in the body of a plant. Animals also need sugar but cannot produce their own, so they must eat other things that contain sugar.

Investigation 10: Movement of Carbon in Ecosystems

1. A

2. B

3. Plants and animals exchange gases with the environment. Animals, like snails, move carbon dioxide from their body to the environment. Plants move more carbon dioxide into their body from the environment than they move from their body into the environment.

Investigation 11: Patterns in Shadows

1. A short shadow pointing down and slightly to the left.

2. A long shadow pointing down and to the right.

3. When the Sun is high in the sky the shadow will be very small, and when the Sun is lower in the sky the shadow will be longer. Shadows go in the opposite direction of the location of the Sun.

Appendix 6

Investigation 12: Daylight and Location

1. A

2. Models show how the components of a system interact with each other. Models A and B include the same components of the Sun-Earth system, but Model A shows how Earth rotates so the Sun rises in the east.

3. C

4. A

5. Cities in the Northern Hemisphere that are farther from the equator would have a shorter duration of daylight during the winter, and cities in the Northern Hemisphere that are closer to the equator would have a longer duration of daylight during the winter.

Investigation 13: Stars in the Night Sky

1. See image below.

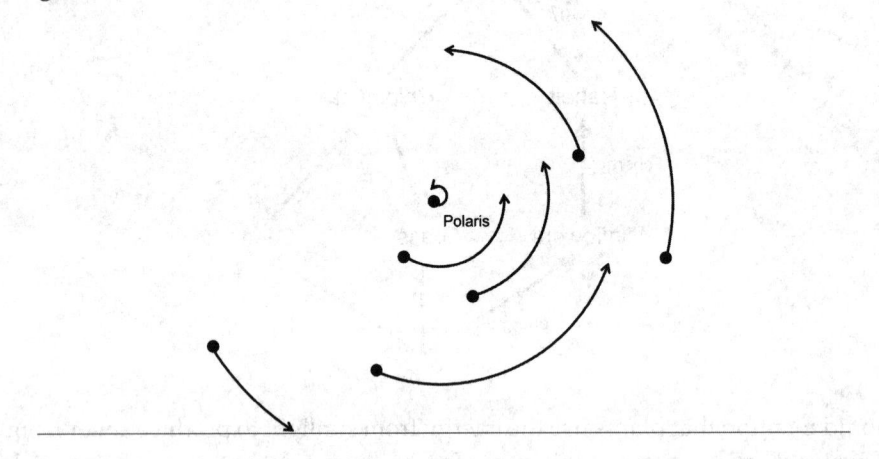

2. See image below.

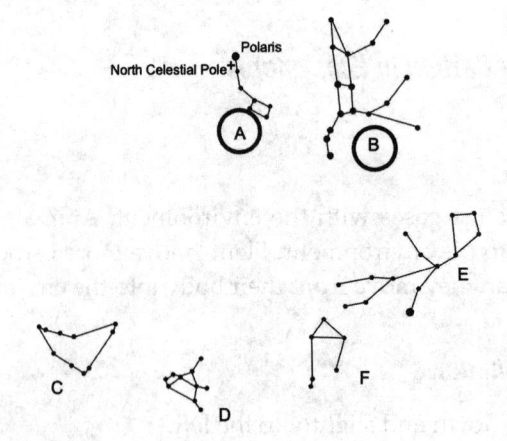

3. We can see different constellations because Earth is moving around the Sun, so the "night" side of Earth is facing different constellations during different seasons. Constellations near the celestial poles are visible all year, and constellations farther away from the celestial poles are visible only during different seasons.

Investigation 14: Star Brightness

1. A
2. D
3. D
4. Stars that are closer to Earth will appear brighter in the sky than stars that are larger and produce more energy but are farther away.

Investigation 15: Geographic Position and Climate

1. B
2. C
3. The mountain ranges, which are part of the geosphere, force clouds, which are part of the atmosphere, to drop lots of rain on the west side of the mountain. The mountains also create a rain shadow on the east side of the mountains. The water that falls on the west side of the mountain creates large rivers and lakes, which are part of the hydrosphere.

Investigation 16: Water Reservoirs

1. A
2. C
3. Nearly all of Earth's available water is in the ocean. Most freshwater is in glaciers or underground; only a tiny fraction is in lakes, rivers, and wetlands. We know this because we use the same units of measurement to described the quantity of water in every reservoir and then compare the amount in each one as a percentage of the total amount.

IMAGE CREDITS

All images in this book are stock photos or courtesy of the authors unless otherwise noted below.

Investigation 1
Figure 1.6: Justin Henry (see Acknowledgments section)

Figure 1.7: Justin Henry (see Acknowledgments section)

Investigation 2
Figure 2.6: BruceBlaus, Wikimedia Commons, CC BY-SA 3.0, *https://commons.wikimedia.org/wiki/File:Blausen_0315_Diffusion.png*

Investigation 3
Figure 3.4: HTO, Wikimedia Commons, Public domain. *https://commons.wikimedia.org/wiki/File:Melting_ice_cream_on_a_bridge.JPG*

Investigation 4
Figure 4.2: Priscilla Shaner (see Acknowledgments section)

Investigation 5
Figure 5.3: Kterha, Wikimedia Commons, CC BY-SA 2.0, *https://commons.wikimedia.org/wiki/File:Baking_soda_and_vinegar.jpg*

Investigation 6
Figure 6.2: Veganbaking.net, Wikimedia Commons, CC BY-SA 2.0, *https://commons.wikimedia.org/wiki/File:All-Purpose_Flour_(4107895947).jpg*

Investigation 7
Figure 7.6: Justin Henry (see Acknowledgments section)

Figure 7.7: Justin Henry (see Acknowledgments section)

Investigation 8
Investigation 8 Checkout Questions: Mokkie, Wikimedia Commons, CC BY-SA 3.0, *https://commons.wikimedia.org/wiki/File:Common_Duckweed_(Lemna_minor).jpg*

Investigation 9
Figure 9.3: Priscilla Shaner (see Acknowledgments section)

Figure 9.5: Fred Hsu, Wikimedia Commons, CC BY-SA 3.0, *https://commons.wikimedia.org/wiki/File:Rockfish_around_kelp_Monterey_Bay_Aquarium.jpg*

Figure 9.6: Jeff Williams/US Fish and Wildlife Service, Public domain. *https://upload.wikimedia.org/wikipedia/commons/2/26/Orca_feeding_at_Bogoslof_Island.jpg*

Investigation 10
Figure 10.3, cow image: DFoidl and DiegoAma, Wikimedia Commons, CC BY-SA 3.0, *https://commons.wikimedia.org/wiki/File:Aurochs_bull_and_cow.svg*

Investigation 12
Figure 12.9: Harman Smith and Laura Generosa (nee Berwin), graphic artists and contractors to NASA's Jet Propulsion Laboratory, Public domain. *https://commons.wikimedia.org/wiki/Solar_System#/media/File:Solar_sys.jpg*

Investigation 13
Figure 13.3, screenshot: Sky & Telescope, *https://skyandtelescope.org/interactive-sky-chart*

Figure 13.5: A. Duro/ESO, Wikimedia Commons, CC BY-SA 4.0, *https://commons.wikimedia.org/wiki/File:All_In_A_Spin_Star_trail.jpg*

Image Credits

Investigation 14
Figure 14.3: NASA, public domain. *www.nasa.gov/sites/default/files/horsehead.jpg*

Investigation 15
Figure 15.1: Wade Greenberg-Brand/Paleontological Research Institution, Wikimedia Commons, CC BY-SA 4.0, *https://upload.wikimedia.org/wikipedia/commons/thumb/b/b4/Rain_Shadow.tif/lossy-page1-1868px-Rain_Shadow.tif.jpg*

Figure 15.3: A McLin, Wikimedia Commons, CC BY-SA 2.0, *www.flickr.com/photos/37486024@N03/5150283634*

Figure 15.4: Thayne Tuason, Wikimedia Commons, CC BY-SA 4.0, *https://commons.wikimedia.org/wiki/File:East_Wenatchee_Washington-_looking_south.jpg*

Investigation 16
Maps of Where Crops Were Grown in the United States in 2017 (on the book's Extras page): United States Department of Agriculture, National Agricultural Statistics Service, Public domain. *www.nass.usda.gov/Charts_and_Maps/Crops_County/index.php#cr*

INDEX

Page numbers printed in **boldface type** refer to figures or tables.

"activity before content" lessons, xxi
agriculture. *See* Water Reservoirs investigation
apparent brightness, 562, 563, 588
    See also Star Brightness investigation
argument
    argumentation session, **4**, 14–18, **14, 15, 16, 18, 28**
    creating a draft argument, **4**, 9–13, **11, 12, 28**
    See also specific investigations
argument-driven inquiry (ADI) model
    about, xxii–xxiv
    described, x–xii
    FAQs, 695–697
    investigation report peer-review guide (elementary school version), **699–700**
    investigations, 33–39
    role of teacher in, 26–27, **27–29**
    safety, 29–30
    stages of, 3–26, **4**
        stage 1: introduction of task and guiding question, 4–6, **4, 5, 27**
        stage 2: designing a method and collecting data, **4**, 6–8, **7, 8, 28**
        stage 3: creating a draft argument, **4**, 9–13, **11, 12, 28**
        stage 4: argumentation session, **4**, 14–18, **14, 15, 16, 18, 28**
        stage 5: reflective discussion, **4**, 19–22, **20, 28**
        stage 6: writing a draft report, **4**, 22–23, **23, 29**
        stage 7: peer review, **4**, 24–25, **29**
        stage 8: revising the report, **4**, 25–26, **29**
assessment, xxv, 38, 39
    See also checkout questions; specific investigations
atmosphere, 628
atoms, 44, 201
attractive force, 112
axis of rotation. *See* Daylight and Location investigation

balloon water dispenser, 44–46, **44–46, 51**, 52, 60, **60**, 71–72, 80–81
Benchmarks for Science Literacy (AAAS), 34
biosphere, 628
brightness. *See* Star Brightness investigation

carbon dioxide (CO_2), 394, 396–397, 425–426
    See also Movement of Carbon in Ecosystems investigation
carnivores, 383
    See also Energy in Ecosystems investigation
Cascade Mountains, 602–603, **614–615**
CCs. *See* crosscutting concepts (CCs)
cellular respiration, 396–397
    See also Movement of Carbon in Ecosystems investigation
checkout questions
    answer guide, 703–707
    for assessment, xxv, 38, 39
    Chemical Reactions investigation, 184, 198–199
    Daylight and Location investigation, 500, 515–518
    Energy in Ecosystems investigation, 378, 392–393
    Geographic Position and Climate investigation, 623, 636–637
    Gravity investigation, 300, 313–316
    Movement of Carbon in Ecosystems investigation, 419, 434–435
    Movement of Matter investigation, 66, 80–81
    Movement of Particles in a Liquid investigation, 106, 120–122
    Patterns in Shadows investigation, 459, 472–473
    Physical and Chemical Properties investigation, 261, 275–276
    Plant Growth investigation, 341, 355
    Reactions and Weight investigation, 224, 238–239
    Star Brightness investigation, 582, 596–597
    Stars in the Night Sky investigation, 544, 559–560
    States of Matter and Weight investigation, 145, 159–160
    Water Reservoirs investigation, 659, 674–675
chemical indicators, 395, 396–397, 425–426
    See also Movement of Carbon in Ecosystems investigation
chemical properties, 162, 163, 241–242, 266–267
    See also Physical and Chemical Properties investigation
chemical reactions
    described, 189–190, 202, 229
    model of, **217**
    reaction of baking soda and vinegar, 206–207, 209, 211, **216**, 229–230
    See also Chemical Reactions investigation; Reactions and Weight investigation
Chemical Reactions investigation, 161–199
    checkout questions, 184, 198–199
    investigation handout
        analyze data, 193
        argumentation session, 194
        collect data, 193
        draft argument, 194
        draft report, 195–197
        guiding question, 191, 192, 195
        introduction, 189–190
        materials, 191
        plan investigation, 191–192
        review report, 197
        safety, 191
        submit final report, 197
        task, 190–191
    teacher notes
        alignment with standards, 184, **185–188**
        background information, 163–164, **164**
        checkout questions, 184
        crosscutting concepts (CCs), 161
        disciplinary core ideas (DCIs), 161
        lesson plan stage 1: introduction of task and guiding question, 161–162
        lesson plan stage 2: designing a method and collecting data, 162–171
        lesson plan stage 3: creating a draft argument, 171–174, **173**
        lesson plan stage 4: argumentation session, 174–177
        lesson plan stage 5: reflective discussion, 177–180, **177**
        lesson plan stage 6: writing a draft report, 180–181
        lesson plan stage 7: peer review, 183–184
        lesson plan stage 8: revising the report, 144–145
        materials and preparation, 165–167, **165–166**
        other concepts, 162
        purpose, 161
        safety, 167
        scientific and engineering practices (SEPs), 161–162
        timeline, 165
        what students figure out, 162–163
circumpolar star trails, **536**
claim, scientific argument components, 9, **10**
climate, 601, 627
    See also Geographic Position and Climate investigation; Water Reservoirs investigation
closed systems
    described, 202, 395, 426
    measurement of weight of, 203, **203**, 238–239
    See also Movement of Carbon in Ecosystems investigation; Reactions and Weight investigation
collisions, states of matter, 72
Common Core State Standards for English Language Arts (CCSS ELA)
    and ADI investigations, 34, **684**
    and argument-driven inquiry (ADI), xi–xii
    Chemical Reactions investigation, **185–187**
    Daylight and Location investigation, **501–503**
    Energy in Ecosystems investigation, **378–381**
    Geographic Position and Climate investigation, **623–626**
    Gravity investigation, **300–303**
    Movement of Carbon in Ecosystems investigation, **419–422**
    Movement of Matter investigation, **67–69**
    Movement of Particles in a Liquid investigation, **107–109**
    Patterns in Shadows investigation, **459–462**
    peer review, 25
    Physical and Chemical Properties investigation, **262–264**
    Plant Growth investigation, **342–344**
    Reactions and Weight investigation, **224–227**
    Star Brightness investigation, **582–585**
    Stars in the Night Sky investigation, **545–547**
    States of Matter and Weight investigation, **146–148**
    Water Reservoirs investigation, **660–662**
    writing a draft report, 23

Index

Common Core State Standards for Mathematics (CCSS Mathematics)
    and ADI investigations, 35, **686–687**
    and argument-driven inquiry (ADI), xii
    creating a draft argument, 9
    Movement of Matter investigation, **69–70**
    Movement of Particles in a Liquid investigation, **109–110**
    Patterns in Shadows investigation, **462**
    Plant Growth investigation, **344–345**
    Reactions and Weight investigation, **227**
    Star Brightness investigation, **585**
    States of Matter and Weight investigation, **148–149**
    Water Reservoirs investigation, **663**
compounds, 44
constellations, 520, 521–523, **522**, **523**, 537, 538, 550–551, **551**
    See also Stars in the Night Sky investigation
consumers, 383
    See also Energy in Ecosystems investigation
contact force, 279, 280, 305
    See also Gravity investigation
crops. *See* Water Reservoirs investigation
crosscutting concepts (CCs)
    and ADI instructional model, xxii–xxiv, 3
    and ADI investigations, 33–34, 35, 37–38, **680**
    argumentation session, 14, 17, 18
    cause and effect, 691
    Chemical Reactions investigation, 161
    creating a draft argument, 9–10
    Daylight and Location investigation, 474
    described, xvii–xviii
    designing a method and collecting data, 6–7
    energy and matter, 691
    Energy in Ecosystems investigation, 356
    Geographic Position and Climate investigation, 600
    Gravity investigation, 278
    and inquiry-based lessons, xxi
    introduction of task and guiding question, 4, 5
    Movement of Carbon in Ecosystems investigation, 394
    Movement of Matter investigation, 42
    Movement of Particles in a Liquid investigation, 82
    patterns, 691
    Patterns in Shadows investigation, 438
    Physical and Chemical Properties investigation, 240
    Plant Growth investigation, 318
    Reactions and Weight investigation, 200
    reflective discussion, 19
    scale, proportion, and quantity, 691
    and science proficiency, ix–x, **xi**
    stability and change, 691
    Star Brightness investigation, 561
    Stars in the Night Sky investigation, 519
    States of Matter and Weight investigation, 123
    structure and function, 691
    systems and system models, 691
    and three-dimensional instruction, xxi–xxii
    Water Reservoirs investigation, 638
    writing a draft report, 22

Daylight and Location investigation, 474–518
    checkout questions, 500, 515–518
    investigation handout
        analyze data, 510
        argumentation session, 511
        collect data, 510
        draft argument, 511
        draft report, 512–514
        guiding question, 508, 509, 512
        introduction, 505–507
        materials, 508
        plan investigation, 508–509
        review report, 514
        safety, 508
        submit final report, 514
        task, 507–508
    teacher notes
        alignment with standards, 500, **501–504**
        background information, 476–478, **476–478**
        checkout questions, 500
        crosscutting concepts (CCs), 474
        disciplinary core ideas (DCIs), 474
        lesson plan stage 1: introduction of task and guiding question, 481–482
        lesson plan stage 2: designing a method and collecting data, 482–484
        lesson plan stage 3: creating a draft argument, 485–587, **487**
        lesson plan stage 4: argumentation session, 488–491
        lesson plan stage 5: reflective discussion, 491–496, **491–493**, **494**
        lesson plan stage 6: writing a draft report, 496–497
        lesson plan stage 7: peer review, 497–499
        lesson plan stage 8: revising the report, 499–500
        materials and preparation, 479–481, **479**, **480**
        other concepts, 475–476
        purpose, 474
        safety, 481
        scientific and engineering practices (SEPs), 474–475
        timeline, 478–479
        what students figure out, 476
daylight. *See* Daylight and Location investigation
DCIs. *See* disciplinary core ideas (DCIs)
designing a method and collecting data, **4**, 6–8, **7**, **8**, **28**
    See also specific investigations
direct instruction and demonstration/hands-on activity, xix–xx
disciplinary core ideas (DCIs)
    and ADI instructional model, xxii–xxiv, 3
    and ADI investigations, 33–34, 35, 37–38, **680**
    argumentation session, 14, 17, 18
    Chemical Reactions investigation, 161
    creating a draft argument, 9–10
    Daylight and Location investigation, 474
    described, xvii
    designing a method and collecting data, 6
    Energy in Ecosystems investigation, 356
    Geographic Position and Climate investigation, 600
    Gravity investigation, 278
    and inquiry-based lessons, xxi
    introduction of task and guiding question, 4, 5
    Movement of Carbon in Ecosystems investigation, 394
    Movement of Matter investigation, 42
    Movement of Particles in a Liquid investigation, 82
    Patterns in Shadows investigation, 438
    Physical and Chemical Properties investigation, 240
    Plant Growth investigation, 318
    Reactions and Weight investigation, 200
    reflective discussion, 19
    and science proficiency, ix–x, **xi**
    Star Brightness investigation, 561
    Stars in the Night Sky investigation, 519
    States of Matter and Weight investigation, 123
    and three-dimensional instruction, xxi–xxii
    Water Reservoirs investigation, 638
    writing a draft report, 22
discipline-specific norms and criteria, scientific argument components, 10, **10**
discussion
    reflective discussion, **4**, 19–22, **20**, **28**
    See also specific investigations
draft report, **4**, 22–23, **23**, **29**
    See also specific investigations
duckweed, 355, **355**
dye added to water, **99**

Earth's place in the universe. *See* Daylight and Location investigation; Patterns in Shadows investigation; Star Brightness investigation; Stars in the Night Sky investigation
Earth's systems. *See* Geographic Position and Climate investigation; Water Reservoirs investigation
ecosystems, 395, 396, 425
    See also Energy in Ecosystems investigation; Movement of Carbon in Ecosystems investigation; Plant Growth investigation
elements, 44
emerging bilingual students and ADI instructional model, xxiv
empirical criteria, scientific argument components, 9–10, **10**
Energy in Ecosystems investigation, 356–393
    checkout questions, 378, 392–393
    investigation handout
        argumentation session, 388
        collect data, 387
        create a model, 387
        draft argument, 388
        draft report, 389–391
        guiding question, 385, 386, 389
        introduction, 382–384

Index

    materials, 385
    plan investigation, 385–386
    review report, 391
    safety, 385
    submit final report, 391
    task, 385
  teacher notes
    alignment with standards, 378, **378–381**
    background information, 358–360, **358**, **359**
    checkout questions, 378
    crosscutting concepts (CCs), 356
    disciplinary core ideas (DCIs), 356
    lesson plan stage 1: introduction of task and guiding question, 361–362
    lesson plan stage 2: designing a method and collecting data, 362–364
    lesson plan stage 3: creating a draft argument, 364–366, **366**, **367**
    lesson plan stage 4: argumentation session, 367–370
    lesson plan stage 5: reflective discussion, 371–373, **371**
    lesson plan stage 6: writing a draft report, 373–374
    lesson plan stage 7: peer review, 375–376
    lesson plan stage 8: revising the report, 376–377
    materials and preparation, 360, **361**
    other concepts, 357
    purpose, 356
    safety, 361
    scientific and engineering practices (SEPs), 356–357
    timeline, 360
    what students figure out, 358
energy. *See* Energy in Ecosystems investigation; Plant Growth investigation
English Language Proficiency (ELP) Standards
  alignment of ADI investigations with, 35, **688–689**
  Chemical Reactions investigation, **188**
  Daylight and Location investigation, **503–504**
  Energy in Ecosystems investigation, **381**
  Geographic Position and Climate investigation, **626**
  Gravity investigation, **303**
  Movement of Carbon in Ecosystems investigation, **422–423**
  Movement of Matter investigation, **70**
  Movement of Particles in a Liquid investigation, **110**
  Patterns in Shadows investigation, **462**
  Physical and Chemical Properties investigation, **264–265**
  Plant Growth investigation, **345**
  Reactions and Weight investigation, **227–228**
  Star Brightness investigation, **585–586**
  Stars in the Night Sky investigation, **547–548**
  States of Matter and Weight investigation, **149**
  Water Reservoirs investigation, **663–664**
evaluating and refining ideas, explanations, or arguments, xxiii, xxiv, 3
evidence, scientific argument components, 9, **10**

firsthand experience, 4
food chains, 357, 358–359, **358**
food webs, 357, 359, **359**
forces
  attractive force, 112
  contact force, 279, 280, 305
  described, 44, 279, 280, 304–305
  non-contact force, 279, 280, 305
  unbalanced forces, 304–305
  See also Gravity investigation; Movement of Matter investigation; Movement of Particles in a Liquid investigation
forest ecosystems, 392–393
Framework for K–12 Science Education
  and ADI investigations, 34
  alignment of ADI investigations with, **679–680**
  and science proficiency, ix–x
  science proficiency as foundation for, xvii
  and three-dimensional instruction, xxi–xxii
freezing point, 125
freshwater, 640, 666, 674

gallery walk format, argumentation session, 15–16, **15**
gas state, 72
  See also States of Matter and Weight investigation
Geographic Position and Climate investigation, 600–637
  checkout questions, 623, 636–637
  investigation handout
    analyze data, 631
    argumentation session, 632
    collect data, 631
    draft argument, 632
    draft report, 632–635
    guiding question, 629, 630, 633
    introduction, 627–628
    materials, 629
    plan investigation, 629–630
    review report, 635
    safety, 629
    submit final report, 635
    task, 629
  teacher notes
    alignment with standards, 623, **623–626**
    background information, 602–603, **603**
    checkout questions, 623
    crosscutting concepts (CCs), 600
    disciplinary core ideas (DCIs), 600
    lesson plan stage 1: introduction of task and guiding question, 605–606
    lesson plan stage 2: designing a method and collecting data, 606–608
    lesson plan stage 3: creating a draft argument, 608–610, **610**
    lesson plan stage 4: argumentation session, 610–614
    lesson plan stage 5: reflective discussion, 614–618, **614–615**
    lesson plan stage 6: writing a draft report, 618–620
    lesson plan stage 7: peer review, 620–621
    lesson plan stage 8: revising the report, 621–623
    materials and preparation, 604–605, **604**
    other concepts, 601
    purpose, 600
    safety, 605
    scientific and engineering practices (SEPs), 600–601
    timeline, 604
    what students figure out, 601
geosphere, 628
Gravity investigation, 278–316
  checkout questions, 300, 313–316
  investigation handout
    analyze data, 308
    argumentation session, 309
    collect data, 308
    draft argument, 309
    draft report, 310–312
    guiding question, 306, 307, 310
    introduction, 304–305
    materials, 306
    plan investigation, 306–307
    review report, 312
    safety, 306
    submit final report, 312
    task, 306
  teacher notes
    alignment with standards, 300, **300–303**
    background information, 280–281, **281**, **282**
    checkout questions, 300
    crosscutting concepts (CCs), 278
    disciplinary core ideas (DCIs), 278
    lesson plan stage 1: introduction of task and guiding question, 283–284
    lesson plan stage 2: designing a method and collecting data, 284–286, **286**
    lesson plan stage 3: creating a draft argument, 287–289, **289**
    lesson plan stage 4: argumentation session, 289–293
    lesson plan stage 5: reflective discussion, 293–295, **294**
    lesson plan stage 6: writing a draft report, 295–297
    lesson plan stage 7: peer review, 297–298
    lesson plan stage 8: revising the report, 299–300
    materials and preparation, 282, **282–283**
    other concepts, 279
    purpose, 278
    safety, 283
    scientific and engineering practices (SEPs), 278–279
    timeline, 282
    what students figure out, 279
guiding question
  introduction of task and guiding question, 4–6, **4**, **5**, **27**
  See also specific investigations

herbivores, 383
  See also Energy in Ecosystems investigation

Index

hot water bath, 127, **127**
hydrosphere, 628
hypothesis, 334
    See also Plant Growth investigation

inquiry and inquiry-based lessons, xx–xxi
Inquiry and the National Science Education Standards (NRC), xx–xxi
introduction of task and guiding question, 4–6, **4**, **5**, **27**
    See also specific investigations
investigating a phenomenon, xxiii, xxiv, 3
investigation reports
    revising the report, **4**, 25–26, **29**
    writing a draft report, **4**, 22–23, **23**, **29**
    See also specific investigations
investigations
    alignment with standards, 34–35
    how to use, 33–35
    instructional materials, 38–39
        checkout questions, 39
        investigation handouts, xxiv–xxv, 38
        peer-review guide and teacher scoring rubric (PRG/TSR), 39
        supplementary materials, 39
    other concepts overview, 35–36
    scientific and engineering practices (SEPs), 35
    teacher notes, 35–38
        checkout questions, 39
        connections to standards, 38
        crosscutting concepts (CCs), 35
        disciplinary core ideas (DCIs), 35
        lesson plan by stage, 37–38
        materials and preparation, 36
        purpose, 35
        safety, 37
        timeline, 36
        what students figure out, 36
    See also specific investigations
irrigation, 640
    See also Water Reservoirs investigation

justification of the evidence, scientific argument components, 9, **10**

kelp and fish, **371**

latitude, 506–507
    See also Daylight and Location investigation
learning and science proficiency, xviii–xix
light. *See* Daylight and Location investigation; Patterns in Shadows investigation; Star Brightness investigation; Stars in the Night Sky investigation
liquid state, 72, 151
    See also Chemical Reactions investigation; Movement of Matter investigation; States of Matter and Weight investigation
listening. *See* speaking and listening
literacy and ADI instructional model, xxiii–xxiv
location. *See* Daylight and Location investigation
longitude, 506–507
    See also Daylight and Location investigation
lumens, 562
    See also Star Brightness investigation
luminosity, 562, 588
    See also Star Brightness investigation
lux, 562
    See also Star Brightness investigation

making sense of a phenomenon, xxiii, xxiv, 3
materials, 241–242, 266–267, 346–347
    See also Physical and Chemical Properties investigation; Plant Growth investigation
mathematics
    and ADI instructional model, xxiii–xxiv
    alignment of ADI investigations with, **686–687**
    Movement of Matter investigation, **69–70**
    Movement of Particles in a Liquid investigation, **109–110**
    Plant Growth investigation, **344–345**
    Reactions and Weight investigation, **227**
    Star Brightness investigation, **585**
    States of Matter and Weight investigation, **148–149**
    Water Reservoirs investigation, **663**
matter
    cow moving matter into and out of its body, **411**
    described, 44, 71–72, 150–151, 202–203, 229, 396
    See also Chemical Reactions investigation; Movement of Carbon in Ecosystems investigation; Movement of Matter investigation; Movement of Particles in a Liquid investigation; Physical and Chemical Properties investigation; Plant Growth investigation; Reactions and Weight investigation; States of Matter and Weight investigation
measurement
    of the angle of a flashlight, **443**
    of light intensity, 565, **565**
    of weight of closed systems, 203, **203**, 238–239
melting point, 124
mixtures. *See* Reactions and Weight investigation
models
    day-night model for inaccurate, nontilted Earth, **476**
    day-night models for accurate, tilted Earth, **477**, **478**, **515**, **517**
    model drawing of the solar system, **494**
    model of a rain shadow, **603**
    model of chemical reactions, **217**
    model of person standing on Earth, 294, **294**, 313
    physical model, 588
    See also Energy in Ecosystems investigation
modified gallery walk format, argumentation session, 16, **16**
molecules, 44
    See also Movement of Particles in a Liquid investigation
motion. *See* Gravity investigation
Movement of Carbon in Ecosystems investigation, 394–435
    checkout questions, 419, 434–435
    investigation handout
        analyze data, 429
        argumentation session, 430
        collect data, 429
        draft argument, 430
        draft report, 431–433
        guiding question, 427, 428, 431
        introduction, 424–426
        materials, 427
        plan investigation, 427–428
        review report, 433
        safety, 427
        submit final report, 433
        task, 427
    teacher notes
        alignment with standards, 419, **419–423**
        background information, 396–397, **398**
        checkout questions, 419
        crosscutting concepts (CCs), 394
        disciplinary core ideas (DCIs), 394
        lesson plan stage 1: introduction of task and guiding question, 401–402
        lesson plan stage 2: designing a method and collecting data, 402–404
        lesson plan stage 3: creating a draft argument, 405–406, **407**
        lesson plan stage 4: argumentation session, 407–410
        lesson plan stage 5: reflective discussion, 411–414, **411**, **413**
        lesson plan stage 6: writing a draft report, 415–416
        lesson plan stage 7: peer review, 416–418
        lesson plan stage 8: revising the report, 418–419
        materials and preparation, 398–400, **399**
        other concepts, 395
        purpose, 394
        safety, 400
        scientific and engineering practices (SEPs), 394–395
        timeline, 398
        what students figure out, 396
Movement of Matter investigation, 42–81
    checkout questions, 66, 80–81
    investigation handout
        analyze data, 75
        argumentation session, 76
        collect data, 75
        draft argument, 76
        draft report, 77–79
        guiding question, 73, 74, 77
        introduction, 71–72
        materials, 73
        plan investigation, 73–74
        review report, 79
        safety, 73
        submit final report, 79
        task, 73
    teacher notes

Index

alignment with standards, 66, **66–70**
background information, 44–46, **44–46**
checkout questions, 66
crosscutting concepts (CCs), 42
disciplinary core ideas (DCIs), 42
lesson plan stage 1: introduction of task and guiding question, 49–50
lesson plan stage 2: designing a method and collecting data, 50–52, **51**
lesson plan stage 3: creating a draft argument, 53–55, **55**
lesson plan stage 4: argumentation session, 55–58
lesson plan stage 5: reflective discussion, 59–61, **60**
lesson plan stage 6: writing a draft report, 62–63
lesson plan stage 7: peer review, 63–65
lesson plan stage 8: revising the report, 65–66
materials and preparation, 47, **48**
other concepts, 43
purpose, 42
safety, 48
scientific and engineering practices (SEPs), 42–43
timeline, 47
what students figure out, 44
Movement of Particles in a Liquid investigation, 82–122
checkout questions, 106, 120–122
investigation handout
analyze data, 115
argumentation session, 116
collect data, 115
draft argument, 116
draft report, 117–119
guiding question, 113, 114, 117
introduction, 111–112
materials, 113
plan investigation, 113–114
review report, 119
safety, 113
submit final report, 119
task, 113
teacher notes
alignment with standards, 106, **106–110**
background information, 84–85, **85**, **86**
checkout questions, 106
crosscutting concepts (CCs), 82
disciplinary core ideas (DCIs), 82
lesson plan stage 1: introduction of task and guiding question, 89–90
lesson plan stage 2: designing a method and collecting data, 90–92
lesson plan stage 3: creating a draft argument, 92–94, **95**
lesson plan stage 4: argumentation session, 94–98
lesson plan stage 5: reflective discussion, 98–101, **99**
lesson plan stage 6: writing a draft report, 101–103
lesson plan stage 7: peer review, 103–105
lesson plan stage 8: revising the report, 105–106
materials and preparation, 86–87, **87**, **88**
other concepts, 83–84
purpose, 82
safety, 88
scientific and engineering practices (SEPs), 82–83
timeline, 86
what students figure out, 84

National Science Education Standards (NRC), 34
nature of scientific inquiry (NOSI)
alignment of ADI investigations with, 34, **683**
concepts overview, **693–694**
purpose subsection of teacher notes, 35
reflective discussion, 19, 21
nature of scientific knowledge (NOSK)
alignment of ADI investigations with, 34, **683**
concepts overview, **692–693**
purpose subsection of teacher notes, 35
reflective discussion, 19, 21
Next Generation Science Standards (NGSS)
alignment of ADI investigations with, 34, **681–682**
Chemical Reactions investigation, **185**
crosscutting concepts overview, **691–692**
Daylight and Location investigation, **501**
Energy in Ecosystems investigation, **378**
Framework as guide to development of, xvii–xviii
Geographic Position and Climate investigation, **623**
Gravity investigation, **300**
Movement of Carbon in Ecosystems investigation, **419**
Movement of Matter investigation, 66
Movement of Particles in a Liquid investigation, **106**
nature of scientific knowledge and scientific inquiry concepts overview, **692–694**
Patterns in Shadows investigation, **459**
Physical and Chemical Properties investigation, **261**
Plant Growth investigation, **341**
Reactions and Weight investigation, **224**
and science proficiency, xi–xii
Star Brightness investigation, **582**
Stars in the Night Sky investigation, **545**
States of Matter and Weight investigation, **145**
Water Reservoirs investigation, **660**
non-contact force, 279, 280, 305
See also Gravity investigation
north celestial pole, 520, 521, 549–550
See also Stars in the Night Sky investigation

omnivores, 383
See also Energy in Ecosystems investigation
orca feeding on a seal, **371**
orcas, 371, 382
See also Energy in Ecosystems investigation

particles. *See* Chemical Reactions investigation; Movement of Particles in a Liquid investigation; Physical and Chemical Properties investigation
Patterns in Shadows investigation, 438–473
checkout questions, 459, 472–473
investigation handout
analyze data, 467
argumentation session, 468
collect data, 467
draft argument, 468
draft report, 469–471
guiding question, 465, 466, 469
introduction, 463–464
materials, 465
plan investigation, 465–466
review report, 471
safety, 465
submit final report, 471
task, 465
teacher notes
alignment with standards, 459, **459–462**
background information, 440
checkout questions, 459
crosscutting concepts (CCs), 438
disciplinary core ideas (DCIs), 438
lesson plan stage 1: introduction of task and guiding question, 442–443, **443**
lesson plan stage 2: designing a method and collecting data, 443–445
lesson plan stage 3: creating a draft argument, 446–448, **447**
lesson plan stage 4: argumentation session, 448–451
lesson plan stage 5: reflective discussion, 451–454, **452**
lesson plan stage 6: writing a draft report, 454–456
lesson plan stage 7: peer review, 456–457
lesson plan stage 8: revising the report, 457–459
materials and preparation, 441, **441**
other concepts, 439–440
purpose, 438
safety, 441
scientific and engineering practices (SEPs), 438–439
timeline, 440–441
what students figure out, 440
peer review
ADI model, 4, 24–25, **29**, **699–700**
investigation report peer-review guide (elementary school version), **699–700**
peer-review guide and teacher scoring rubric (PRG/TSR), 24–25, 39
See also specific investigations
performance expectations, alignment of ADI investigations with *NGSS*, **681–682**
personal protective equipment (PPE), 29–30
Physical and Chemical Properties investigation, 240–276
checkout questions, 261, 275–276
investigation handout
analyze data, 270

Index

argumentation session, 271
collect data, 270
draft argument, 271
draft report, 272–274
guiding question, 267, 269, 272
introduction, 266–267
materials, 268
plan investigation, 268–269
review report, 274
safety, 268
submit final report, 274
task, 267
teacher notes
    alignment with standards, 261, **261–265**
    background information, 241–242
    checkout questions, 261
    crosscutting concepts (CCs), 240
    disciplinary core ideas (DCIs), 240
    lesson plan stage 1: introduction of task and guiding question, 244–245
    lesson plan stage 2: designing a method and collecting data, 245–247
    lesson plan stage 3: creating a draft argument, 247–249, **250**
    lesson plan stage 4: argumentation session, 250–253
    lesson plan stage 5: reflective discussion, 253–256, **254**
    lesson plan stage 6: writing a draft report, 256–258
    lesson plan stage 7: peer review, 258–259
    lesson plan stage 8: revising the report, 260–261
    materials and preparation, 242–243, **243**
    other concepts, 241
    purpose, 240
    safety, 243–244
    scientific and engineering practices (SEPs), 240–241
    timeline, 242
    what students figure out, 241
physical model, 588
physical properties, 162, 163, 190, 241–242, 266–267
    See also Chemical Reactions investigation; Physical and Chemical Properties investigation
Plant Growth investigation, 318–355
checkout questions, 341, 355
investigation handout
    analyze data, 350
    argumentation session, 351
    collect data, 350
    draft argument, 351
    draft report, 352–354
    guiding question, 348, 352
    introduction, 346–347
    materials, 348
    plan investigation, 348–349
    review report, 354
    safety, 348
    submit final report, 354
    task, 348
teacher notes
    alignment with standards, 341, **341–345**
    background information, 320
    checkout questions, 341
    crosscutting concepts (CCs), 318
    disciplinary core ideas (DCIs), 318
    lesson plan stage 1: introduction of task and guiding question, 322–323
    lesson plan stage 2: designing a method and collecting data, 323–325
    lesson plan stage 3: creating a draft argument, 326–327, **328**
    lesson plan stage 4: argumentation session, 328–331
    lesson plan stage 5: reflective discussion, 332–336, **332–335**
    lesson plan stage 6: writing a draft report, 336–338
    lesson plan stage 7: peer review, 338–340
    lesson plan stage 8: revising the report, 340–341
    materials and preparation, 320–322, **321**
    other concepts, 319
    purpose, 318
    safety, 322
    scientific and engineering practices (SEPs), 318–319
    timeline, 320
    what students figure out, 319
prime meridian, 506
    See also Daylight and Location investigation
producers, 383

    See also Energy in Ecosystems investigation
property(ies), 266
    See also Physical and Chemical Properties investigation
rain shadows, 601, 602–603, **603**
    See also Geographic Position and Climate investigation
Reactions and Weight investigation, 200–239
checkout questions, 224, 238–239
investigation handout
    analyze data, 233
    argumentation session, 234
    collect data, 233
    draft argument, 234
    draft report, 235–237
    guiding question, 231, 232, 235
    introduction, 229–230
    materials, 231
    plan investigation, 231–232
    review report, 237
    safety, 231
    submit final report, 237
    task, 231
teacher notes
    alignment with standards, 224, **224–228**
    background information, 202–203, **203**
    checkout questions, 224
    crosscutting concepts (CCs), 200
    disciplinary core ideas (DCIs), 200
    lesson plan stage 1: introduction of task and guiding question, 206–207
    lesson plan stage 2: designing a method and collecting data, 207–209
    lesson plan stage 3: creating a draft argument, 210–212, **212**
    lesson plan stage 4: argumentation session, 213–216
    lesson plan stage 5: reflective discussion, 216–219, **216, 217**
    lesson plan stage 6: writing a draft report, 219–221
    lesson plan stage 7: peer review, 221–222
    lesson plan stage 8: revising the report, 222–224
    materials and preparation, 204–206, **204–205**
    other concepts, 201–202
    purpose, 200
    safety, 205
    scientific and engineering practices (SEPs), 200–201
    timeline, 204
    what students figure out, 202
reading
    alignment of ADI investigations with, **684**
    Chemical Reactions investigation, **185**
    Daylight and Location investigation, **501**
    Energy in Ecosystems investigation, **378–379**
    Geographic Position and Climate investigation, **623–624**
    Gravity investigation, **300–301**
    Movement of Carbon in Ecosystems investigation, **419–420**
    Movement of Matter investigation, **67**
    Movement of Particles in a Liquid investigation, **107**
    Physical and Chemical Properties investigation, **262**
    Plant Growth investigation, **342**
    Reactions and Weight investigation, **224–225**
    Star Brightness investigation, **582–583**
    Stars in the Night Sky investigation, **545**
    States of Matter and Weight investigation, **146**
    Water Reservoirs investigation, **660**
reflective discussion, **4**, 19–22, **20**, **28**
    See also specific investigations
reports
    revising the report, **4**, 25–26, **29**
    writing a draft report, **4**, 22–23, **23**, **29**
    See also specific investigations
rigor and science proficiency, xviii–xix
rotation, 439
    See also Patterns in Shadows investigation
rubrics
    Chemical Reactions investigation, 198, 199
    Daylight and Location investigation, 516, 518
    Energy in Ecosystems investigation, 393
    Geographic Position and Climate investigation, 637
    Gravity investigation, 313, 316
    of Matter investigation, 80, 81
    Movement of Carbon in Ecosystems investigation, 435
    Movement of Matter investigation, 80, 81
    Movement of Particles in a Liquid investigation, 120, 122

Index

Patterns in Shadows investigation, 473
peer-review guide and teacher scoring rubric (PRG/TSR), 24–25, 25–26, 39
Physical and Chemical Properties investigation, 275, 276
Plant Growth investigation, 355
Reactions and Weight investigation, 238, 239
Star Brightness investigation, 597
Stars in the Night Sky investigation, 560
States of Matter and Weight investigation, 159, 160
Water Reservoirs investigation, 675
safety
argument-driven inquiry (ADI) model, 29–30, 37
personal protective equipment (PPE), 29–30
safety acknowledgment form, **701**
See also specific investigations
saltwater, 640, 666, 674
San Diego, CA city information, **492**, **493**
science instruction
direct instruction and demonstration/hands-on activity, xix–xx
inquiry-based lessons, xx–xxi
three-dimensional instruction, xxi–xxii
science proficiency
and ADI instructional model, xxii–xxiii
defined, xvii
importance of, ix
and three-dimensional instruction, xxi–xxii
scientific and engineering practices (SEPs)
and ADI instructional model, xxiii–xxiv, 3
and ADI investigations, 33–34, 35, 37, **679**
argumentation session, 14, 16–17
Chemical Reactions investigation, 161–162
creating a draft argument, 9
Daylight and Location investigation, 474–475
described, xviii
designing a method and collecting data, 6
Energy in Ecosystems investigation, 356–357
Geographic Position and Climate investigation, 600–601
Gravity investigation, 278–279
and inquiry-based lessons, xxi
introduction of task and guiding question, 4, 5
Movement of Carbon in Ecosystems investigation, 394–395
Movement of Matter investigation, 42–43
Movement of Particles in a Liquid investigation, 82–83
Patterns in Shadows investigation, 438–439
peer review, 25
Physical and Chemical Properties investigation, 240–241
Plant Growth investigation, 318–319
Reactions and Weight investigation, 200–201
and science proficiency, ix–x, **xi**
Star Brightness investigation, 561–562
Stars in the Night Sky investigation, 519–520
States of Matter and Weight investigation, 123–124
and three-dimensional instruction, xxi–xxii
Water Reservoirs investigation, 638–639
writing a draft report, 23
scientific argument, creating a draft argument, 9–13, **11**, **12**
Secchi disk, 87, 92, 93
secondhand experience, 4–5
SEPs. *See* scientific and engineering practices (SEPs)
shadows, 463
See also Patterns in Shadows investigation
sky chart, 521–522, **522**, 524–525, **525**
solar luminosity, 563
solar system model drawing, **494**
solid state, 71–72, 150–151
See also Movement of Matter investigation; States of Matter and Weight investigation
south celestial pole, 520, 521, 550
See also Stars in the Night Sky investigation
speaking and listening
alignment of ADI investigations with, **685**
Chemical Reactions investigation, **187**
Daylight and Location investigation, **503**
Energy in Ecosystems investigation, **380–381**
Geographic Position and Climate investigation, **625–626**
Gravity investigation, **302–303**
Movement of Carbon in Ecosystems investigation, **422**
Movement of Matter investigation, **69**
Movement of Particles in a Liquid investigation, **109**
Patterns in Shadows investigation, **461–462**

Physical and Chemical Properties investigation, **263–264**
Plant Growth investigation, **344**
Reactions and Weight investigation, **226–227**
Star Brightness investigation, **584–585**
Stars in the Night Sky investigation, **547**
States of Matter and Weight investigation, **148**
Water Reservoirs investigation, **662**
Spokane, WA city information, **491**, **493**
stability. *See* Gravity investigation
standard units, 123
Star Brightness investigation, 561–597
checkout questions, 582, 596–597
investigation handout
analyze data, 591
argumentation session, 592
collect data, 591
draft argument, 592
draft report, 593–595
guiding question, 589, 590, 593
introduction, 587–588
materials, 589
plan investigation, 589–590
review report, 595
safety, 589
submit final report, 595
task, 589
teacher notes
alignment with standards, 582, **582–586**
background information, 563
checkout questions, 582
crosscutting concepts (CCs), 561
disciplinary core ideas (DCIs), 561
lesson plan stage 1: introduction of task and guiding question, 566
lesson plan stage 2: designing a method and collecting data, 567–568
lesson plan stage 3: creating a draft argument, 569–571, **571**
lesson plan stage 4: argumentation session, 571–575
lesson plan stage 5: reflective discussion, 575–577, **575**
lesson plan stage 6: writing a draft report, 577–579
lesson plan stage 7: peer review, 579–580
lesson plan stage 8: revising the report, 581–582
materials and preparation, 564–565, **564**, **565**
other concepts, 562
purpose, 561
safety, 565
scientific and engineering practices (SEPs), 561–562
timeline, 563–564
what students figure out, 562
Stars in the Night Sky investigation, 519–560
checkout questions, 544, 559–560
investigation handout
analyze data, 554
argumentation session, 555
collect data, 554
draft argument, 555
draft report, 556–558
guiding question, 552, 553, 556
introduction, 549–551
materials, 552
plan investigation, 552–553
review report, 558
safety, 552
submit final report, 558
task, 552
teacher notes
alignment with standards, 545, **545–548**
background information, 521–523, **522**, **523**
checkout questions, 544
crosscutting concepts (CCs), 519
disciplinary core ideas (DCIs), 519
lesson plan stage 1: introduction of task and guiding question, 526–527
lesson plan stage 2: designing a method and collecting data, 527–529
lesson plan stage 3: creating a draft argument, 529–532, **531**
lesson plan stage 4: argumentation session, 532–535
lesson plan stage 5: reflective discussion, 535–540, **536**, **537–538**
lesson plan stage 6: writing a draft report, 540–541
lesson plan stage 7: peer review, 541–543

Index

    lesson plan stage 8: revising the report, 543–544
    materials and preparation, 524–525, **524**, **525**
    other concepts, 520
    purpose, 519
    safety, 525
    scientific and engineering practices (SEPs), 519–520
    timeline, 524
    what students figure out, 521
states of matter, 44, 71–72, 125, 150–151
States of Matter and Weight investigation, 123–160
    checkout questions, 145, 159–160
    investigation handout
        analyze data, 154
        argumentation session, 155
        collect data, 154
        draft argument, 155
        draft report, 156–158
        guiding question, 152, 153, 156
        introduction, 150–151
        materials, 152
        plan investigation, 152–153
        review report, 158
        safety, 152
        submit final report, 158
        task, 152
    teacher notes
        alignment with standards, 145, **145–149**
        background information, 125
        checkout questions, 145
        crosscutting concepts (CCs), 123
        disciplinary core ideas (DCIs), 123
        lesson plan stage 1: introduction of task and guiding question, 128–129, **128**
        lesson plan stage 2: designing a method and collecting data, 129–131
        lesson plan stage 3: creating a draft argument, 131–133, **134**
        lesson plan stage 4: argumentation session, 134–137
        lesson plan stage 5: reflective discussion, 138–140, **138**, **139**
        lesson plan stage 6: writing a draft report, 141–142
        lesson plan stage 7: peer review, 142–144
        lesson plan stage 8: revising the report, 144–145
        materials and preparation, 126–127, **126**, **127**
        other concepts, 124–125
        purpose, 123
        safety, 127–128
        scientific and engineering practices (SEPs), 123–124
        timeline, 125
        what students figure out, 125
student talk goals, reflective discussion, 19–21, **20**
substances, 201, 202, 229
    See also Chemical Reactions investigation; Reactions and Weight investigation
the Sun
    described, 563, 588
    Earth's orbit around, 522–523, **523**
    See also Daylight and Location investigation; Patterns in Shadows investigation
systems
    closed systems, 202, 395, 426
    described, 71
    See also Movement of Matter investigation; Reactions and Weight investigation; States of Matter and Weight investigation

tea bag model, 85, **85**, **86**, 121
teacher scoring rubrics
    Chemical Reactions investigation, 198, 199
    Daylight and Location investigation, 516, 518
    Energy in Ecosystems investigation, 393
    Geographic Position and Climate investigation, 637
    Gravity investigation, 313, 316
    Movement of Carbon in Ecosystems investigation, 435
    Movement of Matter investigation, 80, 81
    Movement of Particles in a Liquid investigation, 120, 122
    Patterns in Shadows investigation, 473
    Physical and Chemical Properties investigation, 275, 276
    Plant Growth investigation, 355
    Reactions and Weight investigation, 238, 239
    Star Brightness investigation, 597
    Stars in the Night Sky investigation, 560
    States of Matter and Weight investigation, 159, 160

Water Reservoirs investigation, 675
teacher talk moves, reflective discussion, 19–21, **20**
temperature, 112, 124–125, 151
theoretical criteria, 10, **10**
thermal energy, 151
three-dimensional instruction, xxi–xxiv
"tool talk," 5–6

unbalanced forces, 304–305
    See also Gravity investigation
units, standard units, 123

water
    dye added to water, **99**
    global water distribution, **640**, **652**, 674
    water usage in the United States (2015), **641**, 666
water particles. *See* Movement of Particles in a Liquid investigation
Water Reservoirs investigation, 638–675
    checkout questions, 659, 674–675
    investigation handout
        analyze data, 669
        argumentation session, 670
        collect data, 669
        draft argument, 670
        draft report, 671–673
        guiding question, 667, 668, 671
        introduction, 665–666
        materials, 667
        plan investigation, 667–668
        review report, 673
        safety, 667
        submit final report, 673
        task, 667
    teacher notes
        alignment with standards, 659, **660–664**
        background information, 640–641, **640**, **641**
        checkout questions, 659
        crosscutting concepts (CCs), 638
        disciplinary core ideas (DCIs), 638
        lesson plan stage 1: introduction of task and guiding question, 643–644
        lesson plan stage 2: designing a method and collecting data, 644–646
        lesson plan stage 3: creating a draft argument, 646–648, **648**
        lesson plan stage 4: argumentation session, 649–652
        lesson plan stage 5: reflective discussion, 652–655, **652**
        lesson plan stage 6: writing a draft report, 655–656
        lesson plan stage 7: peer review, 656–658
        lesson plan stage 8: revising the report, 658–659
        materials and preparation, 642, **642–643**
        other concepts, 639
        purpose, 638
        safety, 643
        scientific and engineering practices (SEPs), 638–639
        timeline, 642
        what students figure out, 640
weather, 601, 602, 627
    See also Geographic Position and Climate investigation
weight. *See* Reactions and Weight investigation; States of Matter and Weight investigation
whole-class presentation format, argumentation session, 14–15, **14**
writing
    alignment of ADI investigations with, 684
    Chemical Reactions investigation, 186–187
    Daylight and Location investigation, **501–502**
    draft report, 4, 22–23, **23**, 29
    Energy in Ecosystems investigation, **379–380**
    Geographic Position and Climate investigation, **624–625**
    Gravity investigation, **301–302**
    Movement of Carbon in Ecosystems investigation, **420–421**
    Movement of Matter investigation, **67–68**
    Movement of Particles in a Liquid investigation, **107–108**
    Physical and Chemical Properties investigation, **262–263**
    Plant Growth investigation, **342–343**
    Reactions and Weight investigation, **225–226**
    revising the report, 25–26
    Star Brightness investigation, **583–584**
    Stars in the Night Sky investigation, **546–547**
    States of Matter and Weight investigation, **146–147**
    Water Reservoirs investigation, **661–662**